APPLIED MATHEMATICAL MODELING AND PROBLEM SOLVING

APPLIED MATHEMATICAL MODELING AND PROBLEM SOLVING

The Consortium for Foundation Mathematics

Ralph Bertelle	*Columbia-Greene Community College*
Judith Bloch	*University of Rochester*
Roy Cameron	*SUNY Cobleskill*
Carolyn Curley	*Erie Community College—South Campus*
Ernie Danforth	*Corning Community College*
Brian Gray	*Howard Community College*
Arlene Kleinstein	*SUNY Farmingdale*
Kathleen Milligan	*Monroe Community College*
Patricia Pacitti	*SUNY Oswego*
Renan Sezer	*Ankara University, Turkey*
Patricia Shuart	*Polk State College—Winter Haven, Florida*
Sylvia Svitak	*Queensborough Community College*

330 Hudson Street, NY NY 10013

Editorial Director: Christine Hoag
Editor in Chief: Michael Hirsch
Assistant Editor: Matthew Summers
Project Manager: Beth Houston
Program Manager: Danielle S. Miller
Project Management Team Lead: Peter Silvia
Project Management Team Lead: Karen Wernholm
Media Producer: Erin Carreiro
TestGen Content Manager: John Flanagan
MathXL Content Developer: Rebecca Williams
Marketing Manager: Alicia Frankel
Marketing Assistant: Kelly Cross

Senior Author Support/Technology Specialist:
  Joe Vetere
Rights and Permissions Project Manager:
  Diahanne Lucas Dowridge
Procurement Specialist: Carol Melville
Associate Director of Design: Andrea Nix
Program Design Lead: Barbara Atkinson
Composition: Cenveo® Publisher Services
Illustrations: Cenveo Publisher Services
Cover Image: Getty Images/Image Hans Elbers
Cover Design: Cenveo Publisher Services

Library of Congress Cataloging-in-Publication Data

Names: Consortium for Foundation Mathematics.
Title: Applied mathematical modeling and problem solving / The Consortium for
  Foundation Mathematics.
Description: Boston : Pearson, [2017]
Identifiers: LCCN 2016022798 | ISBN 9780134654416 (alk. paper)
Subjects: LCSH: Mathematics—Textbooks. | Mathematical models—Textbooks.
Classification: LCC QA11.2 .A6285 2017 | DDC 510—dc23
LC record available at https://lccn.loc.gov/2016022798

SE ISBN 13: 978-0-134-65441-6
ISBN 10:   0-134-65441-2

Contents

8. Measure the strength of the correlation (association) by a correlation coefficient.

9. Recognize that a strong correlation does not necessarily imply a linear, or a cause-and-effect, relationship.

CHAPTER 3

Problem Solving with Quadratic and Variation Function Models   353

CHAPTER 4

Modeling with Exponential and Logarithmic Functions 459

3. Read a bond listing.

4. Calculate the price of bonds.

5. Solve problems involving stocks and bonds.

APPENDICES

To the Instructor

Our Vision

Applied Mathematical Modeling and Problem Solving is intended to give students a deeper understanding of mathematical relationships leading to real-world mathematical literacy and a strong foundation for future study of mathematics and related disciplines. The main goals for this textbook are to teach students how to problem-solve, communicate mathematically, create and interpret mathematical representations and models, and make efficient and appropriate use of technology to solve problems. Additionally, an emphasis can be placed upon developing skills necessary in today's workplace, including organization and a collaborative team approach.

Applied Mathematical Modeling and Problem Solving is based on the principle that students learn mathematics best by doing mathematics within a meaningful context. In keeping with this premise, students solve problems in a series of realistic situations from which the crucial need for mathematics arises. Students will not only develop reasoning skills to problem solve in realistic situations, but will have a deeper understanding of the mathematical content needed to be successful for a non-STEM college math course. *Applied Mathematical Modeling and Problem Solving* guides students toward developing a sense of independence and taking responsibility for their own learning. Students are encouraged to construct, reflect on, apply, and describe their own mathematical models, which they use to solve meaningful problems. We see this as the key to bridging the gap between abstraction and application and as the basis for transfer learning. Appropriate technology is integrated throughout the book, allowing students to interpret real-life data verbally, numerically, symbolically, and graphically.

We expect that by using the *Applied Mathematical Modeling and Problem Solving* book, all students will be able to achieve the following goals:

- Develop mathematical intuition and a relevant base of mathematical knowledge.

- Gain experiences that connect classroom learning with real-world applications.

- Learn to work in groups as well as independently.

- Increase knowledge of mathematics through explorations with appropriate technology.

- Develop a positive attitude about learning and using mathematics.

- Build techniques of reasoning for effective problem solving.

- Learn to apply and display knowledge through alternative means of assessment, such as mathematical portfolios and journal writing.

Our vision for your students is to join the growing number of students using our approach who discover that mathematics is an essential and learnable survival skill for the twenty-first century.

Pedagogical Features

The pedagogical core of *Applied Mathematical Modeling and Problem Solving* is a series of guided-discovery activities in which students work in groups to discover mathematical principles embedded in realistic situations. The key principles of each activity are highlighted and summarized at the activity's conclusion. Each activity is followed by exercises that reinforce the concepts and skills revealed in the activity.

The activities are clustered into sections within each chapter. Each section contains regular activities along with project activities that relate to particular topics. The activities may require more than just paper, pencil, and calculator. They may also require measurements and data collection and are ideal for in-class group work. The project activities are designed to allow students to explore specific topics in greater depth, either individually or in groups. These activities are usually self-contained and have no accompanying exercises. For specific suggestions on how to use the two types of activities, we strongly encourage you to refer to the *Instructor's Resource Manual* that accompanies this text.

Occurring naturally as summary and review are the What Have I Learned? and How Can I Practice? exercises. The What Have I Learned? exercises are designed to help students pull together the key concepts of the section. The How Can I Practice? exercises are designed primarily to provide additional work with the numeric, graphical, and algebraic skills of the section. Taken as a whole, these exercises give students the tools they need to bridge the gaps between abstraction, skills, and application.

Additionally, each chapter ends with a summary that contains a brief description of the concepts and skills discussed in the chapter, plus examples illustrating these concepts and skills. The concepts and skills are also cross-referenced to the activity in which they appear, making the format easier to follow for those students who are unfamiliar with our approach. Each chapter also ends with a Gateway Review, providing students with an opportunity to check their understanding of the chapter's concepts and skills.

Supplements

Teacher Supplements

Annotated Instructor's Edition
ISBN   0-134-65715-2

This special version of the student text provides answers to all exercises directly beneath each problem.

Instructor's Resource Manual
ISBN 0-134-65447-1

This valuable teaching resource includes the following materials:

- Teaching notes for each chapter. These notes are ideal for those using the *Applied Mathematical Modeling and Problem Solving* approach for the first time.

- Extra practice skills worksheets for topics with which students typically have difficulty.

- Sample chapter tests and final exams for in-class and take-home use by individual students and groups.

- Sample journal topics for the students to write comments and observations about the course are included for each chapter.

- A section discussing learning in groups provides questions and answers for teachers trying collaborative learning for the first time.

- Information about incorporating technology in the classroom, including sample graphing calculator assignments.

TestGen®

(Download Only) available from the Instructor Resource Center. To request access, please visit www.pearsoned.com/testgen.

TestGen enables instructors to build, edit, print, and administer tests using a computerized bank of questions developed to cover all the objectives of the text. TestGen is algorithmically based, allowing instructors to create multiple but equivalent versions of the same question or test with the click of a button. Instructors can also modify test bank questions or add new questions. The software and testbank are available for download from Pearson Education's online catalog.

Supplements for Instructors and Students

MathXL® Online Course (access code required)

MathXL® is the homework and assessment engine that runs MyMathLab. (MyMathLab is MathXL plus a learning management system.)

With MathXL, instructors can:

- Create, edit, and assign online homework and tests using algorithmically generated exercises correlated at the objective level to the textbook.

- Create and assign their own online exercises and import TestGen tests for added flexibility.

- Maintain records of all student work tracked in MathXL's online gradebook.

With MathXL, students can:

- Take chapter tests in MathXL and receive personalized study plans and/or personalized homework assignments based on their test results.

- Use the study plan and/or the homework to link directly to tutorial exercises for the objectives they need to study.

- Access supplemental animations and video clips directly from selected exercises.

MathXL is available to qualified adopters. For more information, visit our website at www.mathxl.com, or contact your Pearson representative.

MyMathLab® Online Course (access code required)

MyMathLab from Pearson is the world's leading online resource in mathematics, integrating interactive homework, assessment, and media in a flexible, easy to use format. It provides **engaging experiences** that personalize, stimulate, and measure learning for each student. Moreover, it comes from an **experienced partner** with educational expertise and an eye on the future.

To learn more about how MyMathLab combines proven learning applications with powerful assessment, visit **www.mymathlab.com** or contact your Pearson representative.

Acknowledgments

A special thank-you to our families for their unwavering support and sacrifice, which enabled us to make this text a reality.

The Consortium for Foundation Mathematics

To the Student

The book in your hands is most likely very different from any mathematics textbook you have seen before. In this book, you will take an active role in developing the important ideas of mathematical modeling. It is the belief of the authors that students learn mathematics best when they are actively involved in solving problems that are meaningful to them.

Problem-Solving: The text is primarily a collection of situations drawn from real life. Each situation leads to one or more problems. By answering a series of questions and solving each part of the problem, you will be using and learning one or more mathematical ideas. Sometimes, these will be basic skills that build on your knowledge of arithmetic. Other times, they will be new concepts that are more general and far-reaching. The important point is that you won't be asked to master a skill until you see a real need for that skill as part of solving a realistic application.

Communication: Another important aspect of this text and the course you are taking is the benefit gained by collaborating with your classmates. Much of your work in class will result from being a member of a team. Working in small groups, you will help each other work through a problem situation. While you may feel uncomfortable working this way at first, there are several reasons we believe it is appropriate in this course. First, it is part of the learning-by-doing philosophy. You will be talking about mathematics, needing to express your thoughts in words. This is a key to learning. Secondly, you will be developing skills that will be very valuable when you leave the classroom. Currently, many jobs and careers require the ability to collaborate within a team environment. Your teacher will provide you with more specific information about this collaboration.

Representation and Technology: One more fundamental part of this course is that you will have access to appropriate technology. You will have access to calculators and some form of graphics tool—either a calculator or computer. Technology is a part of our modern world, and learning to use technology and recognizing what the results of the technology represents goes hand in hand with learning mathematics. Your work in this course will help prepare you for whatever you pursue in your working life.

This course will help you develop both the mathematical and general skills necessary in today's workplace, such as organization, problem solving, communication, representation, and collaborative skills. By keeping up with your work and following the suggested organization of the text, you will gain a valuable resource that will serve you well in the future. With hard work and dedication, you will be ready for the next step.

The Consortium for Foundation Mathematics

APPLIED MATHEMATICAL MODELING AND PROBLEM SOLVING

Introduction to Problem Solving and Mathematical Models

The Bookstore

Objectives

1. Practice communication skills.

2. Organize information.

3. Write a solution in sentences.

4. Develop problem-solving skills.

The latest novel by your favorite author has just arrived and by 11:00 A.M., a line has formed outside the crowded bookstore. Knowing you have class at noon, you ask the guard at the door how long you can expect to wait. She provides you with the following information: She is permitted to let 6 people into the bookstore only after 6 people have left; customers are leaving at the rate of 2 customers per minute; and she has just let 6 new customers in. Also, each customer spends an average of 15 minutes browsing at books and 10 minutes waiting in line to check out.

Currently 38 people are ahead of you in line. You know that it is a 10-minute walk to your noon class. Can you buy your book and still expect to make it to your noon class on time? Use the following questions to guide you in solving this problem.

1. What was your initial reaction after reading the problem?

2. Have you ever worked on a problem such as this before?

3. Organizing the information will help you solve the problem.

    a. How many customers must leave the bookstore before the guard allows more to enter?

    b. How many customers per minute leave the bookstore?

    c. How many minutes are there between groups of customers entering the bookstore?

    d. How long will you stand in line outside the bookstore?

e. Now finish solving the problem and answer the question: How early or late for your noon class will you be?

4. In complete sentences, write what you did to solve this problem. Then, explain your solution to a classmate.

SUMMARY: ACTIVITY 1.1

Steps in Problem Solving

1. Sort out the relevant information and organize it.

2. Discuss the problem with others to increase your understanding of the problem.

3. Write your solution in complete sentences to review your steps and check your answer.

EXERCISES: ACTIVITY 1.1

1. Think about the various approaches you and your classmates used to solve Activity 1.1, The Bookstore. Choose the approach that is best for you, and describe it in complete sentences.

2. What mathematical operations and skills did you use?

Activity 1.2

The Classroom

Objectives

1. Organize information.

2. Develop problem-solving strategies.
   - Draw a picture.
   - Recognize a pattern.
   - Do a simpler problem.

3. Communicate problem-solving ideas.

The Handshake

This mathematical modeling course involves working with other students in the class, so form a group of three, four, or five students. Introduce yourself to every other student in your group with a firm handshake. Share some information about yourself with the other members of your group.

1. How many people are in your group?

2. How many handshakes in all were there in your group?

3. Discuss how your group determined the number of handshakes. Be sure everyone understands and agrees with the method and the answer. Write the explanation of the method here.

 Shaking Hands

NUMBER OF STUDENTS IN GROUP	NUMBER OF HANDSHAKES
2	
3	
4	
5	
6	
7	

4. Share your findings with the other groups, and fill in the chart.

5. a. Describe a rule for determining the number of handshakes in a group of seven students.

b. Describe a rule for determining the number of handshakes in a class of n students.

6. If each student shakes hands with each other student, how many handshakes will be needed in your class?

7. Is shaking hands during class time a practical way for students to introduce themselves? Explain.

George Polya's book, *How to Solve It,* **outlines a four-step process for solving problems.**

1. Understand the problem (determine what is involved).

2. Devise a plan (look for connections to obtain the idea of a solution).

3. Carry out the plan.

4. Look back at the completed solution (review and discuss it).

8. Describe how your experiences with the handshake problem correspond with Polya's suggestions.

The Classroom

The tables in your classroom have square tops. Four students can comfortably sit at each table with ample working space. Putting tables together in clusters as shown will allow students to work in larger groups.

9. Construct a table of values for the number of tables and the corresponding total number of students.

NUMBER OF SQUARE TABLES IN EACH CLUSTER	TOTAL NUMBER OF STUDENTS
1	4
2	6

10. How many students can sit around a cluster of 7 square tables?

11. Describe the pattern that connects the number of square tables in a cluster and the total number of students that can be seated. Write a rule in sentences that will determine the total number of students that can sit in a cluster of a given number of square tables.

12. There are 24 students in a science course at your school.

   a. How many tables must be put together to seat a group of six students?

   b. How many clusters of tables are needed?

13. Discuss the best way to arrange the square tables into clusters given the number of students in your class.

SUMMARY: ACTIVITY 1.2

1. Problem-solving strategies include:

   • discussing the problem

   • organizing information

   • drawing a picture

   • recognizing patterns

   • doing a simpler problem

2. George Polya's book, *How to Solve It,* outlines **a four-step process for solving problems**.

   i. Understand the problem (determine what is involved).

   ii. Devise a plan (look for connections to obtain the idea of a solution).

   iii. Carry out the plan.

   iv. Look back at the completed solution (review and discuss it).

EXERCISES: ACTIVITY 1.2

1. At the opening session of the United States Supreme Court, each justice shakes hands with all the others. Select a technique, such as mental math, to determine the number of handshakes.

   a. How many justices are there?

   b. How many handshakes do they make?

2. Identify how the numbers are generated in this triangular arrangement, known as Pascal's triangle. Fill in the missing numbers.

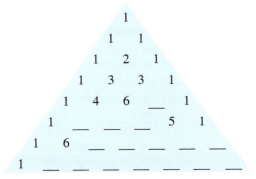

3. An **arithmetic sequence** is a list of numbers in which consecutive numbers share a common difference. Each number after the first is calculated by adding the common difference to the preceding number. For example, the arithmetic sequence 1, 4, 7, 10, . . . has 3 as its common difference. Identify the common difference in each arithmetic sequence that follows.

   a. 2, 4, 6, 8, 10, . . .

   b. 1, 3, 5, 7, 9, 11, . . .

   c. 26, 31, 36, 41, 46, . . .

4. A **geometric sequence** is a list of numbers in which consecutive numbers share a common ratio. Each number after the first is calculated by multiplying the preceding number by the common ratio. For example, 1, 3, 9, 27, . . . has 3 as its common ratio. Identify the common ratio in each geometric sequence that follows.

   a. 2, 4, 8, 16, 32, . . .

   b. 1, 5, 25, 125, 625 . . .

5. The operations needed to get from one number to the next in a sequence can be more complex. Describe a relationship shared by consecutive numbers in the following sequences.

   a. 2, 4, 16, 256, . . .

   b. 2, 5, 11, 23, 47, . . .

   c. 1, 2, 5, 14, 41, 122, . . .

6. In biology lab, you conduct the following experiment. You put two rabbits in a large caged area. In the first month, the pair produces no offspring (rabbits need a month to reach adulthood). At the end of the second month, the pair produces exactly one new pair of rabbits (one male and one female). The result makes you wonder how many male/female pairs you might have if you continue the experiment and each existing pair of rabbits produces a new pair each month, starting after their first month. The numbers for the first four months are calculated and recorded for you in the following table. The arrows in the table illustrate that the number of pairs produced in a given month equals the number of pairs that existed at the beginning of the preceding month. Continue the pattern and fill in the rest of the table.

Hare Today

MONTH	NUMBER OF PAIRS AT THE BEGINNING OF THE MONTH	NUMBER OF NEW PAIRS PRODUCED	TOTAL NUMBER OF PAIRS AT THE END OF THE MONTH
1	1	0	1
2	1	1	2
3	2	1	3
4	3	2	5
5			
6			
7			
8			

The list of numbers in the second column is called the **Fibonacci sequence**. This problem on the reproduction of rabbits first appeared in 1202 in the mathematics text *Liber Abaci*, written by Leonardo of Pisa (nicknamed Fibonacci). Using the first two numbers, 1 and 1, as a starting point, describe how the next number is generated. Your rule should generate the rest of the numbers shown in the sequence in column two.

7. If you shift all the numbers in Pascal's triangle so that all the 1s are in the same column, you get the following triangle.

   a. Add the numbers crossed by each arrow. Put the sums at the tip of the arrow.

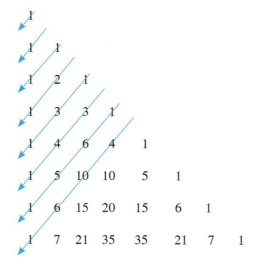

b. What is the name of the sequence formed by these sums?

8. There are some interesting patterns within the Fibonacci sequence itself. Take any number in the sequence and multiply it by itself, and then subtract the product of the number immediately before it and the number immediately after it. What is the result? Pick another number and follow the same procedure. What result do you obtain? Try two more numbers in the sequence.

For example, choose 5.   For example, choose 3.

$5 \cdot 5 = 25$ $3 \cdot 3 = 9$

$3 \cdot 8 = 24$ $2 \cdot 5 = 10$

$25 - 24 = 1$ $9 - 10 = -1$

9. a. Complete the following table by performing the following sequence of operations on the original number: Multiply the number by 4, add 12 to the product, divide the sum by 2, and subtract 6 from the quotient.

ORIGINAL NUMBER	RESULT OF SEQUENCE OF OPERATIONS
2	
5	
10	
15	

b. There is a hidden pattern in the sequence of operations that causes the original number and the result of the sequence of operations to always have the same relationship. What relationship do you observe between the number and the result?

Based on the specific examples in part a, you might conclude that if you follow the given sequence of operations on *any* number, the result would always be twice the original number. Arriving at a general conclusion from specific examples is a type of reasoning called **inductive reasoning** or **induction**. However, this type of reasoning does not prove that the general conclusion is true for all numbers.

c. Generalize the situation in part a by choosing a variable to represent the original number. Use x as your variable and perform the sequence of operations on the variable x. Simplify the resulting algebraic expression.

d. What general conclusion is proven in part c?

The type of reasoning used in parts c and d is called **deductive reasoning** or **deduction**. Deductive reasoning is used to prove conjectures true or false.

10. a. Explore the Internet and obtain data and information regarding the role of team work in the workplace. What are the characteristics of an effective work team?

b. Do you believe that your instructor's use of group work will help prepare you for working in groups in a workplace? Explain.

c. What issues might you experience in class group work that could also possibly arise in the workplace? How can these issues be resolved?

Activity 1.3

Make Me an Offer

Objectives

1. Use the basic steps for problem solving.

2. Translate verbal statements into algebraic equations.

3. Use the basic principles of algebra to solve real-world problems.

4. Use formulas to solve problems.

Many students believe "I can do math. I just can't do word problems." In reality, you learn mathematics so you can solve practical problems in everyday life, and in science, business, technology, medicine, and most other fields.

On a personal level, you solve problems every day by calculating the delivery charge for an online order, remembering a birthday, or figuring out how to get more exercise. Most every-day problems don't require algebra, or even arithmetic, to be solved. But the basic steps and methods for solving any problem can be discussed and understood. In the process you will, with practice, become a better problem solver.

Solve the following problem any way you can, making notes of your thinking as you work. In the space provided, record your work on the left side. On the right side jot down in a few sentences what you are thinking as you work on the problem.

"Night Train" Henson's Contract

1. "Night Train" Henson is negotiating a new contract with his team. He wants $800,000 for the year with an additional $6000 for every game he starts. His team offered $10,000 for every game he starts, but only $700,000 for a base salary. How many games would he need to start in order to make more with the team's offer?

SOLUTION TO THE PROBLEM	YOUR THINKING

2. Talk about and compare your methods for solving the above problem with classmates or your group. As a group, write down the steps you all went through from the very beginning to the end of your problem solving.

> ### Procedure
>
> **Basic Steps for Problem Solving**
>
> Solving any problem generally requires the following four steps.
>
> 1. Understand the problem.
>
> 2. Develop a strategy for solving the problem (Devise a plan).
>
> 3. Execute your strategy to solve the problem (Carry out the plan).
>
> 4. Check your solution for correctness (Look back at the completed solution).

3. How do these four steps compare to your group's solutions? How do your group's answers compare with other groups' answers?

Let's consider each of the steps 1 through 4.

Step 1. Understand the problem.   This step may seem obvious, but many times this is the most important step, and is the one that commonly leads to errors. Read the problem carefully, as many times as necessary. Draw some diagrams to help you visualize the situation. Get an explanation from available resources if you are unsure of any details.

4. List some strategies you could use to make sure you understand the "Night Train" Henson problem.

Step 2. Develop a strategy for solving the problem.   Once you understand the problem, it may not be at all clear what strategy is required. Practice and experience is the best guide. Algebra is often useful for solving a math problem, but is not always required.

5. How many different strategies were tried in your group for solving the "Night Train" problem? Describe one or two of them.

Step 3. Execute your strategy to solve the problem.   Once you have decided on a strategy, carrying it out is sometimes the easiest step in the process. If you really understand the problem, and are clear on your strategy, the solution should happen almost automatically.

6. When you decided on a strategy, how confident did you feel about your solution?

Step 4. Check your solution for correctness.   This last step is *critically important.* Your solution must be reasonable, it must answer the question, and most importantly, it must be correct.

7. Are you absolutely confident your solution is correct? If so, how do you know?

Solution: "Night Train" Henson's Contract

As beginning problem solvers, you may find it useful to develop a consistent strategy. Let's examine one possible way you can solve "Night Train's" problem by applying some algebra tools.

Step 1. Understand the problem.   If you understand the problem, you will realize that "Night Train" needs to make $100,000 more with the team's offer (800,000 − 700,000) to make up for the difference in base salary. So he must start enough games to make up for this difference. Otherwise, he should not want the team's offer.

Recall that a **variable**, usually represented by a letter, is a quantity or quality that may change in value from one particular instance to another. In this situation, you could let the variable x represent the number of games "Night Train" needs to start to make the same amount of money with either offer. Then you note that he would make $4000 more for each game started with the team's offer. So $4000 for each game started times the number of games started must equal $100,000.

Step 3. Execute your strategy for solving the problem.   You can translate the statement above into an equation, then apply the fundamental principles of solving equations to obtain the number of games.

$$4000x = 100,000$$
$$4000x \div 4000 = 100,000 \div 4000$$
$$x = 25$$

Your answer would be: "Night Train needs to start in more than 25 games to make more with the team's offer."

Step 4. Check your solution for correctness.   This answer seems reasonable, but to check it you should refer back to the original statement of the problem. If "Night Train" starts in exactly 25 games his salary for the two offers can be calculated.

Night Train's demand: $800,000 + $6000 \times 25 = $950,000

Team's offer: $700,000 + $10,000 \times 25 = $950,000

If he starts in 26 games, the team's offer is $960,000 and "Night Train's" demand is $956,000. This confirms the solution.

There are other ways to solve this problem. But this gives you an example of how algebra can be used. Use algebra in a similar way to solve the following problems. Be sure to follow the four steps of problem solving.

Additional Problems to Solve

8. You search the Internet for a basic cell phone voice plan for your grandmother. She is interested in being able to have a cell phone for limited use and in case of emergencies. You find two possible plans:

Plan 1: $29 fee per month plus $0.05 for each minute
Plan 2: $10 fee per month plus $0.25 for each minute

How many minutes would you need to make Plan 1 the better deal?

Step 1. Understand the problem.

Step 2. Develop a strategy for solving the problem.

Step 3. Execute your strategy to solve the problem.

Step 4. Check your solution for correctness.

9. The perimeter of a rectangular pasture is 2400 feet. If the width is 800 feet, how long is the pasture?

Step 1. Understand the problem.

Length = ?

Width = 800 ft.   800 ft.

Step 2. Develop a strategy for solving the problem.

Step 3. Execute your strategy to solve the problem.

Step 4. Check your solution for correctness.

Formulas

The solution to Problem 9 involved the relationship between a rectangle's perimeter and its length and width. Stated in words, the perimeter is the sum of twice the width and twice the length. This can be translated into an equation involving letters called a **formula**. If P represents the perimeter, w represents the width, and l represents the length, the stated relationship can be written as

$$P = 2w + 2l.$$

Formulas are useful in many business applications. For Problems 10–12,

    a. First, choose appropriate letters to represent each variable quantity and write what each letter represents.

    b. Use the letters to translate each stated relationship into a symbolic formula.

    c. Use the formula to solve the exercise.

10. Net income is equal to the total revenue from selling an item minus the cost of producing the item. Determine the net income if the revenue from a business is $400,000 and the cost is $156,800.

11. Net pay is the difference between a worker's gross income and his or her deductions. A person's gross income for the year was $65,000 and their total deductions were $12,860. What was their net pay for the year?

12. Annual depreciation equals the difference between the original cost of an item and its remaining value, divided by its estimated life in years. Determine the annual depreciation of a new car that costs $25,000, has an estimated life of ten years, and a remaining value of $2000.

Formulas are also found in the sciences and healthcare. In Problems 13–15, use the given formulas to solve each problem.

13. In general, temperature is measured using either the Celsius (C) or Fahrenheit (F) scales. To convert from degrees Celsius to degrees Fahrenheit, use the following formula.

$$F = (9C \div 5) + 32$$

    a. Use the formula to convert 20°C into degrees F.

    b. 100°C is equal to how many degrees F?

14. Belgian statistician, Adolphe Quetelet, developed the Body Mass Index (BMI) formula in the 1800s. BMI is an internationally used measure to determine your weight status.

BMI	Weight Status
Below 18.5	Underweight
18.5–24.9	Normal
25–29.9	Overweight
30 or greater	Obese

The formula for BMI, B, is

$$B = \frac{703w}{h^2},$$

where w is weight in pounds and h is height in inches.

a. What is your Body Mass Index?

b. Suppose your friend weighs 180 pounds. Substitute this value for w to obtain a formula for B in terms of height h.

c. Complete the following table using the formula for Body Mass Index of a 180-pound person.

h, HEIGHT IN INCHES	60	64	68	72	76	80
B, BODY MASS INDEX						

d. What happens to the Body Mass Index as height increases? Does this make sense in the context of the situation? Explain why or why not.

15. The distance an object travels is determined by how fast it goes and for how long it moves. If an object's speed remains constant, the formula is $d = r \cdot t$, where d is the distance, r is the speed, and t is the time.

a. If you drive at 50 miles per hour for 5 hours, how far will you have traveled?

b. A satellite is orbiting the Earth at the rate of 25 kilometers per minute. How far will it travel in two hours?

Orbits at 25 km per minute

SUMMARY: **ACTIVITY 1.3**

The Four Steps of Problem Solving

Step 1. Understand the problem.

  a. Read the problem completely and carefully.

  b. Draw a sketch of the problem, if possible.

Step 2. Develop a strategy for solving the problem.

  a. Identify and list everything you know about the problem, including relevant formulas. Add labels to the diagram if you have one.

  b. Identify and list what you want to know.

Step 3. Execute your strategy to solve the problem.

  a. Write an equation that relates the known quantities and the unknowns.

  b. Solve the equation.

Step 4. Check your solution for correctness.

  a. Is your answer reasonable?

  b. Is your answer correct? (Does it answer the original question? Does it agree with all the given information?)

EXERCISES: **ACTIVITY 1.3**

In Exercises 1–7, solve each problem by applying the four steps of problem solving. Use the strategy of solving an algebra equation for each problem.

1. In preparing for a family trip, your assignment is to make the travel arrangements. You can rent a car for $75 per day with unlimited mileage. If you have budgeted $600 for car rental, how many days can you drive?

2. You need to drive 440 miles to get to your best friend's wedding. How fast must you drive to get there in 8 hours?

3. Your goal is to save $1200 to pay for next year's books and fees. How much must you save each month if you have 5 months to accomplish your goal?

4. You have enough wallpaper to cover 240 square feet. If your walls are 8 feet high, what wall length can you paper?

5. In your part-time job selling cutlery products, you have two different knife sets available. The better set sells for $35, the cheaper set for $20. Last week you sold more of the cheaper set, in fact twice as many as the better sets. Your receipts for the week totaled $525. How many of the better sets did you sell?

6. A rectangle that has an area of 357 square inches is 17 inches wide. How long is the rectangle?

7. A rectangular field is five times longer than it is wide. If the perimeter is 540 feet, what are the dimensions (length and width) of the field?

$$w \boxed{}$$
$$5w$$

8. In your part-time job, your hours vary each week. Last month you worked 22 hours the first week, 25 hours the second week, 14 hours during week three and 19 hours in week four. Your gross pay for those four weeks was $960.

   a. Let p represent your hourly pay rate. Write the equation for your gross pay over these four weeks.

   b. Solve the equation in part a to determine your hourly pay rate.

9. You are a sales associate in a large retail electronics store. Over the first three months of the year you sold 12, 15, and 21 ProPix digital cameras. The total in sales was $18,960.

   a. Let p represent the price of one ProPix digital camera. Write the equation for your total in sales over these three months.

   b. Solve the equation in part a to find the retail price of the ProPix digital camera.

10. Evaluate each formula for the given value(s).

   a. $d = r \cdot t$, where $r = 35$ and $t = 6$

   b. $F = ma$, where $m = 120$ and $a = 25$

   c. $V = lwh$, where $l = 100$, $w = 5$, and $h = 25$

   d. $F = \dfrac{mv^2}{r}$, where $m = 200$, $v = 25$, and $r = 125$

   e. $A = \dfrac{t_1 + t_2 + t_3}{3}$, where $t_1 = 76$, $t_2 = 83$, and $t_3 = 81$

Activity 1.4

Proportional Reasoning and Scaling

Objectives

1. Use proportional reasoning as a problem-solving strategy.

2. Write a proportion and then solve the resulting proportion.

3. Use scale factors to solve problems.

The following table summarizes Michael Jordan's statistics during the six games of one of his National Basketball Association (NBA) championship series.

GAME	POINTS	FIELD GOALS	FREE THROWS
1	28	9 out of 18	9 out of 10
2	29	9 out of 22	10 out of 16
3	36	11 out of 23	11 out of 11
4	23	6 out of 19	11 out of 13
5	26	11 out of 22	4 out of 5
6	22	5 out of 19	11 out of 12

1. What was his points-per-game average over the six-game series?

2. In which game did he score the most points?

3. In which game(s) did he score the most field goals? The most free throws?

Problem 3 focused on the *actual* number of Jordan's successful field goals and free throws in these six games. Another way of assessing Jordan's performance is to *compare* the number of successful shots to the total number of attempts for each game. This comparison gives you information on the *relative* success of his shooting. For example, in each of games 1 and 2, Jordan made 9 field goals. Relatively speaking, you could argue that he was more successful in game 1 because he made 9 out of 18 attempts; in game 2, he only made 9 out of 22 attempts.

4. Use the free-throw data from the six games to express Jordan's *relative* performance in the given comparison formats (verbal, fraction, division, and decimal). The data from the first game have been entered for you.

Jordan's Relative Free-Throw Performance

	VERBAL	FRACTION	DIVISION	DECIMAL
GAME 1	9 out of 10	$\frac{9}{10}$	$9 \div 10$ or $10\overline{)9}$	0.90
GAME 2				
GAME 3				
GAME 4				
GAME 5				
GAME 6				

5. a. For which of the six games was his relative free-throw performance highest?

   b. Which comparison format did you use to answer part a? Why?

6. For which of the six games was Jordan's *actual* free-throw performance the lowest?

7. For which of the six games was Jordan's *relative* free-throw performance the lowest?

When relative comparisons using quotients are made between different values or quantities of the same kind (e.g., number of baskets to number of baskets), the comparison is called a **ratio**. Ratios can be expressed in several forms—verbal, fraction, division, or decimal, as you saw in Problem 4.

Proportional reasoning is the ability to recognize when two ratios are equivalent, that is, when equivalent ratios represent the same relative performance level.

Definition

Two ratios are said to be **equivalent** if the ratios have equal numerical (e.g., decimal or fraction) values. The mathematical statement that two ratios are equivalent is called a **proportion**. In fraction form, the proportion is written $\dfrac{a}{b} = \dfrac{c}{d}$.

You can determine *equivalent* ratios the same way you determine equivalent fractions. For example, 3 out of 4 is equivalent to 6 out of 8, because $\dfrac{3}{4} \cdot \dfrac{2}{2} = \dfrac{6}{8}$.

8. Fill in the blanks in each of the following proportions.

   a. 3 out of 4 is equivalent to _____ out of 12

   b. 3 out of 4 is equivalent to _____ out of 32

   c. 3 out of 4 is equivalent to _____ out of 100

   d. Write the resulting proportion from part c using a fraction format.

9. a. Explain why the following ratios are equivalent.

      i. 27 out of 75   **ii.** 63 out of 175   **iii.** 36 out of 100

b. Write each ratio in fraction form.

    i.   **ii.**   **iii.**

c. Determine the "reduced" form of the equivalent fractions from part b.

d. With which of the equivalent ratios in part a do you feel most comfortable? Explain.

The number 100 is a very familiar quantity of comparison: There are 100 cents in a dollar and frequently 100 points on a test. Therefore, people feel most comfortable with a ratio such as 70 out of 100 or 36 out of 100.

> The phrase "out of 100" is commonly referred to by its Latin equivalent, *percent*. Per means "division" and cent means "100," so **percent** means "divide by 100."

Therefore, 70 out of 100 can be rephrased as 70 percent and written in the familiar notation 70%, which equals $70 \div 100 = \frac{70}{100} = 0.70$. Similarly, 36 out of 100 can be rephrased as 36 percent and written in the familiar notation $36\% = 36 \div 100 = \frac{36}{100} = 0.36$.

10. Complete the following table using Michael Jordan's field goal data from the beginning of the activity.

Jordan's Relative Field Goal Performance

	VERBAL	FRACTION	DECIMAL	PERCENT
GAME 1	9 out of 18	$\frac{9}{18}$	0.50	50%
GAME 2				
GAME 3				
GAME 4				
GAME 5				
GAME 6				

Solving Proportions

In an effort to increase the education level of their police officers, many municipalities are requiring new recruits to have at least a two-year college degree. A recent survey indicated that 2 out of 5 police officers in your city hold a four-year college degree. There were approximately 3,400 officers when the survey was conducted.

To calculate the number of four-year college degree holders, you can start with the proportion statement

$$2 \text{ out of } 5 = \underline{\ ? \ } \text{ out of } 3,400$$

This proportion can be written in fraction form as

$$\frac{2}{5} = \frac{n}{3,400},$$

where n is the unknown quantity of the proportion.

> One method of solving proportions uses the fact that the two mathematical statements
>
> $$\frac{a}{b} = \frac{c}{d} \quad \text{and} \quad a \cdot d = b \cdot c$$
>
> are equivalent.

Transforming the statement on the left (containing two fractions) into the statement on the right is commonly called **cross-multiplication** because the numerator of the first fraction is multiplied by the denominator of the second and the numerator of the second fraction is multiplied by the denominator of the first.

You can use cross-multiplication to solve proportions as demonstrated in Example 1:

Example 1   *Solve the proportion $\dfrac{3}{4} = \dfrac{n}{1460}$.*

SOLUTION

Original proportion:	$\dfrac{3}{4} = \dfrac{n}{1460}$
Cross multiply:	$3 \cdot 1460 = 4 \cdot n$
Divide both sides by 4:	$\dfrac{3 \cdot 1460}{4} = \dfrac{\cancel{4} \cdot n}{\cancel{4}}$
Simplify:	$1095 = n$

11. Use the cross-multiplication method to solve the proportion $\dfrac{2}{5} = \dfrac{n}{3,400}$.

12. Solve the following proportions.

    a. 2 out of 3 = _____ out of 36

    b. $\dfrac{2}{3} = \dfrac{n}{45}$

Application

In 2013, the Texas legislature passed a bill that reduced the number of required end-of-course examinations for high school students. The hope is that while the testing load will be reduced, graduation rates will increase and the achievement levels of the students will not be compromised. In your cousin's county last year, 8 out of 10 high school seniors actually graduated.

13. If 5400 students in your cousin's county graduated last year, estimate (without actually doing a calculation) the total number of high school seniors in that county last year.

This situation differs from the police officer's situation at the beginning of the activity because the total number is not known. The 5400 represents that part of the total number of high school seniors who graduated. Written as a proportion,

$$8 \text{ out of } 10 = 5400 \text{ out of } \underline{\quad ? \quad}$$

14. a. Rewrite this proportion in fraction form. Let n represent the unknown.

b. To determine the total number of seniors, use cross-multiplication to calculate the unknown denominator in part a.

15. Solve these proportions.

    a. 2 out of 3 = 80 out of _____.   **b.** $\dfrac{2}{3} = \dfrac{216}{n}$

Scale Factors

Architects, engineers, filmmakers, photographers, and artists are among the many professions that use scaling in their occupation. For example, when architects are working on large projects, like skyscrapers, they will often build a scale model before they start construction on the real project. Engineers will often build scale models of new projects to test them to save on the cost of constructing a full-size prototype. In artwork, scale is the size of the art object relative to another object.

> **Definition**
>
> The ratio of the length of the scale drawing to the corresponding length of the actual object is called the **scale factor**.
>
> $$\text{scale factor} = \frac{\text{length of model}}{\text{length of actual}} \text{ or}$$
>
> $$\text{scale factor} = \text{length of model : length of actual}$$

The scale factor is a number used as a multiplier in scaling. A reduction has a scale factor less than 1, and an enlargement has a scale factor greater than 1. For example,

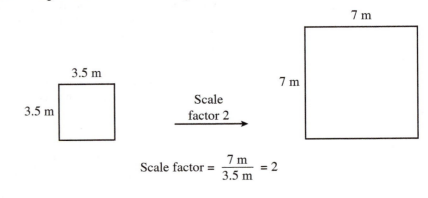

$$\text{Scale factor} = \frac{7 \text{ m}}{3.5 \text{ m}} = 2$$

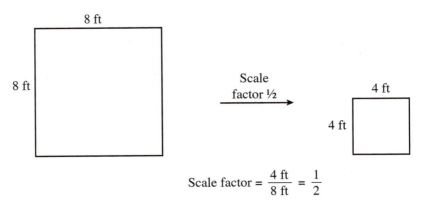

$$\text{Scale factor} = \frac{4 \text{ ft}}{8 \text{ ft}} = \frac{1}{2}$$

Scale factors answer the question: given a model, how many times bigger (or smaller) is the real thing?

> Multiplying the dimensions of a geometric figure by a scale factor results in a figure that is **similar** to the original. Similar figures have the exact same shape, but generally different sizes. The ratios of the lengths of the corresponding sides of similar figures are constant and are equal to the scale factor.

Be careful! In a scale drawing, you might have 1 inch on the drawing corresponding to 15 feet of actual length. The ratio 1 inch to 15 feet or 1:15 is called the **scale**, or **scale ratio**. This ratio has measurements in different units: the scale drawing length is in inches and the actual measurement is in feet. The **scale factor** is always *a ratio between distances measured in the same units*. Therefore, the scale factor of 1 inch corresponds to $(15 \cdot 12) = 180$ inches, written as $\frac{1}{180}$ or 1:180.

The scale factor generally gives you a better sense of the comparison between the model and the actual object. Since the scale factor is $\frac{1}{180}$, the scale drawing lengths are $\frac{1}{180}$th of the actual lengths.

Example 2   *Working with a design engineer, a blueprint has a scale factor of 3:200. If the room dimensions are 30 feet by 9 feet, what lengths would be drawn on the blueprint?*

SOLUTION

The scale factor 3:200 means that the ratio of the corresponding length of the model drawing to the room length as well as the ratio of the corresponding width of the model drawing to the room width must be equal to $\frac{3}{200}$. This is a proportion problem.

Step 1. Let l represent the length of the model drawing, and solve the following proportion:

$$\frac{\text{length of model}}{\text{length of actual}} = \frac{3}{200} = \frac{l}{30}$$

$$3 \cdot 30 = 200 \cdot l$$

$$\frac{3 \cdot 30}{200} = \frac{200 \cdot l}{200}$$

$$0.45 \text{ ft} = l \text{ or } 5.4 \text{ inches}$$

Step 2. Let w represent the width of the model drawing, and solve the corresponding proportion.

$$\frac{\text{width of model}}{\text{width of actual}} = \frac{3}{200} = \frac{w}{9}$$

$$3 \cdot 9 = 200 \cdot w$$

$$\frac{3 \cdot 9}{200} = \frac{200 \cdot w}{200}$$

$$0.135 \text{ ft} = w \text{ or } 1.62 \text{ inches}$$

In Example 2, you determined the scale model dimensions from the actual dimensions. In the following problem, use the same problem-solving strategy to determine the dimension(s) of the actual object from the scale model drawing.

16. Model trains are scale models of actual trains. You recently encountered a 1:87 scale model of the "Glaskasten" or Glass Box. Glaskasten was a lightweight tank locomotive designed to handle local passenger traffic in Bavaria in the early 1900s. These locomotives saw service in Germany, Austria, Switzerland, and Norway. The model is 3.15 inches long. How long was the Glaskasten in feet?

17. A 3×5 inch photo (height $\times$ width) is enlarged 300% in both height and width. Determine the dimensions of the enlargement. Recall that 300% is equal to $\frac{300}{100}$ or 3.

Proportional vs. Non-Proportional

In Example 2, the blueprint had a scale factor of 3:200. Therefore, the ratio of corresponding length of the model drawing to the actual room length was $\frac{3}{200}$. Alternatively, you could state that the 30-foot length of the room multiplied by the scale factor $\frac{3}{200}$ gives the corresponding length of the model drawing.

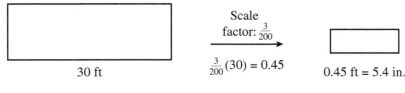

Scale
factor: $\frac{3}{200}$

$\frac{3}{200}(30) = 0.45$

30 ft

0.45 ft = 5.4 in.

Actual Room
Dimensions

Model Drawing

Rather than having a specific length of 30 feet for the room, suppose the length of the actual room is variable and is represented by x. If y represents the length of the model drawing, then the relationship between these two variable quantities can be expressed by the equation

$$y = \frac{3}{200}x.$$

18. a. Use the equation $y = \frac{3}{200}x$ to complete the following table, where x represents the length of the room and y represents the corresponding length of the model drawing.

x (FEET)	30	25	20	15	10
y (FEET)					

   b. Select a pair of values of x and y from the table in part a, and determine the ratio $\frac{y}{x}$. Is this value what you expected? Explain your answer.

In such a case, the variable quantities x and y are said to be **proportional**. In similar fashion, the width of the actual room and the corresponding width of the model drawing are also **proportional**.

Definition

Two quantities Q_1 and Q_2 are **proportional** if they have the same constant ratio $\frac{Q_2}{Q_1}$. If the ratio $\frac{Q_2}{Q_1}$ is not constant, the two quantities are said to be **non-proportional**.

If k represents the constant ratio, then the proportional relationship between Q_1 and Q_2 can be written as

$$\frac{Q_2}{Q_1} = k, \text{ or equivalently as } Q_2 = kQ_1.$$

Therefore, a proportional relationship between two variable quantities Q_1 and Q_2 can be described by an equation of the form $Q_2 = kQ_1$, where k is the scale factor (constant ratio).

19. Due to a change in zoning restrictions, an architect needs to modify his plans for a rectangular space near the building as follows:

   i. double the length of the rectangular space
   ii. increase the width by 10 feet.

   a. Write an equation that expresses the length y of the new rectangular space in terms of the length x of the original rectangular space.

   b. Is the length y of the new rectangular space proportional to the length of the original? Explain.

c. Write an equation that expresses the width of the new rectangular space in terms of the width of the original space.

d. Is the width of the new space proportional to the original? Explain.

SUMMARY: ACTIVITY 1.4

1. When comparisons using quotients are made between different quantities of the same kind, the comparison is called a **ratio**.

2. Ratios can be expressed verbally (4 out of 5), as a fraction $\frac{4}{5}$, as a division $(5\overline{)4})$, as a decimal (0.80), or as a percent (80%).

3. Two ratios are said to be equivalent if the ratios have equal numerical values. The mathematical statement that two ratios are equivalent is called a **proportion**.

4. A proportion expressed in fraction form, $\frac{a}{b} = \frac{c}{d}$, is equivalent to the statement $ad = bc$.

5. Problems that involve **proportional reasoning** usually include a known ratio $\frac{a}{b}$ and a given piece of information, either a "part" or a "total" value, resulting in the proportion:

$$\frac{a}{b} = \frac{part}{total}$$

The missing value can be determined by cross-multiplying and solving the resulting equation.

6. The ratio of the length of the scale drawing to the corresponding length of the actual object is called the **scale factor**.

$$\text{scale factor} = \frac{\text{length of model}}{\text{length of actual}} \text{ or}$$

$$\text{scale factor} = \text{length of model : length of actual}$$

7. Two quantities Q_1 and Q_2 are **proportional** if they have the same constant ratio $\frac{Q_2}{Q_1}$. If the ratio $\frac{Q_2}{Q_1}$ is not constant, the two quantities are said to be **non-proportional**.

EXERCISES: ACTIVITY 1.4

1. Here are your scores on three graded assignments. On which assignment did you perform best?

a. 25 out of 30   b. 30 out of 40   c. 18 out of 25

2. At competitive colleges, the admissions office often compares the number of students accepted to the total number of applications received. This comparison is known as the *selectivity index*. The admissions office also compares the number of students who actually attend to the number of students who have been accepted for admission. This comparison is known as the *yield*. Complete the following table to determine the selectivity index and yield (in percent format) for colleges A, B, and C.

	NUMBER OF APPLICANTS	NUMBER ACCEPTED	NUMBER ATTENDING	SELECTIVITY INDEX	YIELD (AS A %)
COLLEGE A	5500	3500	1000		
COLLEGE B	8500	4800	2100		
COLLEGE C	4200	3200	900		

Which college do you think is the most competitive? The least competitive? Explain.

3. Solve each proportion for the unknown quantity.

   a. $\dfrac{2}{3} = \dfrac{x}{48}$ **b.** $\dfrac{5}{8} = \dfrac{120}{x}$ **c.** $\dfrac{3}{20} = \dfrac{x}{3500}$

In Exercises 4–12, write a proportion that represents the situation and then determine the unknown value in the proportion.

4. A conservationist can estimate the total number of a certain type of fish in a lake by using a technique called "capture-mark-recapture." Suppose a conservationist marks and releases 100 rainbow trout into a lake. A week later she nets 50 trout in the lake and finds 3 marked fish. An estimate of the number of rainbow trout in the lake can be determined by assuming that the proportion of marked fish in the sample taken is the same as the proportion of marked fish in the total population of rainbow trout in the lake. Use this information to estimate the population of rainbow trout in the lake.

5. The school taxes on a house assessed at $210,000 are $2340. At the same tax rate, what are the taxes (to the nearest dollar) on a house assessed at $275,000?

6. According to the local utility company's report, 7 out of every 25 homes are heated by electricity. At this rate, predict how many homes in a community of 12,000 would be heated by electricity.

7. A hybrid car can travel 45 miles on one gallon of gas. Determine the amount of gas needed for a 500 mile trip.

8. A normal 10 cc specimen of human blood contains 1.2 g of hemoglobin. How much hemoglobin would 16 cc of the same blood contain?

9. The ratio of the weight of an object on Mars to the weight of the same object on Earth is 0.4 to 1. How much would a 170 pound astronaut weigh on Mars?

10. The ancient Greeks thought that the most pleasing shape for a rectangle was one for which the ratio of length to width was 8 to 5. This ratio is called the Golden Ratio. If the length of a rectangular painting is 20 inches, determine the width of the painting in order for the length and width of the painting to have the Golden Ratio.

11. The designers of sport shoes assume that the force exerted on the soles of shoes in a jump shot is proportional to the weight of the person jumping. A 140-pound athlete exerts a force of 1960 pounds on his shoe soles when he returns to the court floor after a jump. Determine the force that a 270-pound professional basketball player exerts on the soles of his shoes when he returns to the court floor after shooting a jump shot.

12. Your high school softball team won 80% of the games it played this year. If your team won 20 games, how many games did it play?

13. You are running for president of student government. Your friend has designed a campaign poster that measures 3 feet by 2 feet. She duplicated the poster on a rectangular button measuring 2 inches by $1\frac{1}{3}$ inches. What is the scale factor?

14. Determine the scale factor used to redraw a circle with radius 9 in. so the new circle has the following measurements.

    a. Circle with diameter 24 in.   **b.** Circle with diameter 6 in.

15. Given an original 4 by 6 inch picture, determine the dimensions of a scale drawing of the picture having the following scale factors.

   a. Scale factor 3

   b. Scale factor $\frac{1}{2}$

16. You take a picture of a 3 ft by 3 ft portrait with your Smartphone. If the scale factor is $\frac{1}{18}$, determine the dimensions of the portrait on your phone.

17. The dimensions of a tablet in an advertisement are 2.35 inches in height and 1.65 inches in width. If the scale factor is 0.25, what are the dimensions of the actual tablet?

18. An 8 inch $\times$ 11 inch photo is enlarged by 200%. What are the dimensions of the enlargement?

19. You are looking at a map of the Midwest section of the United States. The scale indicates that 1 inch is equivalent to 100 miles. The distance on the map from St. Louis, Missouri, to Louisville, Kentucky, is $2\frac{5}{8}$ inches. How far is it from St. Louis to Louisville?

20. You land a new job as the advertising agent for motorcycles. The scale factor you will use to make your drawings for the ads is 1:24. If the actual motorcycle is 83 inches long and 33 inches wide, how long and how wide will the motorcycle be in your ad when drawing it to scale?

21. You are going to make a scale drawing of a ladybug for your science class project using a scale factor of 16. If the ladybug is 10 mm long and 6 mm wide, what will be the length and the width of the sketch in centimeters?

22. Model cars, trucks, and motorcycles are popular for many hobbyists and collectors. Models are carefully scaled so that the dimensions are proportional to the actual vehicle.

   a. The following are the approximate dimensions of a model 2010 Ford F-150 STX red pickup truck having a scale of $\frac{1}{27}$: length of 7.5 inches, width of 3 inches, and height of 2.5 inches. Determine the approximate dimensions of the full-size 2010 Ford truck. Give your answer in feet, and round to the thousandths.

b. The following are the approximate dimensions of a model 1973 Ford F-100 style side blue pickup truck having a scale of $\frac{1}{25}$: length of 8.75 inches, width of 3.5 inches, and height of 3 inches. Determine the approximate dimensions of the full-size 1973 Ford truck. Give your answer in feet, and round to the thousandths.

c. Compare the dimensions of the 2010 red Ford pickup from part a with that of the 1973 blue Ford pickup from part b. Which dimension had the biggest percent change?

23. Determine whether each relationship is proportional or non-proportional. If the relationship is proportional, determine the scale factor.

a. A local frozen yogurt shop sells its yogurt for a price that depends on the weight.

SERVING (OUNCES), x	3 oz	5 oz	8 oz	12 oz
PRICE, y	$2.25	$3.25	$4.25	$5.25

b. A gourmet chocolate shop sells its chocolates and truffles by the piece.

NUMBER OF CHOCOLATES AND/OR TRUFFLES, x	6	9	20	36
COST OF SPECIALTY BOX, y	$10.80	$16.20	$36.00	$64.80

Fuel Economy

Objectives

1. Apply rates directly to solve problems.

2. Use proportions to solve problems.

3. Use unit or dimensional analysis to solve problems.

You are excited about purchasing a reliable used car for your commute to school during your senior year. Concerned about the cost of driving, you do some research on the Internet and come across a Web site that lists fuel efficiency, in miles per gallon (mpg), for five cars that you are considering. You record the mpg for city and highway driving in the following table.

1. a. For each of the cars listed in the following table, how many city miles can you travel per week on five gallons of gasoline? Explain the calculation you will do to obtain the answers. Record your answers in the third column of the table.

Fuel for Thought

MAKE/MODEL	CITY MPG	CITY MILES ON 5 GAL OF GAS	HIGHWAY MPG	GAL NEEDED TO DRIVE 304 MILES	FUEL TANK CAPACITY IN GAL
Chevrolet Cruze	32		42		14.0
Ford Focus	27		36		13.2
Honda Civic	29		41		13.2
Hyundai Accent	28		38		11.4
Toyota Corolla	34		41		11.9

b. The round-trip drive to school and back home is 30.4 city miles, which you do five days a week. Which of the cars would get you to school and back home each week on five gallons of gas?

2. Suppose your round-trip commute to a summer job would be 304 highway miles each week. How many gallons of gas would each of the cars require? Explain the calculation you do to obtain the answers. Record your answers to the nearest tenths in the fifth column of the above table.

Rates and Unit Analysis

Miles per gallon (mpg) is an example of a rate. Mathematically, a **rate** is a comparison of two quantities that have different units of measurement. Numerically, you calculate with rates as you would with ratios. Paying attention to what happens to the units of measurement during the calculation is critical to determining the unit of the result.

Definition

A **rate** is a comparison, expressed as a quotient, of two quantities that have different units of measurement.

Unit analysis (sometimes called *dimensional analysis*) uses units of measurement as a guide in setting up a calculation or writing an equation involving one or more rates. When the calculation or equation is set up properly, the result is an answer with the appropriate units of measurement.

There are two common methods using unit analysis to solve problems involving rates. One method is to apply the known rate directly by multiplication or division. The second method is to set up and solve a proportion. In each method, you use the units of measurement as a guide in setting up the calculation or writing the appropriate equation so that all measurement units can be divided out except the measurement unit of the answer. Example 1 demonstrates these two methods.

Example 1

Direct Method: Apply a known rate directly by multiplication to solve a problem.

In Problem 1, the known rate is miles per gallon. You can write miles per gallon in fraction form, $\dfrac{\text{number of miles}}{1 \text{ gal}}$. The fuel economy rating for the Chevrolet Cruze is $\dfrac{32 \text{ miles}}{1 \text{ gal}}$. To determine how many city miles you can travel on five gallons of gasoline, you multiply the known rate by 5 gal as follows:

$$\frac{32 \text{ miles}}{1 \text{ gal}} \cdot 5 \text{ gal} = 160 \text{ miles}.$$

Notice that the unit of measurement, gallon, occurs in both the numerator and the denominator. You divide out common units of measurement in the same way that you divide out common numerical factors.

Proportion Method: Set up and solve a proportion.

Form two fractions, $\dfrac{a}{b}$ and $\dfrac{c}{d}$, set them equal to one another and solve the resulting proportion.

Step 1. The first fraction, $\dfrac{a}{b}$, is the known rate. Its numerator represents one measurement unit and its denominator the other unit. In Problem 1, the known rate is $\dfrac{32 \text{ miles}}{1 \text{ gallon}}$.

Step 2. To determine the second fraction, $\dfrac{c}{d}$, the measurement unit of the numerator c must have the same unit as the numerator of the known rate (miles). The denominator d must have the same measurement unit as the denominator of the known rate (gallons).

You know that you have five gallons of gas. Therefore, five gallons is the denominator d. The unknown is the number of miles that you can drive on five gallons. Represent the unknown in the numerator by a symbol such as x. The numerator c is x miles. Therefore, the fraction $\dfrac{c}{d}$ is

$$\frac{x \text{ miles}}{5 \text{ gallons}}.$$

Step 3. Set the two fractions $\dfrac{a}{b}$ and $\dfrac{c}{d}$ equal to each other to obtain a proportion equation,

$\dfrac{a}{b} = \dfrac{c}{d}$. In Problem 1, the proportion equation is

$$\frac{32 \text{ miles}}{1 \text{ gallon}} = \frac{x \text{ miles}}{5 \text{ gallons}}.$$

Step 4. Solve $\dfrac{32}{1} = \dfrac{x}{5}$ to obtain $x = 160$ miles.

Therefore, in the Chevrolet Cruze, you can drive 160 miles on five gallons of gas.

A rate may be expressed in two ways. For miles and gallons, the rate can be expressed as $\dfrac{\text{miles}}{\text{gallons}}$ or as $\dfrac{\text{gallons}}{\text{miles}}$. Therefore, the highway mpg of the Chevrolet Cruze, 42 miles per gallon, can be written as $\dfrac{42 \text{ miles}}{1 \text{ gal}}$ or $\dfrac{1 \text{ gal}}{42 \text{ miles}}$.

In Problem 2, you were asked to determine how many gallons of gas were required to drive 304 highway miles. For the Chevrolet Cruze, the highway mpg would be expressed as $\dfrac{1 \text{ gal}}{42 \text{ miles}}$ so that the multiplication by 304 miles would result in the unit miles divided out and the unit gallons would remain.

$$304 \text{ miles} \cdot \frac{1 \text{ gal}}{42 \text{ miles}} = \frac{304}{42} \approx 7.2 \text{ gal}$$

You could also have set up a proportion.

$$\frac{42 \text{ miles}}{1 \text{ gallon}} = \frac{304 \text{ miles}}{x \text{ gallons}}$$
$$42x = 304$$
$$x = \frac{304}{42}$$
$$x \approx 7.2 \text{ gal}$$

3. Solve the following problems by both methods demonstrated in Example 1 and compare the results.

   a. The gas tank of a Ford Focus holds 13.2 gallons. How many highway miles can you travel on a full tank of gas?

   b. The Toyota Corolla gas tank holds 11.9 gallons. Is it possible to travel as far on the highway in this car as in the Ford Focus?

4. After you purchase your car, you would like to take a trip to see a good friend in another state. The highway distance is approximately 560 miles. Solve the following problems by the direct method and by the proportion method, and compare your results.

   a. If you bought the Hyundai Accent, how many gallons of gas would you need to make the round-trip?

   b. How many tanks of gas would you need for the trip?

Procedure

Methods for Solving Problems Involving Rates

Direct Method: Multiply and Divide by the Known Rate

1. Identify the unit of the result.

2. Set up the calculation so the appropriate units will divide out, leaving the unit of the result.

3. Multiply or divide the numbers as usual to obtain the numerical part of the result.

4. Divide out the common units to obtain the unit of the answer.

Proportion Method: Set Up and Solve a Proportion Equation

1. Identify the known rate, and write it in fractional form.

2. Identify the given information and the quantity to be determined.

3. Write a second fraction, placing the given information and the quantity x in the same positions as their units in the known rate.

4. Equate the two fractions to obtain a proportion equation.

5. Solve the equation for x, affixing the correct unit to the numerical result.

Applying Consecutive Rates to Solve a Problem

5. You are on a part of a 1500-mile trip where gas stations are far apart. Your car is averaging 40 miles per gallon and you are traveling at 60 miles per hour (mph). The fuel tank holds twelve gallons of gas, and you just filled the tank. How long is it before you have to fill the tank again?

You can solve this problem in two parts, as follows:

a. Determine how many miles you can travel on one tank of gas.

b. Use the result of part a to determine how many hours you can drive before you have to fill the tank again.

Problem 5 is an example of applying consecutive rates to solve a problem. Part a was the first step and part b was the second step. Alternatively, you can also determine the answer in a *single calculation* by considering what the measurement unit of the answer should be. In this case, it is hours per tank. Therefore, you will need to set up the calculation so that you can divide out miles and gallons.

$$\frac{40 \text{ miles}}{1 \text{ gal}} \cdot \frac{12 \text{ gal}}{1 \text{ tank}} \cdot \frac{1 \text{ hour}}{60 \text{ miles}} = \frac{8 \text{ hours}}{1 \text{ tank}}$$

Notice that miles and gallons divide out to leave hours per tank as the measurement unit of the answer.

Procedure

Applying Consecutive Rates to Solve Problems

To apply consecutive rates:

1. Identify the measurement unit of the result.

2. Set up the sequence of multiplications so that the appropriate units divide out, leaving the appropriate measurement unit of the result.

3. Multiply and divide the numbers as usual to obtain the numerical part of the result.

4. Check that the appropriate measurement units divide out, leaving the expected unit for the result.

---

6. You have been driving for several hours and notice that your car's 13.2-gallon fuel tank registers half empty. How many more miles can you travel if your car is averaging 35 miles per gallon?

Unit Conversion

In many countries, distance is measured in kilometers (km) and gasoline in liters. A mile is equivalent to 1.609 kilometers, and a liter is equivalent to 0.264 gallons. Each of these equivalences can be treated as a rate and written in fraction form. Thus, the fact that a mile is equivalent to 1.609 kilometers is written as $\dfrac{1.609 \text{ km}}{1 \text{ mile}}$ or $\dfrac{1 \text{ mile}}{1.609 \text{ km}}$. Using the fraction form, you can convert one measurement unit to another by applying multiplication or division directly, or you can use a proportion equation.

7. Your friend joins you on a trip through Canada, where gasoline is measured in liters and distance in kilometers.

   a. Write the equivalence of liters and gallons in fraction form.

   b. If you bought 20 liters of gas, how many gallons did you buy?

8. To keep track of mileage and fuel needs in Canada, your friend suggests that you convert your car's miles per gallon into kilometers per liter. Your car's highway fuel efficiency is 45 miles per gallon. What is its fuel efficiency in kilometers per liter?

SUMMARY: ACTIVITY 1.5

1. A **rate** is a comparison, expressed as a quotient, of two quantities that have different units of measurement.

2. **Unit analysis** (sometimes called *dimensional analysis*) uses units of measurement as a guide in setting up a calculation or writing an equation involving one or more rates. When the calculation or equation is set up properly, the result is an answer with the appropriate units of measurement.

3. Two common methods to solve problems involving rates:

Direct Method: Multiply directly by rates to solve problems:

1. Identify the measurement unit of the result.

2. Set up the calculation so that the appropriate measurement units will divide out, leaving the unit of the result.

3. Multiply or divide the numbers as usual to obtain the numerical part of the result.

4. Divide out the common measurement units to obtain the unit of the answer.

Proportion Method: Set up and solve a proportion equation

1. Identify the known rate, and write it in fractional form.

2. Identify the given information and the unknown quantity to be determined.

3. Write a second fraction, placing the given information and unknown quantity x in the same position as the measurement units in the known rate.

4. Equate the two fractions to obtain a proportion equation.

5. Solve the equation for x, affixing the correct unit to the numerical result.

4. Applying several rates consecutively:

1. Identify the measurement unit of the result.

2. Set up the sequence of multiplications so the appropriate measurement units divide out, leaving the unit of the result.

3. Multiply and divide the numbers as usual to obtain the numerical part of the result.

4. Check that the appropriate measurement units divide out, leaving the expected unit of the result.

EXERCISES: ACTIVITY 1.5

Use the conversion tables in Appendix B of this textbook for conversion equivalencies.

1. The length of a football playing field is 100 yards between the opposing goal lines. What is the length of the football field in feet?

2. The distance between New York City, New York, and Los Angeles, California, is approximately 4485 kilometers. What is the distance between these two major cities in miles?

3. The aorta is the largest artery in the human body. In the average adult, the aorta attains a maximum diameter of about 1.18 inches where it adjoins the heart. What is the maximum diameter of the aorta in centimeters? In millimeters?

4. In the Himalaya Mountains along the border of Tibet and Nepal, Mt. Everest reaches a record height of 29,035 feet. How high is Mt. Everest in miles? In kilometers? In meters?

5. The average weight of a mature human brain is approximately 1400 grams. What is the equivalent weight in kilograms? In pounds?

6. Approximately 4.5 liters of blood circulates in the body of the average human adult. How many quarts of blood does the average person have? How many pints?

7. How many seconds are in a day? In a week? In a year?

8. Tissues of living organisms consist primarily of organic compounds; that is, they contain carbon molecules known as proteins, carbohydrates, and fats. A healthy human body is approximately 18% carbon by weight. Determine how many pounds of carbon your own body contains. How many kilograms?

9. The following places are three of the wettest locations on Earth. For each site, determine the rainfall in centimeters per year. Record your results in the third column of the table.

Looks Like Rain

LOCATIONS	ANNUAL RAINFALL (IN INCHES)	ANNUAL RAINFALL (IN CM)
Mawsynram, Meghalaya, India	467	
Tutenendo, Colombia	463.5	
Mt. Waialeale, Kauai, Hawaii	410	

10. You have been selected to study for a semester in London next year. You also have received a $1000 grant for books and expenses.

a. What is the value of the 1000 U.S. dollars in British pounds? Many websites (e.g., www.gocurrency.com) provide up-to-the minute currency exchange rates. Use the following exchange rates obtained on January 8, 2014.

	USD	GBP	EUR	JPY
1 USD =	1	0.61	0.734	104.413
1 GBP =	1.64	1	1.204	171.281
1 EUR =	1.363	0.831	1	142.271
1 JPY =	0.01	0.006	0.007	1

b. You plan to visit Paris, France, which uses euros (EUR) as its unit of currency. Determine the value of 100 USD in euros.

c. Convert 75 euros to British pounds.

d. A group of students from Japan is attending your college for the academic year. The unit of currency in Japan is the yen. Convert 1000 yen to U.S. dollars.

e. Which is the largest unit of currency—a U.S. dollar, a British pound, a European euro, or a Japanese yen? Explain.

Activities 1.1–1.5   **What Have I Learned?**

Activities 1.1–1.5 gave you an opportunity to develop some problem-solving strategies. Apply the skills you used in this section to solve the following problems:

1. As you settle down to read a chapter of text for tomorrow's class, you notice a group of students forming a circle outside and beginning to randomly kick an odd-looking ball from person to person. You notice that there are 12 students in the circle and that they are able to keep the object in the air as they kick it. Sometimes they kick it to the person next to them; other times they kick it to someone across the circle.

You later learn that this is a refined version of the original Hacky-Sack game invented in 1972. Hacky-Sack has regained popularity internationally and online with blogs and Facebook groups dedicated to promote a resurgence of the game.

   a. Suppose each student kicks the Hacky-Sack exactly once to each of the others in the circle. How many total kicks would that take? Explain your method.

   b. One student in the circle invites you and another student to join them for a total of 14. How many kicks would it now take if each student kicks the Hacky-Sack exactly once to each of the others? How do you arrive at your answer?

   c. George Polya's book, *How to Solve It,* outlines a four-step process for solving problems.

   i. Understand the problem (determine what is involved).

   ii. Devise a plan (look for connections to obtain the idea of a solution).

   iii. Carry out the plan.

   iv. Look back at the completed solution (review and discuss it).

   Describe how your procedures in parts a and b correspond with Polya's suggestions.

2. You are assigned to read *War and Peace* for your literature class. The edition you have contains 1232 pages. You time yourself and estimate that you can read 12 pages in one hour. You have five days before your exam on this book. Will you be able to finish reading it before the exam?

3. On a 40-question practice test for this course, you answered 32 questions correctly. On the test itself you correctly answered 16 out of 20 questions. Does this mean that you did better on the practice test than you did on the test itself? Explain.

4. In a recent year, Florida's total population was 15,982,378 persons and 183 out of every 1000 residents were 65 years and older. Show how you would determine the actual number of Florida residents who were 65 years or older.

5. Ratios and rates are often encountered in everyday life. For example, the student-faculty ratio at your high school might be 30 to 1, written as $\frac{30}{1}$. Your part-time job might pay an hourly rate of 9.50 dollars per hour, written as $\frac{\$9.50}{1 \text{ hour}}$.

    a. Explain the difference between a ratio and a rate.

    b. Give three examples of ratios used in daily life.

    c. Give three examples of rates used in daily life.

6. Submit two articles from a newspaper, magazine, or the Internet in which ratios and proportions are used. Write a brief summary explaining how the ratios and proportions are used in each article.

Activities 1.1–1.5   How Can I Practice?

I. Solve the following problem by applying the four steps of problem solving. Use the strategy of solving an algebra equation.

Your average reading speed is 160 words per minute. There are approximately 800 words on each page of the textbook you need to read. Approximately how long will it take you to read 50 pages of your textbook?

2. You need to buy grass seed for a new lawn. The lawn will cover a rectangular plot that is 60 feet by 90 feet. Each ounce of grass seed will cover 120 square feet.

 a. How many ounces of grass seed will you need to buy?

 b. If the grass seed you want only comes in one-pound bags, how many bags will you need to buy?

3. Translate each statement into a formula.

 a. The total weekly earnings equal $12 per hour times the number of hours worked.

 b. The total time traveled equals the distance traveled divided by the average speed.

 c. The distance between two cities, measured in feet, is equal to the distance measured in miles times 5280 feet per mile.

4. Use your formulas in Exercise 3 to answer the following questions.

 a. If you make $12 per hour, what will your earnings be in a week when you work 35 hours?

b. If you earned $336 last week (still assuming $12 per hour), how many hours did you work?

c. If you travel 480 miles at an average speed of 40 miles per hour, for how many hours were you traveling?

d. The distance between Dallas and Houston is 244 miles. How far is the distance in feet?

5. What will the temperature be in degrees Fahrenheit when it is 40° Celsius outside? (Recall the formula $F = (9C \div 5) + 32$)

6. Solve the following proportions.

a. 3 out of 5 = _____ out of 20

b. $\dfrac{3}{5} = \dfrac{n}{765}$

c. $\dfrac{3}{5} = \dfrac{27}{n}$

d. $\dfrac{3}{5} = \dfrac{1134}{n}$

In Exercises 7–11, write a proportion that represents the situation and then determine the unknown value in the proportion.

7. You correctly answered two-thirds of the questions on your psychology exam. There were 75 questions on the exam. How many questions did you answer correctly?

8. During a recent infestation by beetles, $\frac{2}{3}$ of the ash trees in a local park were destroyed. If 120 trees were destroyed, how many ash trees were originally in the park?

9. Your high school freshman class consists of 760 students. In recent years, only 4 out of 7 students actually graduated in four years. Approximately how many of your classmates are expected to graduate in four years?

10. Your car averages 380 miles on a 14-gallon tank of gas. You run out of gas on a deserted highway, but have $\frac{1}{2}$ gallon of lawn mower gas with you. Will you be able to reach the nearest gas station 15 miles away?

11. As part of your job as a quality-control inspector in a factory you can check 16 parts in 3 minutes. How long will it take you to check 80 parts?

12. Your car averages about 27 miles per gallon on highways. With gasoline priced at $3.799 per gallon, how much will you expect to spend on gasoline during your 500-mile trip?

13. You currently earn $11.50 per hour. Assuming that you work 40-hour weeks with no raises, what total gross salary will you earn in the next five years?

14. You are traveling at 75 miles per hour on a straight stretch of highway in Nevada. It is noon now. When will you arrive at the next town 120 miles away?

15. To estimate the size of the grizzly bear population in a national park, rangers tagged and released 12 grizzly bears into the park. A few months later, 2 out of 21 grizzly bears sighted had tags. Assuming the proportion of tagged bears is the same for the entire bear population, estimate the number of grizzly bears in this national park.

16. Your friend is on a weight-loss program and wants to reach a final goal weight of 160 pounds. His starting weight was 180 pounds. If he has lost 8% of his body weight so far, how many pounds does he still have to lose to reach his final goal?

17. A company that offers management solutions to companies that sell online announced in a recent consumer survey that nearly 9 out of 10 customers reported problems with transactions online. The survey sampled 2010 adults in the United States, 18 years and older, who had conducted an online transaction in the past year. Estimate the number of adults who reported problems with transactions online.

Activity 1.6

Florida Heat

Objectives

1. Identify input and output in situations involving two variable quantities.

2. Identify a functional relationship between two variables.

3. Identify the independent and dependent variables.

4. Use a table to numerically represent a functional relationship between two variables.

5. Represent a functional relationship between two variables graphically.

6. Identify trends in data pairs that are represented numerically and graphically, including increasing and decreasing.

A key step in the problem-solving process is to look for relationships and connections between the variable quantities in a given situation. Problems encountered in the world around us, including the environment, medicine, economics, and the Internet, are often very complicated and contain several variables. In this text, you will primarily deal with situations that contain two variables. In many of these situations, the variables will have a special relationship called a **function**.

Function

It is July, and you have just moved to Tallahassee, Florida. The area is in the middle of a heat spell. The high temperature for each of the last 5 days has exceeded 100°F. You are curious about the temperatures you will experience while living in Florida. According to the Weather Channel, the average monthly high temperature in Tallahassee Florida, is given in the following table. Note that January is represented by 1, February is represented by 2, and so forth.

MONTH OF THE YEAR	1	2	3	4	5	6	7	8	9	10	11	12
AVERAGE MONTHLY HIGH TEMPERATURE, °F	64	67	74	80	87	91	92	92	89	81	73	66

Source: The Weather Channel

This situation involves two variables, the month and the average monthly high temperature. Typically, one variable is designated as the **input** and the other is called the **output**. The input is the value given first, and the output is the value that corresponds to, or is determined by, the given input value. The input variable is often represented by the letter x. The output variable is frequently represented by the letter y.

1. In the temperature situation, identify the input variable and the output variable.

2. **a.** For an input of 7 (July), what is the average high temperature (output)?

   b. For an input of 1 (January), what is the average high temperature (output)?

c. For each value of input (month), how many different outputs (average high temperature) are there?

The set of data in the table is an example of a mathematical **function**.

> ### Definition
>
> A **function** is a correspondence between an input variable and an output variable that assigns a single, unique output value to each input value. Therefore, for a function, any given input value has exactly one corresponding output value. If x represents the input variable and y represents the output variable, then the function assigns a single, unique y-value to each x-value.

3. Explain how the data in the previous table fit the description of a function.

A functional relationship is stated as follows: "The output variable is a function of the input variable." Using x for the input variable and y for the output variable, then the functional relationship is stated "y is a function of x." Because the input for the temperature function is the month and the output is the average high temperature for that month, you write that the average high temperature is a function of the month.

> **Example 1** *Consider the following table listing the official high temperature (in °F) in the village of Lake Placid, New York, during the first week of January. Note that the date has been designated the input and the high temperature on that date the output. Is the high temperature a function of the date?*

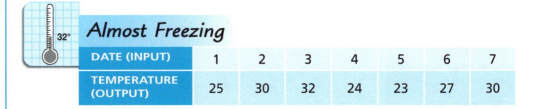

Almost Freezing — 32°

DATE (INPUT)	1	2	3	4	5	6	7
TEMPERATURE (OUTPUT)	25	30	32	24	23	27	30

SOLUTION

From this table, you observe that the high temperature is a function of the date. For each date there is exactly one high temperature. The relationship in this example can be visualized as follows:

DATE (INPUT)	TEMPERATURE (OUTPUT)
1 ⟶	25
2 ⟶	30
3 ⟶	32
4 ⟶	24
5 ⟶	23
6 ⟶	27
7 ⟶	30

If d represents the input (date), and T represents the output (temperature), then T is a function of d.

Example 2 *Determine if the amount of postage for a letter is a function of the weight of the letter. Give a reason for your answer.*

SOLUTION

Yes, the situation does describe a function. The weight of the letter is the input and the amount of postage is the output. Each letter has one weight. This weight determines the postage necessary for the letter. There is only one amount of postage for each letter. Therefore, for each value of input (weight of the letter) there is one output (postage). Note that if w represents the input (weight of the letter), and p represents the output (postage), then p is a function of w.

4. Determine whether or not each situation describes a function. Give a reason for your answer.

   a. The amount of property tax you have to pay is a function of the assessed value of the house.

   b. The weight of a letter in ounces is a function of the postage paid for mailing the letter.

Definition

For all functions, the input variable is called the **independent** variable, and the output variable is called the **dependent** variable. If x represents the input variable and y represents the output variable, then x is the independent variable and y is the dependent variable.

---

5. The independent variable in Example 1 is the date. The dependent variable is the temperature. Identify the independent and dependent variables in Example 2.

Representing Functions Numerically

The input/output pairing in the temperature function on page 48 is presented as a **table of matched pairs**. In such a situation, the function is defined **numerically**. Another way to define a function numerically is as a set of **ordered pairs**.

Definition

An **ordered pair** of numbers consists of two numbers written in the form

$$(x, y),$$

where x represents the independent variable and y represents the dependent variable.

The order in which they are listed *is* significant.

Example 3   *The ordered pair* (3, 4) *is distinct from the ordered pair* (4, 3). *In the ordered pair* (3, 4), *3 is the x-value and 4 is the y-value. In the ordered pair* (4, 3), *4 is the x-value and 3 is the y-value.*

Definition

A function may be defined **numerically** as a set of ordered pairs (x, y) where x represents the input and y represents the output. No two ordered pairs have the same x-value and different y-values.

---

6. Write three other ordered pairs for the Florida temperature function.

Representing Functions Graphically

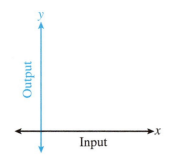

You may have seen an ordered pair before as the coordinates of a point in a **rectangular coordinate system**, typically using ordered pairs of the form (x, y), where x is the input (independent variable) and y is the output (dependent variable). The first value, the horizontal coordinate, indicates the directed distance (right or left) from the vertical axis. The second value, the vertical coordinate, indicates the directed distance (up or down) from the horizontal axis.

The variables may not always be represented by x and y, but the horizontal axis will always be the input axis, and the vertical axis will always be the output axis.

7. Plot each ordered pair in the Florida temperature table on the following grid. Set your axes and scales by noting the smallest and largest values for both input and output. Label each axis by the variable name. *Remember, the scale for the horizontal axis does not have to be the same as the scale for the vertical axis. However, each scale (vertical and horizontal) must be divided into equal intervals.*

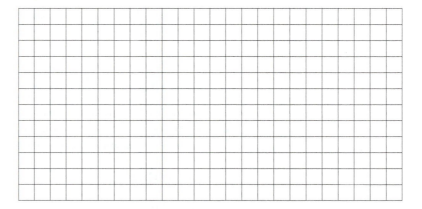

Definition

A set of points in the plane whose coordinate pairs represent input/output pairs of a data set is called a **scatterplot**.

The preceding graph, which consists of a set of labeled axes and 12 points, presents the same information that is in the Florida temperature table on page 47, but in a different way. It shows the information as a graph and therefore defines the function **graphically**.

Note that the graph of the Florida temperature function consists of 12 distinct points that are not connected. The input variable (month of the year) is defined only for whole numbers. This function is said to be **discrete** because it is defined only at isolated, distinct input values. The function is not defined for input values between these particular values.

> **Caution.** In order to use the graph for a relationship such as the Florida temperature situation to make predictions or to recognize patterns, it is convenient to connect the points with line segments. This creates a type of continuous graph. This changes the domain shown in the graph from "some values" to "all values." Therefore, you need to be cautious. Connecting data points may cause confusion when working with real-world situations.

8. In Example 1, the high temperature in Lake Placid is a function of the date.

  a. Convert to ordered pairs all the values in the Almost Freezing table on page 48.

  b. Plot each ordered pair as a point on an appropriately scaled and labeled set of coordinate axes.

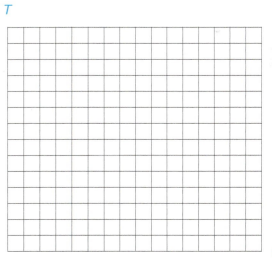

More Graphing

Living on Earth's surface, you experience a relatively narrow range of temperatures. You may know what −20°F (or −28.9°C) feels like on a bitterly cold winter day. Or you may have sweated through 100°F (or 37.8°C) during summer heat-waves. If you were to travel below Earth's surface and above Earth's atmosphere, you would discover a wider range of temperatures. The following graph displays a relationship between the altitude and temperature. Note that the altitude is measured from Earth's surface. That is, Earth's surface is at altitude 0.

The graph is said to be **continuous**, because it is totally connected. There are no gaps or holes in the graph. The function is defined for all input values.

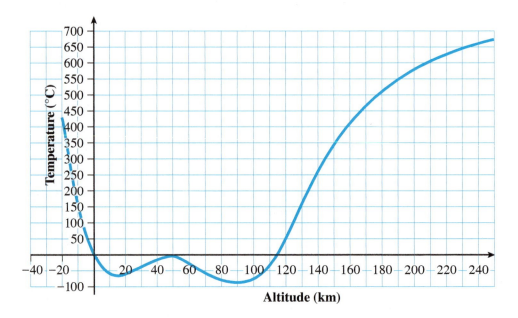

Use this graph to answer Problems 9 and 10.

9. a. With what variable and units of measure is the horizontal axis labeled?

   b. What is the practical significance of the positive values of this quantity?

   c. What is the practical significance of the negative values of this quantity?

   d. How many kilometers are represented between the tick marks on the horizontal axis?

10. a. With what variable and units of measure is the vertical axis labeled?

   b. What is the practical significance of the positive values of this quantity?

   c. What is the practical significance of the negative values of this quantity?

   d. How many degrees Celsius are represented between the tick marks on the vertical axis?

Recall, the two perpendicular coordinate axes divide the plane into four **quadrants**. The quadrants are labeled counterclockwise, using Roman numerals, with quadrant I being the upper-right quadrant.

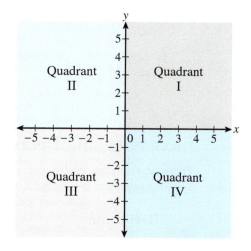

11. The following graph displays eight points selected from the Temperature versus Altitude graph on page 52.

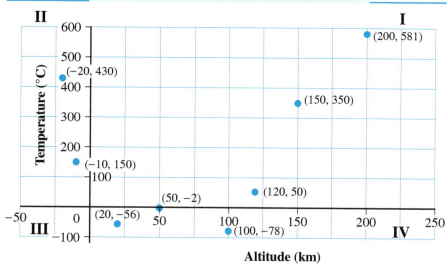

Altitude (km)

a. What is the practical meaning of the point with coordinates (150, 350)?

b. What is the practical meaning of the point with coordinates (100, −78)?

c. What is the practical meaning of the point with coordinates (−20, 430)?

d. In which quadrant are the points (120, 50), (150, 350), and (200, 581) located?

    e. In which quadrant are the points $(-10, 150)$ and $(-20, 430)$ located?

    f. In which quadrant are the points $(20, -56)$ and $(100, -78)$ located?

    g. Are there any points located in quadrant III? What is the significance of your answer?

Increasing and Decreasing Functions

There are many other advantages to having the function in graphical form. For example, you are often interested in determining how the output values change as the input values increase.

> **Definition**
>
> A function is **increasing** if its graph goes up to the right, **decreasing** if its graph goes down to the right, and **constant** if its graph is horizontal. In each case, you are viewing the graph as a point moves along the curve from left to right, that is, as the input values increase.

Example 4

    a. The graph of this function is always increasing. The graph rises from left to right.

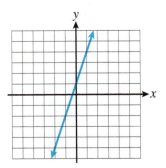

    b. This function is decreasing because its graph falls from left to right.

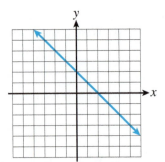

c. This function is constant, because the graph goes neither up nor down. The *y*-value is always 3 no matter what the *x*-value is.

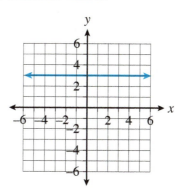

12. Is the following function increasing, decreasing, or constant over the domain displayed in the window?

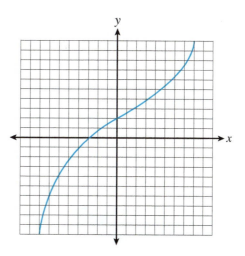

SUMMARY: ACTIVITY 1.6

1. A **variable**, usually represented by a letter, is a quantity that may change in value from one particular instance to another.

2. In a situation involving two variables, one variable is designated the **input** and the other the **output**. The input is the value given first, and the output is the value that corresponds to or is determined by the given input value.

3. A **function** is a correspondence relating an input variable (*independent variable*) and an output variable (*dependent variable*) so that a single, unique output value is assigned to each input value. In such a case, you state that the output variable is a function of the input variable.

4. An **ordered pair** of numbers consists of two numbers written in the form

$$(input\ value, output\ value).$$

The order in which they are listed *is* significant.

5. Functions may be defined **numerically** using ordered pairs of numbers. The ordered pairs are always given in the form (input value, output value). These can be displayed as a table of values or points on a graph. For each input value, there is one and only one corresponding output value.

6. When a function is represented **graphically**, the *x* and *y* data pairs are represented as plotted points on a grid called a *rectangular coordinate system*. The *x*-variable is referenced on the horizontal axis. The *y*-variable is referenced on the vertical axis.

7. The two perpendicular coordinate axes divide the plane into four **quadrants**. The quadrants are labeled counterclockwise, using Roman numerals, with quadrant I being the upper-right quadrant.

$$
\begin{array}{c|c}
\text{II} & \text{I} \\
\hline
\text{III} & \text{IV}
\end{array}
$$

8. A function is **increasing** if its graph rises to the right, **decreasing** if its graph falls to the right, and **constant** if its graph is horizontal.

EXERCISES: ACTIVITY 1.6

1. The weights and heights of six mathematics students are given in the following table:

Weighty Issues

WEIGHT (IN POUNDS)	HEIGHT (IN CENTIMETERS)
165	172
123	157
212	183
175	178
165	163
147	167

a. In the statement, "Height is a function of weight," which variable is the input and which is the output?

b. Is height a function of weight for the six students? Explain using the definition of function.

c. In the statement, "Weight is a function of height," which variable is the input and which is the output?

d. Is weight a function of height for the six students? Explain using the definition of function.

e. For all students, is weight a function of height? Explain.

For Exercises 2–6, determine whether or not each of the situations describes a function. Give a reason for your answer. For each functional relationship, determine the independent and dependent variables.

2. a. The amount of federal income tax you pay is a function of your taxable income.

b. The value of your car is a function of the total mileage on the car.

3. a. The letter grade in this course is a function of your numerical grade.

b. The numerical grade in this course is a function of the letter grade.

4. a. The input is any number and the output is the square of the number.

b. The square of a number is the input and the output is the number.

5. In the following table, elevation is the input and amount of snowfall is the output.

ELEVATION (IN FEET)	SNOWFALL (IN INCHES)
2000	4
3000	6
4000	9
5000	12

6. Number of hours using the Internet is the input variable, and the monthly cost for the Internet service is $39.95, regardless of the number of hours of usage.

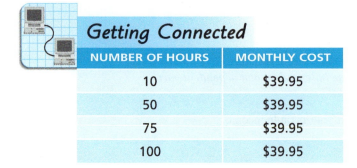

Getting Connected

NUMBER OF HOURS	MONTHLY COST
10	$39.95
50	$39.95
75	$39.95
100	$39.95

7. Each of the following tables defines a relationship between an input x and an output y. Which of the relationships represent functions? Explain your answers.

a.

x	−8	−3	0	6	9	15	24	38	100
y	24	4	9	72	−14	−16	53	29	7

b.

x	−8	−5	0	6	9	15	24	24	100
y	24	4	9	72	14	−16	53	29	7

c.

x	−8	−3	0	6	9	15	24	38	100
y	24	4	9	72	4	−16	53	24	7

8. You work for the National Weather Service and are asked to study the average daily temperatures in Anchorage, Alaska. You calculate the mean of the average daily temperatures for each month. You decide to place the information on a graph in which the date is the input and the temperature is the output. You also decide that January 1950 will correspond to the month 0 as indicated by the dot on the input scale. Determine the quadrant in which you would plot the points that correspond to the following data.

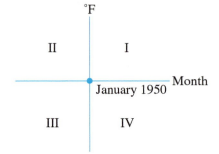

a. The average daily temperature for January 1936 was −15°F.

b. The average daily temperature for July 1963 was 63°F.

c. The average daily temperature for July 1910 was 71°F.

d. The average daily temperature for January 1982 was −21°F.

9. Measurements in wells and mines have shown that the temperatures within the earth generally increase with depth. The following table shows average temperatures for several depths below ground level.

A Hot Topic

DEPTH (km) BELOW GROUND LEVEL	0	25	50	75	100	150	200
TEMPERATURE (°C)	20	600	1000	1250	1400	1700	1800

a. Represent the data from the table graphically on the grid following part d. Place depth along the horizontal axis and temperature along the vertical axis.

b. How many units does each tick mark on the horizontal axis represent?

c. How many units does each tick mark on the vertical axis represent?

d. Explain your reasons for selecting the particular scales that you used.

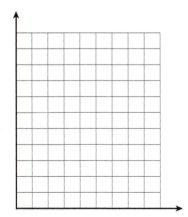

e. Which representation (table or graph) presents the information and trends in this data more clearly? Explain your choice.

10. The following table presents the average recommended weights for given heights for 25- to 29-year-old medium-framed women (wearing 1-inch heels and 3 pounds of clothing). Consider height to be the input variable and weight to be the output variable. As ordered pairs, height and weight take on the form (h, w). Designate the horizontal (input) axis as the h-axis, and the vertical (output) axis as the w-axis. Since all values of the data are positive, the points will lie in quadrant I only.

h, HEIGHT (in.)	58	60	62	64	66	68	70	72
w, WEIGHT (lb)	115	119	125	131	137	143	149	155

Plot the ordered pairs in the height–weight table on the following grid. Note the consistent spacing between tick marks on each axis. The distance between tick marks on the horizontal axis represents 1 inch. On the vertical axis the distance between tick marks represents 5 pounds.

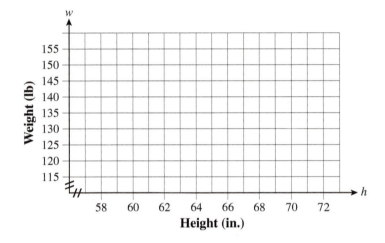

Note: The slash marks (//) near the origin indicate that the interval from 0 to 56 on the horizontal axis and the interval from 0 to 115 on the vertical axis are not shown. That is, only the part of the graph containing the plotted points is shown.

11. On December 3, 1992, the first text message was sent in the United Kingdom from a computer to a mobile device, using the Short Message Service (SMS). The message was "Merry Christmas." In 1999, 100 billion text messages were sent worldwide. In 2005, 1 trillion text messages were sent. In 2012, the number of text messages sent worldwide surpassed 8.5 trillion.

A recent study revealed that persons who text while driving are 23 times more likely to have an accident. Another study concluded that driving while texting is more dangerous than driving while under the influence of alcohol. It is estimated that persons in the age group of 18 to 24 send an average of 71 texts daily.

The following table gives estimates of the average number of text messages (in billions) sent each month in the United States.

YEAR, y	2005	2006	2007	2008	2009	2010	2011	2012
Average number of monthly text messages (in billions), n	7	8	30	86	153	188	212	220

a. Identify the input and output variable quantities.

b. Construct an appropriately scaled and labeled scatterplot of the given data.

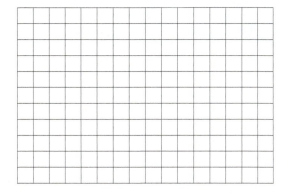

c. Describe any patterns or trends that you observe.

d. Is this function continuous or discrete? Explain.

Activity 1.7

Fill 'er Up

Objectives

1. Determine the equation (symbolic representation) that defines a function.

2. Identify the independent and the dependent variables of a function.

3. Determine the domain and range of a function.

4. Represent functions numerically and graphically using technology.

5. Distinguish between a discrete and a continuous function.

You probably need to fill your car with gas more often than you would like. You commute to high school each day and to a part–time job each weekend. Your car gets good gas mileage, but the recent dramatic fluctuation in gas prices has wreaked havoc on your budget.

1. There are two input variables that determine the cost (output) of a fill-up. What are they? Be specific.

2. Assume you need 12.6 gallons to fill up your car. Now one of the input variables in Problem 1 will become a constant. The value of a constant will not vary throughout the problem. The cost of a fill-up is now dependent on only one variable, the price per gallon.

a. Complete the following table:

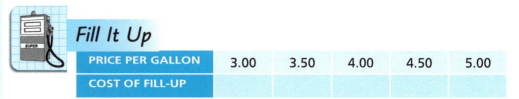

Fill It Up

PRICE PER GALLON	3.00	3.50	4.00	4.50	5.00
COST OF FILL-UP					

b. Is the cost of a fill-up a function of the price per gallon? Explain.

3. a. Write a verbal statement that describes how the cost of a fill-up is determined.

b. Let p represent the price of a gallon of gasoline pumped (input) and c represent the cost of the fill-up (output). Translate the verbal statement in part a into a symbolic statement (an equation) that expresses c in terms of p.

Defining Functions by a Symbolic Rule (Equation)

The symbolic rule (equation) $c = 12.6p$ is an example of a third method of defining a function. Recall that the other methods are numerical (tables and ordered pairs) and graphical.

4. a. Use the given equation to determine the cost of a fill-up at a price of $3.60 per gallon

b. Explain the steps that you used to determine the cost in part a.

Function notation is an efficient and convenient way of representing the output variable. The equation $c = 12.6p$ may be written using the function notation by replacing c with $f(p)$ as follows.

$$f(p) = 12.6p$$

Now, if the price per gallon is $3.60, then the cost of a fill-up can be represented by $f(3.60)$. To evaluate $f(3.60)$, substitute 3.60 for p in $f(p) = 12.6p$ as follows:

$$f(3.60) = 12.6(3.60) = 45.36$$

The results can be written as $f(3.60) = 45.36$ or as the ordered pair $(3.60, 45.36)$. Therefore, at a price of \$3.60 per gallon, the cost of filling your car with 12.6 gallons of gas will be \$45.36.

5. a. Using function notation, write the cost if the price is \$2.85 per gallon and evaluate. Write the result as an ordered pair.

b. Use the equation for the cost-of-fill-up function to evaluate $f(4.95)$, and write a sentence describing its meaning. Write the result as an ordered pair.

Real Numbers

The numbers that you will be using as input and output values in this text will be **real numbers**. A real number is any rational or irrational number.

A **rational number** is any number that can be expressed as the quotient of two integers (negative and positive counting numbers as well as zero) such that the division is not by zero.

Example 1 *Rational numbers include the following.*

$$\frac{3}{4}; -\frac{7}{8}; 2\frac{1}{3} = \frac{7}{3}; 5 = \frac{5}{1}; 0 = \frac{0}{1}; -3\frac{1}{4} = -3.25; \frac{2}{3} = 0.666\ldots = 0.\overline{6}$$

An **irrational number** is a real number that cannot be expressed as a quotient of two integers.

Example 2 *Irrational numbers include* $\sqrt{2}, -\sqrt{7}, \sqrt[3]{5}, \pi$.

All of the numbers in Examples 1 and 2 are real numbers. A real number can be represented as a point on the number line.

Domain and Range

6. Can any number be substituted for the input variable p in the cost of fill-up function? Describe the values of p that make sense, and explain why they do.

Definition

The collection of all possible values of the input or independent variable is called the **domain** of the function. The **practical domain** is the collection of replacement values of the input variable that makes practical sense in the context of the situation.

7. a. Determine the practical domain of the cost-of-fill-up function. Refer to Problem 6.

b. Determine the domain for the general function defined by $c = 12.6p$ with no connection to the context of the situation.

> ### Definition
>
> The collection of all possible values of the output or dependent variable is the **range** of the function. The **practical range** corresponds to the practical domain.

8. a. What is the practical range for the cost function defined by $f(p) = 12.6p$ if the practical domain is 2 to 5?

b. What is the range of this function if it has no connection to the context of the situation?

9. a. Sketch a graph of the cost-of-fill-up function by first plotting the five points from Problem 2a on properly scaled and labeled coordinate axes.

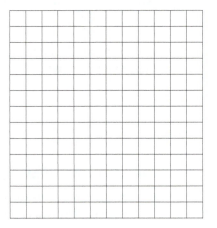

b. Can the five points be connected to form a continuous graph? Is the function discrete or continuous? Explain.

c. Describe any patterns or trends in the graph.

Constructing Tables of Input/Output Values Using Technology

10. Use the symbolic form of the gas cost-of-fill-up function, $f(p) = 12.6p$, to evaluate $f(2), f(2.50), f(3), f(3.50)$, and $f(4)$, and complete the following table. Note that the input variable p increases by 0.50 unit. In such a case, you say the input increases by an **increment** of 0.50 unit.

PRICE PER GALLON, p	2.00	2.50	3.00	3.50	4.00
COST OF FILL-UP, $f(p)$					

A numerical form of the cost-of-fill-up function is a table or a collection of ordered pairs. When a function is defined in symbolic form, you can use technology to generate the table. The TI-83/TI-84 Plus calculator is a function grapher. The y variables Y_1, Y_2, and so on represent function output (dependent) variables. The input, or independent variable, is x. The steps to build tables with the TI-83/TI-84 Plus can be found in Appendix A.

11. Use your graphing calculator to generate a table of values for the function represented by $f(p) = 12.6p$ to check your values in the table in Problem 10. The screens on your graphing calculator should appear as follows.

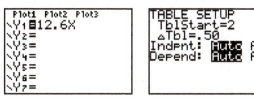

Graphing Functions Using Technology

Graph 1 below can be used to represent the graph of the cost-of-fill-up function defined by $c = 12.6p$ over its practical domain. The points on the graph from $(2, 25.20)$ to $(5, 63)$ can be connected to form a continuous graph in order to help observe trends in the data.

Graph 1

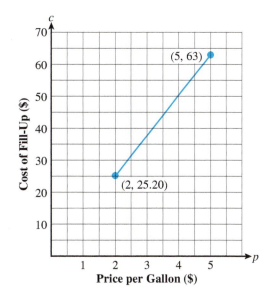

The domain for the general function defined by $c = 12.6p$, with no connection to the context of the situation, is the set of all real numbers, since any real number can be substituted for p in $12.6p$. Following is a graph of $c = 12.6p$ for any real number p.

Graph 2

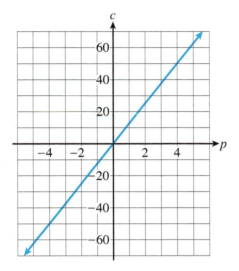

Each of these graphs can be obtained using your graphing calculator as demonstrated in Problems 12 and 13.

Appendix

Recall that the independent variable in your graphing calculator is represented by x and the dependent variable is represented by y. Therefore, the cost-of-fill-up equation $c = 12.6p$ needs to be keyed in the "Y=" menu as $y = 12.6x$. The procedure for graphing a function using the TI-83/TI-84 Plus calculator can be found in Appendix A.

12. a. The viewing window is the portion of the rectangular coordinate system that is displayed when you graph a function. Use the practical domain and practical range of the cost-of-fill-up function to determine X_{min}, X_{max}, Y_{min}, and Y_{max} in the window screen. Key these values into your calculator. Your screens should appear as follows:

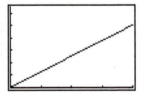

b. Graph the function. Your screen should appear as follows:

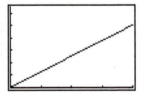

13. a. Determine reasonable window settings to obtain the graph of the general function defined by $c = 12.6p$ for which the domain is any real number.

b. Type in the values determined in part a and graph the function. The screens should appear as follows:

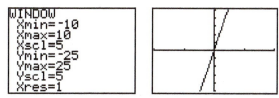

c. How does the graph in part b compare to Graph 2 on the previous page?

Gross Pay Function

14. a. You are working a part-time job. You work between 0 and 25 hours per week. If you earn $10.50 per hour, write an equation to determine the gross pay, g, for working h hours.

b. What is the independent variable? What is the dependent variable?

c. Complete the following table using your graphing calculator.

NUMBER OF HOURS, h	0	5	10	15	20	25
GROSS PAY, g (dollars)						

d. Using f for the name of the function, the output variable g can be written as $g = f(h)$.

Rewrite the equation in part a using the function notation $f(h)$ for gross pay.

e. What are the practical domain and the practical range of the function? Explain.

f. Evaluate $f(14)$ and write a sentence describing its meaning.

g. Plot the ordered pairs determined in part c on an appropriately scaled and labeled set of axes.

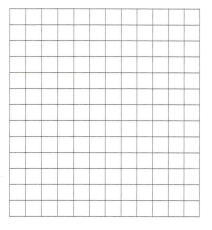

h. Is the graph continuous? Why, or why not? Explain.

SUMMARY: ACTIVITY 1.7

1. **Independent variable** is another name for the input variable of a function.

2. **Dependent variable** is another name for the output variable of a function.

3. The collection of all possible replacement values for the independent variable is called the **domain of the function**. The **practical domain** is the collection of replacement values of the independent variable that makes practical sense in the context of the problem.

4. The collection of all values of the dependent variable of a function is called the **range of the function**. When a function describes a real situation or phenomenon, its range is often called the **practical range** of the function.

5. When a function is represented by an equation, the function may also be written in **function notation**. For example, given $y = 2x + 3$, you can replace y with $f(x)$ and rewrite the equation as $f(x) = 2x + 3$.

6. Functions are **continuous** if they are defined for all input values, and if there are no gaps between consecutive input values.

7. Functions are **discrete** if they are defined only at isolated input values and do not make sense or are not defined for input values between those values.

EXERCISES: ACTIVITY 1.7

In Exercises 1 and 2,

    a. *Identify the independent and dependent variables.*

    b. *Let x represent the independent variable. Use function notation to represent the dependent variable.*

    c. *Translate the written statement into an equation.*

1. Sales tax is a function of the price of an item. The amount of sales tax is 0.08 times the price of the item. Use h to represent the function.

2. A taxi cab from George Bush International Airport charges $12 plus $2 per mile for up to 4 passengers. The cost of the taxi cab is a function of the number of miles. Let c represent the function.

For each function in Exercises 3–5, evaluate $f(2)$, $f(-3.2)$, and $f(a)$.

3. $f(d) = 2d - 5$ **4.** $f(t) = -16t^2 + 7.8t + 12$ **5.** $f(x) = 4$

In Exercises 6–8, construct a table of values of four ordered pairs for the given function. Check your results using the table feature of your grapher.

6. $g(x) = x^2$. Start the independent variable x at 3 and use an increment of 2.

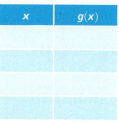

7. $h(x) = \dfrac{1}{x}$. Start the x-values at 10 and use an increment of 10.

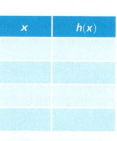

8. $f(x) = 3.5x + 6$. Start the x-values at 0 and use an increment of 5.

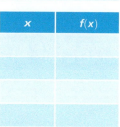

9. **a.** The distance you travel while hiking is a function of how fast you hike and how long you hike at this rate. You usually maintain a speed of three miles per hour while hiking. Write a statement that describes how the distance that you travel is determined.

   b. Identify the independent and dependent variables of this function.

   c. Write the statement in part a using function notation. Let t represent the independent variable. Let h represent the function, and $h(t)$ the dependent variable.

   d. Use the equation from part c to determine the distance traveled in four hours.

   e. Evaluate $h(7)$ and write a sentence describing its meaning. Write the result as an ordered pair.

f. Determine the domain and range of the general function.

g. Determine the practical domain and the practical range of the function.

h. Use your calculator to generate a table of values beginning at zero with an increment of 0.5.

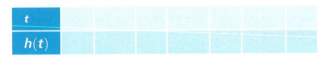

t					
$h(t)$					

10. Determine the domain and range of each function.

a. $\{(-2, 4), (0, 3), (5, 8), (8, 11)\}$

b. $\{(-6, 5), (-2, 5), (0, 5), (3, 5)\}$

11. Plot the ordered pairs of each of the following functions on an appropriately scaled and labeled set of axes. Then determine if a continuous or discrete graph is more appropriate for the situation.

a. In a science experiment, the amount of water displaced in a graduated cylinder is a function of the number of marbles placed in the cylinder. The results of an experiment are recorded in the following table.

NUMBER OF MARBLES	1	2	3	4	5
VOLUME OF WATER DISPLACED (ml)	20	40	60	80	100

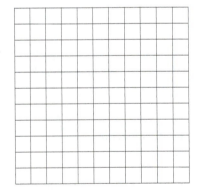

b. Forensic anthropologists predict the height of a male based on the length of his femur. The following table demonstrates this functional relationship.

LENGTH OF FEMUR (inches)	10	12	14	16	18
HEIGHT (inches)	50.9	54.7	58.4	62.2	66.0

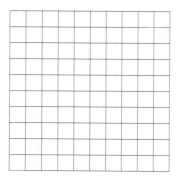

12. The cost to own and operate a car depends on many factors including gas prices, insurance costs, size of car, and finance charges. Using a Cost Calculator found on the Internet, you determine the average cost per mile to own and operate a small sedan in the U.S. is $0.449. The total cost C is a function of the number m of miles driven and can be represented by the function defined by $C = 0.449m$. When you finally take your car to the junkyard, the odometer reads 200,000 miles.

a. Identify the input variable.

b. Identify the output variable.

c. Use f to represent the function and rewrite $C = 0.449m$ using function notation.

d. The average number of miles driven per year is 15,000 miles. Evaluate $f(15,000)$, and write a sentence describing its meaning. Write the result as an ordered pair.

e. Use the table feature of your calculator to create a table of values with a beginning input of 0. Use increments of 50,000 miles.

f. What is the practical domain for this situation?

g. What is the practical range for this situation?

13. To change a Celsius temperature to Fahrenheit, use the formula $F = 1.8C + 32$. You are concerned only with temperatures between freezing and boiling.

a. What is the practical domain of the function?

b. What is the practical range of the function?

14. Your high school service organization has volunteered to help with Spring Cleanup Day at a youth summer camp. You have been assigned the job of supplying paint for the exterior of the bunkhouses. You discover that 1 gallon of paint will cover 400 square feet of flat surface.

a. If n represents the number of gallons of paint you supply and s represents the number of square feet you can cover with the paint, complete the following table:

Painting by Numbers

n, NUMBER OF GALLONS OF PAINT	1	2	4	6
s, SQUARE FEET COVERED BY THE PAINT	400	800		

b. Plot the ordered pairs determined in part a on an appropriately scaled and labeled set of axes.

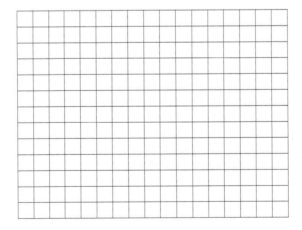

c. Let s be represented by $f(n)$, where f is the name of the function. Determine $f(6)$.

d. Write a sentence explaining the meaning of $f(4) = 1600$.

15. Give an example of a function that you may encounter in your daily life or that describes something about the world around you.

    a. Identify the independent and the dependent variables.

    b. Write the function in the form "output is a function of the input."

    c. Explain how the example fits the definition of a function.

Activity 1.8

Mathematical Modeling

Objectives

1. Identify a mathematical model.

2. Solve problems using formulas as models.

3. Develop a function model to solve a problem.

4. Recognize patterns and trends between two variables using tables as models.

After an automobile accident, the investigating police officers often estimate the speed of the vehicle by measuring the length of the tire skid distance. The following table gives the average skid distances for an automobile with good tires on dry pavement.

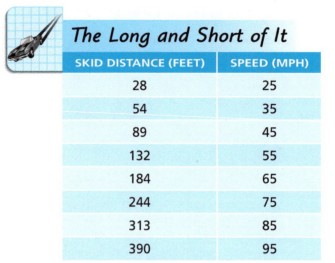

The Long and Short of It

SKID DISTANCE (FEET)	SPEED (MPH)
28	25
54	35
89	45
132	55
184	65
244	75
313	85
390	95

1. Does the table data define speed as a function of skid distance? Explain using the definition of function.

2. Identify the independent variable and the dependent variable.

3. Plot the data as ordered pairs on an appropriately scaled and labeled set of coordinate axes. Remember, the values of the independent variable appear along the horizontal axis, and the values of the dependent variable appear along the vertical axis.

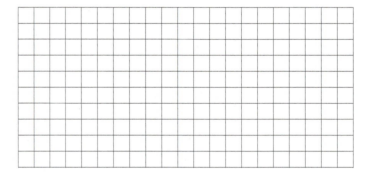

After a particular accident, a skid distance was measured to be 200 feet. From the table, the investigating officer knows that the speed of the vehicle was between 65 and 75 miles per hour. She would, however, like to be able to be more precise in reporting the speed. One way of getting values that are not listed in the table is to use a graph or equation of a function that best fits the actual data.

Note that the points on the graph in Problem 3 do not appear to lie exactly on a specific curve. However, calculators use a special mathematical process to produce an equation that best fits actual data. From the data in the table, The Long and Short of It, the TI-83/84 Plus can be used to generate the following equation:

$$y = -0.00029x^2 + 0.31x + 18.6 \text{ (coefficients are rounded)}, (1)$$

where x represents the skid distance in feet and y represents the speed in miles per hour. The process for generating such equations is covered in later activities.

Appendix

4. a. Enter the function equation above into your calculator. For help with the TI-83/84 Plus, see Appendix A. The screen should appear as follows:

b. The values in the table, The Long and Short of It, can help to set appropriate window values in your calculator to view the graph. Starting at a minimum skid distance of 0, a reasonable maximum value of the skid distance in this situation would be 450 feet. Beginning at a minimum speed of 0, a reasonable maximum speed value would be 100 mph. Using these settings, the window screen should appear as follows:

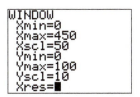

c. Display your graph using the window settings from part b. Your graph should resemble the following:

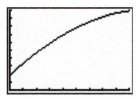

5. a. By pressing the trace button and the left and right arrow keys, you can display the x-y values of points on the graph on the bottom of the display. Use the trace feature of your calculator to approximate the speed of the car when the length of the skid distance is 200 feet. You can obtain the exact value for any x-value between Xmin and Xmax by entering the x-value while in trace mode and pressing (ENTER). Your screen should appear as follows:

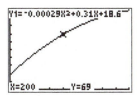

b. Use your calculator to verify the result in part a by evaluating
$-0.00029(200)^2 + 0.31(200) + 18.6$.

The equation $y = -0.00029x^2 + 0.31x + 18.6$ is called a **mathematical model**. A mathematical model is a mathematical form, such as a formula, an equation, a graph, or a table, that fits or approximates the important features of a given situation. Such models can then be used to estimate, make predictions, and draw conclusions about the given situation.

The process of developing a mathematical model to represent a given situation involves several steps. First, you need to identify the problem and develop a well-defined question about what you want to know. For example, in the automobile accident situation, can you determine the speed of the car when the brakes are applied?

Next, you look for relationships and connections between the variables that are involved in the situation. You may also need to make assumptions about the situation before you begin to construct the mathematical model.

6. What are some of the variable quantities present in determining what you want to know in the automobile accident situation? What assumptions did you make?

Using your mathematical and problem solving skills, you can now translate the features and relationships you have identified into a mathematical model. The model can be used to answer the questions that you originally raised.

Finally, you need to verify that the mathematical model you have developed accurately represents the situation under investigation.

7. How well does the mathematical model (equation 1) determine the speed of the car for a given skid distance? Explain how you came to this conclusion.

Note that one way to verify the accuracy of your mathematical model is to collect data and compare the data to the results predicted by the model. In the speed and skid-distance situation, the use of data actually helped you determine the model. By observing patterns revealed by the data and graph, an appropriate model was developed.

The process just described to develop a mathematical model is called **mathematical modeling**. The development and use of mathematical models to help solve simple to extremely complicated problems is a very important use of mathematics.

Modeling with Formulas

In Activity 1.3, you used formulas to model situations in geometry, business, and science. In almost every field of study, you are likely to encounter formulas that relate two or more variables represented by letters. Problems 8 and 9 feature formulas used in the health field.

8. Medical researchers have determined that for exercise to be beneficial, a person's desirable heart rate, R, in beats per minute, can be approximated by the formulas

$$R = 143 - 0.65a \text{ for women}$$
$$R = 165 - 0.75a \text{ for men,}$$

where a represents the person's age.

a. If the desirable heart rate for a woman is 130 beats per minute, how old is she?

b. If the desirable heart rate for a man is 135 beats per minute, how old is he?

9. The basal energy rate is the daily amount of energy (measured in calories) needed by the body at rest to maintain body temperature and the basic life processes of respiration, cell metabolism, circulation, and glandular activity. As you may suspect, the basal energy rate differs for individuals, depending on their gender, age, height, and weight. The formula for the basal energy rate for a female is

$$B = 655.096 + 9.563W + 1.85H - 4.676A,$$

where B is the basal energy rate (in calories), W is the weight (in kilograms), H is the height (in centimeters), and A is the age (in years).

a. A female patient is 70 years old, weighs 55 kilograms, and is 172 centimeters tall. A total daily caloric intake of 1000 calories is prescribed for her. Determine if she is being properly fed.

b. A female is 178 centimeters tall and weighs 84 kilograms. If her basal energy rate is 1500 calories, how old is the female?

Tables as Models

In Activity 1.6, you encountered several tables of matched pairs of values for situations involving two variables. These tables serve as models that help represent the relationship between the quantities involved. Patterns are revealed and often lead to an equation that can model the relationship more completely.

For example, you feel colder on a windy cold day than the air temperature indicates. This is commonly referred to as windchill.

10. Suppose the wind is a constant 20 mph. The following table gives the windchill temperature (how cold your skin feels) for various air temperatures.

AIR TEMPERATURE °F	40	30	20	10	0	−10	−20
WINDCHILL TEMPERATURE (20-MPH WIND)	30	17	4	−9	−22	−35	−48

    a. What relationship do you observe between the windchill temperature and the air temperature?

    b. Estimate the windchill temperature for an air temperature of −30°F.

    c. If the windchill speed is 30 mph, how would you expect the windchill temperatures in the table to change?

11. The following windchill chart was published by the National Weather Service. Note that the formula used to generate the table values is based on scientific knowledge and experiments.

Windchill Chart

Temperature (°F)

Wind (mph)	40	35	30	25	20	15	10	5	0	−5	−10	−15	−20	−25	−30	−35	−40	−45
5	36	31	25	19	13	7	1	−5	−11	−16	−22	−28	−34	−40	−46	−52	−57	−63
10	34	27	21	15	9	3	−4	−10	−16	−22	−28	−35	−41	−47	−53	−59	−66	−72
15	32	25	19	13	6	0	−7	−13	−19	−26	−32	−39	−45	−51	−58	−64	−71	−77
20	30	24	17	11	4	−2	−9	−15	−22	−29	−35	−42	−48	−55	−61	−68	−74	−81
25	29	23	16	9	3	−4	−11	−17	−24	−31	−37	−44	−51	−58	−64	−71	−78	−84
30	28	22	15	8	1	−5	−12	−19	−26	−33	−39	−46	−53	−60	−67	−73	−80	−87
35	28	21	14	7	0	−7	−14	−21	−27	−34	−41	−48	−55	−62	−69	−76	−82	−89
40	27	20	13	6	−1	−8	−15	−22	−29	−36	−43	−50	−57	−64	−71	−78	−84	−91
45	26	20	12	5	−2	−9	−16	−23	−30	−37	−44	−51	−58	−65	−72	−79	−86	−93
50	26	19	12	4	−3	−10	−17	−24	−31	−38	−45	−52	−60	−67	−74	−81	−88	−95
55	25	18	11	4	−3	−11	−18	−25	−32	−39	−46	−54	−61	−68	−75	−82	−89	−97
60	25	17	10	3	−4	−11	−19	−26	−33	−40	−48	−55	−62	−69	−76	−84	−91	−98

Frostbite Times   ■ 30 minutes   ■ 10 minutes   ■ 5 minutes

Windchill (°F) = 35.74 + 0.6215T − 35.75($V^{0.16}$) + 0.4275T($V^{0.16}$)

Where, T = Air Temperature (°F) V = Wind Speed (mph)

    a. In the windchill chart, locate the table values in Problem 10.

    b. If the wind is blowing at 30 mph, use the windchill chart to complete the following table.

AIR TEMPERATURE °F	40	30	20	10	0	−10	−20
WINDCHILL TEMPERATURE (30-MPH WIND)							

c. If the wind speed is 40 mph, what is the windchill temperature if the air temperature is −20°F?

In Chapter 8, you will use similar style tables to help determine the true interest rate on an installment plan and determine monthly mortgage payments.

Modeling with Functions

Many of the mathematical models you will use to solve problems in this course will be function models. As you learned in Activities 1.5 and 1.6, functions can be represented by tables, graphs, and equations. Problem 12 gives you a sample of how to develop a function model to solve a problem.

12. As part of a community service project at your high school, you are organizing a fund-raiser at the neighborhood roller rink. Money raised will benefit a summer camp for children with special needs. The admission charge is $4.50 per person, $2.00 of which is used to pay the rink's rental fee. The remainder is donated to the summer camp fund.

a. State a question that you want answered in this situation.

b. What two variables are involved in this problem?

c. Which variable can best be designated as the dependent variable? As the independent variable?

d. Complete the following table.

INDEPENDENT VARIABLE	1	2	3	4	5	6	7	8
DEPENDENT VARIABLE								

e. State in words the relationship between the independent and independent variables.

f. Use appropriate letters to represent the variables involved and translate the written statement in part e as an equation.

g. If 91 tickets are sold, use the equation model developed in part f to determine the amount donated to the summer camp fund.

h. If the maximum capacity of the rink is 200 people, what is the maximum amount that can be donated?

You will learn more about using graphs as mathematical models in Activity 1.13.

SUMMARY: ACTIVITY 1.8

1. A **mathematical model** is a mathematical form, such as a formula, an equation, a graph, or a table, that fits or approximates the important features of a given situation.

2. **Mathematical modeling** is the process of developing a mathematical model for a given situation.

EXERCISES: ACTIVITY 1.8

1. The total cost, c, of purchasing tickets for a college football game through a ticket service depends on the number of tickets, t, that you purchase. For one particular game, the tickets cost $41.50 apiece plus a $6.00 service charge for the order.

   a. Write an equation that relates the total cost, c, to the number of tickets purchased, t.

   b. If you purchase 6 tickets, what is the total cost of your order?

   c. If your total cost was $296.50, how many tickets did you purchase?

2. The pressure, p, of water (in pounds per square foot) at a depth of d feet below the surface is given by the formula

$$p = 15 + \frac{15}{33}d.$$

   a. If a diver is 100 feet below the surface, what is the pressure of the water on the diver?

   b. On November 14, 1993, Francisco Ferreras reached a record depth for breath-held diving. During the dive, he experienced a pressure of 201 pounds per square foot. What was his record depth?

3. The following formula is used by the NCAA to calculate the quarterback passing efficiency rating:

$$R = \frac{8.4Y + 330T + 100C - 200I}{A},$$

where R = quarterback rating
A = passes attempted
C = passes completed
Y = passing yardage
T = touchdown passes
I = number of interceptions

The Heisman Trophy is awarded annually to the best college football player. In 2013, the award was won by quarterback Jameis Winston of Florida State University. In 2014, the award was given to quarterback Marcus Mariota of the University of Oregon. Consider their statistics for their winning seasons.

PLAYER	PASSES ATTEMPTED	PASSES COMPLETED	PASSING YARDAGE	NUMBER OF TOUCHDOWN PASSES	NUMBER OF INTERCEPTIONS
Jameis Winston 2013	384	257	4057	40	10
Marcus Mariota 2014	445	304	4454	42	4

a. Determine the quarterback rating for Jameis Winston for the 2013 NCAA football season.

b. Determine the quarterback rating for Marcus Mariota for the 2014 NCAA football season.

c. Visit the NCAA website and select stats to obtain the rating of your favorite quarterback.

4. a. You want to invest money in order to receive the best return. You have two options.

Option 1: Invest at 6% simple annual interest for 10 years.

Option 2: Invest at 5% interest compounded annually.

The following table models the growth of $2000 over a 10-year period using the two options. All values are rounded to the nearest dollar.

	NUMBER OF YEARS									
	1	2	3	4	5	6	7	8	9	10
6% SIMPLE ANNUAL INTEREST	2120	2240	2360	2480	2600	2720	2840	2960	3080	3200
5% COMPOUNDED ANNUALLY	2100	2205	2315	2431	2553	2680	2814	2955	3103	3258

Describe any trends or patterns that you observe in the data.

b. The amount of your investment in option 1 can be determined by the following formula

$$A = P + Prt,$$

where A = amount of the investment
P = principal or amount invested
r = annual percentage rate (expressed as a decimal)
t = number of years invested

Use the formula to determine the amount of your $2000 investment in option 1 after 20 years. What is the total amount of interest earned?

c. The amount of your investment in option 2 can be determined by

$$A = P(1 + r)^t,$$

where A = amount of the investment
P = principal
r = annual percentage rate (expressed as a decimal)
t = number of years invested

Use the formula to determine the amount of your $2000 investment in option 2 after 20 years.

d. Which option would you choose? Explain.

Compound interest will be discussed in more detail in Chapter 5.

5. The following table gives the number of people infected by the flu over a given number of months.

NUMBER OF MONTHS	0	1	2	3	4	5
NUMBER OF PEOPLE INFECTED	1	5	13	33	78	180

Describe any trends or patterns that you observe.

6. The value of almost everything you own (assets), such as a car, computer, or house, depreciates (goes down) over time. When an asset's value decreases by a fixed amount each year, the depreciation is called straight-line depreciation.

Suppose a piece of machinery has an initial value of $12,400 and depreciates $820 per year.

a. State a question that you might want answered in this situation.

b. What two variables are involved in this problem?

c. Which variable can best be designated as the dependent variable? As the independent variable?

d. Complete the following table.

INDEPENDENT VARIABLE	1	2	3	4	5
DEPENDENT VARIABLE					

e. State in words the relationship between the independent and dependent variables.

f. Use appropriate letters to represent the variables involved and translate the written statement in part e as an equation.

g. If you plan to keep the machine for 7 years, determine the value of the machine at the end of this period. Explain the process you used.

h. What assumption was made regarding the rate of depreciation of the machine? Does this seem reasonable?

Fund-Raiser Revisited

Objective

Solve an equation numerically and graphically.

In Problem 12, page 79, of Activity 1.8, you developed the following equation to model the fund-raiser situation:

$$y = 2.50x,$$

where x represents the number of admission tickets sold, and y represents the amount of money donated to the summer camp fund.

Since the variable y is written by itself on one side, you say the equation is "solved for y."

Evaluating an Algebraic Expression

In the following example, you are asked to determine a value for the dependent variable y for a given value of the independent variable x.

Example 1   *Use the equation $y = 2.5x$ to determine the amount of money raised if 84 tickets are sold. Stated another way, determine y for a value of $x = 84$.*

SOLUTION

Step 1. Substitute 84 for x in the equation $y = 2.5x$:

$$y = 2.50(84)$$

Step 2. Perform the arithmetic operations on the right-hand side to obtain the y-value:

$$y = \$210$$

Note that the expression $2.50x$ has been **evaluated** for $x = 84$ to obtain the corresponding y-value.

1. Use the equation $y = 2.50x$ to complete the table.

x, NUMBER OF TICKETS SOLD	y, AMOUNT RAISED ($)
26	
94	
278	

Solving Equations Numerically and Graphically

Suppose the goal is to raise a total of at least $200 for the summer camp fund. You can determine how many tickets to sell to raise $200 by replacing y with 200 in $y = 2.50x$ to produce the **equation**

$$200 = 2.50x.$$

The problem now is to solve the equation for x. There are three different methods for solving such equations: numerical, graphical, and algebraic. The numerical and graphical methods are often quite useful, but frequently overlooked. Problems 2 and 3 will illustrate the numerical and graphical solution techniques.

2. a. Use a table of values to estimate how many tickets must be sold to raise $200. Begin by estimating a value for x so that the expression $2.50x$ is close to 200. For example,

$x = 100$ tickets produces a value of 250. In this case, increase or decrease your guesses for x until you obtain 200.

x, NUMBER OF TICKETS SOLD	y, AMOUNT RAISED ($)

This approach, consisting of a guess, check, and repeat, is a **numerical method** for solving the equation $200 = 2.50x$ for x.

b. List at least one disadvantage of solving an equation numerically.

3. Recall that the practical domain for x (number of tickets sold) of the summer camp situation is 0, 1, 2, 3, …, 200. Some of the x-y pairs are shown in the following graph.

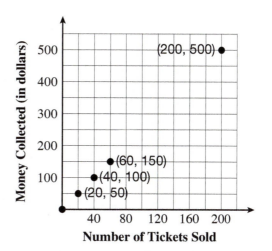

a. The point (60, 150) lies on the graph. Interpret the meaning of the coordinates 60 and 150 in the fund-raiser situation.

b. Is the graph of the fund-raiser continuous or discrete? Explain.

c. Connect the points on the graph in Problem 3 to form a line. Use this line to estimate how many tickets must be sold to raise at least $200. Remember, you need to determine x when $y = 200$.

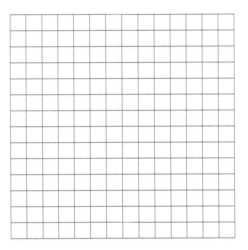

Be careful. In part c, the points of the scatterplot were connected to form a smooth continuous line. This was done as a **graphing aid** to help solve the given equation graphically.

d. Explain the process you used in part c.

The approach in part c is the **graphical method** for solving the equation $200 = 2.50x$ for x.

e. List at least one disadvantage to using a graphing approach to solving an equation.

4. The recommended weight of an adult male can be approximated by the formula
$$w = 5.5h - 220,$$
where w = recommended weight

h = height in inches.

a. What is a possible practical domain for the weight function defined by the given formula?

b. Complete the following table. If you are using a graphing calculator, remember to replace the independent variable h with x and the dependent variable w with y.

h, HEIGHT IN INCHES	60	64	68	72	76	80
w, WEIGHT IN POUNDS						

c. Plot the points determined in part b.

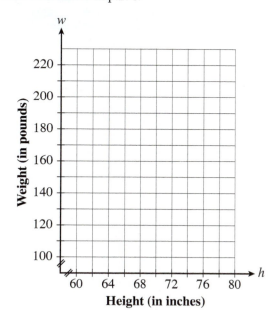

d. Connect the points in the graph to form a continuous line segment from (60, 110) to (80, 220). Are all the points on this line segment on the graph of the weight function? Explain.

e. Write an equation that can be used to determine the height of a male whose recommended weight is 165 pounds.

f. Solve the equation in part e using a numerical approach.

g. Solve the equation in part e using a graphical approach.

5. The sales and use tax collected on taxable items in your city is 8.25%. The tax you must pay depends on the price of the item you are purchasing.

a. Determine the sales tax you must pay on the following items.

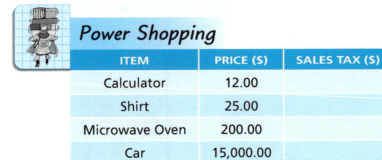

Power Shopping

ITEM	PRICE ($)	SALES TAX ($)
Calculator	12.00	
Shirt	25.00	
Microwave Oven	200.00	
Car	15,000.00	

b. Write a statement that describes how to determine the sales tax for a given price of an item.

c. Translate the statement in part a into an equation, with x representing the price and y representing the sales tax.

d. Write an equation that can be used to determine the price of a DVD player for which you paid a sales tax of $34.65. Solve the equation numerically and graphically.

SUMMARY: ACTIVITY 1.9

1. To solve an equation using a **numerical approach**, use a guess, check, and repeat process.

2. To solve an equation using a **graphical approach**, read the appropriate coordinates on the graph of the equation relating x and y.

3. A **solution** of an equation containing one variable is a replacement value for the variable that produces equal values on both sides of the equation.

4. A graph is said to be **continuous** if you can place your pencil on any point on the graph and then trace the entire graph without lifting the pencil off the paper. The graph has no breaks or holes.

EXERCISES: ACTIVITY 1.9

1. Your favorite gas station is having a super sale and is selling regular unleaded gasoline for three hours at $1.239 per gallon.

a. Select a technique to approximate the number of gallons that you can purchase with $10.

b. Let x be the independent variable representing the number of gallons purchased. Let C be the dependent variable representing the total cost of the fuel. Write an equation relating x and C.

c. What is the practical domain for x?

d. Use the equation from part b to complete the following table.

x, NUMBER OF GALLONS PURCHASED	5	10	15	20
c, TOTAL COST OF THE PURCHASE				

e. You have only $10 with you. Use the equation from part b to write an equation that can be used to determine how many gallons you can purchase.

f. Solve your equation from part e using a numerical approach.

g. The graph of the equation determined in part b is given below. Use the graph to estimate the number of gallons you can purchase with $10.

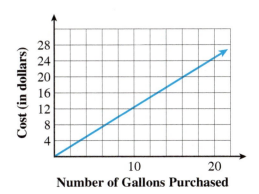

2. You are president of the high school band booster club. You have arranged for the school's jazz band to perform at a local coffee shop for three hours. In exchange for the performance, the booster club will receive three-quarters of the shop's gross receipts during that three-hour period.

a. Let x be the independent variable representing the gross receipts of the coffee shop during the performance. Let y be the dependent variable representing the share of the gross receipts that the coffee shop will donate to the band boosters. Write an equation relating x and y.

b. Use the equation from part a to complete the following table.

x, TOTAL GROSS RECEIPTS ($)	250	500	750	1000
y, BOOSTERS' SHARE ($)				

c. If the coffee shop presents you with a check for $650, what were the gross receipts during the performance? Estimate your answer using a numerical approach.

d. The graph of the equation determined in part a is given below. Use a graphical approach to estimate the gross receipts from part c.

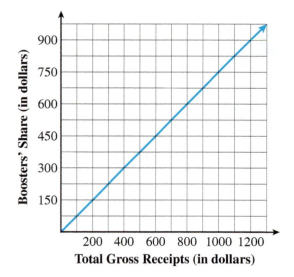

e. Which tool would you select, the table or the graph, to approximate the gross receipts if the coffee shop presented you with a check for $450? Explain.

f. Use the tool you selected in part e to approximate the gross receipts if the coffee shop presented you with a check for $450.

3. A tennis ball is dropped from the top of the Willis Tower in Chicago. The following table gives the ball's distance from ground level at a given time after it is dropped.

TIME (SEC)	1	2	3	4	5	6	7
DISTANCE (FT)	1398	1350	1270	1158	1014	838	630

The ball was dropped from a height of 1414 feet. Use the information in the table to approximate the time when the ball will have fallen half the height of the building.

4. **a.** The depth (inches) of water that accumulates in the spring soil from melted snow can be determined by dividing the cumulative winter snowfall (inches) by 12. Translate this statement into an equation, using *I* for the accumulated inches of water and *n* for the inches of fallen snow.

b. Determine the amount of water that accumulates in the soil if 25 inches of snow falls.

c. Write an equation that can be used to determine the total amount of winter snow that accumulates 6 inches of water in the soil.

d. Solve the equation in part c using a numerical approach.

e. Solve the equation in part c using a graphical approach.

Activity 1.10

Leasing a Copier

Objectives

1. Develop a mathematical model in the form of an equation.

2. Solve an equation of the form $ax + b = c$, where $a \neq 0$, using an algebraic approach.

As part of a special summer program at your high school, you are an intern in a law office. You are asked by the office manager to gather some information about leasing a copy machine for the office. The sales representative at Western Supply Company recommends a 50-copy/minute copier to satisfy the office's copying needs. The copier leases for $300 per month, plus 1.5 cents a copy. Maintenance fees are covered in the monthly charge. The lawyers would own the copier after 39 months.

1. **a.** The total monthly cost depends upon the number of copies made. Identify the independent and dependent variables.

   b. Write a statement in words to determine the total monthly cost in terms of the number of copies made during the month.

   c. Complete the following table.

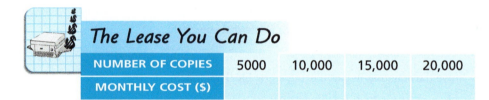

NUMBER OF COPIES	5000	10,000	15,000	20,000
MONTHLY COST ($)				

   d. Translate the statement in part b into an equation. Let n represent the number of copies made during the month and c represent the total monthly cost.

2. **a.** Use the result obtained in Problem 1d to determine the monthly cost if 12,000 copies are made.

   b. Is 12,000 a replacement value for the independent or for the dependent variable?

In Problem 2, you determined the cost by evaluating the expression $0.015n + 300$ for $n = 12,000$.

3. Suppose the monthly budget for leasing the copier is $800.

   a. Is 800 a value for n or a value for c?

   b. If the monthly budget is $800, write an equation that can be used to determine the number of copies in the month.

Notice in Problem 3 that n is not isolated by itself on one side of the equation. To isolate n, you need to undo two operations: the addition of 300 and the multiplication by 0.015. You do this by reversing the sequence of operations, undoing each operation by its inverse.

Example 1   *The following example illustrates a systematic algebraic procedure by undoing two operations to solve an equation.*

Solve for x: $120 = 3x + 90$

$$120 = 3x + 90$$

$$\underline{-90 \qquad\quad -90}$$

$$30 = 3x$$

Step 1: To undo the addition by 90, subtract 90 from each side of the equation.

$$\frac{30}{3} = \frac{3x}{3}$$

Step 2: To undo the multiplication of 3, divide each side of the equation by 3.

$$10 = x$$

The variable x has been isolated on the right side with coefficient 1; the solution is 10.

Check: $120 \overset{?}{=} 3(10) + 90$

$120 \overset{?}{=} 30 + 90$

$120 = 120$

Note that the algebraic procedure emphasizes that an equation can be thought of as a scale whose arms are in balance. The equals sign can be thought of as the balancing point.

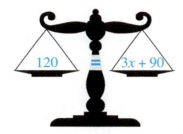

As you perform the appropriate inverse operation to solve the equation $120 = 3x + 90$ for x, you must maintain the balance as you perform each step in the process. If you subtract 90 from one side, then you must subtract 90 from the other side. Similarly, if you divide one side by 3, then you must divide the other side by 3.

4. a. Solve the equation in Problem 3 using the algebraic procedure outlined in Example 1.

   b. Verify your answer in part a using either a numerical or graphical approach.

5. For each of the following, substitute the given value of y and use an algebraic approach to solve the resulting equation for x.

   a. If $y = 3x - 5$ and $y = 10$, determine x.

   b. If $y = 30 - 2x$ and $y = 24$, determine x.

    c. If $y = \dfrac{3}{4}x - 21$ and $y = -9$, determine x.

    d. If $-2x + 15 = y$ and $y = -3$, determine x.

6. You need to purchase a camera and lenses for your photography class. There is a special promotion for students in which you can purchase an \$850 complete camera outfit by making monthly payments for 2 years at 0% interest. You are required to make a 10% down payment, with the balance to be paid in 24 equal monthly payments.

    a. Determine the amount of down payment required.

    b. Determine the amount owed after the down payment.

    c. What are the monthly payments?

    d. Let A represent the total amount paid after n payments made. What is the amount A when $n = 0$? What does this value of A represent in this situation?

    e. Complete the following table.

PAYMENT NUMBER, n	AMOUNT PAID, A (\$)
0	
1	
2	
3	

    f. Write an equation that represents the relationship between the amount paid, A, and the number n of payments.

    g. If you leave college, you are required to pay the balance due. What is the amount due after 3 payments?

    h. Write an equation for the balance due, B, in terms of A.

SUMMARY: ACTIVITY 1.10

1. The goal of **solving an equation** of the form $ax + b = c$, where $a \neq 0$, for the variable x is to isolate the variable x (with coefficient 1) on one side of the equation.

2. Use **inverse operations** to isolate the variable. Apply the inverse operations in the following order:

Step 1. Undo the addition of b by subtracting b from each side of the equation; undo the subtraction of b by adding b to each side of the equation.

Step 2. Undo the multiplication of the variable x by the nonzero coefficient a by dividing each side of the equation by a.

EXERCISES: ACTIVITY 1.10

1. You are planning an extended trip to Mexico. You have a new smartphone to keep in touch with your friends but are concerned about roaming charges, because you know you will be outside the home network service area. Your plan costs $35 per month plus $0.25 per minute for roaming charges.

  a. Write a statement to determine your total monthly cost for your smartphone per month during your extended stay in Mexico.

  b. Translate the statement in part a into an equation, using c to represent the total monthly cost and n to represent the total roaming minutes for the month.

  c. Determine your monthly cost, if 250 minutes of roaming charges are added to your bill.

  d. If you budget $75 per month for your smartphone, how many minutes of roaming charges can you incur?

2. You are considering taking some courses at a nearby community college. For out of district students, the tuition is $246 for each credit hour taken. All students, part-time or full-time, must pay a $15 parking fee for the semester. This fixed fee is added directly to your tuition and is included in your total bill.

  a. Complete the following table, where t represents the total bill and n represents the number of credit hours taken.

NUMBER OF CREDIT HOURS, n	1	2	3	4	5	6
TOTAL BILL, t($)						

  b. Write an equation to determine the total bill, t, for an out of district student. Use n to represent the number of hours taken for the semester.

c. Determine the total bill if you take nine credit hours.

d. Suppose you have $1000 to spend on the total bill. How many credit hours can you carry for the semester?

3. The senior year class is organizing an entertainment night to benefit charities in the community. You are a member of the budget committee for the class project. The committee suggests a $10 per person admission donation for food, nonalcoholic beverages, and entertainment. The committee determines that the fixed costs for the event (food, drinks, posters, tickets, etc.) will total $2100. The high school principal is offering the use of the gymnasium for the evening.

a. The total revenue (gross income before expenses are deducted) depends on the number, n, of students who attend. Write an expression in terms of n that represents the total revenue if n students attend.

b. Profit is the net income after expenses are deducted. Write an equation expressing the profit, p, in terms of the number, n, of students who attend.

c. If the gymnasium holds a maximum of 700 people, what is the maximum amount of money that can be donated to charity?

d. Suppose that the members of the committee want to be able to donate at least $1500 to community charities. How many students must attend in order to have a profit of $1500?

4. The method of straight-line depreciation is often used by companies to reduce the value of their assets by a fixed amount each year. The amount of the reduction depends on the useful life of the asset. This is generally determined by the company. Suppose a business selling computer hardware has obtained a fleet of new hybrid cars for its sales force. Each car costs $26,700.

a. The business estimates that the useful life of each vehicle is 5 years. Using the straight-line method, determine the amount of depreciation per year.

b. Determine the value of each car after 3 years.

c. In this situation, the value of each car, v, is a function of the age of the car, a. Identify the independent and dependent variables.

d. Use function notation to write a function rule for the value $v(a)$ of each car in terms of its age a.

e. What is the practical domain and practical range of the depreciation function?

f. Determine $v(4)$. Interpret the result in the context of this situation.

g. Use the function rule to determine the age of each car if the value is $16,020.

5. To help pay for college, you have accepted a summer position as a sales associate in a family-owned men's and women's specialty clothing store. You are offered a weekly salary of $200 plus 5% commission on all of your sales.

   a. Identify the input and output variables in this situation.

   b. Write a statement that describes how the output in part a is determined for a given input value.

   c. Does the relationship in part b define a function? Explain.

   d. Let $s(t)$ represent the weekly salary and t represent total sales. Write a function rule for $s(t)$ in terms of t.

   e. What are the practical domain and practical range of the salary function?

   f. Evaluate $s(2000)$, and write a sentence explaining the meaning of $s(2000)$ in the context of the situation.

   g. Use the function rule to determine total sales needed to earn a weekly salary of $400.

6. In the summer of 2009, during a period of economic woes for the United States, President Barack Obama's "Cash for Clunkers" program was wildly popular. The program was intended to aid the auto industry, thereby boosting the economy. It was also a "green" initiative, because consumers replaced their cars with cars having better fuel economy. Your 1995 Jeep Grand Cherokee 4WD had a combined city/highway fuel economy of 15 miles per gallon. In return for it, you would receive a $4500 credit toward the purchase of a new car provided that you chose one with fuel economy at least 10 miles per gallon more than your Jeep. This incentive, in addition to the 0% interest loans that were being offered by the troubled auto industry, was too much for you to resist. Your car payment is $420 per month.

    a. Write an equation to determine the total amount, p, paid on the car after n months, including the $4500 credit.

    b. Use the equation you wrote in part a to complete the following table.

NUMBER OF MONTHS, n	1	2	3	4	5	6
TOTAL AMOUNT PAID, p ($)						

    c. Determine the amount, p, paid on the car after 9 months.

    d. The car you chose was a 2009 Ford Escape Hybrid, with combined fuel economy of 30 miles per gallon, and the price of your car after negotiation, including taxes and all fees, was $29,700. How many years will it take you to pay off your loan?

7. Archaeologists and forensic scientists use the length of human bones to estimate the height of individuals. A person's height, h, in centimeters, can be determined from the length of the femur, f (the bone from the knee to the hip socket), in centimeters, using the following formulas:

        Man: $h = 69.089 + 2.238f$

        Woman: $h = 61.412 + 2.317f$

    a. A partial skeleton of a man is found. The femur measures 50 centimeters. How tall was the man?

b. What is the length of the femur for a woman who is 150 centimeters tall?

8. Let p represent the perimeter of an isosceles triangle that has two equal sides of length a, and a third side of length b. The formula for the perimeter is $p = 2a + b$. Determine the length of the equal sides of an isosceles triangle having perimeter of $\frac{3}{4}$ yard and third side measuring $\frac{1}{3}$ yard.

9. The recommended weight of an adult male is given by the formula $w = \frac{11}{2}h - 220$, where w represents his recommended weight in pounds and h represents his height in inches. Determine the height of a man whose recommended weight is 165 pounds.

10. Housing prices in your neighborhood have been increasing steadily since you bought your home in 2006. The relationship between the market value of your home and the length of time you have been living there can be expressed algebraically by the rule

$$V = 130{,}000 + 3500x,$$

where x is the length of time (in years) in your home and V is the market value (in dollars).

a. Complete the following table:

YEAR	x	MARKET VALUE
2006		
2011		
2014		

b. Determine the value of your home in 2016.

c. In which year will the value of your home reach $186,000?

11. Your local utility company offers you a free energy audit of your home. The report is based on a temperature difference of 68°F inside and 28°F outside. The resulting report indicates a wall of the house containing 40 square feet of glass (single pane) and 120 square feet of standard insulated (R-11) wall lost 2035 BTU per hour. If the heat loss per square foot per hour for a standard wall is 3.7 BTU/h, determine the heat lost per square foot per hour for glass (single pane).

12. According to a 2012 U.S. Census Bureau Report, a person with a bachelor's degree could expect to have an annual salary that is $21,528 more than that of someone who has only a high school diploma.

a. Let x represent the annual salary of a person with a high school diploma in 2012. Write an algebraic expression in terms of x that represents the average annual salary of a person with a bachelor's degree in 2012.

b. Together, the individuals would have a total annual salary of $89,336. Write an equation that represents the relationship between the two average salaries in part a.

c. Solve the equation in part b to determine the average annual salary of a person at each level of education in 2012.

13. Use an algebraic approach to solve each of the following equations for x.

a. $10 = 2x + 12$

b. $-27 = -5x - 7$

c. $3x - 26 = -14$

d. $24 - 2x = 38$

e. $5x - 15 = 15$

f. $-4x + 8 = 8$

g. $12 + \dfrac{1}{5}x = 9$

h. $\dfrac{2}{3}x - 12 = 0$

i. $0.25x - 14.5 = 10$

j. $5 = 2.5x - 20$

Activity 1.11

Comparing Energy Costs

Objectives

1. Develop mathematical models to solve problems.

2. Write and solve equations of the form $ax + b = cx + d$, where $a \neq 0$ and $c \neq 0$.

3. Use the distributive property to solve equations involving grouping symbols.

4. Solve formulas for a specified variable.

An architect is hired to design a 3,000 square foot home. She obtains installation and annual operating costs for two types of heating and cooling systems: solar and electric.

TYPE OF HEATING AND COOLING SYSTEM	INSTALLATION COST	OPERATING COST PER YEAR
Solar	$40,000	$100
Electric	$9000	$1200

You will use this information to compare the cost for each system over a period of years. The following questions will guide you in making this comparison.

1. a. If you select the solar system, you would be eligible to receive 30% off the total installation cost from the federal government in the form of a Federal Solar Tax Credit. Determine the amount of credit.

b. What is the new installation cost of the solar system after the credit is applied?

2. a. Determine the total cost of the solar heating system after five years of use.

b. Write a statement for the total cost of the solar heating system in terms of the number of years of use.

c. Let x represent the number of years of use and c represent the total cost of the solar heating system. Translate the statement in part b into an equation.

d. Use the equation from part c to complete the following table:

NUMBER OF YEARS IN USE, x	5	10	15	20
TOTAL COST, c				

3. a. Determine the total cost of the electric heating system after five years of use.

b. Write a statement for the total cost of the electric heating system in terms of the number of years of use.

c. Let x represent the number of years of use and c represent the total cost of the electric heating system. Translate the statement in part b into an equation.

d. Use the equation from part c to complete the following table:

NUMBER OF YEARS IN USE, x	5	10	15	20
TOTAL COST, c				

The installation cost of solar heating is much more than that of the electric system, but the operating cost per year of the solar system is much lower. Therefore, it is reasonable to think that the total cost for the electric system will eventually "catch up" and surpass the total cost of the solar system.

4. Compare the table values for total heating costs in Problems 2d and 3d. Estimate in what year the total cost for electric heating will "catch up" and surpass the total cost for solar heating. Explain.

5. The year in which the total costs of the two heating systems are equal can be determined algebraically. Write an equation you can solve to determine when the total heating costs are the same by setting the expressions in the symbolic rules in Problems 2c and 3c equal to each other.

The following example demonstrates a systematic algebraic procedure that you can use to solve equations similar to the equation in Problem 5. Remember that your goal is to isolate the variable on one side of the equation with coefficient 1 by applying the appropriate inverse operations.

Example 1   *Solve for x: $2x + 14 = 8x + 2$*

SOLUTION

First, add and/or subtract terms appropriately so that all terms involving the variable are on one side of the equals sign and all other terms are on the other side.

$$2x + 14 = 8x + 2 \qquad \text{Subtract } 8x \text{ from both sides and combine like terms.}$$
$$\underline{-8x \qquad\quad -8x}$$
$$-6x + 14 = 2$$
$$\underline{\qquad -14 = -14} \qquad \text{Subtract 14 from each side and combine like terms.}$$
$$-6x \qquad = -12$$
$$\frac{-6x}{-6} = \frac{-12}{-6} \qquad \text{Divide each side by } -6, \text{ the coefficient of } x.$$
$$x = 2$$

Check: $2(2) + 14 = 8(2) + 2$
$$4 + 14 = 16 + 2$$
$$18 = 18$$

6. In Example 1, the variable terms are combined on the left side of the equation. Solve the equation $2x + 14 = 8x + 2$ by combining the variable terms on the right side.

7. a. Solve the equation in Problem 5 for x.

    b. Interpret what your answer in part a represents in the context of the heating system situation.

8. Your architect tells you that there could actually be a surplus of $500 per year by using the solar system. Redo your calculations and report your results.

Purchasing a Car

You are interested in purchasing a new car and have narrowed the choice to a Honda Accord LX (4 cylinder) and a Passat GLS (4 cylinder). Being concerned about the value of the car depreciating over time, you search the Internet and obtain the following information:

Driven Down

MODEL OF CAR	MARKET SUGGESTED RETAIL PRICE (MSRP) ($)	ANNUAL DEPRECIATION ($)
Accord LX	20,925	1730
Passat GLS	24,995	2420

9. a. Complete the following table:

YEARS THE CAR IS OWNED	VALUE OF ACCORD LX ($)	VALUE OF PASSAT GLS ($)
1		
2		
3		

b. Could the value of the Passat GLS ever be lower than the value of the Accord LX? Explain.

c. Let v represent the value of the car after x years of ownership. Write a symbolic rule to determine v in terms of x for the Accord LX.

d. Write a symbolic rule to determine v in terms of x for the Passat GLS.

e. Write an equation to determine when the value of the Accord LX will equal the value of the Passat GLS.

f. Solve the equation in part e.

NBA Basketball Court

You and your friend are avid professional basketball fans and discover that your mathematics instructor shares your enthusiasm for basketball. During a mathematics class, your instructor tells you that the perimeter of an NBA basketball court is 288 feet and the length is 44 feet more than its width. He challenges you to use your algebra skills to determine the dimensions of the court. To solve the problem, you and your friend use the following plan.

10. a. Let w represent the width of the court. Write an expression for the length in terms of the width, w.

b. Use the formula for perimeter of a rectangle, $P = 2l + 2w$, and the information given to obtain an equation containing just the variable w.

c. Solve the equation you obtained in part b by first applying the distributive property and then combining like terms in the expression involving w.

d. What are the dimensions of an NBA basketball court?

Solving Formulas for a Specified Variable

11. Windchill temperature, w, produced by a 30-mile-per-hour wind at various Fahrenheit degrees can be modeled by the formula

$$w = 1.6t - 49.$$

a. Use the formula $w = 1.6t - 49$ to determine the windchill temperature if the air temperature is 7°F.

b. On a cold day in Michigan, the wind is blowing at 30 miles per hour. If the windchill temperature is reported to be -18°F, then what is the air temperature on that day? Use the formula $w = 1.6t - 49$.

The formula $w = 1.6t - 49$ is said to be solved for w in terms of t because the variable w is isolated by itself on one side of the equation. To determine the windchill temperature, w, for air temperature $t = 7$°F (Problem 11a), you simply substitute 7 for t in $1.6t - 49$ and do the arithmetic:

$$w = 1.6(7) - 49 = -37.8°F$$

In Problem 11b, you needed to determine the air temperature, t, for a given windchill temperature. In this case, you substitute -18 for w in $w = 1.6t - 49$, and then solve the resulting equation for t.

If you had to determine the corresponding t-value for several different w-values, you would have to solve several equations for t. It is much more convenient and efficient to solve the original formula $w = 1.6t - 49$ for t and then use the new equation to determine values of t.

Example 2   *Solving the formula $w = 1.6t - 49$ for t is similar to solving the equation $-18 = 1.6t - 49$ for t.*

$$-18 = 1.6t - 49$$

$$\underline{+49 \qquad\qquad +49}$$

$$31 = 1.6t$$

$$\frac{31}{1.6} = \frac{1.6t}{1.6}$$

$$19.4 \approx t$$

$$w = 1.6t - 49$$

$$\underline{+49 \qquad\qquad +49} \qquad \text{Add 49 to each side.}$$

$$w + 49 = 1.6t$$

$$\frac{w + 49}{1.6} = \frac{1.6t}{1.6} \qquad \text{Divide each side by the coefficient 1.6.}$$

$$\frac{w + 49}{1.6} = t$$

The new formula is $t = \dfrac{w + 49}{1.6}$ or $t = \dfrac{w}{1.6} + \dfrac{49}{1.6}$, which is equivalent to $t = 0.625w + 30.625$.

To solve the equation $w = 1.6t - 49$ for t means to isolate the variable t (with coefficient 1) on one side of the equation, with all other terms on the opposite side.

12. Redo Problem 11b using the new formula derived in Example 2 that expresses t in terms of w.

13. a. If the wind speed is 15 miles per hour, the windchill can be approximated by the formula $w = 1.4t - 32$, where t is the air temperature in degrees Fahrenheit. Solve the formula for t.

  b. Use the new formula from part a to determine the air temperature, t, if the windchill temperature is $-10°F$.

14. Solve each of the following formulas for the specified variable.

  a. $A = lw$ for w

  b. $p = c + m$ for m

  c. $P = 2l + 2w$ for l

  d. $R = 165 - 0.75a$ for a

  e. $V = \pi r^2 h$ for h

  f. $P = 2l + 2w$ for w

  g. $V(P + a) = k$ for P

Additional Practice

15. Solve each of the following equations for x. Remember to check your result in the original equation.

a. $2x + 9 = 5x - 12$

b. $21 - x = -3 - 5x$

c. $2(x - 3) = -8$

d. $2(x - 4) + 6 = 4x - 7$

SUMMARY: ACTIVITY 1.11

1. General strategy for solving equations for a variable, such as x.

- Remove parentheses, if necessary, by applying the distributive property.

- Combine like terms that appear on the same side of the equation.

- Write the equation so that the product of the coefficient and the variable is on one side and all other numbers and variables are on the other side. This is generally accomplished by adding and/or subtracting terms so that all terms involving the variable you are solving for are on one side of the equation and all terms involving numbers and other variables appear on the other side.

- Solve for the variable by dividing each side of the equation by the nonzero coefficient of the variable.

- Check the result in the original equation to be sure that the value of the variable produces a true statement.

2. To solve a **formula** for a variable, isolate that variable (with coefficient 1) on one side of the equation with all other terms on the "opposite" side.

EXERCISES: ACTIVITY 1.11

1. You need to rent a van so you contact two local rental companies and acquire the following information for the one-day cost of renting a van:

   Company 1: $60 per day, plus $0.75 per mile
   Company 2: $30 per day, plus $1.00 per mile

Let x represent the number of miles driven in one day and C represent that total daily rental cost ($).

   a. Write an equation that represents the total cost of renting the van for one day from company 1.

   b. Write an equation that represents the total cost of renting the van for one day from company 2.

   c. Write an equation to determine for what mileage the one-day rental cost would be the same.

   d. Solve the equation you obtained in part c.

   e. For which mileages would company 2 have the lower price?

2. You are considering installing a basic security system in your new house. You gather the following information from the Internet about similar security systems from two local home security dealers:

   Lease: $99 to install and $35 per month monitoring fee
   Purchase: $450 to purchase equipment and install and $15 per month for monitoring

You see that the initial cost of purchasing the security system is much higher than that of leasing the system, but that the monitoring fee is lower. You want to determine which is the better system for your needs.

Let x represent the number of months that you have the security system, and C represent the total cost ($).

   a. Write an equation that represents the total cost of leasing the system in terms of the number of months you have the system.

   b. Write an equation that represents the total cost of purchasing the system in terms of the number of months you have the system.

   c. Write an equation to determine the number of months for which the total cost of the systems will be equal.

d. Solve the equation in part c.

e. If you plan to live in the house and use the system for five years, which system would be less expensive?

3. You are able to get three summer jobs to help save for college expenses. In your job as a cashier, you work twenty hours per week and earn $9.50 per hour. Your second and third jobs are both at a local hospital. There you earn $9.00 per hour as a payroll clerk and $7.00 per hour as an aide. You always work ten hours less per week as an aide than you do as the payroll clerk. Your total weekly salary depends on the number of hours that you work at each job.

a. Determine the independent and dependent variables for this situation.

b. Explain how you calculate the total amount earned each week.

c. If x represents the number of hours that you work as a payroll clerk, represent the number of hours that you work as an aide in terms of x.

d. Write an equation that describes the total amount you earn each week. Use x to represent the input variable and y to represent the output variable. Simplify the equation as much as possible.

e. If you work twelve hours as a payroll clerk, how much will you make in one week?

f. What is a practical domain for x? Would eight hours at your payroll job be a realistic replacement value? What about fifty hours?

g. When you don't work as an aide, what is your total weekly salary?

h. If you plan to earn a total of $505 in one week from all jobs, how many hours would you have to work at each job? Is the total number of hours worked realistic? Explain.

i. Solve the equation in part d for x in terms of y. When would it be useful to have the equation in this form?

4. A florist sells roses for $1.50 each and carnations for $0.85 each. Suppose you purchase a bouquet of one dozen flowers consisting of roses and carnations.

a. Let x represent the number of roses purchased. Write an expression in terms of x that represents the number of carnations purchased.

b. Write an expression that represents the cost of purchasing x roses.

c. Write an expression that represents the cost of purchasing the carnations.

d. What does the sum of the expressions in parts b and c represent?

e. Suppose you are willing to spend $14.75. Write an equation that can be used to determine the number of roses that can be included in a bouquet of one dozen flowers consisting of roses and carnations.

f. Solve the equation in part e to determine the number of roses and the number of carnations in the bouquet.

5. The viewing window of a certain calculator is in the shape of a rectangle.

    a. Let w represent the width of the viewing window in centimeters. If the window is 5 centimeters longer than it is wide, write an expression in terms of w for the length of the viewing window.

    b. Write a symbolic rule that represents the perimeter, P, of the viewing window in terms of w.

    c. If the perimeter of the viewing window is 26 centimeters, determine the dimensions of the window.

Solve the equations in Exercises 6–17.

6. $5x - 4 = 3x - 6$ **7.** $3x - 14 = 6x + 4$

8. $0.5x + 9 = 4.5x + 17$ **9.** $4x - 10 = -2x + 8$

10. $0.3x - 5.5 = 0.2x + 2.6$ **11.** $4 - 0.025x = 0.1 - 0.05x$

12. $5t + 3 = 2(t + 6)$ **13.** $3(w + 2) = w - 14$

14. $21 + 3(x - 4) = 4(x + 5)$

15. $2(x + 3) = 5(2x + 1) + 4x$

16. $500 = 0.75x - (750 + 0.25x)$

17. $1.5x + 3(22 - x) = 70$

18. The National Weather Service reports the daily temperature in degrees Fahrenheit. The scientific community, as well as Canada and most of Europe, reports temperature in degrees Celsius. The Celsius, C, and Fahrenheit, F, temperature readings are related by the formula

$$F = 1.8C + 32$$

a. Determine the Fahrenheit reading corresponding to the temperature at which water boils, 100°C.

b. Solve the formula $F = 1.8C + 32$ for C.

c. Use the new formula from part b to answer part a.

19. In 1966, the U.S. Surgeon General's health warnings began appearing on cigarette packages. The following data seems to demonstrate that public awareness of the health hazards of smoking has had some effect on consumption of cigarettes.

	YEAR							
	1997	1999	2001	2003	2005	2007	2009	2011
% OF TOTAL POPULATION 18 AND OLDER WHO SMOKE	24.7	23.5	22.8	21.6	20.9	19.8	20.6	19.0

Data Source: U.S. National Center for Health Statistics.

The percent P of the total population (18 and older) who smoke can be modeled by the formula

$$P = -0.38t + 24.3,$$

where t is the number of years since 1997.

a. Use the formula to predict in which year 18% of the total population (18 and older) will smoke.

b. Use the formula to predict the percent of the total population 18 and older that will smoke in 2020.

c. How confident are you in the prediction in parts a and b? Explain.

Solve each of the following formulas for the specified variable.

20. $E = IR$ for I

21. $C = 2\pi r$ for r

22. $P = 2a + b$ for b

23. $P = R - C$ for R

24. $P = 2l + 2w$ for l

25. $R = 143 - 0.65a$ for a

26. $A = P + Prt$ for r

27. $y = mx + b$ for m

28. $m = g - vt^2$ for g

29. $m = g - vt^2$ for v

You will be attending college this fall, so it is important for you to find a summer job that pays well. Luckily, the classified section of your newspaper lists numerous summer job opportunities in sales, road construction, and food service. The advertisements for all these positions welcome applications from high school students. All positions involve the same period during the summer.

Sales

A new electronics store opened recently. There are several sales associate positions that pay an hourly rate of $7.50 plus a 5% commission based on your total weekly sales. You would be guaranteed at least 30 hours of work per week, but not more than 40 hours.

Construction

Your state's highway department hires high school students every summer to help with road construction projects. The hourly rate is $11.75 with the possibility of up to 10 hours per week in overtime, for which you would be paid time and a half. Of course, the work is totally dependent on good weather, and so the number of hours that you would work per week could vary.

Restaurants

Local restaurants experience an increase in business during the summer. There are several positions for waitstaff. The hourly rate is $4.90, and the weekly tip total ranges from $200 to $750. You are told that you can expect a weekly average of approximately $450 in tips. You would be scheduled to work five dinner shifts of 6.5 hours each for a total of 32.5 hours per week. However, on slow nights you might be sent home early, perhaps after working only 5 hours. Thus, your total weekly hours might be fewer than 32.5.

All of the jobs would provide an interesting summer experience. Your personal preferences might favor one position over another. Keep in mind that you will have a lot of college expenses.

I. At the electronics store, sales associates average $7000 in sales each week.

    a. Based on the expected weekly average of $7000 in sales, calculate your gross weekly paycheck (before any taxes or other deductions) if you worked a full 40-hour week in sales.

    b. Use the average weekly sales figure of $7000 and write an equation for your weekly earnings, s, where x represents the total number of hours you would work.

    c. What would be your gross paycheck for the week if you worked 30 hours and still managed to sell $7000 in merchandise?

    d. You are told that you would typically work 35 hours per week if your total electronic sales do average $7000. Calculate your typical gross paycheck for a week.

    e. You calculate that to pay college expenses for the upcoming academic year, you need to gross at least $550 a week. How many hours would you have to work in sales each week? Assume that you would sell $7000 in merchandise.

2. In the construction job, you would average a 40-hour workweek.

    a. Calculate your gross paycheck for a typical 40-hour workweek.

    b. Write an equation for your weekly salary, s, for a week with no overtime. Let x represents the total number of hours worked.

    c. If the weather is ideal for a week, you can expect to work 10 hours in overtime (over and above the regular 40-hour workweek). Determine your total gross pay for a week with 10 hours of overtime.

    d. The equation in part b can be used to determine the weekly salary, s, when x, the total number of hours worked, is less than or equal to 40 (no overtime). Write an equation to determine your weekly salary, s, if x is greater than 40 hours.

    e. Suppose it turns out to be a gorgeous summer and your supervisor says that you can work as many hours as you want. If you are able to gross $800 a week, you will be able to afford to buy a computer. How many hours would you have to work each week to achieve your goal?

3. The restaurant job involves working a maximum of five dinner shifts of 6.5 hours each.

    a. Calculate what your gross paycheck would be for an exceptionally busy week of five 6.5-hour dinner shifts and $750 in tips.

    b. Calculate what your gross paycheck would be for an exceptionally slow week of five 5-hour dinner shifts and only $200 in tips.

    c. Calculate what your gross paycheck would be for a typical week of five 6.5-hour dinner shifts and $450 in tips.

    d. Use $450 as your typical weekly total for tips, and write an equation for your gross weekly salary, s, where x represents the number of hours.

    e. Calculate what your gross paycheck would be for a 27-hour week and $450 in tips.

f. During the holiday week of July 4, you would be asked to work an extra dinner shift. You are told to expect $220 in tips for that night alone. Assuming a typical workweek for the rest of the week, would working that extra dinner shift enable you to gross at least $850?

4. You would like to make an informed decision in choosing one of the three positions. Based on all the information you have about the three jobs, fill in the following table.

	LOWEST WEEKLY GROSS PAYCHECK	TYPICAL WEEKLY GROSS PAYCHECK	HIGHEST WEEKLY GROSS PAYCHECK
SALES ASSOCIATE			
CONSTRUCTION WORKER			
WAITSTAFF			

5. Money may be the biggest factor in making your decision. But it is summer, and it would be nice to enjoy what you are doing. Discuss the advantages and disadvantages of each position. What would your personal choice be? Why?

6. You decide that you would prefer an indoor job. Use the algebraic rules you developed for the sales job in Problem 1b and for the restaurant position in Problem 3d to calculate how many hours you would have to work in each job to receive the same weekly salary.

Activity 1.13

Graphs Tell Stories

Objectives

1. Describe in words what a graph tells you about a given situation.

2. Sketch a graph that best represents the situation described in words.

3. Identify increasing, decreasing, and constant parts of a graph.

4. Identify minimum and maximum points on a graph.

5. Use the vertical line test to determine whether a graph represents a function.

The expression "A picture is worth a thousand words" is a cliché, but it is true. Mathematical models are often easier to understand when presented in visual form. To understand such pictures, you need to practice going back and forth between graphs and word descriptions.

Every graph shows how the independent and dependent variables change in relation to one another. As you read a graph from left to right, the *x*-variable is increasing in value. The graph indicates the change in the *y*-values (increasing, decreasing, or constant) as the *x*-values increase.

Graphs are always constructed so that as you read the graph from left to right, the *x*-variable increases in value. The graph shows the change (increasing, decreasing, or constant) in the output values as the input values increase.

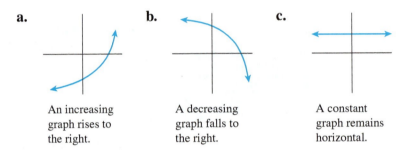

a. An increasing graph rises to the right.

b. A decreasing graph falls to the right.

c. A constant graph remains horizontal.

If a function increases and then decreases, the point where the graph changes from rising to falling is called a **maximum point**. The *y*-value of this point is called a **local maximum value**. If a function decreases and then increases, the point where the graph changes from falling to rising is called a **minimum point**. The *y*-value of this point is called a **local minimum value**.

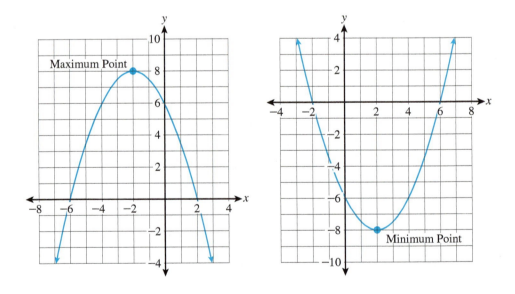

> Example 1
>
> *The following graph describes your walk from the parking lot to the library.*

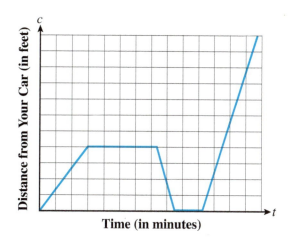

As you read this graph from left to right, it shows how your distance from your car changes as time passes. One possible scenario this graph describes is:

You leave your car and walk at a steady pace toward the library. You meet some friends and stop to chat for a while. You realize that you forgot something and quickly return to your car. After rummaging around for a while, you hurry off to the library.

How did anyone come up with this from the graph? Look at the graph in sections.

a. The first increasing line segment indicates you are moving away from your car because the time and the distance are increasing.

b. The first horizontal section indicates that your distance from the car is constant, so you are standing still.

c. The decreasing line segment indicates your distance from the car is decreasing. When it reaches the horizontal axis, it tells you that you are back at your car.

d. The second horizontal segment indicates you stay at your car for a time.

e. The final increasing segment is steeper and longer than the first, so you are moving away from the car faster and farther than in the first segment.

Graphs to Stories

The graphs in Problems 1–4 present visual images of several situations. Each graph shows how the *y*-values change in relation to the *x*-values. In each situation, identify the independent variable and the dependent variable. Then, interpret the situation; that is, describe, in words, what the graph is telling you about the situation. Indicate whether the graph rises, falls, or is constant and whether the graph reaches either a minimum (smallest) or maximum (largest) *y*-value.

1. A person's core body temperature (°F) in relation to time of day

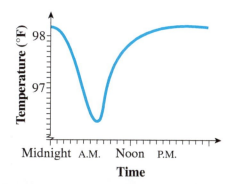

a. Independent: _____   Dependent: _____

b. Interpretation:

2. Performance of a simple task in relation to interest level

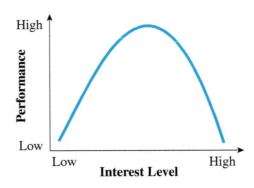

a. Independent: _____   Dependent: _____

b. Interpretation:

3. Net profit of a particular business in relation to time given quarterly

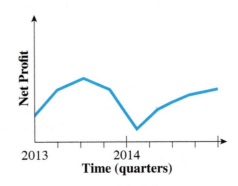

a. Independent: _____   Dependent: _____

b. Interpretation:

4. Annual gross income in relation to number of years

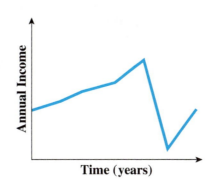

a. Independent: _____   Dependent: _____

b. Interpretation:

Stories to Graphs

In Problems 5 and 6, sketch a graph that best represents the situation described. Note that in many cases, the actual values are unknown, so you will need to estimate what seems reasonable to you. Remember to label your axes and identify the independent/dependent variables. Provide numerical scales when appropriate.

5. You drive to visit your parents, who live 100 miles away. Your average speed is 50 miles per hour. On arrival, you stay for five hours and then return home, again at an average speed of 50 miles per hour. Graph your distance from home, from the time you leave until you return home.

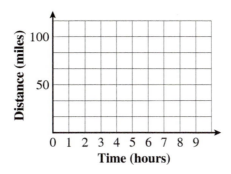

6. Your small business started slowly, losing money in each of its first 2 years, then break-
ing even in year three. In the fourth year, you made as much as you lost in the first year
and then doubled your profits each of the next 2 years. Graph your profit as the output
and the number of years since starting your business as the input.

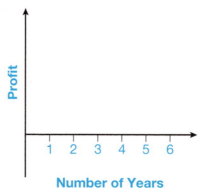

Representing Functions Graphically

Although each of the graphs in Problems 1–6 has a different shape and different properties,
they all reflect a functional relationship between the independent and dependent variables.
In the following problem, you will explore this special relationship or correspondence more
closely.

7. a. Use the graph in Problem 1 to complete the following table. Estimate the temperature for each value of time.

TIME OF DAY	1 A.M.	5 A.M.	7 A.M.	10 A.M.	12 NOON	1 P.M.
BODY TEMPERATURE (°F)						

b. How many corresponding temperatures are assigned to any one particular time?

For any specific time of day, there is one and only one corresponding temperature. Therefore,
the body temperature is a function of the time of day.

8. The following graph shows the distance from home over a 9-hour period as described in
Problem 5.

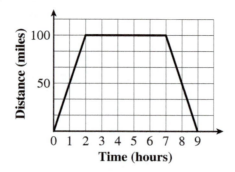

a. Use the graph to complete the following table.

TIME (hours)	1	3	4	8	9
DISTANCE FROM HOME (miles)					

b. Is the distance from home a function of the time on the trip? Explain.

9. A unit circle is a circle having a center at the origin and a radius of 1. Such circles are used in the study of trigonometry. The graph of a unit circle is

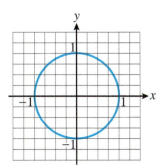

Does the graph of a unit circle represent a function? Explain how you can determine from the graph whether or not it represents a function.

Vertical Line Test

You can determine directly from a graph whether or not there is a functional relationship between the two variables.

10. **a.** Refer to the graph in Problem 1 on page 119, and select any specific time of day along the horizontal (x) axis. Then, move straight up or down (vertically) from that value of time to locate the corresponding point on the graph. How many points do you locate for any given time?

b. If you locate only one point in part a, explain why this would mean that body temperature is a function of the time of day.

c. If you had located more than one point on the graph in part a, explain why this would mean that body temperature is not a function of time.

The procedure in Problem 10 is referred to as the **vertical line test**.

> **Definition**
>
> In the **vertical line test**, a graph represents a function if any vertical line drawn through the graph intersects the graph no more than once.

11. Use the vertical line test on the graphs in Problems 2 through 4 to verify that each graph represents a function.

12. Use the vertical line test to determine which of the following graphs represent functions. Explain.

a.

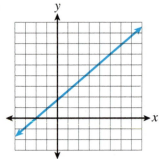

b.

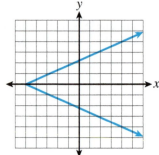

c.

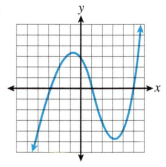

SUMMARY: ACTIVITY 1.13

1. In the **vertical line test**, a graph represents a function if any vertical line drawn through the graph intersects the graph no more than once.

2. If a graph increases and then decreases, the point where the graph changes from rising to falling is called a **maximum point**. The y-value of this point is called a **local maximum value**. If a graph decreases and then increases, the point where the graph changes from falling to rising is called a **minimum point**. The y-value of this point is called a **local minimum value**.

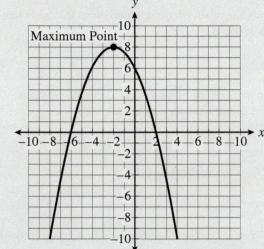

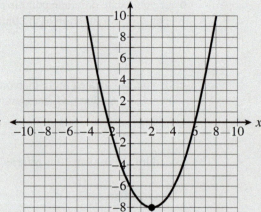

EXERCISES: ACTIVITY 1.13

1. You are a technician at the local power plant, and you have been asked to prepare a report that compares the output and efficiency of the six generators in your sector. Each generator has a graph that shows output of the generator as a function of time over the previous week, Monday through Sunday. You take all the paperwork home for the night (your supervisor wants this report on his desk at 7:00 A.M.), and to your dismay your cat scatters your pile of papers out of the neat order in which you left them. Unfortunately, the graphs for generators A through F were not labeled (you will know better next time!). You recall some information and find evidence elsewhere for the following facts.

- Generators A and D were the only ones that maintained a fairly steady output.
- Generator B was shut down for a little more than two days during midweek.
- Generator C experienced a slow decrease in output during the entire week.
- On Tuesday morning, there was a problem with generator E that was corrected in a few hours.
- Generator D was the most productive over the entire week.

Match each graph with its corresponding generator. Explain in complete sentences how you arrive at your answers.

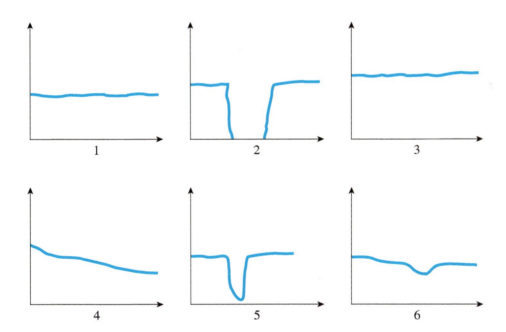

In Exercises 2 and 3, identify the independent variable and the dependent variable. Then interpret the situation being represented. Indicate whether the graph rises, falls, or is constant and whether the graph reaches either a minimum (smallest) or maximum (largest) output value.

2. Time required to complete a task in relation to number of times the task is attempted

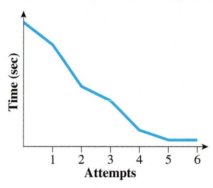

a. Independent: _____   Dependent: _____

b. Interpretation:

3. Number of units sold in relation to selling price

a. Independent: _____   Dependent: _____

b. Interpretation:

4. You leave home on Friday afternoon for your weekend getaway. Heavy traffic slows you down for the first half of your trip, but you make good time by the end. Express your distance from home as a function of time.

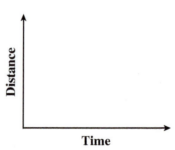

5. You decide to invest in the stock market for the first time, and you buy 10 shares of a local company that you admire. For the first two months, your stock steadily drops in price. If you had to sell, it would be at a loss. During the third month, the stock makes up lost ground and ends the month at the same price you paid for it originally. During months 4 and 5, it continues to rise, but not quite as quickly. In the sixth month, it rises as fast as it did during month 3. At this point, you sell and make a nice profit. Graph your profit as the output and time as the input.

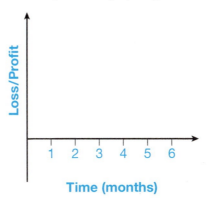

6. Use the vertical line test to determine which of the following graphs represents a function. Explain your answer.

a.

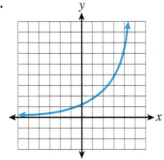

b.

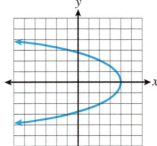

c.

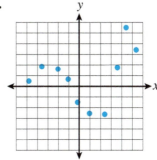

Activity 1.14

Heating Schedule

Objectives

1. Obtain a new graph from an original graph using a vertical shift.

2. Obtain a new graph from an original graph using a horizontal shift.

3. Identify vertical and horizontal shifts.

4. Write a new formula for a function for which its graph has been shifted vertically or horizontally.

The cost of fuel oil is rising and is affecting the school district's budget. In order to save money, the building maintenance supervisor of the high school has decided to keep the building warm only during school hours.

1. Sketch a graph that best represents the following building heating schedule, which begins at midnight.

At midnight, the building temperature is 55°F. This temperature remains constant until 4 A.M., at which time the temperature of the building steadily increases. By 7 A.M., the temperature is 68°F. The temperature is maintained at a constant 68°F until 7 P.M., when the temperature begins a steady decrease. By 10 P.M., the temperature is back to 55°F.

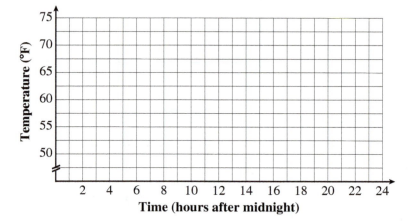

Now, suppose the building maintenance supervisor is instructed by the Board of Education to increase the temperature of the building by a constant 3°F.

2. **a.** On the graph in Problem 1, sketch a graph that represents the new building heat schedule over a 24-hour period, starting at midnight.

   b. Describe how the graph in part a can be obtained from the original graph of the building heating schedule in Problem 1.

The graph of the new heating schedule in Problem 2 represents a **vertical shift** of the original graph. The graph has been shifted vertically upward by 3 units.

3. The heating schedule for a warehouse is represented by the following graph.

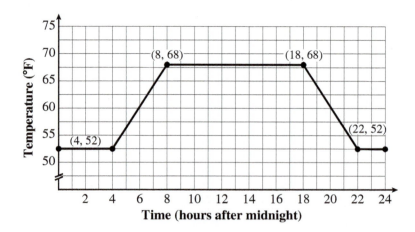

a. As a cost saving measure, the building superintendent decides to drop the temperature of the warehouse by 2°F. Using the given graph, sketch a graph that represents the new heat schedule.

b. Describe how the graph of the new heating schedule can be obtained from the original graph.

Horizontal Shift

4. The following graph represents the heating schedule in an office building.

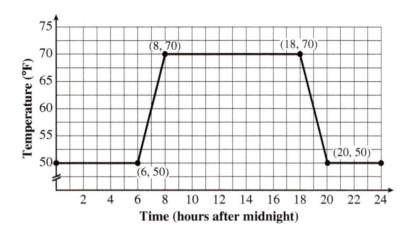

The building superintendent needs to change the heating schedule by moving everything two hours earlier.

a. On the given graph, sketch a graph that represents the new heating schedule. Describe the new schedule in words.

b. Describe how the graph of the new heating schedule can be obtained from the original graph.

The graph of the new heating schedule in Problem 4 represents a **horizontal shift** of the original graph. The original graph has been shifted two units horizontally to the left.

5. a. The heating schedule for the warehouse in Problem 3 is delayed by 2 hours. Sketch a graph that represents the new heating schedule.

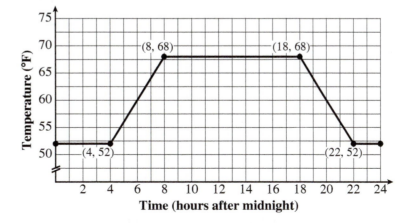

b. Describe how the graphs of the new and original heating schedule are related.

Function Formulas for a Graph Resulting from Vertical or Horizontal Shift

6. a. Use a graphing calculator to graph the squaring function, defined by $y = x^2$, in a standard window. The screen should appear as follows:

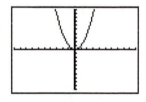

   Note: The squaring function is an example of a quadratic function. Quadratic functions are studied in detail in Chapter 3.

   b. Graph $y_2 = x^2 + 2$ on the same coordinate axis as $y = x^2$ in part a.

   c. Describe how the graph for $y_2 = x^2 + 2$ can be obtained from the graph of $y = x^2$.

   d. Now, let $y_3 = x^2 - 3$. Use a graphing calculator to graph the function defined by this equation. Describe how the graph can be obtained from the graph of $y = x^2$.

   e. The graph of $y = x^2 + c$ represents a vertical shift of the graph of $y = x^2$. How do you determine if the graph is shifted vertically upward or vertically downward?

7. a. Use a graphing calculator to sketch a graph of $y_2 = (x + 2)^2$ on the same coordinate axis as $y = x^2$. The screen should appear as follows:

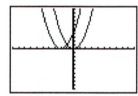

b. Does the graph of $y_2 = (x + 2)^2$ represent a shift of the graph $y = x^2$? If yes, describe the type of shift.

c. Use a graphing calculator to sketch a graph of $y_3 = (x - 4)^2$ on the same coordinate axis as $y = x^2$. Describe how the graph of $y_3 = (x - 4)^2$ can be obtained from the graph of $y = x^2$.

d. The graph of $y = (x + c)^2$ represents a horizontal shift of the graph of $y = x^2$. How do you determine if the shift is to the right or left?

In Problem 6b, adding 2 to the output (the square of a number) results in a vertical shift of the graph of the squaring function. Using function notation, if $y = f(x) = x^2$, then

$$y = x^2 + 2 = f(x) + 2 \text{ is the upward vertical shift of } y = f(x) = x^2.$$

Since $y = f(x) + 2$ involves a change to the output value, $f(x)$, vertical shifts result from "outside" changes to the function.

In Problem 7b, adding 2 to the input value x **before** the squaring is done results in a horizontal shift of the graph of the squaring function. Using function notation, if $y = f(x) = x^2$, then

$$y = (x + 2)^2 = f(x + 2) \text{ is a horizontal shift to the left of } y = f(x) = x^2.$$

Since $y = f(x + 2)$ involves a change to the input value, x, horizontal shifts result from "inside" changes to the function.

The results from Problems 6 and 7 can be generalized as follows:

In general,

1. If a function is defined by $y = f(x)$, and c is a constant, then the graph of $y = f(x) + c$ is the graph of $y = f(x)$ shifted vertically $|c|$ units. If $c > 0$, then the shift is upward. If $c < 0$, then the shift is downward.

2. If a function is defined by $y = f(x)$, and c is a constant, then the graph of $y = f(x + c)$ is the graph of $y = f(x)$ shifted horizontally $|c|$ units. If $c > 0$, then the shift is to the left. If $c < 0$, the shift is to the right.

A vertical or horizontal shift of the graph of a function is called a **translation**. A translation does not change the shape of the graph. It translates (moves) the graph into a new position in the plane.

8. The graph of the absolute value function, defined by $y = |x|$, is

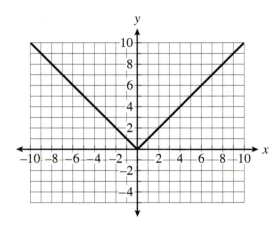

Determine the graph of each of the following translations of the absolute value function. Describe the shift in words. Verify your results using a graphing calculator.

a. $y = |x| + 5$ **b.** $y = |x + 5|$

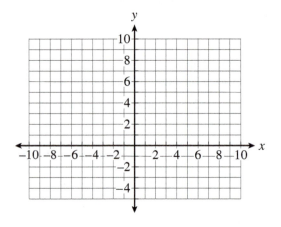   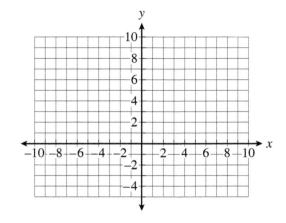

c. $y = |x| - 2$ **d.** $y = |x - 2|$

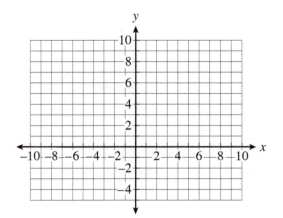   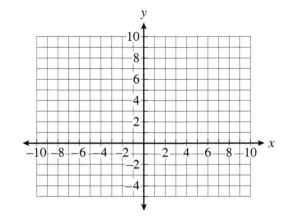

9. The graph of the squaring function, defined by $y = x^2$, is translated (shifted) in parts a–d. Write the equation of the resulting graph.

   a. vertical shift of the graph 3 units upward

   b. vertical shift of the graph 5 units downward

   c. c horizontal shift of the graph 2 units to the left

   d. horizontal shift of the graph 4 units to the right

10. The graph of $y = (x - 3)^2 + 5$ involves a vertical and horizontal shift of the graph of $y = x^2$. Use the patterns discussed in this activity to sketch the graph. Verify using a graphing calculator.

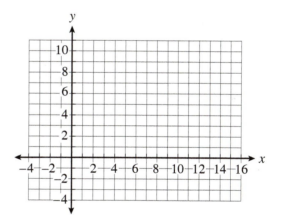

SUMMARY: **ACTIVITY 1.14**

1. In a **vertical shift** of a graph, a positive constant is added to (upward shift) or subtracted from (downward shift) each output value. As a result, the graph is moved vertically upward or vertically downward.

2. In a **horizontal shift** of a graph, a positive constant is first added to (shift left) or subtracted from (shift right) each input value. As a result, the graph is moved horizontally to the left or horizontally to the right.

3. In general, if a function is defined by $y = f(x)$, and c is a constant, then the graph of
   i. $y = f(x) + c$ is the graph of $y = f(x)$ shifted vertically $|c|$ units. If $c > 0$, then the shift is upward. If $c < 0$, then the shift is downward.
   ii. $y = f(x + c)$ is the graph of $y = f(x)$ shifted horizontally $|c|$ units. If $c > 0$, then the shift is to the left. If $c < 0$, the shift is to the right.

4. A vertical or horizontal shift of the graph of a function is called a **translation.** A translation does not change the shape of the graph. It translates (moves) the graph into a new position in the coordinate plane.

EXERCISES: ACTIVITY 1.14

1. The graph of the absolute value function is given in Problem 8 on page 132.

Match each of the graphs in parts a–d with the corresponding shift of the graph of the absolute value function from i–iv.

i. vertical shift of the graph 3 units upward

ii. vertical shift of the graph 3 units downward

iii. horizontal shift of the graph 3 units to the right

iv. horizontal shift of the graph 3 units to the left

a.

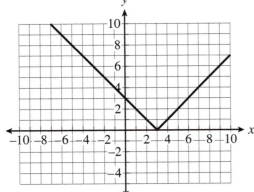

b.

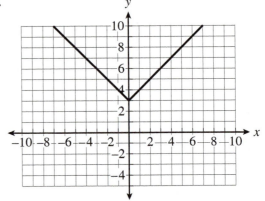

c.

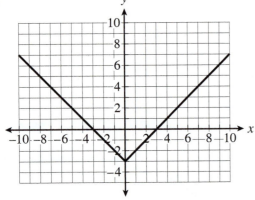

d.

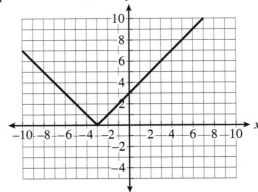

2. In parts a–d, sketch a new graph by performing the specified shift on the given graph.

   a. Vertical shift of (the graph 3 units downward.

   b. Horizontal shift of the graph 2 units to the right.

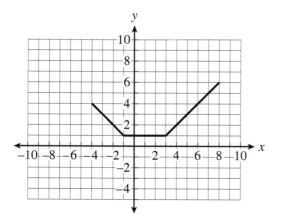

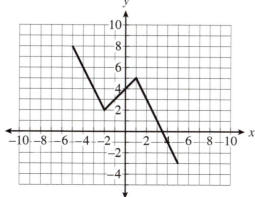

   c. Vertical shift of the graph 4 units upward.

   d. Horizontal shift of the graph 3 units to the left.

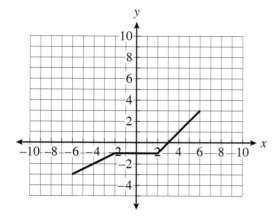

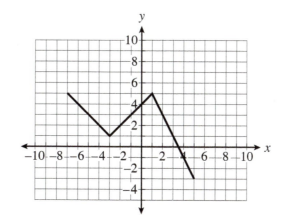

3. The graphs below are horizontal translations of the basic squaring function, $f(x) = x^2$. Write the equation for each function. Check your answer by graphing on your calculator.

   a.

   b.

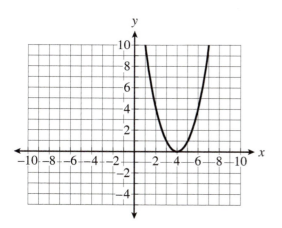

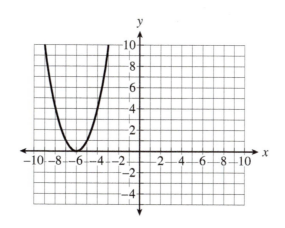

4. Given the graph of $y = f(x)$, translate as indicated by the given equation to graph a new function in parts a–c.

a. $y = f(x) - 3$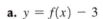

b. $y = f(x + 2)$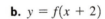

c. $y = f(x) + 2$

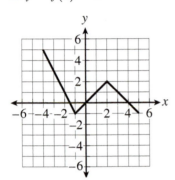

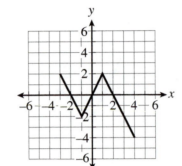

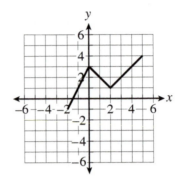

5. Describe in words how the graph of the basic absolute value function will be translated in each case. Check by graphing on your calculator.

a. $y = |x - 5|$

b. $y = |x| - 7$

c. $y = 2 + |x + 4|$

6. Write the equation for each described function. Check by graphing on your calculator.

a. The squaring function shifted down 8 units.

b. The absolute value function shifted to the left 4 units.

c. The linear function $y = 3x$ shifted to the right 5 units.

d. The squaring function shifted up 3 units and to the left 6 units.

7. Each of the following graphs represents the translation of the indicated basic function. Write the equation for each graph. Check by graphing on your calculator.

a. Translation of $y = x^2$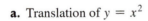

b. Translation of $y = |x|$

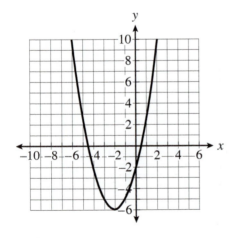

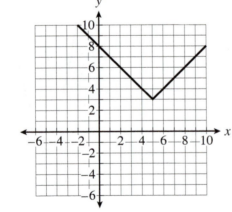

8. A computer technician charges a flat fee of $50 for an on-site service call and $75 per hour thereafter. The following graph represents the total cost c as a function of the number of hours.

  a. What is the equation for this function?

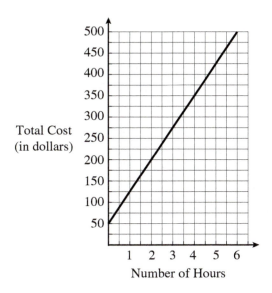

  b. Suppose the flat fee for an on-site service increases by 20 dollars. How does the increase affect the cost?

  c. Sketch a graph of the new cost function.

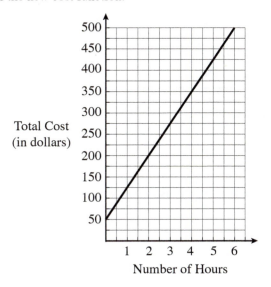

  d. Describe how the new graph can be obtained from the original graph.

  e. Write an equation for the new cost function in part c.

9. The speed limits over a four mile stretch of country road are represented by the graph to the right. If all the speed limits are to be reduced by 5 MPH, sketch the new graph and describe how it is obtained from the original graph.

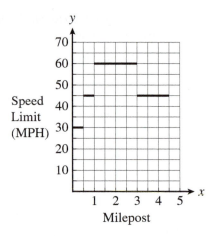

10. A new car depreciates in value over time. Assuming a straight-line depreciation, the car will decrease in value the same amount every year. If a car is purchased in 2014 for $32,000 and decreases in value $2000 every year, the graph shows the car's value, where x represents the years since purchase.

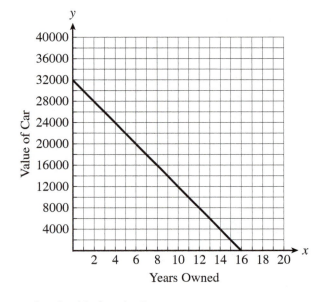

a. What is the equation for this function?

b. If you want the independent variable x to be the actual year instead of the number of years since 2014, how would the graph be shifted?

c. Write the equation for this translated function, where x represents the actual year.

Activities 1.6–1.14   **What Have I Learned?**

1. What is the mathematical definition of a function? Give a real-life example, and explain how this example satisfies the definition of a function.

2. Describe how you can tell from a graph when a function is increasing and when it is decreasing.

3. Identify four different ways that a function can be defined. Give an example of each.

COST OF ITEM (dollars)	10	20	30	40	50	100	150	200	300	500	1000
SALES TAX (dollars)	0.60	1.20	1.80	2.40	3.00	6.00	9.00	12.00	18.00	30.00	60.00

4. Use your newspaper to find at least 4 examples of functions and report your findings back to the class. For each example, you should do the following:

   a. Explain how the example satisfies the definition of a function.

   b. Describe how the function is defined (see Problem 3).

   c. Identify the independent and dependent variables.

   d. Determine the domain of the function.

5. The notation $g(t)$ represents the weight (in grams) of a melting ice cube t minutes after being removed from the freezer. Interpret the meaning of $g(10) = 4$.

6. The graph of $y = |x + 2| - 3$ involves a horizontal and vertical shift of the graph of the function defined by $y = |x|$.

   a. Sketch a graph of $y = |x + 2| - 3$ by first doing a horizontal shift of the graph of $y = |x|$, followed by a vertical shift of the graph resulting from the horizontal shift.

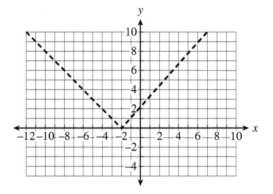

   b. Now, sketch a graph of $y = |x + 2| - 3$ by first doing the vertical shift of the graph of $y = |x|$, followed by the horizontal shift.

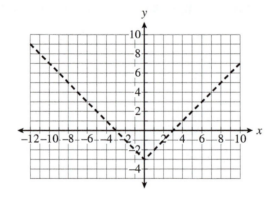

   c. Compare the graphs in parts a and b.

Activities 1.6–1.14   **How Can I Practice?**

1. You bought a company in 2011 and have tracked the company's profits and losses from its beginning in 2005 to the present. You decide to graph the information where the number of years since 2011 is the x-variable and profit or loss for the year is the y-variable. Note that the year 2011 corresponds to 0 on the x-axis. Determine the quadrant or axis on which you would plot the points that correspond to the following data. If your answer is on an axis, indicate between which quadrants the point is located.

a. The loss in 2008 was $1500.

b. The profit in 2012 was $6000.

c. The loss in 2014 was $1000.

d. In 2007, there was no profit or loss.

e. The profit in 2005 was $500.

f. The loss in 2011 was $800.

2. Fish need oxygen to live, just as you do. The amount of dissolved oxygen (D.O.) in water is measured in parts per million (ppm). Trout need a minimum of 6 ppm to live.

The data in the table shows the relationship between the temperature of the water and the amount of dissolved oxygen present.

TEMP (°C)	11	16	21	26	31
D.O. (in ppm)	10.2	8.6	7.7	7.0	6.4

a. Represent the data in the table graphically. Place temperature along the horizontal axis and dissolved oxygen along the vertical axis.

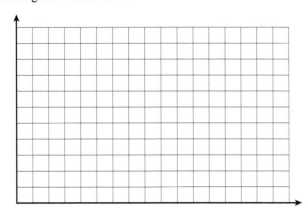

b. What general trend do you notice in the data?

c. In which of the 5-degree temperature intervals given in the table does the dissolved oxygen content change the most?

d. Which representation (table or graph) presents the information and trends more clearly?

3. When you were born, your uncle invested $1000 for you in a local bank. Your bank has compounded the interest continuously at a rate of 6%. The following graph shows how your investment grows.

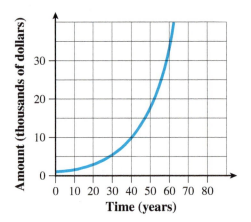

a. Which variable is the independent variable?

b. How much money did you have when you were 10 years old?

c. Estimate in what year your original investment will have doubled.

d. If your college bill is estimated to be $2600 in the first year of college, will you have enough to pay the bill with these funds? (Assume that you attend when you are eighteen.) Explain.

4. You decide to lose weight and will cut down on your calories to lose 2 pounds a week. Suppose that your present weight is 180 pounds. Sketch a graph covering twenty weeks showing your projected weight loss. Describe your graph. If you stick to your plan, how much will you weigh in three months (13 weeks)?

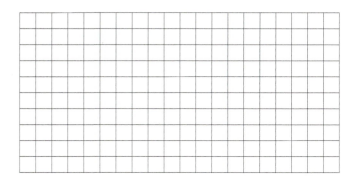

5. A taxicab driver charges a flat rate of $2.50 plus $1.50 per mile. The fare F (in dollars) is a function of the distance driven, x (in miles). The driver wants to display a table for her customers to show approximate fares for different locations within the city.

   a. Write an equation for F in terms of x.

   b. Use the equation to complete the following table.

x	0.25	0.5	0.75	1.0	1.5	2.0	3.0	5.0	10.0
f									

6. Interpret each situation represented by the following graphs. That is, describe, in words, what the graph is saying about the relationship between the variables. Indicate what occurs when either a minimum or maximum value is reached.

   a. Hours of daylight per day in relation to time of year

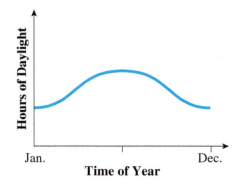

b. Population of fish in a pond in relation to the number of years since stocking

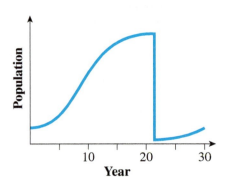

c. Amount of money saved per year in relation to amount of money earned

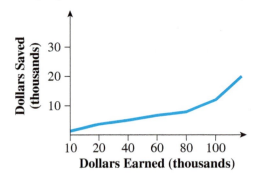

7. In the following situations, determine the independent and dependent variables, and then sketch a graph that best describes the situation. Remember to label the axes with the names of the variables.

  a. Sketch a graph that approximates average temperatures where you live as a function of the number of months since January.

  Independent variable: _____

  Dependent variable: _____

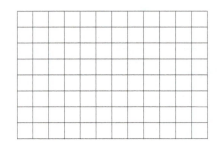

b. If you don't study, you would expect to do poorly on the next test. If you study several hours, you should do quite well, but if you study for too many more hours, your test score will probably not improve. Sketch a graph of your test score as a function of study time.

Independent variable: _____   Dependent variable: _____

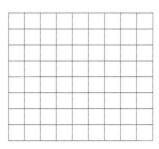

In Exercises 8–15, solve each equation for x.

8. $2(x + 4) = 12$

9. $3x + 2(x - 6) = x + 4$

10. $\frac{1}{3}(2x + 3) = 3(x + 1)$

11. $(2x - 5) - (4 - 3x) = 11$

12. $\frac{1}{2}x + \frac{1}{3}x + 5 = x$

13. $1.50x + 0.85(12 - x) = 14.75$

14. $-7 = -3(x - 7) + 2x + 1$

15. $2[6x - 7(x - 1)] = 5x - 21$

16. The cost of printing a brochure to advertise your lawn-care business is a flat fee of $25 plus $0.55 per copy. Let C represent the total cost of printing and x represent the number of copies.

   a. Write an equation that models the relationship between C and x.

   b. Complete the following table. Begin with 100 copies, increase by increments of 100, and end with 500 copies.

NUMBER OF COPIES, x	TOTAL COST ($), C
100	

   c. Graph the data obtained in part b. Use a straightedge to connect the points and extend the graph. Scale the axes appropriately so that you can plot the ordered pair corresponding to 1,000 copies.

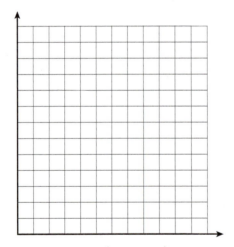

   d. What is the total cost of printing 800 copies?

   e. You have $100 to spend on advertising. How many copies can you have printed for that amount?

17. Solve each of the following equations for the given variable:

a. $d = rt$ for r

b. $P = a + b + c$ for b

c. $A = P + Prt$ for t

d. $y = 4x - 5$ for x

e. $w = \dfrac{4}{7}h + 3$ for h

18. A volatile stock began the last week of the year worth x dollars per share. The following table shows the changes during that week. If you own 30 shares, write an equation that represents the total value, V, of your stock at the end of the week.

DAY	1	2	3	4	5
CHANGE IN VALUE/SHARE	Doubled	Lost 10	Tripled	Gained 12	Lost half its value

19. A triathlon includes swimming, long-distance running, and cycling.

a. Let x represent the number of miles the competitors swim. If the long-distance run is 10 miles longer than the distance swum, write an expression that represents the distance the competitors run in the event.

b. The distance the athletes cycle is 55 miles longer than they run. Use the result in part a to write an expression in terms of x that represents the cycling distance of the race.

c. Write an expression that represents the total distance of all three phases of the triathlon. Simplify the expression.

d. If the total distance of the triathlon is 120 miles, write and solve an equation to determine x. Interpret the result.

e. What are the lengths of the running and cycling portions of the race?

20. You planned a trip with your best friend. You had only four days for your trip and planned to travel x hours each day.

The first day, you stopped for sightseeing and lost two hours of travel time. The second day, you gained one hour because you did not stop for lunch. On the third day, you traveled well into the night and doubled your planned travel time. On the fourth day, you traveled only a fourth of the time you planned because your friend was sick. You averaged 45 miles per hour for the first two days and 48 miles per hour for the last two days.

a. How many hours, in terms of x, did you travel the first two days?

b. How many hours, in terms of x, did you travel the last two days?

c. Express the total distance, D, traveled over the four days as an equation in terms of x. Simplify the equation. Recall that distance = average rate · time.

d. Write an equation that expresses the total distance, y, you would have traveled had you traveled exactly x hours each day at the average speeds indicated above. Simplify the equation.

e. If you anticipated traveling for seven hours each day, how many miles did you actually go on your trip?

f. How many miles would you have gone had you traveled exactly seven hours each day?

21. As a prospective employee in a furniture store, you are offered a choice of salary. The following table shows your options.

OPTION 1	$350 per week	Plus 30% of all sales
OPTION 2	$400 per week	Plus 15% of all sales

a. Write an equation to represent the total salary, S, for option 1 if the total sales are x dollars per week.

b. Write an equation to represent the total salary, S, for option 2 if the total sales are x dollars per week.

c. Write an equation that you could use to determine how much you would have to sell in a week to earn the same salary under both plans.

d. Solve the equation in part c. Interpret your result.

e. What is the common salary for the amount of sales found in part d?

f. Graph the two equations from parts a and b on the following grid. Locate the point of the common salary and use the graph to determine which option provides the larger salary for furniture sales more than $333 per week.

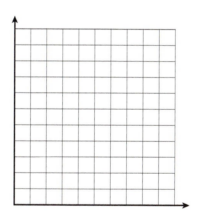

22. The following table shows the total number of points accumulated by each student and the numerical grade in the course.

STUDENT	TOTAL POINTS	NUMERICAL GRADE
TOM	432	86.4
JEN	394	78.8
KATHY	495	99
MICHAEL	330	66
BRADY	213	42.6

a. Is the numerical grade a function of the total number of points? Explain.

b. Is the total number of points for these five students a function of the numerical grade? Explain.

c. Using the total points as the input and the numerical grade as the output, write the ordered pairs that represent each student. Call this function f, and write it as a set of ordered pairs.

d. Plot the ordered pairs determined in part c on an appropriately scaled and labeled set of axes.

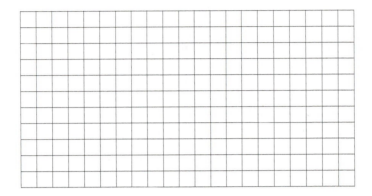

e. What is the value of $f(394)$?

f. What is the practical meaning of $f(394)$?

g. What is the value of $f(213)$?

h. What is the practical meaning of $f(213)$?

i. Determine the numerical value n, given that $f(n) = 66$.

23. In parts a–i, determine which of the given relationships represent functions.

a. The money you earn at a fixed hourly rate is a function of the number of hours you work.

b. Your heart rate is a function of your level of activity.

c. The cost of daycare depends on the number of hours a child stays at the facility.

d. The number of children in a family is a function of the parents' last name.

e. $\{(2, 3), (4, 3), (5, -5)\}$

f. $\{(-3, 4), (-3, 6), (2, 6)\}$

g.

x	−3	5	7
f(x)	0	−5	9

h.

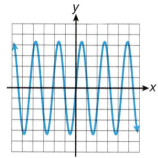

i.

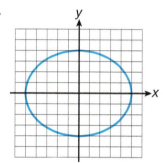

24. a. For a part-time student taking less than 12 hours, the cost of college tuition is a function of the number of hours for which the student is registered. Write an equation to represent the tuition cost, *c,* if the cost per credit hour is $245 and *h* represents the number of credit hours taken for the current semester.

b. Use $f(h)$ to represent the cost, and rewrite the equation in part a using function notation.

c. Complete the table.

h	2	4	7	8	11
f(h)					

d. Evaluate $f(3)$, and write a sentence describing its meaning. Write the results as an ordered pair.

e. Given $f(h) = 1225$, determine the value of *h*.

f. Which variable is the output? Explain.

g. Which is the independent variable? Explain.

h. Explain (using the table in part c) how you know that the data represents a function.

i. What is the practical domain for this function?

j. Plot the ordered pairs on an appropriately scaled and labeled coordinate system. Which axis represents the input values?

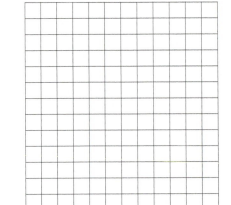

k. Explain from the graph how you know that *f* is a function.

l. Use your graphing calculator to verify your answers to parts c and j.

m. Use the trace and table features to determine the cost of 9 credit hours.

25. According to the U.S. Centers for Disease Control and Prevention, the average life expectancy from birth for males in the United States may be modeled by the function $f(x) = 0.18x + 65.6$, where *x* is the number of years since 1950.

a. Use your calculator to complete the following table. Round your results to the nearest tenth.

	YEAR						
	1950	**1960**	**1970**	**1985**	**2000**	**2010**	**2020**
x, YEARS SINCE 1950	0	10	20	35	50	60	70
f(x), LIFE EXPECTANCY							

b. Evaluate $f(30)$, and explain its practical meaning in this situation.

c. Use the table values to set appropriate window values to view the graph of *f*. Graph the function on your calculator. Identify the window you used.

d. Is the graph increasing, decreasing, or constant? Explain.

e. Use the trace feature of your calculator to determine $f(30)$. Compare your answer to your result in part b.

26. Determine the domain and the range of each of the following functions.

a. $\{(3, 5), (4, 5), (5, 8), (6, 10)\}$

b.

YEAR	2005	2006	2007	2008	2009	2010	2011
ONLINE MUSIC SALES (BILLIONS OF DOLLARS)	1.1	2.0	2.8	3.7	4.3	4.9	5.3

27. The following graph is a sketch of the step function. Note that an open circle indicates that the point is not included in the graph.

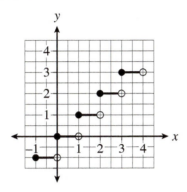

In parts a–d, sketch a new graph of the step function by performing the specified translation.

a. Vertical shift of the graph 1 unit upward.

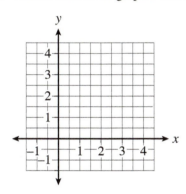

b. Vertical shift of the graph 1 unit downward.

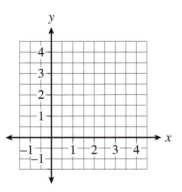

c. Horizontal shift of the graph 1 unit to the right.

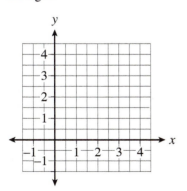

d. Horizontal shift of the graph 1 unit to the left.

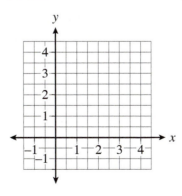

28. The graph of the function defined by $y = f(x)$ is

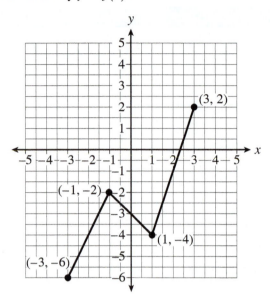

Sketch a graph of each function defined by the given equation. Describe the type of shift that was performed on the original graph.

a. $y = f(x) + 3$ **b.** $y = f(x + 3)$

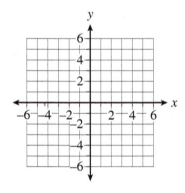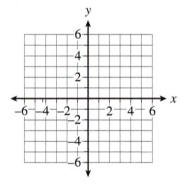

The bracketed numbers following each concept indicate the activity in which the concept is discussed.

CONCEPT/SKILL	DESCRIPTION	EXAMPLE
Problem solving [1.2], [1.3]	Problem-solving strategies include: • discussing the problem • organizing information • drawing a picture • recognizing patterns • doing a simpler problem	You drive for 2 hours at an average speed of 47 mph. How far do you travel? $d = rt$ $d = \dfrac{47 \text{ mi}}{\text{hr}} \cdot 2 \text{ hr} = 94 \text{ mi}$
Variable [1.2], [1.3]	A quantity or quality that may change in value from one particular instance to another, usually represented by a letter. When a variable describes an actual quantity, its values must include the unit of measurement of that quantity.	The number of miles you drive in a week is a variable. Its value may (and usually does) change from one week to the next. x and y are commonly used to represent variables.
Formulas [1.3]	A formula shows the arithmetic relationship between two or more quantities; each quantity is represented by a letter or symbol.	$F = ma$ for $m = 120$ and $a = 25$ $F = 120 \cdot 25 = 3000$
Applying a known ratio to a given piece of information [1.4]	Total $\times$ known ratio = unknown part Part $\div$ known ratio = unknown total	Forty percent of the 350 children play an instrument; 40% is the known ratio; 350 is the total. $350 \times 0.40 = 140$ A total of 140 children play an instrument. Twenty-four children, constituting 30% of the marching band, play the saxophone; 30% is the known ratio, 24 is the part. $24 \div 0.30 = 80$ A total of 80 children are in the marching band.
Solving a proportion [1.4]	$\dfrac{a}{b} = \dfrac{c}{d}$ Equivalently, $a \cdot d = b \cdot c$	$\dfrac{7}{30} = \dfrac{x}{4140}$ $30 \cdot x = 7 \cdot 4140$ $x = \dfrac{7 \cdot 4140}{30} = 966$
Scale factor [1.4]	The ratio of the length of the scale drawing to the corresponding length of the actual object is called the **scale factor**. scale factor $= \dfrac{\text{length of model}}{\text{length of actual}}$ or scale factor = length of model: length of actual	See Example 2 on page 24.

CONCEPT/SKILL	DESCRIPTION	EXAMPLE
Proportional relationship [1.4]	Two quantities Q_1 and Q_2 are **proportional** if they have the same constant ratio $\dfrac{Q_2}{Q_1}$. If the ratio $\dfrac{Q_2}{Q_1}$ is not constant, the two quantities are said to be **non-proportional**.	See Problem 19 on page 26.
Using unit (or dimensional) analysis to solve conversion problems [1.5]	1. Identify the unit of the result. 2. Set up the sequence of multiplications so the appropriate units cancel, leaving the unit of the result. 3. Multiply and divide the numbers as usual to obtain the numerical part of the result. 4. Check that the appropriate units cancel, leaving the expected unit of the result.	To convert your height of 70 inches to centimeters: $70 \text{ in.} \cdot \dfrac{2.54 \text{ cm}}{1 \text{ inch}} = 177.8 \text{ cm}$
Input variable [1.6]	The input variable, often represented by x, is the value given first in a relationship.	In the relationship between the perimeter and the side of a square, $P = 4s$, s is the input variable.
Output variable [1.6]	The output, often represented by y, is the value that corresponds to or is determined by the given input value (x).	In the relationship between the perimeter and the side of a square, $P = 4s$, P is the output variable.
Function [1.6]	A function is a correspondence between an input (x) variable and an output (y) variable that assigns a single, unique output value (y) to each input value (x).	See Example 1 in Activity 1.6.
Ordered pair [1.6]	An ordered pair of numbers consists of two numbers written in the form (x, y). The order in which they are listed is significant.	(2, 3) is an ordered pair. In this pair, 2 is the input and 3 is the output.
Verbally defined function [1.6]	A function is defined verbally when it is defined using words.	The high temperature in Houston, Texas, is a function of the day of the year, because for each day there is one high temperature.
Graphically defined function [1.6]	A function is defined graphically when the x-variable is represented on the horizontal axis and the y-variable on the vertical axis.	
Numerically defined function [1.6]	A function is defined numerically using ordered pairs, or presented as a table of matched pairs.	See Problem 8 on page 49.

CONCEPT/SKILL	DESCRIPTION	EXAMPLE
Horizontal axis [1.6]	In graphing an x-y relationship, the variable x is referenced on the horizontal axis (often called the x-axis).	
Vertical axis [1.6]	In graphing an x-y relationship, the variable y is referenced on the vertical axis (often called the y-axis).	
Rectangular coordinate system [1.6]	Allows every point in the plane to be identified by an ordered pair of numbers, determined by the distance of the point from two perpendicular number lines (called coordinate axes) that intersect at their respective 0 values, the origin.	
Scaling [1.6]	Setting the same distance between each pair of adjacent tick marks on an axis.	
Quadrants [1.6]	Two perpendicular coordinate axes divide the plane into four quadrants, labeled counterclockwise with quadrant I being the upper-right quadrant.	
Point in the plane [1.6]	A point that is identified by an ordered pair of numbers (x, y) in which x represents the horizontal distance from the origin and y represents the vertical distance from the origin.	$(2, 30)$ are the coordinates of a point in the first quadrant located 2 units to the right and 30 units above the origin.
Discrete [1.6]	Functions are discrete if they are defined only at isolated input values and do not make sense or are not defined for input values between those values.	See page 51 in Activity 1.6.
Continuous [1.6]	Functions are continuous if they are defined for all input values, and if there are no gaps between consecutive input values. Their corresponding graphs are totally connected.	See page 52 in Activity 1.6.
Independent variable [1.7]	Independent variable is another name for the x-variable of a function.	See Example 4 of Activity 1.7.

CONCEPT/SKILL	DESCRIPTION	EXAMPLE
Dependent variable [1.7]	Dependent variable is another name for the y-variable of a function.	See Example 4 of Activity 1.7.
Domain of a function [1.7]	The domain of the function is the collection of all replacement values for the independent variable.	See page 63 in Activity 1.7.
Practical domain of a function [1.7]	The practical domain is the collection of replacement values of the independent variable that makes practical sense in the context of the situation.	See page 64 in Activity 1.7.
Range of a function [1.7]	The range of a function is the collection of all dependent variable values of a function.	See page 64 in Activity 1.7.
Practical range of a function [1.7]	The practical range is the collection of all dependent variable values that make practical sense in the context of the situation.	See page 64 in Activity 1.7.
Mathematical model [1.8]	A mathematical model is a mathematical structure such as a formula, a graph, or a table, that fits or approximates the important features of a given situation.	See Problems 8 and 10 in Activity 1.8.
Solution of an equation [1.9]	The solution of an equation is a replacement value for the variable that makes both sides of the equation equal in value.	3 is a solution of the equation $4x - 5 = 7$.
Solve an equation using a numerical approach [1.9]	To solve an equation using a numerical approach use a guess, check and repeat process. This process can be automated using the table feature of a graphing calculator.	$4x - 7 = 5$ try $x = 2, 4(2) - 7 = 1$ (too low) try $x = 4, 4(4) - 7 = 9$ (too high) try $x = 3, 4(3) - 7 = 5$ (This is it!)
Solve an equation graphically [1.9]	To solve an equation graphically, first graph the equation. Then locate the point that has the desired value as its y-coordinate. The value of the x-coordinate of that point is the solution of the equation.	Solve $-2x + 3 = 5$ The output value is 5 when the input value is -1. Therefore, -1 is the solution.
Evaluate an algebraic expression [1.9]	To evaluate an algebraic expression replace the variable(s) by its (their) assigned value(s) and perform the indicated arithmetic operation(s).	Evaluate $3x^2 - 2x + 4$ when $x = 2$. $3(2)^2 - 2(2) + 4$ $3 \cdot 4 - 4 + 4$ $12 - 4 + 4$ 12

CONCEPT/SKILL	DESCRIPTION	EXAMPLE
Isolate a variable in an equation [1.9], [1.10]	Write the equation so that the variable you are solving for is on one side and all other numbers and variables are on the other side.	The variable y is isolated in the equation $y = 3x + 25$.
How to solve an equation for a given variable. [1.9], [1.10]	Add and/or subtract appropriate terms so that all terms involving the variable you are solving for are on one side of the equation and all terms involving numbers and other variables appear on the other side. Then, combine like terms, and finally divide and/or multiply by the coefficient of the variable.	$4x + 7 + 5x = 7 + 2x - 13$ $9x + 7 = 2x - 6$ $7x = -13$ $x = -13/7$
Solve an equation algebraically for a specified variable [1.9], [1.10]	An algebraic approach to solving an equation is complete when the variable of interest is alone on one side of the equation with coefficient 1, and the other side is simplified.	$4x + 7 + 5x = 7 + 2x - 13$ $9x + 7 = 2x - 6$ $7x = -13$ $x = -13/7$
Solve a formula for a given variable [1.11]	To solve a formula, collect the term containing the variable of interest on one side of the equation, with all other expressions on the other side. Then, divide both sides of the equation by the coefficient of the variable.	$2(a + b) = 8 - 5a$ for a $2a + 2b = 8 - 5a$ $2a + 5a = 8 - 2b$ $7a = 8 - 2b$ $a = \dfrac{8 - 2b}{7}$
Distributive property [1.11]	$\underbrace{a \cdot (b + c)}_{\text{factored form}} = \underbrace{a \cdot b + a \cdot c}_{\text{expanded form}}$	$4(x + 6) = 4 \cdot x + 4 \cdot 6$ $= 4x + 24$
General strategy for solving equations in one variable algebraically [1.11]	General strategy for solving an equation algebraically: 1. Simplify the expressions on each side of the equation (as discussed above). 2. Collect the variable terms on one side of the equation. 3. Solve for the variable by dividing each side of the equation by the coefficient of the variable. 4. Check your result in the original equation.	$3(2x - 5) + 2x = 9$ $6x - 15 + 2x = 9$ $8x - 15 = 9$ $8x = 24$ $x = 3$ Check: $3(2 \cdot 3 - 5) + 2 \cdot 3 =$ $3(6 - 5) + 6 =$ $3 \cdot 1 + 6 =$ 9

CONCEPT/SKILL	DESCRIPTION	EXAMPLE
Vertical line test [1.13]	In the **vertical line test**, a graph represents a function if any vertical line drawn through the graph intersects the graph no more than once.	This graph is not a function.
Increasing function [1.13]	The graph of an increasing function rises to the right.	
Decreasing function [1.13]	The graph of a decreasing function falls to the right.	
Constant function [1.13]	The graph of a constant function is a horizontal line.	
Vertical shift [1.14]	A positive constant is added to (upward shift) or subtracted from (downward shift) each output value. As a result, the graph is moved vertically upward or vertically downward.	See Problem 6 on page 130.
Horizontal shift [1.14]	A positive constant is first added to (shift left) or subtracted from (shift right) each input value. As a result, the graph is moved horizontally to the left or horizontally to the right.	See Problem 7 on pages 130–131

1. Determine whether each of the following is a function.

    a. The loudness of the stereo system is a function of the position of the volume dial.

    b. $\{(2, 9), (3, 10), (2, -9)\}$

    c.

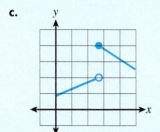

2. For an average yard, the fertilizer costs $20. You charge $8 per hour to do yard work. If x represents the number of hours worked on the yard, and y represents the total cost, including fertilizer, complete the following table:

x	0	2	3	5	7
y					

    a. Is the total cost a function of the hours worked? Explain.

    b. What is the independent variable?

    c. Which is the dependent variable?

    d. Which value(s) of the domain would not be realistic for this situation? Explain.

3. As a real estate salesperson, you earn a small salary, plus a percentage of the selling price of each house you sell. If your salary is $100 a week, plus 3.5% of the selling price of each house sold, what must your total annual home sales be for you to gross $60,000 in one year? Assume that you work 50 weeks per year.

4. Use the formula $I = Prt$ to evaluate I, given the following information.

    a. $P = \$2000, r = 5\%, t = 1$

    b. $P = \$3000, r = 6\%, t = 2$

5. Use the formula $P = 2(w + l)$, to evaluate P for the following information.

    a. $w = 2.8$ and $l = 3.4$

    b. $w = 7\dfrac{1}{3}$ and $l = 8\dfrac{1}{4}$

6. Solve the given equations, and check your results.

    a. $4(x + 5) - x = 80$ **b.** $-5(x - 3) + 2x = 6$

    c. $38 = 57 - (x + 32)$ **d.** $-13 + 4(3x + 5) = 7$

    e. $5x + 3(2x - 8) = 2(x + 6)$ **f.** $-4x - 2(5x - 7) + 2 = -3(3x + 5) - 4$

    g. $-32 + 6(3x + 4) = -(-5x + 38) + 3x$

7. You visit some relatives in Austin, Texas. You want to drive to San Antonio to see your uncle. The cost of renting a car for a day is \$25, plus 15 cents per mile.

    a. Identify the independent variable.

    b. Identify the dependent variable.

    c. Use x to represent the independent variable and y to represent the dependent variable. Write an equation that models the daily rental cost in terms of the number of miles driven.

    d. Complete the following table.

x	100	200	300	400	500
y					

e. Plot the points from the table in part d. Then draw the line through all five points.

(Make sure you label your axes and use appropriate scaling.)

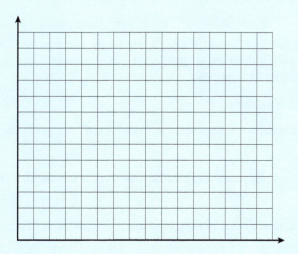

f. The distance from Austin to San Antonio is 80 miles. Estimate from the graph in part e how much it will cost to travel from Austin to San Antonio and back.

g. Use the mathematical model in part c to determine the exact cost of this trip.

h. Use the mathematical model in part c to determine the exact number of miles you can travel in the day and not exceed a budget of $115.

i. What are realistic replacement values for the independent variable if you rent the car for only one day?

8. The proceeds from your high school talent show to benefit a local charity totaled $950. Since the seats were all taken, you know that 500 people attended. The cost per ticket was $1.50 for students and $2.50 for adults. Unfortunately, you misplaced the ticket stubs that would indicate how many students and how many adults attended. You need this information for accounting purposes and future planning.

a. Let n represent the number of students who attended. Write an expression in terms of n to represent the number of adults who attended.

b. Write an expression in terms of n that will represent the proceeds from the student tickets.

c. Write an expression in terms of n that will represent the proceeds from the adult tickets.

d. Write an equation that indicates that the total proceeds from the student and adult tickets totaled $950.

e. How many student tickets and how many adult tickets were sold?

9. You have an opportunity to be the manager of a day camp for the summer. You know that your fixed costs for operating the camp are $690 per week, even if there are no campers. Each camper who attends costs the management $11.50 per week. The camp charges each camper $45 per week.

Let x represent the number of campers.

a. Write an equation in terms of x that represents the total cost, C, of running the camp per week.

b. Write an equation in terms of x that represents the total income (revenue), R, from the campers per week.

c. Write an equation in terms of x that represents the total profit, P, from the campers per week.

d. How many campers must attend for the camp to break even with revenue and costs?

e. The camp would like to make a profit of $600. How many campers must enroll to make that profit?

f. How much money would the camp lose if only 10 campers attend?

g. Use your graphing calculator to graph the revenue and cost equations. Compare the break-even point from the graph (point of intersection) with your answer in part d.

10. Solve each of the following equations for the specified variable.

a. $I = Prt$, for P **b.** $f = v + at$, for t **c.** $2x - 3y = 7$, for y

11. You contact the local print shop to produce a commemorative booklet for your high school theater group. It is the group's twenty-fifth anniversary, and in the booklet you want a short history plus a description of all the theater productions for the past 25 years in the booklet. The set up charges for the booklet are $150. It costs 40 cents for each copy produced.

  a. Write a mathematical model that gives the total cost, C, in terms of the number, x, of booklets produced.

  b. Use the equation from part a to determine the total cost of producing 500 booklets.

  c. How many booklets can be produced for $1000?

  d. Suppose the booklets are sold for $1.50 each. Write an equation for the total revenue, R, from the sale of x booklets.

  e. How many booklets must be sold to break even? That is, for what value of x is the total cost of production equal to the total amount of revenue?

  f. How many booklets must be sold to make a $200 profit?

12. You are designing a cylindrical container as new packaging for a popular brand of coffee. The current package is a cylinder with a diameter of 4 inches and a height of 5.5 inches. The volume of a cylinder is given by the formula

$$V = \pi r^2 h,$$

where V is the volume (in cubic inches), r is the radius (in inches), and h is the height (in inches).

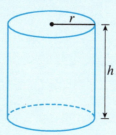

How much coffee does the current container hold? Round your answer to the nearest tenth of a cubic inch.

13. You have been asked to alter the dimensions of the container in Problem 12 so that the new package will contain less coffee. To save money, the company plans to sell the new package for the same price as before.

You will do this in one of two ways:

    i. By increasing the diameter and decreasing the height by 1/2 inch each (resulting in a slightly wider and shorter can), or,

    ii. By decreasing the diameter and increasing the height by 1/2 inch each (resulting in a slightly narrower and taller can).

a. Determine which new design, if either, will result in a package that holds less coffee than the current one.

b. By what percent will you have decreased the volume?

14. In parts a–d, sketch a graph of the new function resulting from the specified translation of the given graph.

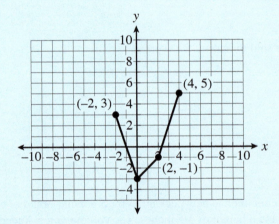

a. Vertical shift of the graph 3 units downward.

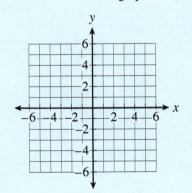

b. Vertical shift of the graph 5 units upward.

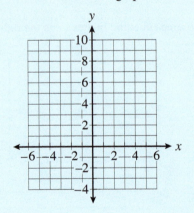

166

c. Horizontal shift of the graph 2 units to the left.

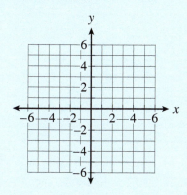

d. Horizontal shift of the graph 4 units to the right.

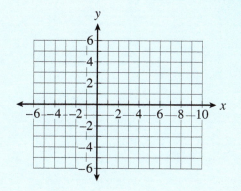

15. The graph of the function defined by $y = g(x)$ is

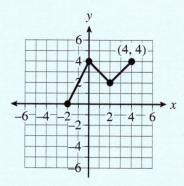

(4, 4)

Sketch a graph of each of the following functions defined by the given equation. Describe the type of shift that you performed on the original graph.

a. $y = g(x) - 2$

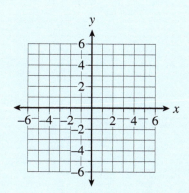

b. $y = g(x - 2)$

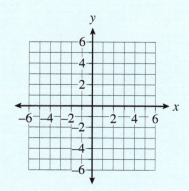

Linear Function Models and Problem Solving

You are a member of a health and fitness club. The club's registered dietitian and your personal trainer helped you develop a special eight-week diet and exercise program. The data in the following table represents your weight, w, as a function of time, t, over an eight-week period.

Time Weights for No One

Time (weeks)	0	1	2	3	4	5	6	7	8
Weight (lb)	140	136	133	131	130	127	127	130	126

1. a. Plot the data points using ordered pairs of the form (t, w). For example, $(3, 131)$ is a data point that represents your weight at the end of the third week.

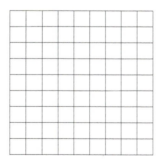

b. What is the practical domain of this function?

    c. What is the practical range of this function?

2. a. What was your weight at the beginning of the program?

    b. What was your weight at the end of the first week?

3. To see how the health program is working for you, you analyze your weekly weight changes during the eight-week period.

    a. During which week(s) does your weight increase?

    b. During which week(s) does your weight decrease?

    c. During which week(s) does your weight remain unchanged?

Average Rate of Change

4. Your weight decreases during each of the first five weeks of the program.

    a. Determine the actual change in your weight over the first five weeks of the program by subtracting your initial weight from your weight at the end of the first five weeks.

    b. What is the sign (positive or negative) of your answer? What is the significance of this sign?

    c. Determine the change in the t-value over the first five weeks; that is, from $t = 0$ to $t = 5$.

    d. Write the ratio of the change in weight from part a to the change in time in part c. Interpret the meaning of this ratio.

The change in weight describes how much weight you have gained or lost, but it does not tell how quickly you shed those pounds. That is, a loss of 3 pounds in one week is more impressive than a loss of 3 pounds over a month's time. Dividing the change in weight by the change in time gives an average rate at which you have lost weight over this period of time. Its units are weight units per time unit—in this case, pounds per week.

The ratio in Problem 4d,

$$\frac{-13 \text{ lb}}{5 \text{ wk}} = -2.6 \text{ pounds per week},$$

is called the **average rate of change** of weight with respect to time over the first five weeks of the program. It can be interpreted as an average loss of 2.6 pounds each week for the first five weeks of the program.

Delta Notation

The change in the values of a variable is so important that special symbolic notation has been developed to denote it.

The uppercase Greek letter delta Δ is used with a variable's name to represent a change in the value of the variable from a starting point to an ending point.

For example in Problem 4, the notation Δw represents the change in value of the dependent variable, weight w. It is calculated by subtracting the initial value of w, denoted by w_1, from the final value of w, denoted by w_2. Symbolically, this change is represented by

$$\Delta w = w_2 - w_1.$$

In the case of Problem 4a, the change in weight over the first five weeks can be calculated as:

$$\Delta w = w_2 - w_1 = 127 - 140 = -13 \, \text{lb}.$$

In the same way, the change in value of the independent variable, time (t), is written as Δt and is calculated by subtracting the initial value of t, denoted by t_1, from the final value of t, denoted by t_2. Symbolically, this change is represented by

$$\Delta t = t_2 - t_1.$$

In Problem 4c, the change in the number of weeks can be written using Δ notation as follows:

$$\Delta t = t_2 - t_1 = 5 - 0 = 5 \, \text{weeks}$$

The average rate of change of weight over the first five weeks of the program can now be symbolically written as follows:

$$\frac{\Delta w}{\Delta t} = \frac{-13}{5} = -2.6 \, \text{lb per week}$$

Note that the symbol Δ for delta is the Greek version of d, for difference, the result of a subtraction that produces the change in value.

5. Using Δ notation, determine the average rate of change of weight over the last four weeks of the program.

Definition

The ratio $\dfrac{\Delta w}{\Delta t}$ is called the **average rate of change** of weight, w, with respect to time, t.

In general, the average rate of change is

$$\frac{\Delta w}{\Delta t} = \frac{w_2 - w_1}{t_2 - t_1},$$

where (t_1, w_1) is the initial point, (t_2, w_2) is the final point, and $t_1 \neq t_2$.

Graphical Interpretation of the Average Rate of Change

6. a. On the graph in Problem 1, connect the points $(0, 140)$ and $(5, 127)$ with a line segment. Does the line segment rise, fall, or remain horizontal as you follow it from left to right?

b. Recall that the average rate of change over the first five weeks was -2.6 pounds per week. What does the average rate of change tell you about the line segment drawn in part a?

7. a. Determine the average rate of change of your weight over the time period from $t = 5$ to $t = 7$ weeks. Include the appropriate sign and units.

b. Interpret the rate in part a with respect to your diet.

c. On the graph in Problem 1, connect the points $(5, 127)$ and $(7, 130)$ with a line segment. Does the line segment rise, fall, or remain horizontal as you follow it from left to right?

d. How is the average rate of change of weight over the given two-week period related to the line segment you draw in part c?

8. a. At what rate is your weight changing during the sixth week of your diet; that is, from $t = 5$ to $t = 6$?

b. Interpret the rate in part a with respect to your diet.

c. Connect the points $(5, 127)$ and $(6, 127)$ on the graph with a line segment. Does the line segment rise, fall, or remain horizontal as you follow it from left to right?

d. How is the average rate of change in part a related to the line segment drawn in part c?

9. a. What is the average rate of change of your weight over the period from $t = 4$ to $t = 7$ weeks?

b. Explain how the rate in part a reflects the progress of your diet over those three weeks.

10. At the beginning of the diet and exercise program, and once a week thereafter, you are tested on the treadmill. The test consists of how many minutes it takes you to walk, jog, or run 3 miles on the treadmill. The following data gives your time, t, over an eight-week period.

End of Week, w	0	1	2	3	4	5	6	7	8
Time, t (in minutes)	45	42	40	39	38	38	37	39	36

Note that $w = 0$ corresponds to the first time on the treadmill, $w = 1$ is the end of the first week, $w = 2$ is the end of the second week, and so on.

a. Plot the data points using ordered pairs of the form (w, t).

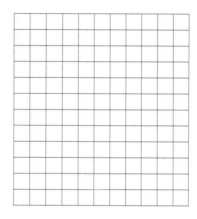

b. Determine the average rate of change of t with respect to w during the sixth and seventh weeks (from the point where $w = 5$ to the point where $w = 7$).

c. What is the significance of the positive sign of the average rate of change in this situation?

d. Connect the data points (5, 38) and (7, 39) on your graph from part a, using a line segment. Is the output increasing, decreasing, or constant on the interval?

e. At what average rate did your time change during the fifth week (from $w = 4$ to $w = 5$?)

f. Interpret your answer from part e.

g. Connect the data points (4, 38) and (5, 38) on the graph from a using a line segment. Is the output increasing, decreasing, or constant over this interval?

SUMMARY: ACTIVITY 2.1

1. Let y_1 represent the corresponding value for x_1 and y_2 represent the corresponding value for x_2. As the variable x changes in value from x_1 to x_2,

    a. the change in x is represented by $\Delta x = x_2 - x_1$

    b. the change in y is represented by $\Delta y = y_2 - y_1$

2. The quotient $\dfrac{\Delta y}{\Delta x} = \dfrac{y_2 - y_1}{x_2 - x_1}$, $x_1 \neq x_2$, is called the **average rate of change** of y with

respect to x over the x-interval from x_1 to x_2. The units of measurement of the quantity $\dfrac{\Delta y}{\Delta x}$

are *y-units* per *x-unit*.

3. The line segment connecting the points (x_1, y_1) and (x_2, y_2)

    a. rises to the right if $\dfrac{\Delta y}{\Delta x} > 0$

    b. falls to the right if $\dfrac{\Delta y}{\Delta x} < 0$

    c. is horizontal if $\dfrac{\Delta y}{\Delta x} = 0$

EXERCISES: ACTIVITY 2.1

The following table of data from the U.S. Bureau of the Census gives the median age of an American man at the time of his first marriage:

Year	1910	1920	1930	1940	1950	1960	1970	1980	1990	2000	2010
Median Age	25.1	24.6	24.3	24.3	22.8	22.8	23.2	24.7	26.1	26.8	27.5

Use this data to answer Exercises 1–6.

1. a. Determine the average rate of change in median age per year from 1950 to 2010.

    b. Describe what the average rate of change in part a represents in this situation.

2. Determine the average rate of change in median age per year from 1930 to 1960.

3. What is the average rate of change over the 100-year period described in the table?

4. During what ten-year period did the average age increase the most?

5. a. What does it mean in this situation if the average rate of change is negative?

b. Determine at least one ten-year period when the average rate of change is negative.

c. What trend would you observe in the graph if the average rate of change were negative? That is, would the graph go up, go down, or remain constant?

6. a. Is the average rate of change zero over any ten-year period? If so, when?

b. What does a rate of change of zero mean in this situation?

c. What trend would you observe in the graph during this period? That is, would the graph go up, go down, or be horizontal?

7. The total amount of rainfall in a given community can vary widely from year to year. The following table gives information on the total rainfall received over a recent seven-year period for an area in southern New York.

ANNUAL RAINFALL							
Year, t	1	2	3	4	5	6	7
Rainfall, r (inches)	45.49	41.88	39.63	32.91	37.47	50.08	37.54

a. Plot the data points using ordered pairs of the form (t, r).

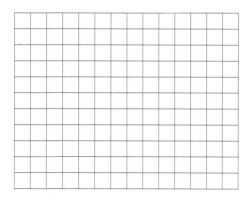

b. Determine the average rate of change of rainfall in the area from year 4 to year 7.

c. Determine the average rate of change of rainfall in the area from year 1 to year 4.

d. Compare the average rate of change from year 1 to year 4 to the average rate of change from year 4 to year 7.

e. When the average rate of change is negative, what trend will you observe in the graph? What does that mean in this situation?

8. Between 1960 and 2010, the size and shape of automobiles in the United States has changed almost annually. The amount of fuel consumed by these vehicles has also changed. The following table describes the average fuel consumed per year per passenger car in gallons of gasoline.

Year, t	1960	1970	1980	1990	1995	2000	2005	2010
Gallons Consumed per Passenger Car (average), g	668	760	576	520	530	547	567	453

a. Determine the average rate of change, in gallons of gas per year, from 1960 to 1970.

b. Determine the average rate of change, in gallons of gas per year, from 1960 to 1990.

c. Determine the average rate of change, in gallons of gas per year, from 1995 to 2005.

d. Determine the average rate of change, in gallons of gas per year, between 1960 and 2010.

e. What does the result in part d mean in this situation?

9. The National Weather Service recorded the following temperatures one February day in Chicago:

Time of Day	10 A.M.	12 NOON	2 P.M.	4 P.M.	6 P.M.	8 P.M.	10 P.M.
Temperature (°F)	30	35	36	36	34	30	28

a. Determine the average rate of change (including units and sign) of temperature with respect to time over the entire twelve-hour period given in the table.

b. Over which period(s) of time is the average rate of change zero? What, if anything, can you conclude about the actual temperature fluctuation within this period?

c. What is the average rate of change of temperature with respect to time over the evening hours from 6 P.M. to 10 P.M.? Interpret this value (including units and sign) in a complete sentence.

d. Write a brief paragraph describing the temperature and its fluctuations during the twelve-hour period in the table.

Activity 2.2

The Snowy Tree Cricket

Objectives

1. Identify linear functions by a constant rate of change.

2. Interpret slope as an average rate of change.

3. Determine the slope of the line drawn through two points.

4. Identify increasing and decreasing linear functions using slope.

5. Identify parallel lines using slope.

6. Identify perpendicular lines using slope.

One of the more familiar late-evening sounds during the summer is the rhythmic chirping of a male cricket. Of particular interest is the snowy tree cricket, sometimes called the temperature cricket. It is very sensitive to temperature, speeding up or slowing down its chirping as the temperature rises or falls. An entomologist collected the following data, which shows how the number of chirps per minute of the snowy tree cricket is related to temperature.

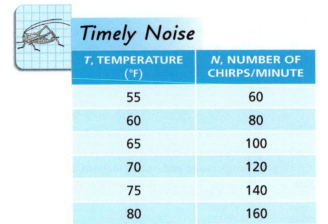

Timely Noise

T, TEMPERATURE (°F)	N, NUMBER OF CHIRPS/MINUTE
55	60
60	80
65	100
70	120
75	140
80	160

1. Crickets are usually silent when the temperature falls below 55°F. What is a possible practical domain for the snowy tree cricket function?

2. **a.** Determine the average rate of change of the number of chirps per minute with respect to temperature as the temperature increases from 55°F to 60°F.

   b. What are the units of measure of this rate of change?

3. **a.** How does the average rate of change determined in Problem 2 compare with the average rate of change as the temperature increases from 65°F to 80°F?

   b. Determine the average rate of change of number of chirps per minute with respect to temperature for the temperature intervals given in the following table. The results from Problems 1 and 2 are already recorded. Add several more of your own choice. List all your results in the table.

TEMPERATURE INCREASES	AVERAGE RATE OF CHANGE (number of chirps/minute with respect to temperature)
From 55° to 60°F	4
From 60° to 80°F	
From 55° to 75°F	
From 65° to 80°F	4

    c. What can you conclude about the average rate of increase in the number of chirps per minute for any particular increase in temperature?

4. For any 7° increase in temperature, what is the expected increase in chirps per minute?

5. Plot the data pairs (temperature, chirps per minute) from the table preceding Problem 1. What type of graph is suggested by the pattern of points?

    Note: Use a double slash (//) to indicate that the horizontal axis has been compressed between 0 and 50, and the vertical axis between 0 and 60.

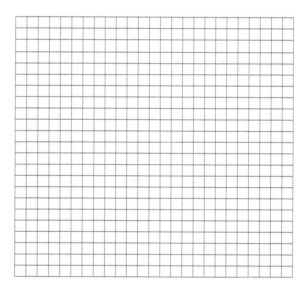

Linear Functions

If the average rate of change in *x* with respect to *y* remains constant (stays the same) for *any* two points in a data set, the points will lie on a straight line. That is, *y* is a **linear function** of *x*. Conversely, if all the points of a data set lie on a straight line when graphed, the average rate of change of *y* with respect to *x* will be constant for any two data points.

6. From the graph in Problem 5, would you conclude that the number of chirps per minute is a linear function of the temperature? Explain.

As you will see in the following activities, many situations in the world around us can be modeled by linear functions.

Slope of a Line

You have seen that the average rate of change between any two points on a line is always the same constant value. This value has a geometric interpretation as well—it describes the "steepness" of the line and is called the **slope** of the line.

   7. a. What is the slope of the line in the snowy tree cricket situation?

   b. Because the slope of this line is positive, what can you conclude about the direction of the line as the independent variable (temperature) increases in value?

   c. What is the practical meaning of slope in this situation?

The slope of a line is often denoted by the letter m. It can be determined by selecting *any* two points on the line and calculating the rate of change between them. That is, if (x_1, y_1) and (x_2, y_2) represent two points on a line, then the slope of the line is calculated by the following formula:

$$m = slope = \frac{\Delta y}{\Delta x} = \frac{y_2 - y_1}{x_2 - x_1}, \text{ and where } x_1 \neq x_2$$

Example 1   *Determine the slope of the line containing the points* $(1, -2)$ *and* $(3, 8)$.

SOLUTION

Let $x_1 = 1$, $y_1 = -2$, $x_2 = 3$, and $y_2 = 8$; so

$$m = \frac{\Delta y}{\Delta x} = \frac{y_2 - y_1}{x_2 - x_1} = \frac{8 - (-2)}{3 - 1} = \frac{10}{2} = \frac{5}{1} = 5.$$

   8. Determine the slope of the line in the snowy tree cricket situation using delta notation. Select any two points from the table preceding Problem 1 for (x_1, y_1) and (x_2, y_2).

On a graph, slope is the ratio of two distances. For example, in the snowy tree cricket situation, the slope of the line between the points (55, 60) and (56, 64) as well as between (56, 64) and (57, 68) is $\frac{4}{1}$, as shown on the following graph. The change in x-value, 1, represents a horizontal distance (the run) in going from one point to another point on the same line. The change in the y-value, 4, represents a vertical distance (the rise) between the same points. The graph below illustrates that a horizontal distance (run) of 2 and a vertical distance (rise) of 8 from (55, 60) will also locate the point (57, 68) on the line.

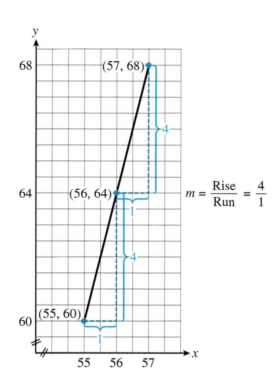

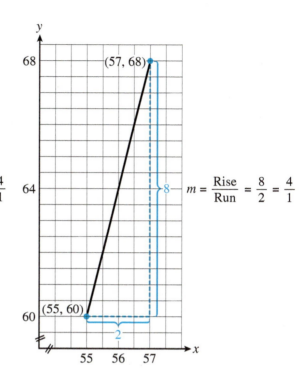

$$m = \frac{\text{Rise}}{\text{Run}} = \frac{4}{1}$$

$$m = \frac{\text{Rise}}{\text{Run}} = \frac{8}{2} = \frac{4}{1}$$

Definition

From its geometric meaning, slope may be determined from a graph by the formula:

$$m = \text{slope} = \frac{\text{rise}}{\text{run}} = \frac{\text{distance up } (+) \text{ or down } (-)}{\text{distance right } (+) \text{ or left } (-)}$$

The slope can be used to locate additional points on a graph.

Note that a positive slope of $\dfrac{5}{3}$ can be interpreted as $\dfrac{+5}{+3} = \dfrac{\text{up } 5}{\text{right } 3}$ or $\dfrac{-5}{-3} = \dfrac{\text{down } 5}{\text{left } 3}$ or any multiple of either, such as $\dfrac{-10}{-6}$.

Example 2   *Use the geometric interpretation of slope to locate two additional points in the snowy cricket situation.*

a. Locate a point below (55, 60).

$$m = \frac{\Delta y}{\Delta x} = \frac{4}{1} = \frac{-4}{-1} = \frac{-12}{-3} = \frac{\text{down } 12}{\text{left } 3} \qquad \begin{array}{l}\text{Moving down 12 and left 3 from (55, 60)} \\ \text{locates the point (52, 48).}\end{array}$$

b. Locate a point between (55, 60) and (56, 64).

$$m = 4 = \frac{4}{1} = \frac{+2}{+0.5} = \frac{\text{up } 2}{\text{right } 0.5} \qquad \begin{array}{l}\text{Moving up 2 and right 0.5 from (55, 60)} \\ \text{locates the point (55.5, 62).}\end{array}$$

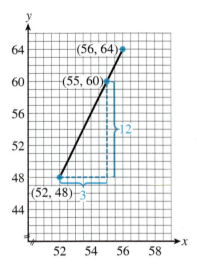
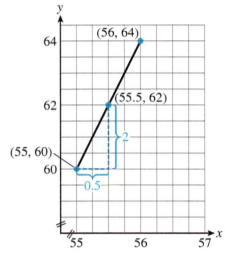

9. a. Consider the line containing the point $(-6, 4)$ and having slope $\dfrac{3}{4}$. Plot the point and then determine two additional points on the line. Draw a line through these points.

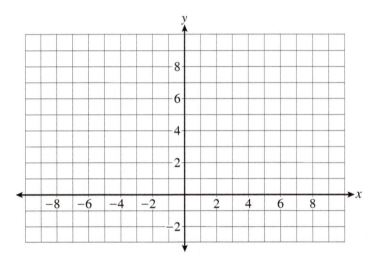

b. Which line increases more rapidly: a line with slope 3 or a line having slope $\dfrac{3}{4}$? Explain.

The graph of a line having positive slope rises from left to right. Such a line represents an increasing function.

Decreasing Linear Function

While on a trip, you notice that the video screen on the airplane, in addition to showing movies and news, records your altitude (in kilometers) above the ground. As the plane starts its descent (at time $x = 0$), you record the following data:

Cleared for a Landing

TIME, x (min)	ALTITUDE, y (km)
0	12
2	10
4	8
6	6
8	4
10	2

10. a. What is the average rate of change in the altitude of the plane from two to six minutes into the descent? Pay careful attention to the sign of this rate of change.

b. What are the units of measurement of this average rate of change?

c. Determine the average rate of change over several other input intervals.

d. What is the significance of the signs of these average rates of change?

e. Based on your calculation in parts a and c, do you think that the data lie on a single straight line? Explain.

f. What is the practical meaning of slope in this situation?

g. By how much does the altitude of the plane change for each three-minute change in time during the descent?

I I. **a.** Plot the data points from the table preceding Problem 10, and verify that the points lie on a line. What is the slope of the line?

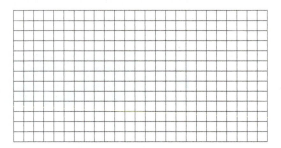

b. Explain to a classmate the method you used to determine the slope in part a.

12. The slope of the line for the descent function in this activity is negative. What does this tell you about how the altitude changes as the time increases in value?

Parallel Lines

13. **a.** Two lines, denoted by l_1 and l_2, contain the given points. Sketch a graph of each line on the following coordinate axes.

$l_1: (-2, -3), (2, 5)$ $l_2: (-1, -5), (3, 3)$

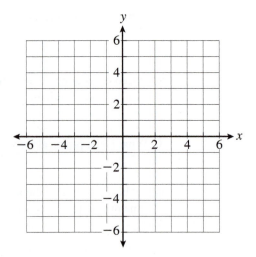

b. What can you say about the graphs of the two lines?

c. Determine and then compare the slopes of the two lines.

d. By inspection of the graph in part a, estimate the points where the two lines intersect the *y*-axis. How do they compare?

Definition

Two lines are **parallel** if they never intersect, no matter how far you extend the lines in either direction. Parallel lines have equal slopes but different *y*-intercepts.

---

14. Are the two lines in Problem 13 parallel? Explain.

Note that if two lines cross the *y*-axis at the same point and have equal slopes, the graphs are exactly the same and the two lines are said to **coincide**.

Perpendicular Lines

15. Consider two lines with slopes $m_1 = -2$ and $m_2 = \dfrac{1}{2}$. Are the lines parallel? Explain.

If the slopes of the two lines are different, the lines will intersect at some point. The two lines in Problem 15 intersect in a special way to form four angles of equal size. Such an angle is called a **right angle** and measures 90°. In a diagram, right angles are usually designated as follows:

Perpendicular Lines

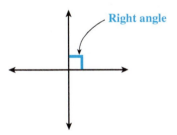

Two lines that intersect at right angles are called **perpendicular lines**. The slopes of perpendicular lines are related in a special way. Problem 16 demonstrates this relationship.

16. Multiply the slopes of the two lines in Problem 15.

If the product of two numbers is -1, the numbers are called **negative reciprocals.**

Definition

Two lines are perpendicular if their slopes are negative reciprocals. Stated symbolically, if m_1 and m_2 represent the slopes of two lines l_1 and l_2, respectively, then l_1 is perpendicular to l_2 if $m_1 \cdot m_2 = -1$.

---

17. Determine if the lines having slopes $m_1 = \dfrac{-4}{5}$ and $m_2 = \dfrac{5}{4}$ are perpendicular:

SUMMARY: ACTIVITY 2.2

1. A **linear function** is one whose average rate of change of y with respect to x from any one data point to any other data point is always the same (constant) value.

2. The **graph** of a linear function is a line whose slope is the constant rate of change of the function.

3. The **slope of a line segment** joining two points (x_1, y_1) and (x_2, y_2) is denoted by m and can be calculated using the formula $m = \dfrac{\Delta y}{\Delta x} = \dfrac{y_2 - y_1}{x_2 - x_1}$, where $x_1 \neq x_2$. Geometrically, Δy represents a vertical distance (rise), and Δx represents a horizontal distance (run).

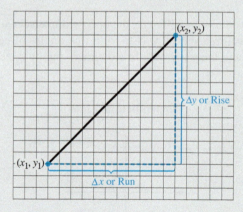

Therefore, $m = \dfrac{\Delta y}{\Delta x} = \dfrac{\text{rise}}{\text{run}}$.

4. The graph of every linear function with **positive slope** is an increasing line, rising to the right. The function is an **increasing function**.

5. The graph of every linear function with **negative slope** is a decreasing line, falling to the right. The function is a **decreasing function**.

6. Two lines are **parallel** if they never intersect, no matter how far you extend the lines in either direction. Parallel lines have equal slopes but intersect the y-axis at different points.

7. If the product of two numbers is -1, then the numbers are called **negative reciprocals**.

8. Two lines are **perpendicular** if their slopes are negative reciprocals. Stated symbolically, if m_1 and m_2 represent the slopes of two lines l_1 and l_2, respectively, then l_1 is perpendicular to l_2 if $m_1 \cdot m_2 = -1$.

EXERCISES: ACTIVITY 2.2

1. In a science lab, you collect the following sets of data. Which of the four data sets are linear functions? If linear, determine the slope.

a.

Time (sec)	0	10	20	30	40
Temperature (°C)	12	17	22	27	32

b.

Time (sec)	0	10	20	30	40
Temperature (°C)	41	23	5	−10	−20

c.

Time (sec)	3	5	8	10	15
Temperature (°C)	12	16	24	28	36

d.

Time (sec)	3	9	12	18	21
Temperature (°C)	25	23	22	20	19

2. a. A special diet and exercise program has been developed for you by a registered dietitian and a personal trainer. You weigh 181 pounds and would like to lose 2 pounds every week. Complete the following table of values for your desired weight each week:

N, Number of Weeks	0	1	2	3	4
W, Desired Weight (lb)					

b. Plot the data points.

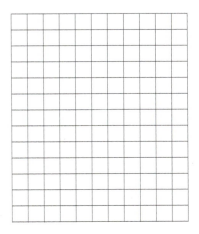

c. Explain why your desired weight is a linear function of time. What is the slope of the line containing the five data points?

d. What is the practical meaning of slope in this situation?

e. How long will it take to reach your ideal weight of 168 pounds?

3. Your aerobics instructor informs you that to receive full physical benefit from exercising, your heart rate must be maintained at a certain level for at least twelve minutes. The proper exercise heart rate for a healthy person, called the target heart rate, is determined by the person's age. The relationship between these two quantities is illustrated by the data in the following table.

A, Age (Years)	20	30	40	50	60
R, Target Heart Rate (beats per minute)	140	133	126	119	112

a. Does the data in the table indicate that the target heart rate is a linear function of age? Explain.

b. What is the slope of the line for this data? What are the units?

c. What is the practical domain for age, A?

d. Plot the data points on coordinate axes where both axes start with zero. Connect the points with a line.

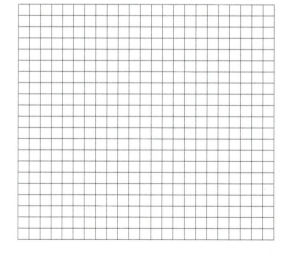

4. Determine the slope of each of the following lines:

i.

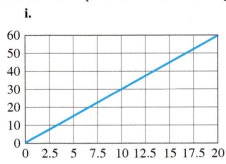

ii.

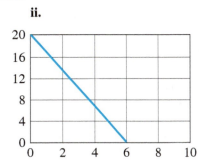

5. Each question refers to the graph that accompanies it. The graphed line in each grid represents the total distance a car travels as a function of time (in hours).

   a. How fast is the car traveling? Explain how you obtained your result.

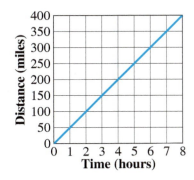

   b. How can you determine visually from the following graph which car is going faster? Verify your answer to the first question by calculating the speed of each car.

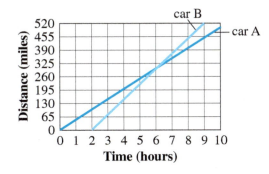

c. Describe in words the movement of the car that is represented by the following graph:

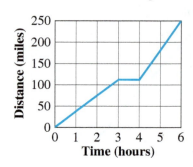

6. a. Determine the slope of each of the following lines:

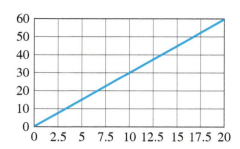 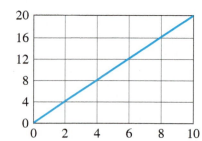 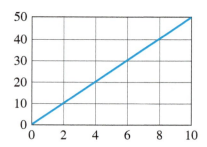

b. At first glance, the three graphs in part a may appear to represent the same line. Do they?

7. For each of the following, determine two additional points on the line. Then sketch a graph of the line.

a. A line contains the point $(-5, 10)$ and has slope $\dfrac{2}{3}$.

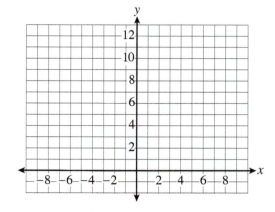

b. A line contains the point $(3, -4)$ and has slope 5.

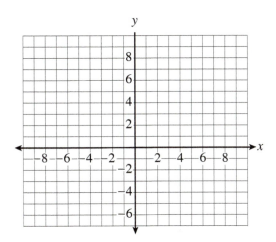

8. The concept of slope arises in many practical applications. When designing and building roads, engineers and surveyors need to be concerned about the grade of the road. The grade, usually expressed as a percent, is one way to describe the steepness of the finished surface of the road. For example, a 5% grade means that the road has a slope (rise over run) of $0.05 = \dfrac{5}{100}$.

5 ft

100 ft

a. If a road has a 5% upward grade over a 1000-foot run, how much higher will you be at the end of that run than at the beginning?

b. What is the grade of a road that rises 26 feet over a distance of 500 feet?

9. An architect is designing a wheelchair ramp for a new apartment complex. The American National Standards Institute (ANSI) requires that the slope for a wheelchair ramp does not exceed $\dfrac{1}{12}$.

a. Does a ramp that is 160 in. long and 10 in. high meet the requirements of ANSI? Explain.

  b. A ramp for a wheelchair must be 20 inches high. Determine the minimum length of the ramp so that it meets the ANSI requirement.

10. Determine if the given pair of lines is parallel. Verify your conclusion by sketching a graph of each pair of lines.

  a. l_1: $(-1, 1), (1, 7)$

  l_2: $(-2, -5), (1, 4)$

  b. l_1: $(-2, -5), (1, 4)$

  l_2: $(-1, 7), (2, 1)$

  c. l_1: $(-2, 2), (4, -1)$

  l_2: $(4, -5), (2, -4)$

  d. l_1: $(2, -3), (1, -1)$

  l_2: $(-1, 3), (2, -3)$

11. Determine if the lines having given slopes are perpendicular.

  a. $m_1 = \dfrac{1}{3}$ and $m_2 = -3$

  b. $m_1 = 2$ and $m_2 = \dfrac{1}{2}$

Activity 2.3

Depreciation

Objectives

1. Identify whether a situation can be modeled by a linear function.

2. Identify the practical meaning of *x*- and *y*-intercepts.

3. Develop the slope/intercept model of an equation of a line.

4. Use the slope/intercept formula to determine *x*- and *y*-intercepts.

5. Determine the zeros of a function.

You have decided to buy a new Honda Civic, but you are concerned about the value of the car depreciating over time. You search the Internet and obtain the following information.

- Suggested retail price $20,905
- Depreciation per year $1,750 (assume constant)

1. a. Complete the following table in which *V* represents the value of the car after *n* years of ownership.

Accordingly

n, YEARS	*V*, VALUE IN DOLLARS
0	
1	
2	
3	
5	
8	

b. Is the value of the car a function of the number of years of ownership? Explain.

c. What is the independent variable? What is the dependent variable?

2. a. Select two ordered pairs of the form (*n*, *V*) from the table in Problem 1 and determine the average rate of change.

b. What are the units of measure of the average rate of change?

c. What is the practical meaning of the sign of the average rate of change?

d. Select two different ordered pairs and compute the average rate of change.

e. Select two ordered pairs not used in parts a or d, and compute the average rate of change.

f. Using the results in parts a, d, and e, what can you infer about the average rate of change over any interval of time?

3. a. Is the value, V, of the car a linear function of the number of years, n, of ownership? Explain using the definition of linear function.

b. Is this function increasing, decreasing, or constant?

Graph of a Linear Function

4. Consider the ordered pairs of the form (n, V), and plot each ordered pair in Problem 1 on an appropriately scaled and labeled set of axes. Connect the points to see if there is a pattern.

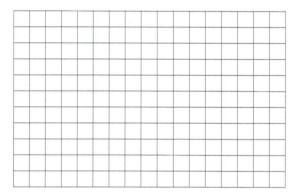

The graph of a linear function is a nonvertical line. Remember that the constant average rate of change is called the **slope** of the line and is denoted by the letter m.

5. a. What is the slope of the line graphed in Problem 4?

b. What is the relationship between the slope of the line and the average rate of change?

c. What is the practical meaning of slope in this situation?

Vertical Intercept (y-intercepts)

Definition

In general, the **vertical intercept** is the point where the graph crosses, or intercepts, the vertical axis. If the dependent variable is represented by y, the vertical intercept is called the **y-intercept**. The corresponding x-value of the y-intercept is always zero.

6. a. Using the table of data in Problem 1 or the graph in Problem 4, determine the V-intercept.

b. What is the practical meaning of the v-intercept in this situation? Include units.

Slope-Intercept Form of a Linear Equation

7. a. Review how you determined the value, V, of the car in Problem 1 for a given number of years, n, of ownership. Write an equation for V in terms of n that models this situation.

b. Use your graphing calculator to sketch a graph of this equation. Use the window $\text{Xmin} = -2$, $\text{Xmax} = 16$, $\text{Ymin} = -500$, and $\text{Ymax} = 25{,}000$.

c. How does this graph compare to your graph in Problem 4?

8. Recall that the slope of your line is $m = -1750$ and the v-intercept is $(0, 20{,}905)$. How is this information contained in the equation of the line you determined in Problem 7a?

Definition

The coordinates of all points (x, y) on the line with slope m and y-intercept $(0, b)$ satisfy the equation

$$y = mx + b \quad \text{or} \quad y = b + mx.$$

This is called the **slope-intercept form** of the equation of a line.

Note that the coefficient of x, which is m, is the *slope* of the line. The constant term, b, is the *y-coordinate of the y-intercept*. If $f(x)$ replaces y, the equation $y = mx + b$ can be written as

$$f(x) = mx + b.$$

Example 1 *The slope-intercept form of the equation of the line with slope 3 and vertical intercept $(0, -6)$ is $y = 3x - 6$. Using function notation and replacing y with $f(x)$, the equation becomes $f(x) = 3x - 6$.*

9. Identify the slope and y-intercept of the line whose equation is given. Write the vertical intercept as an ordered pair.

a. $y = -2x + 5$

b. $s = \dfrac{3}{4}t + 2$

c. $q = 2 - r$

d. $y = \dfrac{5}{6} + \dfrac{x}{3}$

Horizontal Intercepts (x-intercepts)

Definition

In general, a **horizontal intercept** of a graph is a point where the graph meets or crosses the horizontal axis. If the independent variable is represented by x, the horizontal intercept is called the **x-intercept**. The y-value of the x-intercept is always zero.

Example 2   *Consider the equation $y = 2x - 10$.*

a. The y-intercept occurs where the line crosses the vertical axis, i.e., where $x = 0$. Letting $x = 0$, $y = 2(0) - 10$ or $y = -10$. The y-intercept is $(0, -10)$.

b. The x-intercept occurs where the line crosses the horizontal axis, that is, where $y = 0$.

Letting $y = 0$,

$$0 = 2x - 10$$
$$\underline{+\ 10 \qquad\quad +\ 10} \quad \text{Add 10 to each side.}$$
$$10 = 2x$$

$$\frac{10}{2} = \frac{2x}{2} \qquad \text{Divide each side by 2.}$$

$$5 = x$$

Therefore, the x-intercept is $(5, 0)$.

c. You can now sketch a graph of $y = 2x - 10$ by plotting the x- and y-intercepts and connecting the points.

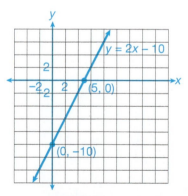

10. a. Determine the *n*-intercept of the graph of the car value equation
$$V = -1750n + 20{,}905.$$

b. What is the practical significance of the *n*-intercept? Include units.

Converting Celsius to Fahrenheit

11. The temperature on a warm summer day may be 27°. Your refrigerator is set at 5°. You set your oven to 180° to cook a turkey. If these temperatures seem strange, it is because they are measured on the Celsius scale. If the TV weather forecaster says it is going to be 65°F tomorrow, you know what that temperature will feel like and you can dress accordingly. How should you dress if the reporter says it will be 10°C? Most of us need to convert Celsius to Fahrenheit so we can better understand the temperature. The equation that relates Celsius measure to Fahrenheit measure is:

$$F = 1.8C + 32,$$

where *C* is the Celsius measure and *F* is the Fahrenheit measure.

a. Use the equation to complete the following table:

C	−10	0	5	10	27	100
F						

b. Sketch a graph of the equation $F = 1.8C + 32$ by plotting the ordered pairs in part a.

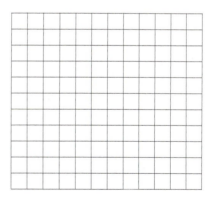

c. Determine the average rate of change of *F* as *C* increases from 0 to 100.

d. What is the slope of the line? How does the slope compare to the average rate of change in part c?

e. Is the function increasing, decreasing, or constant?

f. Determine the intercepts of the graph using the equation of the line. Verify your results using the graph of the line.

g. What is the practical significance of each intercept in this situation?

h. When does F = C?

Zeros of a Function

A value, a, is a zero of a function f if $f(a) = 0$. In other words, a zero of a function is an input value that makes the output zero.

> **Example 3**   *If $f(x) = 2x - 6$, then $f(3) = 2(3) - 6 = 0$. So 3 is a zero of f.*

12. If the value, 3, is a zero of a function f, then $f(3) = 0$.

a. Write the ordered pair corresponding to $f(3) = 0$.

b. Describe the location of the point whose ordered pair is given in part a.

Zeros and x-intercepts

If a value, a, is a zero of a function f, then $f(a) = 0$ and $(a, 0)$ is an x-intercept of the graph of $y = f(x)$.

13. Determine the zero of $f(x) = 3x + 12$ by setting $f(x) = 0$ and solving for x.

14. What is the x-intercept of the graph of $f(x) = 3x + 12$?

SUMMARY: ACTIVITY 2.3

1. A function for which the average rate of change between any pair of points remains constant is called a **linear function**.

2. The graph of a linear function is a nonvertical line. The constant average rate of change is called the **slope** and is denoted by the letter m (from the French verb *monter*, "to climb" or "rise").

3. The **slope** of a line segment joining two points (x_1, y_1) and (x_2, y_2) is denoted by m and is defined by $m = \dfrac{y_2 - y_1}{x_2 - x_1}$, where $x_1 \neq x_2$.

4. The **y-intercept** $(0, b)$ of a graph is the point where the graph crosses the y-axis.
   The **x-intercept** $(a, 0)$ of a graph is the point, where the graph crosses the x-axis.

5. The slope-intercept form of the equation of a line is $y = mx + b$.

6. To determine the b-value of the y-intercept $(0, b)$ from $y = mx + b$, set $x = 0$ and solve for y.

7. To determine the a-value of the x-intercept $(a, 0)$ from $y = mx + b$, set $y = 0$ and solve for x.

8. If the **slope** of a linear function is **positive**, the graph of the function rises to the right.

9. If the **slope** of a linear function is **negative**, the graph of the function falls to the right.

10. Determine the **zero** of $f(x) = mx + b$ by setting $f(x) = 0$. If the value, a, is a zero of the function, then $f(a) = 0$ and $(a, 0)$ is an x-intercept of the graph of $y = f(x)$.

EXERCISES: ACTIVITY 2.3

1. Determine whether the following functions are linear. If they are, determine the constant average rate of change, called the slope of the line.

a.

x	y
0	−1
1	9
2	19

b.

x	2	4	6
y	11	8	−1

c. $\{(-2, 18), (2, 9), (6, 0)\}$

2. Determine whether the following tables contain data that represent a linear function. Assume that the first row of the table is the x-variable and the second row is the y-variable. Explain your reasoning.

a. You owe your uncle $1000. He paid for your first semester at community college. The conditions of the loan are that you must pay him back the whole amount in one payment using a simple interest rate of 2% per year. He doesn't care in which year you pay him. The table contains input and output values to represent how much money you will owe your uncle 1, 2, 3, or 4 years later.

Year	1	2	3	4
Amount Owed (in $)	1020	1040	1060	1080

b. You are driving along the highway, and just before you reach the crest of a hill, you notice a sign that indicates the elevation at the crest is 2250 feet. As you proceed down the hill, your elevation is as given by the following table.

Distance Traveled From the Crest of The Hill (in feet)	0	1000	3000	6000	10,000
Elevation (in feet)	2250	2180	2040	1830	1550

c. You decide to invest $1000 of your 401k funds into an account that pays 3.5% interest compounded continuously. The table contains input and output values that represent the amount an initial investment of $1000 is worth at the end of each year.

Year	1	2	3	4	5
Value of Investment (in $)	1036	1073	1111	1150	1191

d. For a fee of $40 per month you may have breakfast (all you can eat) in the college snack bar each day. The table contains input and output values that represent the total number of breakfasts consumed each month and the amount you pay each month.

Number of Breakfasts	10	22	16	13
Cost (in $)	40	40	40	40

3. You and your friends are enrolled in the community center's weight-loss program. The charts contain data that represent the weight over a four-week period for you and your two friends. Determine which charts contain data that is linear and explain why.

a.

Week	1	2	3	4
Weight (in lb)	150	147	144	141

b.

Week	1	2	3	4
Weight (in lb)	183	178	174	171

c.

Week	1	2	3	4
Weight (in lb)	160	160	160	160

4. Consider the equation $y = -2x + 5$.

    a. Construct a table of five ordered pairs that satisfy the equation.

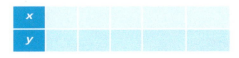

    b. What is the slope of the line represented by the equation?

    c. What is the y-intercept?

    d. What is the x-intercept?

    e. Sketch a graph of the line using each of the following methods:

    Method 1: Plot the five ordered pairs.

    Method 2: Plot one point and use the slope to obtain additional points on the line.

    Method 3: Plot the intercepts.

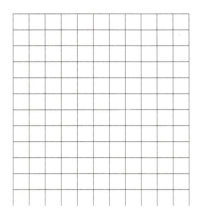 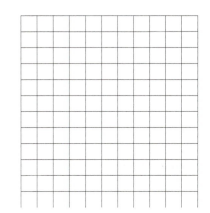 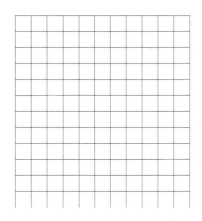

5. a. Determine the slope of the line that goes through the points $(2, -7)$ and $(0, 5)$.

    b. Determine the y-intercept of this line.

    c. What is the equation of the line through these points? Write the equation in function notation.

    d. What is the x-intercept?

6. A car is traveling on a highway. The distance (in miles) from its destination and the time (in hours) is given by the equation $d = 420 - 65t$.

  a. What is the d-intercept of the line?

  b. What is the practical meaning of the d-intercept?

  c. What is the slope of the line represented by the equation?

  d. What is the practical meaning of the slope determined in part c?

  e. What is the t-intercept? Interpret the meaning of this intercept in this situation.

  f. What is the practical domain of this function?

  g. Graph the equation, both by hand and with your graphing calculator, to verify your answers.

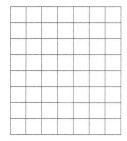

7. The following table gives a jet's height above the ground (in feet) as a function of time (in seconds) as the jet makes a fast descent.

t (in seconds)	0	5	10	15	20	25
h (in feet)	3500	3000	2500	2000	1500	1000

  a. Is this function linear? Explain.

  b. Calculate the slope using the formula $m = \dfrac{\Delta h}{\Delta t}$.

  c. What is the significance of the sign of the slope in part b?

  d. Determine where the graph crosses the h-axis.

e. Write the equation in slope-intercept form.

f. Determine the t-intercept. What is its significance in this situation?

8. Determine the x-intercept of the line whose equation is given.

a. $y = 4x + 2$

b. $y = \dfrac{x}{2} - 3$

9. a. Use your graphing calculator to graph the linear functions defined by the following equations: $y = 2x - 3, y = 2x, y = 2x + 2, y = 2x + 5$. Discuss the similarities and the differences of the graphs.

b. Use your graphing calculator to graph the linear functions defined by the following equations: $y = x - 2, y = 2x - 2, y = -x - 2, y = -2$. Discuss the similarities and the differences in the graphs.

c. Use your graphing calculator to graph the linear functions defined by the following equations: $y = 3x, y = -2x, y = \dfrac{1}{2}x, y = -5x$. Discuss the similarities and differences in the graphs.

10. Determine the zero of each of the following:

a. $f(x) = 2x - 8$

b. $g(x) = \dfrac{2}{3}x + 4$

Activity 2.4

Skateboard Heaven

Objectives

1. Write an equation of a line in general form $Ax + By = C$.

2. Write the slope-intercept form of a linear equation given the general form.

3. Identify lines that have zero or undefined slopes.

4. Determine the equation of a horizontal line.

5. Determine the equation of a vertical line.

Your town has just authorized funding to build a new ramp and pathways for skateboarding. For security, the ramp and pathways must have a rectangular fence surrounding them. The money allocated in the budget for fencing will be enough to purchase 350 feet of fence. The only stipulation is that the width must be between 35 and 60 feet to properly enclose the new ramp. The length will depend on the width you choose. Your task is to determine the length and width of the rectangular region so that you use all of the fencing.

1. **a.** What does the value of 350 represent with regard to the rectangular region?

   b. Using x to represent the width and y to represent the length, write an equation for the perimeter of this rectangular region.

   c. The linear equation in part b should be in the form $Ax + By = C$. Identify the values of the constants A, B, and C in the equation.

Definition

When a linear equation is written in the form $Ax + By = C$, it is said to be in **general form**, where A, B, and C represent constants and A and B cannot both be zero.

Example 1   *Sometimes it is advantageous to rewrite a linear equation given in general form as its equivalent slope-intercept form. For example, consider the equation $3x + 7y = 5$. To write this equation in slope-intercept form, you need to solve for y as follows:*

$$3x + 7y = 5$$
$$\underline{-3x \qquad\quad -3x}$$ Add $-3x$ to each side of the equation.
$$7y = -3x + 5$$
$$\frac{7y}{7} = \frac{-3x}{7} + \frac{5}{7}$$ Divide each side of the equation by 7, the coefficient of y.
$$y = -\frac{3}{7}x + \frac{5}{7}$$

2. a. Rewrite the equation from Problem 1b in slope-intercept form by solving the equation for y in terms of x.

b. What is the slope of the line represented by the equation you wrote in part a? What is the practical meaning of the slope in this situation?

c. What is the vertical intercept of the line represented by the equation you determined in part a? What is the practical meaning of the vertical intercept in this situation?

d. What is the practical domain and range of this function?

3. a. Determine the corresponding y-value for $x = 35$ using $2x + 2y = 350$ (general form of the equation of a line).

b. Determine the corresponding y-value for $x = 35$ using $y = -x + 175$ (slope-intercept form of the equation of a line).

c. Was it more convenient to use the standard form or the slope-intercept form to determine a y-value given $x = 35$? Explain.

d. Complete the table to determine some possible lengths and widths for the rectangular region.

x	35	40	50	60
y				

4. Use function notation to write the length as a function of the width, letting f represent the function.

5. You are working in the purchasing department for an electronics retailer. Your job this month is to stock up on GPS units and tablet PCs. Your supervisor informs you that you have a budget of $12,000 this month. You know that the average wholesale cost of a GPS system is $125 and the average wholesale cost of a tablet PC is $400.

a. If g represents the number of GPS systems you can purchase, write an expression that represents the amount you can spend on GPS systems.

b. If c represents the number of tablet PCs you can purchase, write an expression that represents the amount you can spend on tablet PCs.

c. Write a linear equation in general form that relates the number of GPS systems and tablet PCs you can expect to purchase with your budget.

d. Solve your equation in part c for *c*. In other words, express the number of tablet PCs you can purchase as a function of the number of GPS systems you can purchase.

e. What is the horizontal intercept for this function? What is its meaning in this situation?

f. What is the vertical intercept for this function? What is its meaning in this situation?

g. What is the slope of this function? What is the significance of the slope in this situation?

h. What are the practical domain and the practical range in this situation?

Equation of a Horizontal Line

If either A, the coefficient of x, or B, the coefficient of y, in the general form $Ax + By = C$ equals zero, a special situation arises. You will explore this in the following problems.

6. a. When the ramp for skateboarding in the community park is complete, you and your friends will be able to pay a monthly fee of $12.50 to use the ramp for as many hours as you wish. Complete the table of values below where x is the number of hours per month that each person who pays the fee uses the ramp and y is the total cost per month for each person.

x, Time (in hours)	5	10	15	20
y, Cost (in $)				

b. Plot the data points from the table in part a.

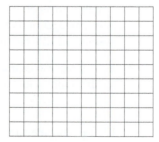

c. The data points in part b are contained on a line. Describe this graph in words.

d. Choose two ordered pairs from part b and determine the slope, *m*, of the line drawn through them.

e. What are the intercepts of the line (if they exist)?

f. Write the equation of this line in slope-intercept form, $y = mx + b$.

g. Write the equation $y = 12.50$ in general form, $Ax + By = C$. What is the value of A?

h. The equation of the horizontal line containing the data points in part b is $y = 12.50$. Does this make sense? Explain.

Definition

A graph in which the output y is a constant or, equivalently, $f(x)$ is a constant, is a **horizontal line**. The equation of a horizontal line is $y = c$ (or $f(x) = c$), where c is some fixed real number. The slope of any horizontal line is zero.

---

7. Determine three ordered pairs that satisfy each of the following equations. Then sketch the graph of each constant function on the same coordinate axes.

a. $y = -2$

b. $f(x) = 1$

c. $g(x) = \dfrac{5}{2}$

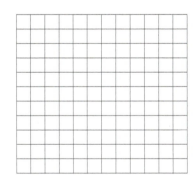

Equation of a Vertical Line

8. You apply for a part-time job at the skateboarding rink to help cover your weekly expenses. The following table gives your weekly salary, x, and corresponding weekly expenses, y, for a typical month.

x, Weekly Salary in Dollars	70	70	70	70
y, Weekly Expenses in Dollars				

a. Sketch a graph of the given data points.

b. The data points in part a are contained on a line. Describe this graph in words.

c. Choose two ordered pairs from the table and determine the slope, m, of the line drawn through them.

d. What are the intercepts of the line (if they exist)?

e. Is y a function of x? Explain using the definition of a function.

f. Do the ordered pairs in the table satisfy the equation $x = 70$? Explain why or why not.

g. Can you graph the equation $x = 70$ using your graphing calculator? Explain.

h. Write the equation of the vertical line in general form, $Ax + By = C$. What is the value of B?

Definition

A graph in which x is a constant is a **vertical line**. The equation of a vertical line is $x = a$, where a is some fixed real number. The slope of a vertical line is **undefined**.

9. Determine three ordered pairs that satisfy each of the following equations, and then sketch each graph.

a. $x = -2$ **b.** $x = 4$ **c.** $x = \dfrac{5}{2}$

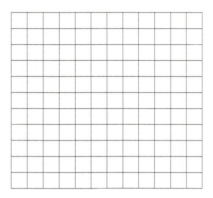

SUMMARY: ACTIVITY 2.4

1. The **general form** of a linear equation is $Ax + By = C$, where A, B, and C are constants and A and B are not both zero.

2. The graph of $y = c$ or $f(x) = c$ is a **horizontal line**. In this case, f is called a constant function. Every point on this line has a y-coordinate equal to c. A horizontal line has slope of zero.

3. The graph of $x = a$ is a **vertical line**. Every point on this line has an x-coordinate equal to a. The slope of a vertical line is undefined.

EXERCISES: ACTIVITY 2.4

1. Write the following linear equations in slope-intercept form. Determine the slope and x-intercept of each line.

a. $2x - y = 3$ **b.** $x + y = -2$

c. $2x - 3y = 7$ **d.** $-x + 2y = 4$

e. $0x + 3y = 12$

2. a. Sketch the graph of the horizontal line through the point $(-2, 3)$.

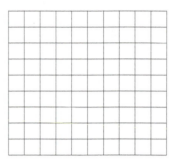

   b. Write the equation of a horizontal line through the point $(-2, 3)$.

   c. What is the slope of the line?

   d. What are the vertical and horizontal intercepts of the line?

   e. Does the graph represent a function? Explain.

3. a. Sketch the graph of the vertical line through $(-2, 3)$.

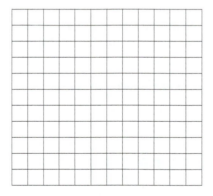

   b. Does the graph represent a function? Explain.

   c. Write the equation of a vertical line through the point $(-2, 3)$.

   d. What is the slope of the line?

4. Explain the difference between a line with a zero slope and a line with an undefined slope.

5. You are retained as a consultant for a major computer company. You receive $2000 per month as a fee no matter how many hours you work.

    a. Using x to represent the number of hours you work each month, write an equation to represent the total amount received from the company each month.

    b. Complete the following table of values.

x, Hours Worked per Month	15	25	35
y, Fee per Month (in $)			

    c. Use your graphing calculator to sketch the graph of this function.

    d. What is the slope of the line? What is the practical meaning of the slope in this situation?

    e. Describe the graph of the function.

6. You are now working in the purchasing department of an appliance retailer. This month you are stocking up on washers and dryers. Your supervisor informs you that your budget this month is $10,000. You know that the average wholesale cost of the washer over the past year has been $250, while the average wholesale cost of a dryer has been $200.

    a. If w represents the number of washers you can purchase, write an expression that represents the amount you can spend on washers.

    b. If d represents the number of dryers you can purchase, write an expression that represents the amount you can spend on dryers.

    c. Write a linear equation in general form that relates the number of washers and dryers you can expect to purchase with your budget.

    d. Solve your equation in part c for d. In other words, express the number of dryers you can expect to purchase as a function of the number of washers you can expect to purchase.

    e. What is the horizontal intercept for this function? What is its practical meaning in this situation?

    f. What is the vertical intercept for this function? What is its practical meaning in this situation?

g. What is the slope of this function? What is its significance in this situation?

h. Use your graphing calculator to graph the function in part d. What part of this graph is relevant to this situation?

i. What are the practical domain and range of this situation?

Activity 2.5

Family of Functions

Objectives

1. Identify the effect of changes in the equation of a line on its graph.

2. Identify the effect of changes in the graph of a line on its equation.

3. Identify the change in the graph and equation of a basic function as a translation, reflection, or vertical stretch or shrink.

A primary objective of this textbook is to help you develop a familiarity with the graphs, equations, and properties of a variety of functions, including linear, quadratic, and exponential. You will group these functions into families and identify the similarities within a family and the differences between families.

The family of functions being investigated in this chapter is linear functions. As you have discovered in previous activities, there are strong relationships between the equation of a line and its corresponding graph. Recognizing such relationships will help you identify patterns in the equations and graphs of other families of functions.

Vertical and Horizontal Shifts Revisited

1. Determine the slope and intercepts of the line having equation $y = 2x$.

2. **a.** Using a graphing calculator, sketch a graph in the standard window of each of the following. What do you observe?

$$y_1 = 2x, y_2 = 2x + 2, y_3 = 2x + 6$$

b. Now, sketch a graph of each of the following. What do you observe?

$$y_1 = 2x, y_2 = 2x - 3, y_3 = 2x - 5$$

c. The equations in parts a and b are of the form $y = 2x + c$, where c is a constant. What is the effect of c on the graph of $y = 2x$?

d. The graph of $y = 2x + 6$ is a vertical shift of the graph of $y = 2x$ upward by 6 units. As a result, which intercept of the graph is changed by exactly 6 units?

e. Does adding a constant c to the output value have any effect on the slope of the line?

3. a. Use a graphing calculator to sketch a graph of $y_1 = 2x$, $y_2 = 2(x + 1)$, $y_3 = 2(x + 4)$, and $y_4 = 2(x + 5.5)$. What do you observe?

b. The equations in part a are of the form $y = 2(x + c)$, where $c > 0$. What is the effect of c on the graph of $y = 2x$?

c. Describe how the graph of $y = 2(x - 3)$ can be obtained from the graph of $y = 2x$. Verify using a graphing calculator.

d. The graph of $y = 2(x + 4)$ is a horizontal shift of the graph $y = 2x$ horizontally to the left 4 units. As a result, which intercept of the graph of $y = 2x$ changed by 4 units?

e. Does adding a constant c to the input x have any effect on the slope of the line $y = 2x$?

Now, look carefully at the graphs of $y_1 = 2x$, $y_2 = 2x + 6$, $y_3 = 2(x + 3)$ from Problems 2 and 3. Note that a horizontal shift 3 units to the left of the graph of $y = 2x$ results in the same graph as a vertical shift upward 6 units. Is this true for all functions? Problem 4 uses the squaring function to demonstrate that this is not generally true.

4. a. Using a graphing calculator, sketch a graph of each of the following in the standard window:

$$y_1 = x^2, y_2 = x^2 + 6, y_3 = (x + 3)^2$$

b. For the squaring function, does a horizontal shift 3 units to the left of the graph result in the same graph as a vertical shift 6 units upward?

Reflections

5. a. Consider the linear function defined by $y = x$. Using a graphing calculator, graph the equation $y = -x$ in the same standard window as $y = x$. What do you observe?

> The graph of $y = -x$ is a **reflection** of the graph of $y = x$ across the x-axis.

    b. When a graph is reflected across the x-axis, the x-value of each point remains the same. What happens to the corresponding y-values?

> In general, if the graph of $y = f(x)$ is reflected across the x-axis, then the equation of the resulting graph is $y = -f(x)$. The graphs of $y = f(x)$ and $y = -f(x)$ are mirror images across the x-axis. The y-values of the graph of $y = f(x)$ change in sign. The x-values remain the same.

6. Reflect each of the following graphs across the x-axis. If a formula is given for the graph, write the equation of the reflected graph. Where possible, verify using a graphing calculator.

    a. $y = 2x$ **b.** $y = x^2$

c.

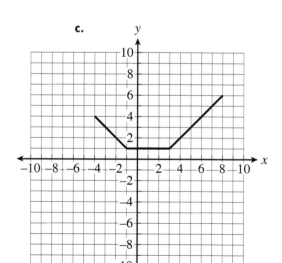

d.

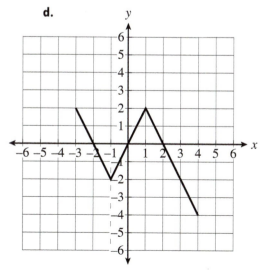

7. a. Determine the slope and intercepts of the graph of $y = f(x) = 3x + 6$.

b. Graph the function defined by $y = 3x + 6$. Verify using a graphing calculator.

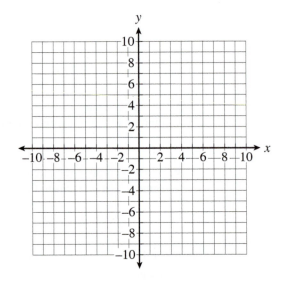

c. On the graph in part b, reflect the graph of $y = 3x + 6$ across the x-axis.

d. Write the equation of the reflected graph in part c.

e. Use a graphing calculator to verify the graph in part d.

f. Determine the slope and intercepts of the reflected graph.

g. Compare the slopes and intercepts of the graph of $y = 3x + 6$ and its reflection across the x-axis.

8. Consider the graph of the function defined by $y = f(x) = 3x + 6$ in Problem 7.

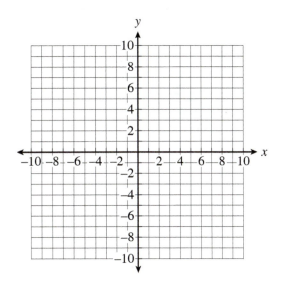

a. Determine an equation for $y_2 = f(-x)$. That is, replace x with $-x$ in the equation for f.

b. Determine the slope and intercepts of the graph of $y_2 = f(-x)$.

c. Sketch a graph of $y = f(-x)$ on the same coordinate axis as $y = f(x)$.

d. How can the graph of $y = f(-x)$ be obtained from the graph of $y = f(x)$?

The graph of $y_2 = f(-x) = -3x + 6$ is a reflection of the graph of $y = f(x) = 3x + 6$ across the y-axis.

e. When a graph is reflected across the y-axis, the y-values of each point remain the same. What happens to the corresponding x-values?

f. Compare the slopes and intercepts of the graph of $f(x) = 3x + 6$ and its reflection across the y-axis $f(-x) = -3x + 6$.

9. Reflect each of the following across the y-axis. If a formula is given for the graph, write the equation of the reflected graph. Where possible, verify using a graphing calculator.

a. $y = 2x - 8$

b. $y = x^2$

c.

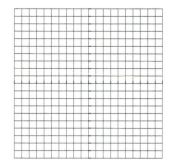

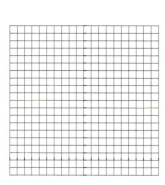

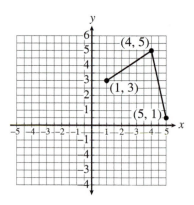

Vertical Stretch and Shrink of a Graph

You have just purchased a stereo system that includes an amplifier. The purpose of the amplifier is to take the signal from a component part (such as the compact disc player) and boost the signal in order to power a set of speakers.

The following graph shows a situation where a sound has been amplified 5 times.

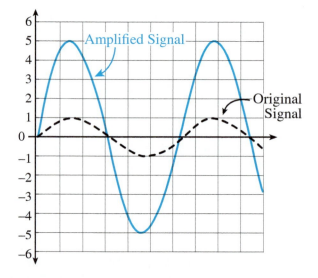

The transformation of the graph of the original signal into the graph of the amplified signal is called a **vertical stretch**. Note that the height of the graph of the amplified signal is 5 times as high as the original signal. The number 5 is called the **stretch factor**. Therefore, if $y = g(x)$ is the equation of the original signal, the equation of the amplified graph is $y = 5\,g(x)$.

> In general, if the graph of $y = f(x)$ is vertically stretched by a factor of a, then the equation of the stretched graph is $y = af(x)$. The number a is called the stretch factor. The y-value of each point in the graph of $y = f(x)$ is multiplied by a factor of a. The x-values remain fixed.

How does a vertical stretch of the graph of a line affect the properties of the graph?

10. a. Using a graphing calculator, sketch a graph of each of the following in a standard window. What do you observe?

$$y_1 = x, y_2 = 2x, y_3 = 5x$$

b. How does the vertical stretch of the graph of $y = x$ affect its graph?

11. a. Determine the slope and intercepts of the graph of $y = g(x) = 2x + 3$.

b. Graph the function defined by $y = 2x + 3$. Verify using a graphing calculator.

c. Write an equation that represents a vertical stretch of the graph of $y = 2x + 3$ by a factor of 3.

d. Graph the line having equation in part c on the same coordinate axis as the graph of $y = 2x + 3$. Verify using a graphing calculator.

e. Determine the slope and intercepts of the vertically stretched graph.

f. Compare the slopes and intercepts with the graph of the original function.

12. a. Using a graphing calculator, sketch a graph of each of the following in a standard window:

$$y_1 = x, y_2 = 0.5x$$

b. Is the graph of $y = 0.5x$ a vertical stretch of the graph of $y = x$? Explain.

In general, if a is a positive constant, then the graph of $y = a f(x)$ is the graph of $y = f(x)$

i. vertically shrunk by a factor of a, if $0 < a < 1$.
ii. vertically stretched by a factor of a, if $a > 1$.

Vertical and horizontal shifts (also called translations), reflections, and vertical stretches and shrinks are all examples of **transformations**. These are ways in which the graph of a given function can be changed or transformed.

13. If the graph of a line is vertically stretched or shrunk, explain why the x-intercept of the line is not affected? See Problem 11f.

14. The graph of $y = -2x + 1$ can be interpreted as a combination of transformations of the graph of $y = x$. Sketch a graph of $y = -2x + 1$ by performing the following changes in the graph of $y = x$: First, a vertical stretch; second, a reflection; last, a vertical shift. Each transformation should be performed in the natural order of operations. Verify using a graphing calculator.

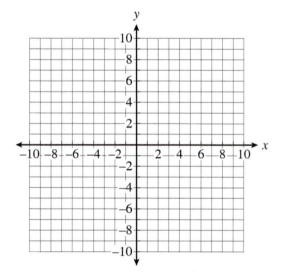

The transformations discussed in this activity can be applied to the graph of any function, as you will see in subsequent activities, as well as some of the exercises that follow.

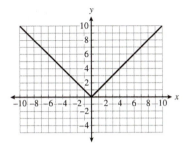

15. Consider the basic absolute value function, $f(x) = |x|$, which is then changed by four consecutive transformations described in parts a–d. To determine the final equation for this function, you will consider one transformation at a time. The graph of $f(x) = |x|$ appears to the left.

a. Graph the basic absolute value function that has been shifted 3 units to the left. Write the equation for the resulting graph.

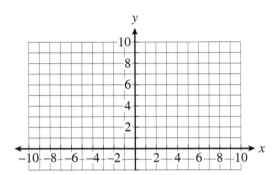

b. Now, vertically stretch the graph in part a by a factor of 2. Graph and write the equation for this function.

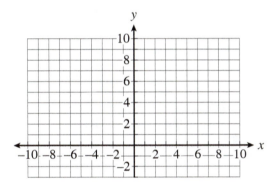

c. Next, reflect the graph in part b across the *x*-axis. Graph and write the equation for this function.

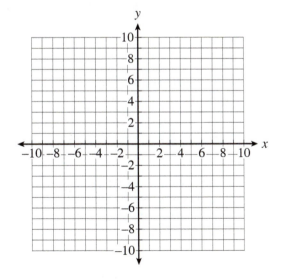

d. Finally, shift the graph in part c upward 5 units. Graph and write the equation for this final graph.

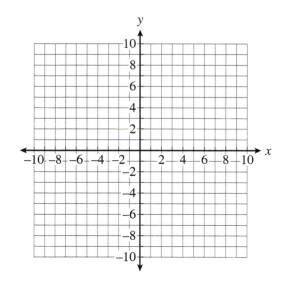

e. Check your final function by graphing the equation on your calculator.

SUMMARY: ACTIVITY 2.5

1. In a **vertical shift** of a graph, a positive constant is added to (upward shift) or subtracted from (downward shift) each output value. As a result, the graph is moved vertically upward or vertically downward.

2. In a **horizontal shift** of a graph, a positive constant is first added to (shift left) or subtracted from (shift right) each input value. As a result, the graph is moved horizontally to the left or horizontally to the right.

3. In general, if a function is defined by $y = f(x)$, and c is a constant, then the graph of

i. $y = f(x) + c$ is the graph of $y = f(x)$ shifted vertically $|c|$ units. If $c > 0$, then the shift is upward. If $c < 0$, then the shift is downward.

ii. $y = f(x + c)$ is the graph of $y = f(x)$ shifted horizontally $|c|$ units. If $c > 0$, then the shift is to the left. If $c < 0$, the shift is to the right.

4. The graph of $y = -x$ is a **reflection** of the graph of $y = x$ across the x-axis.

5. If the graph of $y = f(x)$ is reflected across the x-axis, then the equation of the resulting graph is $y = -f(x)$. The graphs of $y = f(x)$ and $y = -f(x)$ are mirror images across the x-axis. The y-value of each point in the original graph changes sign. The x-value of each point remains the same.

6. If the graph of $y = f(x)$ is reflected across the y-axis, then the equation of the resulting graph is $y = f(-x)$. The graphs of $y = f(x)$ and $y = f(-x)$ are mirror images across the y-axis. The x-value of each point in the original graph changes sign. The y-value of each point remains the same.

7. If the graph of $y = f(x)$ is vertically stretched by a factor of $a > 0$, then the equation of the stretched graph is $y = af(x)$. The number a is called the **stretch factor**. The y-value of each point in the graph of $y = f(x)$ is multiplied by a factor of a. The x-values remain fixed.

8. In general, if a function is defined by $y = f(x)$, and a is a positive constant, then the graph of

  i. $y = af(x)$ is the graph of $y = f(x)$ vertically shrunk by a factor of a, if $0 < |a| < 1$.

  ii. $y = af(x)$ is the graph of $y = f(x)$ vertically stretched by a factor of a, if $a > 1$.

9. Vertical and horizontal shifts (also called translations), reflections, and vertical stretches and shrinks are all examples of **transformations**.

EXERCISES: ACTIVITY 2.5

1. Describe in words how the basic linear function, $f(x) = x$, is transformed by each equation.

  a. $y = 3x$

  b. $y = x - 4$

  c. $y = -x$

  d. $y = \dfrac{1}{2}x + 3$

2. Write the equation for the basic squaring function, $f(x) = x^2$, that has been transformed by

  a. reflecting across the x-axis.

  b. shifting down 6 units.

  c. stretching vertically by a factor of 4.

  d. shifting right 3 units and shrinking vertically by a factor of $\dfrac{1}{3}$.

Check each equation by graphing on your calculator.

3. For each function graphed below, sketch the graph of the new function after it is transformed as described.

  a. Shift left 3 and reflect across x-axis.

  b. Stretch vertically by a factor of 2, then shift down 3 units.

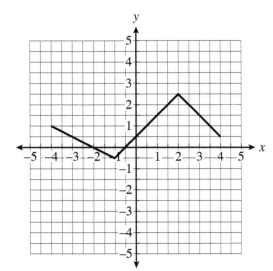

4. For each of the following linear function graphs, describe how $f(x) = x$ is transformed, write the equation for the line, and then check by graphing on your calculator.

a.

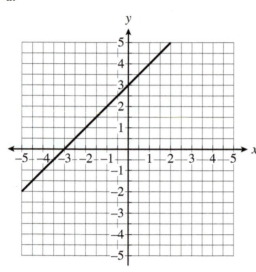

b.

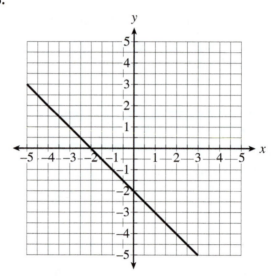

c.

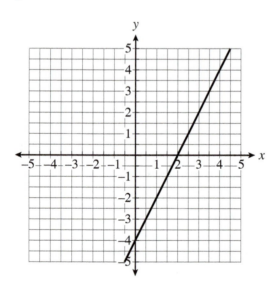

d.

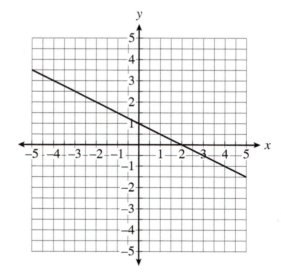

5. If the point $(3, -5)$ is on the graph of $y = f(x)$ and the graph is transformed with a vertical stretch by a factor of 5, then reflected across the x-axis, what must be the coordinates of one point on the new graph?

6. If the point $(-2, 8)$ is on the graph of $y = f(x)$ and the graph is transformed with a vertical shrink by a factor of $\frac{1}{2}$ then shifted down 5 units, what must be the coordinates of one point on the new graph?

7. If the point $(-5, 2)$ is on the graph of $y = f(x)$ and the graph is transformed with a horizontal shift of 3 units to the right, then a reflection across the x-axis, what must be the coordinates of one point on the new graph?

Activity 2.6

Predicting Population

Objectives

1. Write an equation for a linear function given its slope and y-intercept.

2. Write linear functions in slope-intercept form.

3. Determine the relative error in a measurement or prediction using a linear model.

4. Interpret the slope and y-intercept of linear functions in contextual situations.

5. Use the slope-intercept form of linear equations to solve problems.

According to the U.S. Bureau of the Census, the population of the United States was approximately 132 million in 1940 and 151 million in 1950.

For this activity, assume that the rate of change of population with respect to time is a constant value over the decade from 1940 to 1950. This means you are assuming that the relationship between the independent variable time and the dependent variable population is linear.

1. a. Write the given data as ordered pairs of the form $(t, P(t))$, where t is the number of years since 1940 and $P(t)$ is the corresponding population, in millions.

b. Plot the two data points, and draw a straight line through them. Label the horizontal axis from 0 to 25 and the vertical axis from 130 to 180.

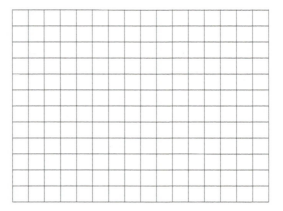

2. a. What is the average rate of change of population from $t = 0$ (1940) to $t = 10$ (1950)?

b. What is the slope of the line connecting the two points in part a? What is the practical meaning of the slope in this situation?

3. What is the P-intercept of this line? What is the practical meaning of this intercept in this situation?

4. a. Use the slope and P-intercept from Problems 2 and 3 to write an equation for the line.

b. Assume that the average rate of change you determined in Problem 2 stays the same through 1960. Use the equation in part a to predict the U.S. population in 1960. Also estimate the population in 1960 from the graph.

The **error** in a prediction is the difference between the observed value (actual value that was measured) and the predicted value. The **relative error** of a prediction is the ratio of the error to the observed value. That is,

$$\text{relative error} = \frac{\text{observed value} - \text{predicted value}}{\text{observed value}} = \frac{\text{error}}{\text{observed value}}.$$

5. a. The relative error is usually reported as a percent. The actual U.S. population in 1960 was approximately 179 million. What is the *relative error* (expressed as a percent) in your prediction?

b. What do you think was the cause of your prediction error?

6. A demographer wants to develop a population model based on more recent data. The U.S. population was approximately 281 million in 2000 and 309 million in 2010.

a. Plot these data points using ordered pairs of the form $(t, P(t))$, where t is the number of years since 2000 (now, $t = 0$ corresponds to 2000). Draw a line through the points.

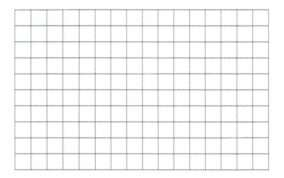

b. Determine the slope of the line in part a. What is the practical meaning of the slope in this situation? How does this slope compare with the slope in Problem 2?

c. In which decade, 1940–1950 or 2000–2010, did the U.S. population increase more rapidly? Explain your answer in terms of slope.

d. Determine the *P*-intercept of the line in part a.

e. Write the equation of the line in part a.

7. a. Use the linear model $P(t) = 2.8t + 281$ developed in Problem 6 to predict the population in 2020. What assumptions do you make about the average rate of change of the population in this prediction?

b. According to the linear model $P(t) = 2.8t + 281$, in what year will the population be 350 million?

8. The following data was obtained from the 2000 and 2010 U.S. censuses for the U.S. population by gender.

Year	2000	2010
m, Number of Males (in millions)	138	151.7
f, Number of Females (in millions)	143	156.9

Data Source: U.S. Census Bureau

a. Write the data for the male population as ordered pairs of the form $(t, m(t))$, where *t* is the number of years since 2000 and $m(t)$ is the corresponding population in millions.

b. Assume that the relationship between time and the U.S. male population is linear from 2000 to 2010. Determine the slope of the line containing the two points in part a. What are its units of measure? Interpret the meaning of the slope in the context of this situation.

c. What is the vertical intercept of the line containing the two points in part a? Interpret the meaning of the vertical intercept in the context of this situation.

d. Write a linear model for $m(t)$ in terms of t.

e. Assume that the relationship between time and the U.S. female population is also linear from 2000 to 2010. Write the data for the female population as ordered pairs of the form $(t, f(t))$, and determine the slope and vertical intercept of the line containing these points.

f. Write a linear model for $f(t)$ in terms of t.

g. The population of which gender group, male or female, is growing more rapidly? Explain.

SUMMARY: ACTIVITY 2.6

1. The **error** in a prediction is the difference between the observed value and the predicted value. The **relative error** in prediction is the ratio of the error to the observed value. That is,

$$\text{relative error} = \frac{\text{observed value} - \text{predicted value}}{\text{observed value}} = \frac{\text{error}}{\text{observed value}}.$$

Relative error is usually reported as a percent.

EXERCISES: ACTIVITY 2.6

1. a. Use the linear model $P(t) = 2.8t + 281$ developed in Problem 6 to predict the U.S. population in the year 2012. What assumptions are you making about the rate of change of the population in this prediction? Recall that t is the number of years since 2000.

b. The actual U.S. population in 2012 was approximately 314 million. What is the relative error in your prediction?

c. How accurate was the prediction?

2. a. According to the U.S. Bureau of the Census, the population of California in 2010 was approximately 37.33 million and was increasing at an approximately constant rate of 350,000 people per year. Let $P(t)$ represent the California population (in millions) and t represent the number of years since 2010. Complete the following table.

t	$P(t)$ (in millions)
0	
1	
2	

b. What information in part a indicates that the California population growth is linear with respect to time? What are the slope and vertical intercept of the graph of the population data?

c. Write a linear function rule for $P(t)$ in terms of t.

d. Use the linear function in part c to estimate the population of California in 2014.

e. Population data from the State of California Department of Finance is used by California state agencies in developing their programs and policies. The Department of Finance estimated that the population of California in 2014 was approximately 38.34 million. Determine the relative error (as a percent) between your prediction in part d and the estimate actually used by the state of California.

f. Use the linear model from part c to predict the population of California in 2020.

g. Do you believe your prediction in part f will be too high, close, or too low? Explain your answer.

3. a. The population of Waco, Texas, was 113.7 thousand in 2000 and 124.8 thousand in 2010. This information is summarized in the accompanying table, where *t* is the number of years since 2000 and $P(t)$ represents the population (in thousands) at a given time *t*.

t	P(t)
0	113.7
10	124.8

Assume that the average rate of change of the population over this ten-year period is constant. Determine this average rate.

b. Plot the two data points on an appropriately scaled and labeled coordinate, and draw a line through them.

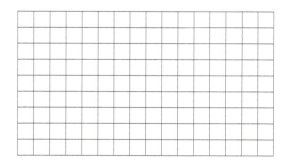

c. Determine the slope and vertical intercept of the line in part b.

d. Write an equation to model Waco's population, $P(t)$ (in thousands), in terms of *t*.

e. Use this linear model to predict Waco's population in 2020.

4. a. According to the U.S. Census Bureau, the population of Detroit, Michigan, was 936.9 thousand in 2000 and 701.5 thousand in 2012. If *t* represents the number of years since 2000, and $P(t)$ represents the population (in thousands) at a given time, *t*, summarize the given information in the accompanying table.

t	P(t)

b. Plot the two data points on appropriately scaled and labeled coordinate axes. Draw a line connecting the points.

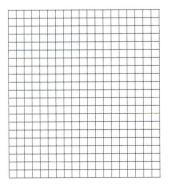

c. What is the slope of the line? What is the practical meaning of the slope in this situation?

d. What is the *P*-intercept of the line? What is the practical meaning of the intercept in this situation?

e. Write an equation to model Detroit's population, $P(t)$, in terms of t.

f. Use this linear model to predict the population of Detroit in 2025.

5. In each part, determine the equation of the line for the given information:

    a. Two points on the line are (0, 4) and (7, 18). Use the points to first determine the slope and *y*-intercept. Then write the equation of the line.

    b. The graph has *y*-intercept (0, 6) and contains the point (2, 1).

6. a. In 2010, the rate of change of the world population was approximately 0.07792 billion per year (or approximately 1 million people every five days). The world population was estimated to be 6.9 billion in 2010. Write an equation to model the population, *P* (in billions), in terms of t, where t is the number of years since 2010 ($t = 0$ corresponds to 2010).

    b. Use the linear model to predict the world population in 2020.

c. According to the model, when will the population of the world be double the 2010 population?

7. According to the National Oceanic and Atmospheric Administration, the Earth's average surface temperature has increased from 56.58°F in 1900 to 58.09°F in 2013. Many experts believe that this increase in global temperature is largely due to the increase in the amount of carbon dioxide in the atmosphere. The pre-industrial concentration of atmospheric carbon dioxide was 289 parts per million (ppm). The following table shows the atmospheric concentration of carbon dioxide in parts per million, measured at the Mauna Loa Observatory in Hawaii.

Year	1960	2012
c, Carbon Dioxide Concentration (parts per million)	317	394

a. Let t represent the number of years since 1960. Determine the average rate of increase in the atmospheric concentration c of carbon dioxide from 1960 to 2012. Interpret the rate in the context of this situation.

b. What is the slope and c-intercept of the line containing the two data points? Interpret each in the context of this situation.

c. Write an equation for c in terms of t.

d. Use the equation in part c to project the atmospheric concentration of carbon dioxide in 2050.

e. Do you have confidence in the projection in part d? Explain your answer.

Activity 2.7

A New Camera

Objectives

1. Write a linear equation given two points, one of which is the vertical intercept.

2. Write a linear equation given two points, neither of which is the vertical (y-) intercept.

3. Use the point-slope form, $y = y_0 + m(x - x_0)$, to write the equation of a non-vertical line.

A camera and lens you want are now available at a cost of $1200. There is a special promotion for students that allows you to make monthly payments for 2 years at 0% interest. You are required to make a 20% down payment, with the balance to be paid in 24 equal monthly payments. This is an opportunity to get the camera you really want, so you investigate to see if you can afford to take advantage of this promotion.

1. a. Determine the amount of the down payment required.

b. Determine the amount owed after the down payment.

c. What are the monthly payments?

A monthly payment of $40 is something you can afford, so you decide to go for it.

At any time during the next 2 years, the total amount, A, paid is a function of the number, n, of payments made.

2. a. What is the amount A when $n = 0$? What does this value of A represent in this situation?

b. What is the total amount paid, A, when $n = 1$ (after the first payment)?

c. What is the total amount paid, A, when $n = 2$?

d. Record the results from parts 2a through 2c to complete the following table.

PAYMENT NUMBER, n	TOTAL AMOUNT PAID, A ($)
0	
1	
2	

e. What is the average rate of change from $n = 0$ to $n = 1$?

f. What is the average rate of change from $n = 1$ to $n = 2$?

g. What type of function is this? Explain.

Summary: To write an equation of a linear function in the slope-intercept form, $y = mx + b$, you need the values of the slope, m, and the vertical intercept, b. The slope is the average rate of change, which is constant for a linear function. The vertical intercept is the output when the input is zero. This output value is often referred to as the **initial value**.

3. a. What is the slope, *m*, of the payment function?

b. What is the initial value, *b*, of the payment function?

c. Write the linear equation that gives the total amount paid, *A,* as a function of the number of payments made, *n.* Note that the equation will have the general form $A = mn + b.$

d. Graph the equation from part c.

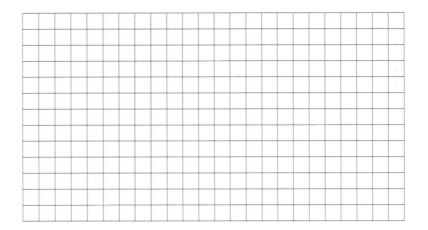

e. What is the practical meaning of the slope in this situation?

f. What is the practical meaning of the vertical intercept in this situation?

Point-Slope Form of an Equation of a Line

In Problem 3 you were given two points, one of which had an input value of zero. You determined the slope and produced a linear equation in slope-intercept form. In the next situation you are given two points, neither of which has an input value of zero, and are asked to determine a linear equation.

4. The cost for a part-time student at the local community college is determined by a fixed activity fee plus a fixed tuition amount per credit. The cost for a student taking 6 credits is $770. The cost for a student taking 9 credits is $1130. The total cost, *C,* is a function of the number of credits taken, *n.*

a. Write two ordered pairs of the form (*n, C*) for the college cost function.

b. Use the result from part a to complete the following table.

NUMBER OF CREDITS, n	TOTAL COST, C ($)
6	
9	

c. Determine the average rate of change for the college cost function.

d. Interpret the average rate of change in the context of the college costs.

e. Is the average rate of change constant? Explain.

The cost, C, is a linear function of the number of credits, n. Therefore, the relationship between C and n can be modeled by an equation written in the slope-intercept form, $C = b + mn$.

5. a. What is the value of the slope, m?

b. Write the equation $C = b + mn$, replacing m with its value.

c. Use the ordered pair (6, 770) from Problem 5a and rewrite the equation in part b by replacing C and n with the appropriate values in the ordered pair.

d. Solve the equation in part c for b.

e. Interpret the value of b in the college cost function.

f. Finally, rewrite the equation $C = b + mn$, replacing b and m with their respective values.

In Problems 4 and 5, you determined a linear equation in the slope-intercept form, using two ordered pairs that satisfy the equation. There is a general method for writing a linear equation from two ordered pairs, based on the method of Problems 4 and 5.

Let (x_0, y_0) be a known point on a line and let (x, y) be any other point on the line. The slope of the line is determined by the two ordered pairs, (x_0, y_0) and (x, y).

$$m = \frac{\Delta y}{\Delta x} = \frac{y - y_0}{x - x_0}, \text{ so}$$

$$y - y_0 = m(x - x_0) \quad \text{or} \quad y = y_0 + m(x - x_0)$$

The last equation, $y = y_0 + m(x - x_0)$, gives a linear equation in **point-slope form**.

Example 1 *Write an equation of the linear function whose graph is a line containing the points $(-1, 3)$ and $(2, 9)$.*

SOLUTION

Step 1. Determine the slope of the line.

$$m = \frac{9 - 3}{2 - (-1)} = \frac{6}{3} = 2$$

Step 2. Choose any point on the line and substitute for x_0 and y_0. Using the point-slope form $y = y_0 + m(x - x_0)$ with $m = 2$ and $(x_0, y_0) = (-1, 3)$, you have

$$y = 3 + 2(x - (-1))$$
$$y = 3 + 2(x + 1)$$
$$y = 3 + 2x + 2, \text{ or } y = 2x + 5$$

Note that in Problem 4 you determined two ordered pairs, $(6, 770)$ and $(9, 1130)$, for the college cost function and then used the two ordered pairs to determine the slope, $m = 120$. Since the input and output variables are n and C, respectively, the point-slope form is $C = C_0 + m(n - n_0)$.

6. a. Use the ordered pair $(6, 770)$ for (n_0, C_0) and $m = 120$ to write the point-slope form of the college cost function.

b. Simplify the right side of the point-slope form of the equation, solve for C, and compare the result to the slope-intercept form in Problem 5.

c. Repeat Problems 6a and 6b using the ordered pair $(9, 1130)$ for (n_0, C_0).

Problem 6 demonstrates that you obtain the same equation of the line regardless of which point you use for (n_0, C_0).

7. Follow the steps below to write the point-slope form of the equation of the line containing the points $(2, 4)$ and $(5, -2)$.

a. Use the given points to determine the slope, m.

b. Use the point $(2, 4)$ for (x_0, y_0) to write the point-slope form of the equation.

8. The basal energy requirement is the daily number of calories that a person needs to maintain basic life processes. For a 20-year-old male who weighs 75 kilograms and is 190.5 centimeters tall, the basal energy requirement is 1952 calories. If his weight increases to 95 kilograms, he will require 2226 calories.

The given information is summarized in the following table.

20-YEAR-OLD MALE, 190.5 CENTIMETERS TALL		
w, Weight (kg)	75	95
B, Basal Energy Requirement (cals)	1952	2226

a. Assume that the basal energy requirement, *B*, is a linear function of weight, *w* for a 20-year-old male who is 190.5 cm tall. Determine the slope of the line containing the two points indicated in the above table.

b. What is the practical meaning of the slope in the context of this situation?

c. Using the point-slope form, determine an equation that expresses *B* in terms of *w* for a 20-year-old, 190.5 cm tall male.

d. Does the *B*-intercept have any practical meaning in this situation? Determine the practical domain of the basal energy function.

SUMMARY: ACTIVITY 2.7

1. To determine the equation of a line when two points on the line are known:

Step 1. Determine the **slope** of the line: $m = \dfrac{\Delta y}{\Delta x}$.

Step 2. If the *x*-value of one of the points is zero, then *b* is the *y*-value of this point (*y*-intercept) and you can write the equation in **slope-intercept form**, $y = mx + b$.

Step 3. If neither point has an *x*-value of zero, then choose one of the points as (x_0, y_0) and write the equation in **point-slope form**, $y - y_0 = m(x - x_0)$ or $y = y_0 + m(x - x_0)$. Then solve for *y*.

EXERCISES: ACTIVITY 2.7

1. Federal income tax paid by an individual single taxpayer is a function of taxable income. The following table represents the federal tax in a recent year for various taxable incomes.

i, Taxable Income ($)	15,000	16,500	18,000	19,500	21,000	22,500	24,000
t, Federal Tax ($)	1889	2114	2339	2564	2789	3014	3239

a. Plot the data points, with taxable income i as input and the federal tax t as output, on an appropriately scaled and labeled coordinate axis. Explain why the relationship is linear.

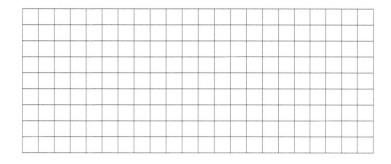

b. Determine the slope of the line. What is the practical meaning of the slope?

c. Use the point-slope form to write an equation to model this situation. Use the variable i to represent the taxable income and the variable t to represent the federal tax owed.

d. What is the t-intercept? Does it make sense?

e. Use the equation from part c to determine the federal tax owed by a college student having a taxable income of $8600 in this particular year.

f. Use the equation from part c to determine the taxable income of a single person who paid $1686 in federal taxes.

In Exercises 2–8, determine the equation of the line that has the given slope and passes through the given point. Then sketch a graph of the line.

2. $m = 3$, through the point $(2, 6)$

3. $m = -1$, through the point $(5, 0)$

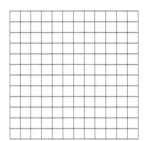

4. $m = 7$, through the point $(-3, -5)$

5. $m = 0.5$, through the point $(8, 0.5)$

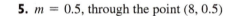

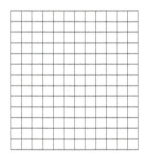
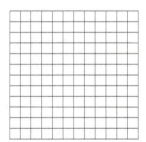

6. $m = 0$, through the point $(5, 2)$

7. $m = -4.2$, through the point $(-4, 6.8)$

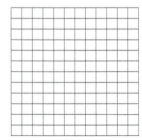
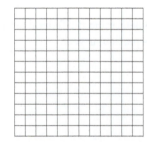

8. $m = -\dfrac{2}{7}$, through the point $\left(5, \dfrac{4}{7}\right)$

In Exercises 9–14, determine the equation of the line that passes through the given points. Use the table feature of a graphing calculator to confirm that the coordinates of both points satisfy your equation.

9. $(2, 6)$ and $(4, 16)$

10. $(-5, 10)$ and $(5, -10)$

11. $(3, 18)$ and $(8, 33)$

12. $(0, 6)$ and $(-10, 0)$

13. $(10, 2)$ and $(-3, 2)$

14. $(3.5, 8.2)$ and $(2, 7.3)$

15. You own a kayak company and open only during the summer months. You discover that if you sell a certain type of kayak for $400, your sales per day average $5200. If you raise the price of the kayak to $450, the sales fall to approximately $3600 per day.

    a. Assume that the sales per day is a function of the price of the kayak. Write two ordered pairs that describe this situation.

    b. Assume that the sales per day is a linear function of the price of the kayak. Write an equation describing this relationship.

    c. You cannot make enough profit if you sell the kayak for less than $375. What would be the average sales per day if you change the price to $375?

16. Your architect will charge you a flat fee of $5000 for the plans for your home. The cost of your home is estimated by the square footage. The following table gives the total estimated cost of your home, c, including the architect's fees, as a function of the square footage, h. Assume that the total cost is a linear function of square footage.

TOTAL SQUARE FEET, h	TOTAL COST c ($)
0	5000
3000	380,000

a. What is the vertical intercept of the line containing these points? Explain how you determined this intercept.

b. Using the data in the table and the formula $m = \dfrac{\Delta c}{\Delta h}$, calculate the slope.

What is the practical meaning of the slope in this situation?

c. Use the results from parts a and b to write the equation of the line in slope-intercept form that can be used to determine the cost for any given square footage.

d. You decide that you cannot afford a house with 3000 square feet. Using the equation from part c, determine the cost of your home if you decrease its size to 2500 square feet.

Activity 2.8

Body Fat Percentage

Objectives

1. Construct scatterplots from sets of data pairs.

2. Recognize when patterns of points in a scatterplot have a linear form.

3. Recognize when the pattern in the scatterplot shows that the two variables are positively related or negatively related.

4. Identify individual data points, called outliers, that fall outside the general pattern of the other data.

5. Estimate and draw a line of best fit through a set of points in a scatterplot.

6. Determine residuals between the actual value and the predicted value for each point in the data set.

7. Use a graphing calculator to determine a line of best fit by the least-squares method.

8. Measure the strength of the correlation (association) by a correlation coefficient.

9. Recognize that a strong correlation does not necessarily imply a linear, or a cause-and-effect, relationship.

Your body fat percentage is simply the percentage of fat your body contains. If you weigh 150 pounds and have a 10% body fat, your body consists of 15 pounds of fat and 135 pounds of lean body mass (bone, muscle, organs, tissue, blood, etc.).

A certain amount of fat is essential to bodily functions. Fat regulates body temperature, cushions and insulates organs and tissues, and is the main form of the body's energy storage. The American Council on Exercise has established the following categories for females and males based upon body fat percentage.

CLASSIFICATION	FEMALE (% fat)	MALE (% fat)
Essential Fat	10–12%	2–4%
Athletes	14–20%	6–13%
Fitness	21–24%	14–17%
Acceptable	25–31%	18–25%
Obese	32% plus	25% plus

Hydrostatic measurement (body composition analysis through immersion of a person's body in water) is considered the most accurate method of determining body fat percentage. However, in most clinical or applied situations, underwater weighing is not practical.

A group of researchers is searching for alternative methods to measure body fat percentage. They first investigate if there is an association between body fat percentage and a person's weight. The body fat percentage of 19 male subjects is accurately determined using hydrostatic weighing method. Then, each subject is weighed, using a traditional scale. The results are recorded in the following table.

w, Weight (pounds)	175	181	200	159	195	192	205	173	187	188	240	175	168	246	160	215	155	146	219
y, Body Fat %	16	21	25	6	22	30	32	21	25	19	15	22	9	38	14	27	12	10	30

I. Plot the data points as ordered pairs of the form (w, y) on an appropriately scaled and labeled coordinate axis.

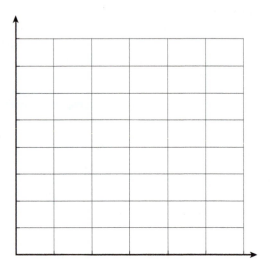

In this situation, you are interested in determining if body fat percentage is related to body weight. The **scatterplot** is an ideal tool to identify patterns in the relationship between two variables.

2. a. Does there appear to be a linear relationship between weight w and body fat percentage y?

b. As body weight increases, what is the general trend in the values of the body fat percentage?

c. Identify any data points that fall outside the general trend or pattern of the other data points. Such data points are called **outliers**. Explain.

3. a. Use a straightedge to draw a single line through the points (175, 16) and (200, 25). A line that lies within the "middle" of a linear pattern of points in a scatterplot is sometimes called a **line of best fit**.

b. Determine the slope of the line in part a.

c. Determine the equation of the line.

d. Use the linear model in part c to predict the body fat percentage of a male weighing 192 pounds. How does the predicted value compare to the actual value of 30%?

Residuals

In almost every situation in which predictions are made from data, it is important to investigate the residuals. **Residuals** are the difference between the actual value and the predicted value for each point in your data. Symbolically stated, you have

$$E = y - \hat{y},$$

where E represents the residual, y is the actual value of the dependent variable, and $\hat{y}$, read "y-hat," is the predicted value for a given x.

Residuals represent the error between the actual value and the value predicted by the model.

4. a. Determine the residual in Problem 3d.

b. If a residual is positive, what does that tell you about the predicted value?

5. a. Determine the residual for a body weight of 168 pounds.

b. If the residual is negative, what does that tell you about the prediction?

c. What if the residual error is zero? What must be true about the predicted value of body fat percentage?

6. The residuals can be used to determine how well a particular line fits the data.

a. Determine all the residuals for the body fat versus weight data using the linear model from Problem 3. Use your graphing calculator to create a table to easily display the values. On the TI-83/84 Plus, enter the body fat percentage data by first pressing (STAT) and choosing (EDIT). Then, enter the weight data (w-values) in L_1 and the body fat % data in L_2. Your equation can be entered in L_3, and then using $L_4 = L_2 - L_3$, the residuals will appear in L_4. (See Appendix A for TI-83/84 Plus procedures.)

Appendix

Record the predicted body fat % values and corresponding residuals in the following table.

WEIGHT (pounds) w	BODY FAT % y	PREDICTED BODY FAT % $\hat{y}$	RESIDUAL $E = y - \hat{y}$	E^2
175	16			
181	21			
200	25			
159	6			
195	22			
192	30			
205	32			
173	21			
187	25			
188	19			
240	15			
175	22			
168	9			
246	38			
160	14			
215	27			
155	12			
146	10			
219	30			

b. If the line is a good estimate of the linear pattern, there should be approximately half the points above the line and half below. Is that the case with your line?

c. The sum of all the residuals should be close to zero, if the line is a reasonably good fit. What is the sum of the residuals in this case? (Use sum(L_4) on TI-83/84 Plus.)

Regression Line

There is an established method for determining the line of best fit. The method depends on calculus to find the line that will have the smallest possible sum of all the squares of the residuals. The line that is determined by this method is called the **least squares regression line**.

7. a. Determine the squares of the residuals in Problem 6. You can use the TI-83/84 Plus again by entering the list $L_5 = L_4{}^2$. Record your results in the last column of the table in Problem 6. Then use sum (L_5) to determine the sums of the squares of the residuals.

Appendix

b. The regression line can be determined with your calculator by pressing (STAT), choosing CALC and option 4: LinReg($ax + b$), rounding the coefficients to four decimal places. Your screen should appear as follows:

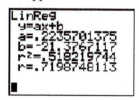

c. When you use the calculator to find the regression line, the residuals have already been calculated and stored for you. They are in a list, which can be found under LIST (2nd (STAT)) then scroll down under the NAMES menu until you see RESID. Determine the sum of the residuals and the sum of the squares of the residuals for the regression line model. Then compare with the results in Problems 6c and 7a.

Note that the sum of the squares of the residuals for any other linear model will be greater than what you found for the regression line. Hence the name, least squares regression line.

Appendix

8. a. Produce the scatterplot for the body fat versus weight data, along with the graph of the regression line and the original line from Problem 3. (See Appendix A for TI-83/84 Plus procedures.) Your screen should appear as follows:

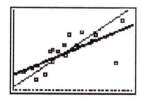

b. Describe how the two lines compare. What do you think might be the cause of any significant differences?

c. Use the regression line model to predict the body fat percentage for a male who weighs 225 pounds.

d. Interpret the practical meaning of the slope of the regression line in this context.

Linear Correlation

When a linear pattern is evident in a scatterplot, there is said to be a linear correlation between the two variables. (When the word correlation appears by itself, it usually means **linear** correlation.) Your calculator window in Problem 7b should have included a value for r, called the **correlation coefficient** as shown below. (If it didn't, see Appendix A to enter DiagnosticOn.)

$$r = 0.7198748113 \text{ or } r = 0.72$$

Note that the correlation coefficient is generally rounded to the nearest hundredth.

The value of the correlation coefficient indicates how strong a linear relationship exists between the two variables under consideration. The value of r ranges between -1 and 1. When the value of r is close to zero, you would conclude that there is little or no linear correlation. The closer r is to either 1 or -1 the stronger the linear correlation between the two variables.

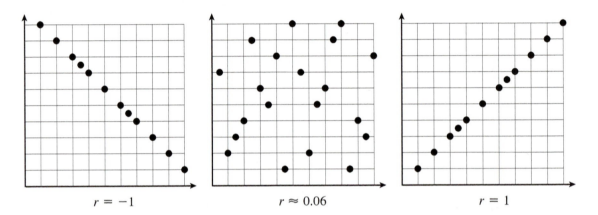

$r = -1$ $r \approx 0.06$ $r = 1$

If $r = 1$, then all the data points lie exactly on a straight line with positive slope. In this case there is a perfect positive correlation. If $r = -1$, then all the decreasing data points lie exactly on a straight line with negative slope. In this case there is a perfect negative correlation. The strength of the correlation is generally best described by both the correlation coefficient and a visual interpretation of the scatterplot. The following scatterplots with their respective correlation coefficients should help make this clear.

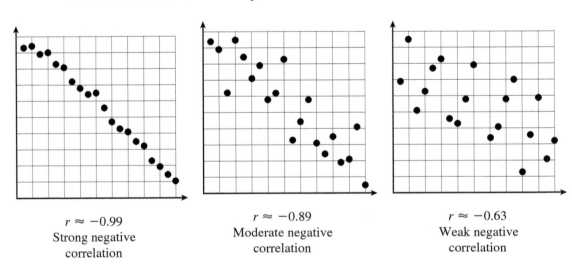

$r \approx -0.99$ $r \approx -0.89$ $r \approx -0.63$
Strong negative   Moderate negative   Weak negative
correlation   correlation   correlation

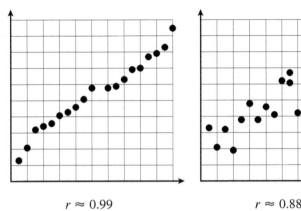

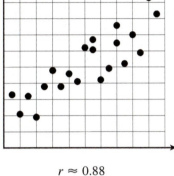

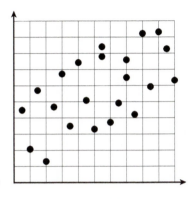

$r \approx 0.99$
Strong positive
correlation

$r \approx 0.88$
Moderate positive
correlation

$r \approx 0.57$
Weak positive
correlation

9. For each of the following, match the correlation coefficient r with the corresponding scatterplot of data points.

a. $r = 0.84$ **b.** $r = 0.53$ **c.** $r = -0.45$

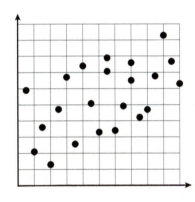

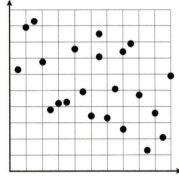

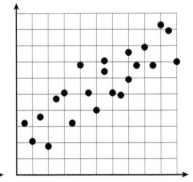

Be careful. The correlation coefficient can be dramatically affected by outlying values, as demonstrated in Problem 10.

10. Recall from Problem 2c that the outlier for the body fat versus weight data was the point (240,15). The correlation coefficient from Problem 7b was $r = 0.72$.

  a. Recalculate the regression line and correlation coefficient when the outlier is removed from the data. What significant differences do you observe?

  b. Use your new regression line to predict the body fat percentage of a 225-pound male, and compare with your result from Problem 8c, when you used the old regression line model.

Cause-and-Effect Relationships

A strong correlation between two variables does not necessarily mean that a cause-and-effect relationship exists between the two variables. For example, there is a strong positive correlation between height and weight. However, an increase in height does not necessarily cause an increase in weight. There are many other variables that could cause the increase in weight such as gender, age, or body type. Such variables are often called **lurking variables**, hidden in the background and not explicitly measured.

11. a. Suppose your home is heated with natural gas. There is a strong negative correlation between the amount of natural gas usage and the outside temperature. Is there a cause-and-effect relationship between these two variables? Explain.

b. In a 2002 publication, *The Natural History of the Rich,* the author claimed a positive correlation between people with two cars and a longer life span. Is there a cause-and-effect relationship between the number of cars you own and length of life? Explain.

SUMMARY: ACTIVITY 2.8

1. A **scatterplot** is a graph of individual data points, useful in determining visually how two variables may be related.

2. An **outlier** is a data point that lies far outside the general pattern of points in a scatterplot.

3. A line that lies within the middle of a linear pattern of data points in a scatterplot is sometimes called a **line of best fit**.

4. A **correlation coefficient**, r, measures how strongly two related variables follow a linear pattern. The value of r ranges between -1 and 1. When the value of r is close to zero, you would conclude that there is little or no linear correlation. The closer r is to either 1 or -1, the stronger the linear correlation between the two variables. If $r < 0$, a negative correlation, then the linear pattern follows a negative slope. If $r > 0$, a positive correlation, then the linear pattern follows a positive slope.

5. A **residual** is the vertical distance between a data point and a best-fit line. In other words, it is the difference between an actual data value and a value predicted by the linear model, $E = y - \hat{y}$, where E represents the residual, y is the actual value of the dependent variable, and $\hat{y}$, read "y-hat," is the predicted value for a given x.

6. A **regression line** is considered to be the best fit line for paired data. It is determined by the least squares method, meaning the sum of the squares of the residuals is as small as possible for the regression line, when compared to any other line.

EXERCISES: ACTIVITY 2.8

1. For each of the following, determine if there is a positive or negative correlation between the independent variable x and the dependent variable y.

a. $x =$ age of a child, $y =$ height of a child

b. $x =$ age of a car, $y =$ resale value of the car

2. During the spring and summer, a concession stand at a community Little League baseball field sells soft drinks and other refreshments. To prepare for the season, the concession owner refers to the previous year's files, in which he had recorded the daily soft drink sales (in gallons) and the average daily temperature (in Fahrenheit degrees). The data is shown in the table.

t, Temperature (°F)	52	55	61	66	72	75	77	84	90	94	97
g, Soft Drink Sales (gal)	35	42	50	53	66	68	72	80	84	91	95

a. Plot the data points as ordered pairs of the form (t, g).

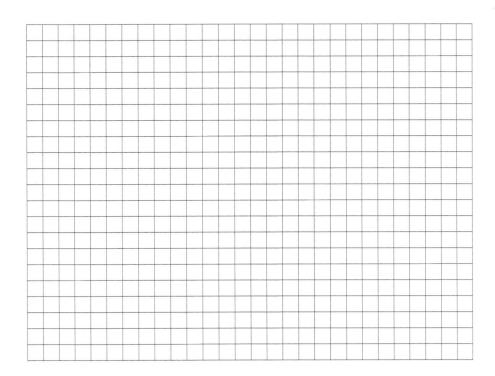

b. Does there appear to be a linear trend between the temperature, t, and the soft drink sales, g? Is there an exact linear fit?

c. Use a straightedge to draw a line of best fit through the points (84, 80) and (66, 53).

d. Use the coordinates of two points in part c to determine the slope of your line.

e. What is the equation of this line?

f. Complete the following table. Compute the sum of the squares of the residuals, as done in Problem 7c.

t, TEMPERATURE	ACTUAL SALES	g, PREDICTED SALES	RESIDUALS
52	35		
55	42		
61	50		
66	53		
72	66		
75	68		
77	72		
84	80		
90	84		
94	91		
97	95		

3. a. Use your graphing calculator's statistics menu to determine the equation for the regression line in the situation in Exercise 2.

b. Determine the residuals for the regression line in part a. Proceed as you did in Exercise 2f.

t, TEMPERATURE	ACTUAL SALES	G(t), PREDICTED SALES	RESIDUALS
52	35		
55	42		
61	50		
66	53		
72	66		
75	68		
77	72		
84	80		
90	84		
94	91		
97	95		

c. Compare the residual errors of your line of best fit in Problem 2f with the error of the least-squares regression line in part b.

d. Determine the correlation coefficient.

4. The research group is trying to determine a better predictor of body fat percentage. In addition to the weight of each of the 19 male subjects, the waist size of each subject was measured and recorded.

w, Waist (inches)	32	36	38	33	39	40	41	35	38	33	40	36	32	44	33	41	34	34	44
y, Body Fat %	16	21	25	6	22	30	32	21	25	19	15	22	9	38	14	27	12	10	30

a. Plot the data points as ordered pairs of the form (w, y) on an appropriately scaled and labeled coordinate axis.

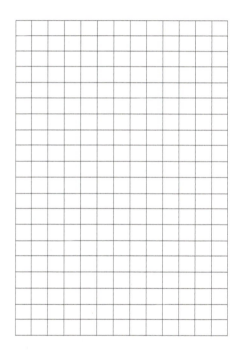

b. Describe any patterns you observe in the scatterplot.

c. Use the linear regression feature of a graphing calculator to determine the equation of the regression line in this situation. What does the slope of this line tell you?

d. Determine the value of the correlation coefficient in this situation. What does it tell you about the relationship between waist size and body fat percentage?

e. Is waist size or body weight a more reliable predictor of body fat percentage? Explain.

5. A correlation coefficient value of $r = 0.85$ suggests a moderately strong linear association between two variables. However, as demonstrated in the following exercise, a line may not be the best model or representation of the data.

   a. Produce a scatterplot for the following table of values. How would you describe the association between x and y?

x	0.5	0.5	1	1	1.5	2	2.5	3	3.5	4	4.5	5	5	5.5	5.5	6	6.5	7	7	7.5	8
y	3.2	2.9	2.5	2.1	2.6	1.8	2.6	2.2	3.2	2.9	3.1	3.6	4	4.1	3.1	3.9	4.9	4.6	5.3	6.3	7.4

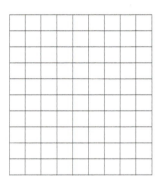

   b. Determine the regression line equation and correlation coefficient. Does the coefficient confirm what you said in part a?

   c. Graph the regression line with the scatterplot. Do you think a different graph might better describe the pattern seen in the scatterplot?

Activity 2.9

College Tuition

Objectives

1. Use a graphing calculator to determine the equation of a line of best fit by the least-squares method.

2. Use the regression equation to interpolate and extrapolate.

The following table contains the average tuition and room and board for full-time matriculated students at four-year colleges as published by the U.S. Department of Education, National Center for Education Statistics.

COLLEGE COSTS						
ENDING YEAR	**2001**	**2003**	**2005**	**2007**	**2009**	**2011**
Years Since 2001	0	2	4	6	8	10
Average Cost	15,996	17,175	18,666	19,611	20,606	21,657

(Ending year means the year in which the school year ended, i.e., the 2008–2009 school year is represented by 2009.)

1. Let t, the number of years since 2001, represent the input variable and c, the average cost, the output variable. Determine an appropriate scale, and plot the data points from the accompanying table.

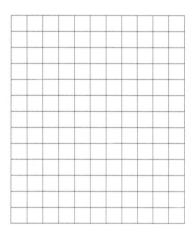

2. **a.** Does there appear to be a linear relationship between the years since 2001 and the average cost of tuition and room and board?

   b. As the years since 2001 increase, what is the general trend in the college costs?

Appendix

3. **a.** Enter the college cost data into your calculator by pressing (STAT) and choosing (EDIT). If necessary, see Appendix A for help operating the TI 83/84 Plus graphing calculator. Your screens should appear as follows.

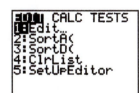

   b. Enter the year data in L1 and the tuition data in L2.

c. Determine the regression line by pressing ⬚STAT⬚ and choosing ⬚CALC⬚ and option 4:LinReg(ax + b). Your screens should appear as follows.

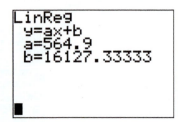

d. Write the result, rounding the coefficients to two decimal places.

4. Now use the regression equation you determined in Problem 3 to model the college costs data. Use the Table feature on your graphing calculator to determine the model's predicted output in the following table.

INPUT, t	ACTUAL OUTPUT, c	MODEL'S PREDICTED OUTPUT
0	15,996	
2	17,175	
4	18,666	
6	19,611	
8	20,606	
10	21,657	

5. a. Use the regression equation to predict the average tuition and room and board at 4-year colleges in the year ending in 2004.

Definition

Using a regression model to predict an output within the boundaries of the input values of the given data is called **interpolation**. Using a regression model to predict an output outside the boundaries of the input values of the given data is called **extrapolation**. In general, interpolation is more reliable than extrapolation.

---

b. Use the regression equation to predict the average tuition and room and board at 4-year colleges in the years ending in 2015 and 2020.

c. Which prediction do you believe would be more accurate? Explain.

d. Graph the regression equation on a graphing calculator. Use the graph to estimate the year in which average tuition and room and board will be at least $30,000.

6. Over the past quarter century, the number of bachelor's degrees conferred by degree-granting institutions has steadily increased. The following table contains data on the number of bachelor's degrees (in thousands) earned by women in a given year. The input *t* represents the number of years since 1990.

	YEAR						
	1990	1995	2000	2005	2007	2008	2009
t, Number of Years Since 1990	0	5	10	15	17	18	19
f(t), Number of Degrees in Thousands	558	634	708	838	886	906	928

a. Sketch a scatterplot of the given data.

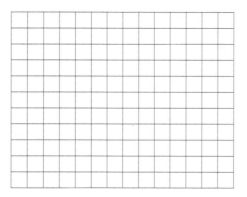

b. Enter the data into your graphing calculator, and determine a linear regression model to represent the data. Write the result here.

c. What is the slope of the line? What is the practical meaning of the slope in this situation?

d. Use the regression model to predict the number of bachelor's degrees that will be earned by women in the year 2020.

SUMMARY: ACTIVITY 2.9

1. The **linear regression equation** is the linear equation that "best fits" a set of data.

2. The **regression line** is a mathematical model for the data.

3. **Interpolation** is the process of using a regression equation to predict a value of output for an input value that lies within the boundaries of the given data.

4. **Extrapolation** is the process of using a regression equation to predict a value of output for an input value that lies outside the boundaries of the given data.

EXERCISES: ACTIVITY 2.9

1. a. Plot the following data.

x	0	3	6	9	12	15	18
f(x)	−0.8	6.3	13.1	19.6	27.0	33.5	40.8

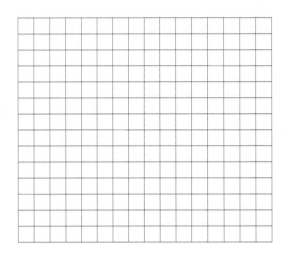

b. With a straightedge, draw a line that you think looks like the line of best fit. Does this data appear to be linear?

c. Use your graphing calculator to determine the equation of the regression line. Write the result below.

d. Use your equation from part c to predict the value of y when $x = 10$.

e. Use your equation from part c to predict the value of y when $x = 25$.

f. Which prediction, $f(10)$ or $f(25)$, would be more accurate? Explain.

2. Worldwide PC shipments have stayed relatively flat from 2009 to 2012, while tablet sales sky-rocketed. The following table gives worldwide PC shipments (in millions of units) from 2009 to 2012 where $t = 0$ corresponds to the year 2009. People want mobility, so Microsoft-powered PC products are competing against enormously popular smartphones and tablets.

t (in years since 2009)	0	1	2	3
c (in millions of units)	33.5	37.1	35.4	36.6

a. Plot the data.

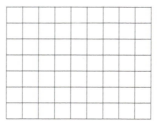

b. Using your graphing calculator, determine the equation of the regression line.

c. What is the slope of the line? What is the practical meaning of slope in this situation?

d. Use the regression line to determine the number of PCs shipped in 2011 ($t = 2$). Compare your result with the actual value of 35.4 in 2011.

e. Use the regression equation to predict the number of PC shipments in 2015.

f. What is the process called to make the prediction in part e?

g. What is the correlation coefficient for this data? Discuss the meaning of this coefficient.

3. In 1966, the U.S. Surgeon General's health warnings began appearing on cigarette packages. The following data seems to demonstrate that public awareness of the health hazards of smoking has had some effect on consumption of cigarettes.

	YEAR, t							
	1997	1999	2001	2003	2005	2007	2009	2011
% of Total Population 18 and Older Who Smoke, P	24.7	23.5	22.8	21.6	20.9	19.8	20.6	19.0

Data Source: U.S. National Center for Health Statistics.

a. Plot the given data as ordered pairs of the form (t, P), where t is the number of years since 1997 and P is the percent of the total population (18 and older) who smoke. Appropriately scale and label the coordinate axes.

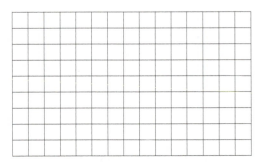

b. Determine the equation of the regression line that best represents the data.

c. Use the equation to predict the percent of the total population 18 and older that will smoke in 2020.

4. Per capita personal income is calculated by taking the total income of a population and dividing it by the total number of people in that population. The per capita income of the United States as reported by the Department of Commerce for the given years is located in the following table.

Year	1997	1999	2001	2003	2005	2007	2009	2011
Per Capita Income	$25,924	$28,546	$30,574	$31,484	$34,757	$38,611	$38,846	$41,633

a. Plot the data points on appropriately scaled and labeled axes. Let x represent the number of years since 1997.

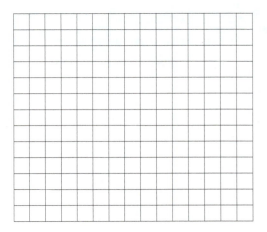

b. Use your graphing calculator to determine the equation of the regression line.

c. What is the slope of the line in part b? What is the practical meaning of the slope?

d. Use the linear model from part b to determine when the per capita income will reach $50,000. How confident are you in the prediction? Explain your answer.

Collecting and Analyzing Data

5. One measure of how well the American economy is doing is the number of new houses built in a given year. Your economics teacher has asked the class to investigate trends in new home construction in recent years and to predict the number of new homes that will be constructed in 2020.

Use the U.S. Bureau of the Census[*] as a resource to obtain the latest data available on new home construction from 2010 to 2014. Work with a group and be prepared to give an oral presentation of your findings to the class. The presentation should include visual displays showing any tables, scatterplots, regression equations, and graphs used in the problem-solving process.

Each member of the group should submit a report of the group's findings. The report should contain the following information:

a. The source(s) of the data.

b. A description of the dependent and independent variables.

c. A table and scatterplot of the data.

d. Identification of any patterns in the scatterplot.

e. A linear regression equation and correlation coefficient for the data.

f. A prediction of the number of new homes to be constructed in 2020 and a description of the level of confidence in this prediction.

g. State a conclusion of the data analysis process.

---

[*]The U.S. Bureau of the Census gathers large amounts of data about the United States and its population. This information is published in the *Statistical Abstract of the United States*. The publication is available on the Internet as well as in print form at most libraries. This is an excellent source of data on a variety of topics.

Project 2.10

Measuring Up

Objectives

1. Collect and organize data in a table.

2. Plot data in a scatterplot.

3. Recognize linear patterns in paired data.

4. Determine a linear regression equation.

Variables arise in many common measurements. Your height is one measurement that has probably been recorded frequently from the day you were born. In this project, you are asked to pair up and make the following body measurements: height (h); arm span (a), the distance between the tips of your two middle fingers with arms outstretched; wrist circumference (w); foot length (f); and neck circumference (n). For consistency, measure the lengths in inches.

1. Gather the data for your entire class, and record it in the following table:

Inch by Inch

STUDENT	HEIGHT (h)	ARM SPAN (a)	WRIST (w)	FOOT (f)	NECK (n)

2. What are some relationships you can identify, based on a visual inspection of the data? For example, how do the heights relate to the arm spans?

3. Construct a scatterplot for heights versus arm span on the grid below, carefully labeling the axes and marking the scales. Does the scatterplot confirm what you may have guessed in Problem 2?

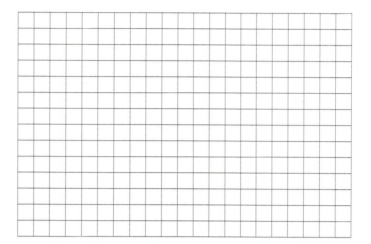

4. Use your calculator to create scatterplots for the following pairs of data, and state whether or not there appears to be a linear relationship. Comment on how the scatterplots either confirm or go against the observations you made in Problem 2.

VARIABLES	LINEAR RELATIONSHIP
Height versus Foot Length	
Arm Span versus Wrist	
Foot Length versus	
Neck Circumference	

5. Determine a linear regression equation to represent the relationship between the two variables in Problems 3 and 4 that show the strongest linear pattern.

Predicting Height from Bone Length

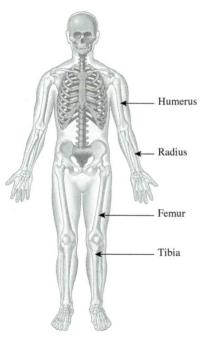

An anthropologist studies human physical traits, place of origin, social structure, and culture. Anthropologists are often searching for the remains of people who lived many years ago. A forensic scientist studies the evidence from a crime scene in order to help solve a crime. Both of these groups of scientists use various characteristics and measurements of the human skeletal remains to help determine physical traits such as height, as well as racial and gender differences.

In the average person, there is a strong relationship between height and the length of two major arm bones (the humerus and the radius), as well as the length of the two major leg bones (the femur and the tibia).

Anthropologists and forensic scientists can closely estimate a person's height from the length of just one of these major bones.

6. Each member of the class should measure his or her leg from the center of the kneecap to the bone on the outside of the hip. This is the length of the femur. Record the results in the appropriate place in the last column of the table in Problem 1.

a. If you want to predict height from the length of the femur, which variable should represent the independent variable? Explain.

b. Make a scatterplot of the data on a carefully scaled and labeled coordinate axes.

c. Describe any patterns you observe in the scatterplot.

d. Determine the correlation coefficient. Describe any correlation between the two variables.

e. Determine the equation of the regression line for the data.

f. Use the equation of the regression line in part e to predict the height of a person whose femur measures 17 inches.

g. Anthropologists have developed the following formula to predict the height of a male based on the length of his femur:

$$h = 1.888L + 32.010$$

where h represents the height in inches and L represents the length of the femur in inches. Use the formula to determine the height of a male whose femur measures 17 inches.

h. Compare your results from parts f and g. What might explain the difference between the height you obtained using the regression formula in part f and the height using the formula in part g?

i. Determine the regression line equation for femur length vs. height using just the male data from Problem 1. How does this new regression line equation compare with the formula $h = 1.888L + 32.010$ used by anthropologists?

7. a. Determine the linear regression equation (femur length vs. height) for the female data in Problem 1.

b. Compare your results to the formula used by anthropologists:

$$h = 1.945x + 28.679$$

where h represents height in inches and x represents femur length in inches.

8. The work of Dr. Mildred Trotter (1899–1991) in skeletal biology led to the development of formulas used to estimate a person's height based on bone length. Her research also led to discoveries about the growth, racial and gender differences, and aging of the human skeleton. Write a brief report on the life and accomplishments of this remarkable scientist.

Activities 2.1–2.10   **What Have I Learned?**

1. What must be true about the average rate of change between any two points on the graph of an increasing function?

2. You are told that the average rate of change of a particular function is always negative. What can you conclude about the graph of that function and why?

3. Describe a step-by-step procedure for determining the average rate of change between any two points on the graph of a function. Use the points represented by $(85, 350)$ and $(89, 400)$ in your explanation.

4. A line is given by the equation $y = -4x + 10$.

   a. Determine its x- and y-intercepts algebraically from the equation.

   b. Use your graphing calculator to confirm these intercepts.

5. a. Does the slope of the line having the equation $4x + 2y = 3$ have a value of 4? Why or why not?

   b. Solve the equation in part a for y so that it is in the form $y = mx + b$.

   c. What is its slope?

6. Explain the difference between a line with zero slope and a line with an undefined slope.

7. Describe how you recognize that a function is linear when it is given

  a. graphically.

  b. as an equation involving x and y.

  c. numerically in a table.

8. Do vertical lines represent functions? Explain.

9. If you know the slope and the vertical intercept of a line, how do you write the equation of the line? Use an example to demonstrate.

10. Demonstrate how you would change the equation of a linear function such as $5y - 6x = 3$ into slope-intercept form. Explain your method.

11. What assumption are you making when you say that the cost, c, of a rental car (in dollars) is a linear function of the number, n, of miles driven?

12. Explain the difference between the slopes of lines that are parallel and the slopes of lines that are perpendicular.

13. When a scatterplot of input-output values from a data set suggests a linear relationship, you can determine a line of best fit. Why might this line be useful in your analysis of the data?

14. Explain how you would determine a line of best fit for a set of data. How would you estimate the slope and y-intercept?

15. Suppose a set of data pairs suggests a linear trend. The input values range from a low of 10 to a high of 40. You use your graphing calculator to calculate the regression equation in the form $y = ax + b$.

    a. Do you think that the equation will provide a good prediction of the output value for an input value of $x = 20$? Explain.

    b. Do you think that the equation will provide a good prediction of the output value for an input value of $x = 60$? Explain.

Activities 2.1–2.10 How Can I Practice?

1. You and your friends have joined an after school fitness program. You record your weight at the beginning of the program and each month thereafter. The following data gives your weight, w, over a five-month period:

Months (t)	0	1	2	3	4	5
Weight, w (lb)	196	183	180	177	174	171

a. Sketch a graph of the data on appropriately scaled and labeled axes.

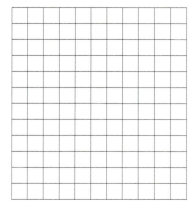

b. Determine the average rate of change of your weight during the first three months. Be sure to include the units of measurement of this rate.

c. Determine the average rate of change during the five-month period.

d. On the graph in part a, connect the points $(0, 196)$ and $(3, 177)$ with a line segment. Does the line segment rise, fall, or remain horizontal as you follow the line left to right?

e. What is the practical meaning of the average rate of change in this situation?

f. What can you say about the average rate of change of weight during any of the time intervals in this situation?

2. A function is linear because the rate of change of y with respect to x from point to point is constant. Use this idea to determine the missing x and y values in each table, assuming that each table represents a linear function.

a.

x	y
1	4
2	5
3	

b.

x	y
1	4
3	8
5	

c.

x	y
0	4
5	9
10	

d.

x	y
−1	3
0	8
	13
2	

e.

x	y
−3	11
0	8
3	
	2

f.

x	y
−2	−5
0	−8
	−11
4	

g. Explain how you used the idea of constant rate of change to determine the values in the tables.

3. The pitch of a roof is an example of slope in a practical setting. The roof slope is usually expressed as a ratio of rise over run. For example, in the building shown, the pitch is 6 to 24 or in fraction form as $\frac{1}{4}$.

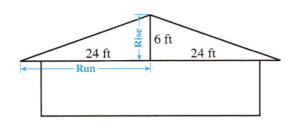

a. If a roof has a pitch of 5 to 16, how high will the roof rise over a 24-foot run?

b. If a roof's slope is 0.25, how high will the roof rise over a 16-foot run?

c. What is the slope of a roof that rises 12 feet over a run of 30 feet?

4. Determine whether any of the following tables contain data that represent a linear function. In each case, give a reason for your answer.

a. You make an investment of $100 at 5% interest compounded semiannually. The following table represents the amount of money you will have at the end of each year.

TIME (years)	AMOUNT ($)
1	105.06
2	110.38
3	115.97
4	121.84

b. A cable-TV company charges a $45 installation fee and $28 per month for basic cable service. The table values represent the total usage cost since installation.

Number of Months	6	12	18	24	36
Total Cost ($)	213	381	549	717	1053

c. For a fee of $20 a month, you have unlimited video rental. Values in the table represent the relationship between the number of videos you rented each month and the monthly fee.

Number of Rentals	10	15	12	9	2
Cost ($)	20	20	20	20	20

5. After stopping your car at a stop sign, you accelerate at a constant rate for a period of time. The speed of your car is a function of the time since you left the stop sign. The following table shows your speedometer reading every third second for the next 21 seconds:

t, Time (sec)	0	3	6	9	12	15	18	21
S, Speed (mph)	0	11	22	33	44	55	55	55

a. Graph the data using ordered pairs of the form (t, s).

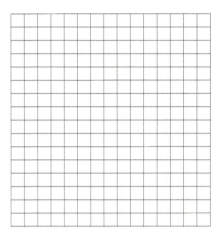

b. For what values of t is the graph increasing?

c. What is the slope of the line segment during the period of acceleration?

d. What is the practical meaning of the slope in this situation?

e. For what values of t is the speed a constant? What is the slope of the line connecting the points of constant speed?

6. a. The three lines shown in the following graphs appear to be different. Calculate the slope of each line.

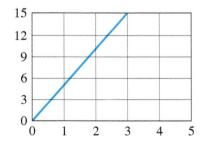
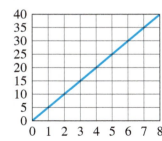
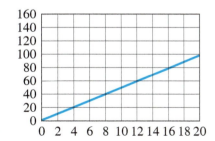

b. Do the three graphs represent the same linear model? Explain.

7. a. Determine the slope of the line $y = -3x - 2$.

b. Determine the slope of the line $2x - 4y = 10$.

c. Determine the slope of the line from the following graph:

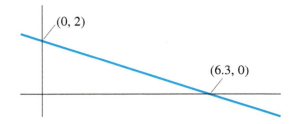

(0, 2)

(6.3, 0)

d. Determine whether the lines are parallel, perpendicular, or neither.

i. $y = -5x + 4$ **ii.** $y = \dfrac{2}{5}x + 1$ **iii.** $3x - y = -2$

$y = 5x - 2$ $y = -\dfrac{5}{2}x - 3$ $6x - 2y = 8$

8. Determine the x- and y-intercepts for the graph of each of the following:

a. $y = 2x - 6$ **b.** $y = -\dfrac{3}{2}x + 10$

c. $y = 10$

9. Determine the equation of each line.

   a. The line passes through the points $(2, 0)$ and $(0, -5)$.

   b. The slope is 7, and the line passes through the point $\left(0, \dfrac{1}{2}\right)$.

   c. The slope is 0, and the line passes through the point $(2, -4)$.

10. Sketch a graph of each of the following. Use your graphing calculator to verify your graphs.

   a. $y = 3x - 6$ 　　　　　　　　　　　　 **b.** $y = -2x + 10$

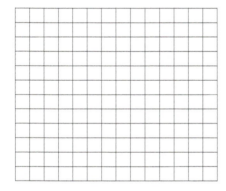
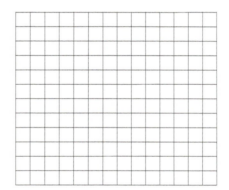

11. Write each equation in slope-intercept form to discover what the graphs have in common. Use your graphing calculator to verify your graphs.

   a. $y = 3x - 4$ 　　　　 **b.** $y - 3x = 6$ 　　　　 **c.** $3x - y = 0$

In Problems 12–15, determine the slope and the x- and y-intercepts of each line.

12. $y = 2x + 1$ 　　　　　　　　　　 **13.** $y = 4 - x$

14. $y = -2$

15. $-\dfrac{3}{2}x - 5 = y$

16. Write an equation in slope-intercept form for the line that passes through the points $(0, 4)$ and $(-5, 0)$.

17. Identify the independent and dependent variables, and write a linear equation model for each of the following situations. Then give the practical meaning of the slope and vertical intercept in each situation.

    a. You make a down payment of \$50 and pay \$10 per month for your new computer.

    b. You pay \$16,000 for a new car whose value decreases by \$1500 each year.

18. Suppose you enter Interstate 90 in Montana and drive at a constant speed of 75 miles per hour.

    a. Write a linear model that represents the total distance, $d(t)$ traveled on the highway as a function of time, t, in hours.

b. Sketch a graph of the function. What are the slope and vertical intercept of the line? What is the practical meaning of the slope?

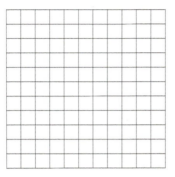

c. How long would you need to drive at 75 miles per hour to travel a total of 400 miles?

d. You start out at 10:00 A.M. and drive for three hours at a constant speed of 75 miles per hour. You are hungry and stop for lunch. One hour later you resume your travel, driving steadily at 60 miles per hour until 6 P.M., when you reach your destination. How far will you have traveled? Sketch a graph that shows the distance traveled as a function of time.

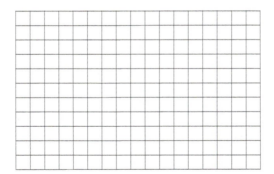

19. Determine the equation of the line passing through each pair of points.

    a. $(0, 6)$ and $(4, 14)$

    b. $(-9, -7)$ and $(-7, -3)$

20. Determine the equation of the line shown on the following graph:

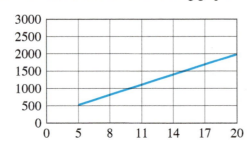

21. On an average winter day in Chicago, the Auto Club receives 125 calls from people who need help starting their cars. The number of calls varies, however, depending on the temperature. Here is some data giving the number of calls as a function of the temperature (in degrees Celsius).

Temperature (°C), t	−12	−6	0	4	9
Number of Auto Club Service Calls, N	250	190	140	125	100

    a. Sketch the given data on an appropriately scaled and labeled coordinate axes.

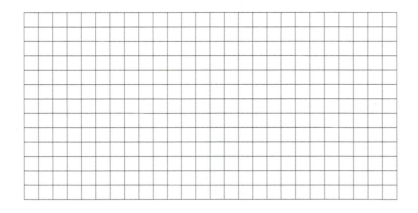

    b. Use your graphing calculator to determine the equation of the regression line for the data in the preceding table.

    c. Use the regression equation from part b to determine how many service calls the Auto Club can expect if the temperature drops to −20°C.

Activity 2.11

Smartphone Plan
Options

Objectives

1. Solve a system of
two linear equations
numerically.

2. Solve a system of
two linear equations
graphically.

3. Solve a system of two
linear equations using
the substitution method.

4. Recognize the connections
between the three methods
of solution.

5. Interpret the solution to
a system of two linear
equations in terms of the
problem's content.

You have researched the Internet to compare the various types of voice, texting, and data plans that are available for your smartphone. You have decided to select a plan with unlimited text, picture, and video messaging but no voice minutes. You will pay a separate monthly fee for voice minutes used. There are two service providers that offer similar plans:

Plan 1: $55.99 per month for Internet access and unlimited text, picture, and video messaging, including 2GB data allowance. There is a voice cost-per-use fee of $0.25 per minute.

Plan 2: $69.99 per month for the same features of Plan 1 but with a voice cost-per-use fee of $0.15 per minute.

Assume that you will not exceed the monthly 2GB data limit.

1. Although the average number of voice minutes used by the average smartphone user is decreasing, in 2013 the average smartphone user used more than 500 voice minutes per month. If you plan on spending about 50 total minutes talking on your smartphone each month, which plan will be more economical?

2. Let m represent the number of voice minutes used in a given month. Write a function rule for the monthly cost, C, in terms of m for Plan 1.

3. Write a function rule for the monthly cost C in terms of m for Plan 2.

4. **a.** Complete the following table for each plan, showing the monthly cost for 100, 125, 150, 200, 250, 300, and 500 voice minutes. Estimate the number of minutes for which the two plans come the closest to being equal in cost. If you have a graphing calculator, use the table feature to complete the table.

Number of Voice Minutes	100	125	150	200	250	300	500
Cost of Plan 1	$80.99	$87.24					
Cost of Plan 2	$84.99	$88.74					

b. Use the table feature of your graphing calculator to determine the m-value that produces identical C-values. What is that value?

5. a. Graph the cost equation for each plan on the same coordinate axes. Plot the data points, and then use a straightedge to draw a line connecting each set of points. Be sure to properly scale and label the axes.

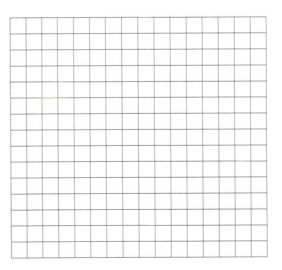

b. Estimate the coordinates of the point where the lines intersect. What is the significance of this point?

c. Verify your results from part b using your graphing calculator. Use the trace or intersect feature of the graphing calculator. See Appendix A for the procedure for the TI-84 Plus. Your final screens should appear as shown on the left.

c.

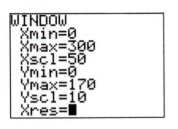

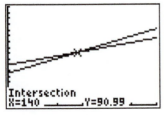

System of Two Linear Equations

Two linear equations that relate the same two variables are called a **system of linear equations**. The two cost equations from Problems 2 and 3 form a system of two linear equations,

$$C = 55.99 + 0.25m$$
$$C = 69.99 + 0.15m.$$

The **solution** of a system is the set of all ordered pairs that satisfy both equations. If the system has exactly one solution, the system is called **consistent**. The solution to the cost system is $(140, 90.99)$. This solution represents the specific number of voice minutes ($x = 140$) that produce identical costs in both accounts ($90.99).

In Problem 4, you solved the cost system **numerically** by completing a table and noting the value of the input that resulted in the same output. In Problem 5, you solved the cost system **graphically** by determining the coordinates of the point of intersection.

You can also determine an exact solution by solving the system of equations **algebraically**. In Problem 6, you will explore one method—the **substitution method**—for solving systems of linear equations algebraically.

Substitution Method for Solving a System of Two Linear Equations

Consider the following system of two linear equations:

$$y = 3x - 10$$
$$y = 5x + 14$$

To solve this system, you need to determine for what value of x are the corresponding y-values the same. The goal of the **substitution method** is to write a single equation involving just one variable.

6. a. Use the two linear equations in the preceding system to write a single equation involving just one variable.

b. Solve the equation in part a for the variable.

c. Use the result in part b to determine the corresponding value for the other variable in the system.

d. Write the solution to this system as an ordered pair.

e. Verify this solution numerically as well as graphically using your graphing calculator.

7. The algebraic process used in Problem 6 is called the substitution method for solving a system of two linear equations. Write a summary of this procedure.

8. a. Using the substitution method, solve the following system of smartphone data plan cost functions.

$$C = 55.99 + 0.25m$$
$$C = 69.99 + 0.15m$$

b. Compare your result with the answers obtained using a numerical approach (Problem 4) and a graphing approach (Problem 5).

c. Summarize your results by describing under what circumstances the basic data Plan 1 is preferable to Plan 2.

9. U.S. online retail sales are forecast to reach 434.2 billion dollars by 2017. Amazon.com is the most popular and well-known example of online shopping. Founded in 1995, the Seattle-based site started out as an online bookstore, but it has since considerably expanded its product line. Amazon.com offers a prime membership to its customers. For a yearly $79 fee, one-day shipping on eligible items for Amazon Prime members is $3.99. There is no minimum order size. Without the prime membership, the cost of one-day shipping is $13.99 per shipment. You do occasionally order from Amazon using one-day shipping and wonder if the annual prime membership fee is worth it.

a. This situation involves two variables: the number of one-day shipments you expect to order in one year and the total cost of the shipping. Identify the input and output.

b. Determine the total cost of shipping 5 orders in one year:

  i. with the prime membership.

  ii. without the prime membership.

c. Write a function rule for the cost c of making n one-day shipments in a year for each of the following:

  i. with Prime Membership.

  ii. without Prime Membership.

d. Solve the system of two linear equations in part c. What does the solution represent in this situation?

e. When would it be worth purchasing the Amazon Prime membership?

Does every system of equations have exactly one solution? Attempt to solve the system in Problem 10 algebraically. Explain your result using a graphical interpretation.

10. $y = 2x + 2$

$y = 2x - 1$

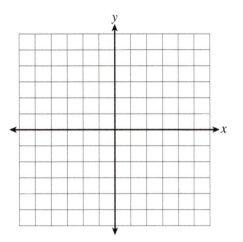

SUMMARY: ACTIVITY 2.11

1. Two equations that relate the same variables are called a **system of equations**. The solution of a system of equations is the set of all ordered pairs that satisfy both equations.

2. If x is the independent variable and y the dependent variable, then a system of two linear equations is often written in the form

$$y = ax + b$$
$$y = cx + d.$$

3. There are three standard methods for solving a system of equations:

1. Numerical method: Make a table of values for both equations. Identify or estimate the x-value that produces the same y-value for both equations.

2. Graphical method: Graph both equations on the same grid. If the two lines intersect, the coordinates of the point of intersection represent the solution of the system.

3. Substitution method: Replace (or *substitute*) the variable y in one equation with its algebraic expression in x from the other equation. Solve for x. Substitute this x-value into either function rule to determine the corresponding y-value.

EXERCISES: ACTIVITY 2.11

1. Two companies sell software products. In 2015, company A had total sales of $17.2 million. Its marketing department projects that sales will increase $1.5 million per year for the next several years. Company B had total sales of $9.6 million of software products in 2015 and projects that its sales will increase an average of $2.3 million each year.

   Let n represent the number of years since 2015.

   a. Write an equation that represents the total sales (in millions of dollars), s, of company A since 2015.

   b. Write an equation that represents the total sales (in millions of dollars), s, of company B since 2015.

   c. The two equations in parts a and b form a system. Solve this system to determine the year in which the total sales of both companies will be the same.

2. Your brother has a total of $10,700 in student loans. Part of the loan was made at a local credit union at 6.50%. The remainder was a Stafford Loan made at 3.86%. After one year, the total amount of interest would accumulate to $531.82.

   a. Let c represent the amount borrowed at the local credit union, and let s represent the amount of the Stafford Loan. Does the sum $c + s$ equal 10,700 or 531.82? Explain.

   b. Write an algebraic expression that represents the amount of interest on the amount c borrowed at the credit union. Then, write an expression that represents the amount of interest on the amount s of the Stafford Loan.

   c. Write an equation for the sum of the two amounts in part b.

   d. Use the results from parts a and c to write a system of equations involving the variables c and s.

e. Solve the system in part d. What does your solution represent in this situation?

3. You need to replace the heating system in your house. A conventional heating system will cost $5000 with a yearly fuel cost of $5400. A modern heating system will cost $8000 with a yearly fuel cost of $4500.

a. Write an equation for the total cost, C, of a conventional system for t years. The total cost is the cost of the system plus the fuel cost.

b. Write an equation for the total cost, C, of a modern system for t years.

c. Solve the system of equations from (a) and (b) to find the number of years it takes for the total cost of the conventional system to equal the total cost of the modern system.

Use substitution to determine algebraically the exact solution to each system of equations in Exercises 4–7. Use the table feature or graphing capability of your calculator to verify these solutions.

4. $p = q - 2$

$p = -1.5q + 3$

5. $n = -2m + 9$

$n = 3m - 11$

6. $y = 1.5x - 8$

$y = -0.25x + 2.5$

7. $z = 3w - 1$

$z = -3w - 1$

8. Attempt to solve the following system algebraically. Explain your result using a graphical interpretation.

$y = -3x + 2$

$y = -3x + 3$

9. You like to drive to a special store to buy your favorite gourmet jelly beans for a low price of $1.30 a pound. However, your best friend points out that you are spending about $3 on gas every time you drive to that store and that you might as well buy your jelly beans at your local supermarket for $2.10 a pound. You decide to show him that you are getting enough of those jelly beans to make it worth the trip.

  a. Let x represent the number of pounds of jelly beans you purchase. Write an equation to determine the total cost, C, if you purchase the jelly beans at the specialty store. Remember to include the cost of the gas.

  b. Write an equation to determine the cost, C, if you purchase the jelly beans at the local supermarket.

c. The two equations in parts a and b form a system of two linear equations. Solve this system using an algebraic approach.

d. Verify your solution using a numerical (table) and graphing approach.

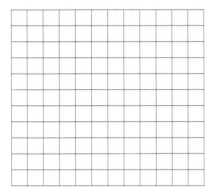

e. How many pounds of jelly beans must you buy to make it worth the trip to the specialty store?

10. Your mother wants to hire someone to prune the trees and shrubs. One service she calls charges a $15 consultation fee plus $8 an hour for the actual work. A neighbor's son says he does not include a consulting fee, but he charges $10 an hour for the work.

a. Write an equation that describes the pruning service's charge, C, as a function of h, the number of hours worked.

b. Write an equation that describes the neighbor's charge, C, as a function of h, the number of hours worked.

c. Whom should your mother hire for a 3-hour job?

d. When, if at all, would it be more economical to hire the other service? Set up and solve a system of equations to answer this question.

e. Use your graphing calculator to verify your results.

11. You and your friend are going rollerblading at a local park. There is a 5-mile path along the lake that begins at the concession stand. You rollerblade at a rate of 10 miles per hour, and your friend rollerblades at 8 miles per hour. You start rollerblading at the concession stand. Your friend starts farther down the path, 0.5 mile from the concession stand.

   a. Write an equation that models your distance from the concession stand as a function of time. What are the units of the input variable? output variable? (Recall that distance $=$ rate $\cdot$ time.)

   b. Write an equation that models your friend's distance from the concession stand as a function of time.

   c. How long will it take you to catch up to your friend? In that time, how far will you have rollerbladed?

   d. Use your graphing calculator to verify your results.

Collecting and Analyzing Data

12. The women's world record time in the 400-meter run is decreasing at a faster rate than the men's record time. A similar occurrence is happening in many events, including the men's and women's Olympic 500-meter speed skating and world record times in the 1500-meter run.

Choose one of these competitions and use linear functions to model the data for the women's event and for the men's event. Use the models to estimate when the women's record time will equal the men's record time.

Work in a group and be prepared to give a presentation to the class. The presentation should include visual displays showing the tables, scatterplots, and equations used in the problemsolving process. Be sure to identify the source of your data.

Project 2.12

Modeling a Business

Objectives

1. Solve a system of two linear equations by any method.

2. Determine the break-even point of a linear system algebraically and graphically.

3. Interpret break-even points in contextual situations.

You are an electrical engineer employed by a company that manufactures solar collector panels. To remain competitive, the company must consider many variables and make many decisions. Two major concerns are those variables and decisions that affect operating expenses (or costs) of making the product and those that affect the gross income (or revenue) from selling the product.

Costs such as rent, insurance, and utilities for the operation of the company are called **fixed costs**. These costs generally remain constant over a short period of time and must be paid whether or not any items are manufactured. Other costs, such as materials and labor, are called **variable costs**. These expenses depend directly on the number of items produced.

1. The records of the company show that fixed costs over the past year have averaged $8000 per month. In addition, each panel manufactured costs the company $95 in materials and $55 in labor. Write a function rule for the total cost, $C(n)$, of producing n solar collector panels in one month.

2. A marketing survey indicates that the company can sell all the panels it produces if the panels are priced at $350 each. The revenue (gross income) is the amount of money collected from the sale of the product. Write a function rule for the revenue, $R(n)$, from selling n solar collector panels in one month.

3. **a.** Complete the following table:

n, Number of Solar Panels	0	10	20	30	40	50	60
C(n), Total Cost ($)							
R(n), Total Revenue ($)							

b. Sketch a graph of the cost and revenue functions using the same set of coordinate axes.

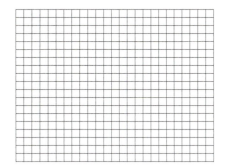

4. The point at which the cost and revenue functions are equal is called the **break-even point**.

a. Estimate the break-even point on the graph.

b. What system of equations must be solved to determine the break-even point for your company?

    c. Solve the system algebraically to determine the exact break-even point.

    d. Does your graph confirm the algebraic solution in part c?

5. Revenue exceeds costs when the graph of the revenue function is above the graph of the cost function. For what values of n is $R(n)$ greater than $C(n)$? What do these values represent in this situation?

6. Profit is defined as revenue minus cost.

    a. Determine the profit when 25 panels are sold. What does the sign of your answer signify?

    b. Determine the profit when 60 panels are sold.

    c. Write a function rule for the profit, $P(n)$, made by selling n solar panels in one month.

    d. Use your result in part c to compute the profit for selling 25 and 60 panels. How do your results compare with the answers in parts a and b?

7. a. Sketch a graph of the profit function $P(n) = 200n - 8000$.

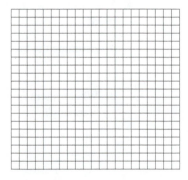

b. Determine the intercepts of the graph. What is the meaning of each intercept in this situation?

c. What is the slope of the line? What is the practical meaning of slope in this situation?

8. Suppose you are the manager of a small company producing interlocking paving pieces, called pavers, for driveways. You sell the pavers in bundles that cost $200; each bundle contains 144 pavers. The total cost in dollars, $C(x)$, of producing x bundles of pavers is modeled by $C(x) = 160x + 1000$.

a. Write a function rule for the revenue in dollars, $R(x)$, from the sale of the pavers.

b. Determine the slope and the C-intercept of the cost function. Explain the practical meaning of each in this situation.

c. Determine the slope and the R-intercept of the revenue function. Explain the practical meaning of each in this situation.

d. Graph the two functions from parts b and c on the same set of axes. Estimate the break-even point from the graph. Express your answer as an ordered pair, giving units. Check your estimate of the break-even point by graphing the two functions on your graphing calculator.

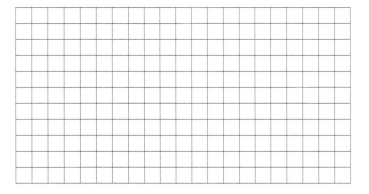

e. Determine the exact break-even point algebraically. If your algebraic solution does not approximate your answer from the graph in part d, explain why.

f. How many bundles of pavers does the company have to sell for it to break even?

g. What is the total cost to the company when you break even? Verify that the cost and revenue values are equal at the break-even point.

h. For what values of x will your revenue exceed your cost?

i. As manager, what factors do you have to consider when deciding how many pavers to make?

j. If you knew you could sell only 30 bundles of pavers, would you make them? Consider how much it would cost you and how much you would make. What if you could sell only 20?

Activity 2.13

Healthy Lifestyle

Objectives

1. Solve a 2 × 2 linear system algebraically using the substitution method and the addition method.

2. Solve equations containing parentheses.

You are trying to maintain a healthy lifestyle. You eat a well-balanced diet and follow a regular schedule of exercise. One of your favorite exercise activities is a combination of walking and jogging in a nearby park.

On one particular day, it takes you 1.3 hours to walk and jog a total of 5.5 miles in the park. You are curious about the amount of time you spent walking and the amount of time you spent jogging during the workout.

Let x represent the time you walked and y represent the time you jogged.

1. Write an equation, using x and y, for the total time of your walk/jog workout in the park.

2. a. If you walk at 3 miles per hour, write an expression that represents the distance you walked.

   b. If you jog at 5 miles per hour, write an expression that represents the distance you jogged.

   c. Write an equation for the total distance you walked/jogged in the park.

The situation just described can be represented by the following system.

$$x + y = 1.3$$
$$3x + 5y = 5.5$$

Note that each equation in this system is in standard form $Ax + By = C$, where A, B, and C are constants, A and B both not zero. One approach to solving this system is to solve each equation for one variable in terms of the other and then use the substitution method.

3. a. Solve each of the equations in the system above for y.

   b. Solve the system in part a using the substitution method.

   c. Check your answer graphically using your graphing calculator. You may want to use the window Xmin $= -2.5$, Xmax $= 2.5$, Ymin $= -2.5$, and Ymax $= 2.5$.

Addition Method

Sometimes it is more convenient to leave each equation in the linear system in standard form $(Ax + By = C)$ rather than solving for one variable in terms of the other. Look again at the original system.

$$x + y = 1.3 \quad \text{(equation 1)}$$
$$3x + 5y = 5.5 \quad \text{(equation 2)}$$

If you apply the addition principle of algebra by adding the two equations (left side to left side and right side to right side), you may be able to obtain a single equation containing only one variable.

$$x + y = 1.3 \quad \text{(equation 1)}$$
$$\underline{3x + 5y = 5.5} \quad \text{(equation 2)}$$
$$4x + 6y = 6.8$$

In this case, adding the equations does not eliminate a variable. But if you multiply both sides of equation 1 by -5, the coefficients of y will be opposite, and the variable y can be eliminated.

$$-5(x + y) = -5(1.3) \rightarrow -5x - 5y = -6.5 \qquad \text{Add the corresponding sides of the equation.}$$

$$3x + 5y = 5.5 \qquad \qquad \underline{3x + 5y = 5.5}$$
$$-2x + 0 = -1$$

Solving the resulting equation, $-2x = -1$, for x, you have

$$-2x = -1$$

$$x = \frac{1}{2} = 0.5.$$

Substituting for x in $x + y = 1.3$, you have

$$0.5 + y = 1.3$$
$$\text{or } y = 0.8.$$

This method of solving systems algebraically is called the **addition method.**

> To solve a 2 $\times$ 2 linear system by the addition method.
>
> **1.** line up the like terms in each equation vertically;
>
> **2.** if necessary, multiply one or both equations by constants so that the coefficients of one of the variables are opposites;
>
> **3.** add the corresponding sides of the two equations.

4. Solve the following system again using the addition method. Multiply the appropriate equation by the appropriate factor to eliminate x and solve for y first.

$$x + y = 1.3$$
$$3x + 5y = 5.5$$

Not all systems will have convenient coefficients, and you may need to multiply one or both equations by a factor that will produce coefficients of the same variable that are additive inverses, or opposites.

5. Consider solving the following system with the addition method.

$$-2x + 5y = -16$$
$$3x + 2y = 5$$

a. Identify which variable you wish to eliminate. Multiply the appropriate equation by the appropriate factor so that the coefficients of your chosen variable are opposite. Show the two equations after you multiply by the factor. (Remember to multiply both sides of the equation by the factor.)

b. Add the two equations to eliminate the chosen variable.

c. Solve the resulting linear equation.

d. Determine the complete solution. Remember to check by substituting into both of the original equations.

Substitution Method Revisited

The substitution method of solving a 2 $\times$ 2 system of linear equations is generally used when each equation in the system is solved for one variable in terms of the other. However, it can be convenient to use the substitution method when only one of the equations is solved for a variable. Example 1 demonstrates this process.

Example 1   *In the walk-jog system,*

$$x + y = 1.3 \quad \textbf{(equation 1)}$$
$$3x + 5y = 5.5 \quad \textbf{(equation 2)}$$

you could solve equation 1 for y and then substitute for y in equation 2 as follows:

Step 1. Solve $x + y = 1.3$ for y.

$$y = 1.3 - x$$

Step 2. Substitute $1.3 - x$ for y in equation 2.

$$3x + 5(1.3 - x) = 5.5$$

Step 3. Solve the resulting equation for x.

$$3x + 5(1.3 - x) = 5.5$$ Remove the parentheses by applying the distributive property.

$$3x + 6.5 - 5x = 5.5$$ Collect like terms on the same side.

$$-2x + 6.5 = 5.5$$

$$\underline{ -6.5 = -6.5}$$ Add the opposite of 6.5 to both sides.

$$-2x = -1$$ Divide each side by -2.

$$\frac{-2x}{-2} = \frac{-1}{-2}$$

$$x = 0.5$$

Step 4. From equation 1, $y = 0.8$ as before.

6. a. Solve the following linear system using the substitution method in which you solve only one of the equations for a variable.

$$x - y = 5$$
$$4x + 5y = -7$$

b. Check your answer in part a by solving the system using the addition method.

SUMMARY: ACTIVITY 2.13

1. There are two methods for solving a 2 × 2 system of linear equations algebraically:

   a. the substitution method and
   b. the addition method

2. To solve linear systems by substitution,

   Step 1. solve one or both equations for a variable;

   Step 2. substitute the expression that represents the variable in one equation for that variable in the other equation;

   Step 3. solve the resulting equation for the remaining variable;

   Step 4. substitute the value from step 3 into one of the original equations, and solve for the other variable.

3. To solve linear systems by addition with equations written in the form $Ax + By = C$.

Step 1. multiply one equation or both equations by the number(s) that will make the coefficients of one of the variables opposites;

Step 2. add the two equations to eliminate one variable and solve the resulting equation;

Step 3. substitute the value from step 2 into one of the original equations, and solve for the other variable.

EXERCISES: **ACTIVITY 2.13**

1. Solve each of the following equations.

a. $5(x + 3) + 4 = 6x - 1 - 5x$

b. $-3(1 - 2x) - 3(x - 4) = -5 - 4x$

c. $2(x + 3) - 4x = 5x + 2$

2. Solve the following systems algebraically using the substitution method.

a. $y = 3x + 1$
$y = 6x - 0.5$

b. $y = 3x + 7$
$2x - 5y = 4$

c. $2x + 3y = 5$

$-2x + y = -9$

d. $4x + y = 10$

$2x + 3y = -5$

3. Solve the following systems algebraically using the addition method.

a. $y = 2x + 3$

$y = -x + 6$

b. $2x + y = 1$

$-x + y = -5$

4. Solve the system both graphically and algebraically.

$$3x + y = -18$$
$$5x - 2y = -8$$

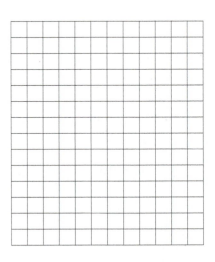

5. A catering service placed an order for eight centerpieces and five glasses, and the bill was $106. For the wedding reception, they were short one centerpiece and six glasses and had to reorder. This order came to $24. Let x represent the cost of one centerpiece, and let y represent the cost of one glass.

   a. Write a system of equations that represents both orders.

   b. Solve the system using the substitution method. Interpret your solution.

   c. Check your result in part b using the addition method.

   d. Use your graphing calculator to solve the system.

Activity 2.14

How Long Can You Live?

Objectives

1. Solve linear inequalities numerically and graphically.

2. Use properties of inequalities to solve linear inequalities algebraically.

3. Solve compound inequalities algebraically and graphically.

Life expectancy in the United States is steadily increasing, and the number of Americans aged 100 or older will exceed 850,000 by the middle of this century. Medical advancements have been a primary reason for Americans living longer. Another factor has been the increased awareness of maintaining a healthy lifestyle.

The life expectancies at birth for men and women born after 1980 in the United States can be modeled by the following functions,

$$W(x) = 0.101x + 77.5$$
$$M(x) = 0.192x + 70.0,$$

where $W(x)$ represents the life expectancy for women, $M(x)$ represents the life expectancy for men, and x represents the number of years since 1980 that the person was born. That is, $x = 0$ corresponds to the year 1980, $x = 5$ corresponds to 1985, and so forth.

1. a. Complete the following table:

Counting the Years	YEAR						
	1980	1985	1990	1995	2000	2005	2010
x, Year Since 1980	0	5	10	15	20	25	30
W(x)							
M(x)							

b. For people born between 1980 and 2005, do men or women have the greater life expectancy?

c. Is the life expectancy of men or women increasing more rapidly? Explain using slope.

You would like to determine in what years the life expectancy of men is greater than that of women. The phrase "greater than" indicates a mathematical relationship called an **inequality**. Symbolically, the relationship can be represented by

$$M(x) \qquad > \qquad W(x).$$

life expectancy   Is greater   life expectancy
for men   than   for women

Other commonly used phrases that indicate inequalities are given in the following example:

Example 1

STATEMENT, WHERE x REPRESENTS A REAL NUMBER	TRANSLATION TO AN INEQUALITY
x is greater than 10	$x > 10$ or $10 < x$
x is less than 10	$x < 10$ or $10 > x$
x is at least 10	$x \geq 10$ (also read, x is greater than or equal to 10)
x is at most 10	$x \leq 10$ (also read, x is less than or equal to 10)

2. Substitute for $M(x)$ and $W(x)$ to obtain an inequality involving x that can be used to determine the birth years for which the life expectancy of men is greater than that of women.

Solving Inequalities in One Variable Numerically and Graphically

Definition

Solving an inequality is the process of determining the values of the variable that make the inequality a true statement. These values are called the **solutions** of the inequality.

3. Solve the inequality in Problem 2 numerically. That is, continue to construct a table of values (see Problem 1) until you determine the values of the years x for which $0.192x + 70.0 > 0.101x + 77.5$. Use the table feature of your graphing calculator.

Therefore, if the trends given by the equations for $M(x)$ and $W(x)$ continue, the approximate solution to the inequality $M(x) > W(x)$ is $x > 82.4$. That is, according to the models, after the year 2062, men will live longer than women.

4. Now, solve the inequality $0.192x + 70.0 > 0.101x + 77.5$ graphically.

  a. Use your graphing calculator to sketch a graph of $M(x) = 0.192x + 70.0$ and $W(x) = 0.101x + 77.5$ on the same coordinate axis.

  b. Determine the point of intersection of the two graphs using the intersect feature of your graphing calculator. What does the point represent in this situation?

To solve the inequality $M(x) > W(x)$ graphically, you need to determine the values of x for which the graph of $M(x) = 0.192x + 70.0$ is above the graph of

$$W(x) = 0.101x + 77.5.$$

   c. Use the graph to solve $M(x) > W(x)$. How does your solution compare to the solution in Problem 3?

5. a. Write an inequality to determine the birth years of women whose life expectancy is at least 85.

   b. Solve the inequality numerically, using the table feature of your graphing calculator.

   c. Use your graphing calculator to solve this inequality graphically.

Solving Inequalities Algebraically

The process of solving an inequality algebraically is very similar to solving an equation algebraically. Your goal is to isolate the variable on one side of the inequality symbol. You isolate the variable in an **equation** by performing the same operations to both sides of the equation so as not to upset the **balance**. You isolate the variable in an **inequality** by performing the same operations to both sides so as not to upset the **imbalance**.

6. a. Write the statement "15 is greater than 6" as an inequality.

   b. Add 5 to each side of $15 > 6$. Is the resulting inequality a true statement? (That is, is the left side still greater than the right side?)

   c. Subtract 10 from each side of $15 > 6$. Is the resulting inequality a true statement?

   d. Multiply each side of $15 > 6$ by 4. Is the resulting inequality true?

   e. Multiply each side of $15 > 6$ by -2. Is the left side still greater than the right side?

   f. Reverse the direction of the inequality symbol in part e. Is the new inequality a true statement?

Problem 6 demonstrates two very important properties of inequalities.

Property 1: If $a < b$ represents a true inequality, then if
     i. The same quantity is added to or subtracted from both sides

or

     ii. both sides are multiplied or divided by the same positive number, then the resulting inequality remains a true statement and the direction of the inequality symbol remains the same.

For example, because $-4 < 10$, then

    i. $-4 + 5 < 10 + 5$ or $1 < 15$ is true.

        $-4 - 3 < 10 - 3$ or $-7 < 7$ is true.

    ii. $-4(6) < 10(6)$ or $-24 < 60$ is true.

        $\dfrac{-4}{2} < \dfrac{10}{2}$ or $-2 < 5$ is true.

Property 2: If $a < b$ represents a true inequality, then if both sides are multiplied or divided by the same *negative number*, then the inequality symbol in the result-ing inequality statement must be reversed ($<$ to $>$ or $>$ to $<$) in order for the resulting statement to be true.

For example, because $-4 < 10$, then $-4(-5) > 10(-5)$ or $20 > -50$.

Because $-4 < 10$, then $\dfrac{-4}{-2} > \dfrac{10}{-2}$ or $2 > -5$.

These properties will be true if $a < b$ is replaced by $a \leq b$, $a > b$, or $a \geq b$.

The following example demonstrates how properties of inequalities can be used to solve an inequality algebraically.

Example 2   *Solve $3(x - 4) > 5(x - 2) - 8$.*

SOLUTION

$3(x - 4) > 5(x - 2) - 8$ **Apply the distributive property.**

$3x - 12 > 5x - 10 - 8$ **Combine like terms on the right side.**

$3x - 12 > 5x - 18$

$\underline{-5x \qquad\qquad -5x}$ **Subtract 5x from both sides; the direction of**
                                    the inequality symbol remains the same.

$-2x - 12 > -18$

$\underline{+ 12 \quad + \quad 12}$ **Add 12 to both sides; the direction of the inequality**
                                      does not change.

$\dfrac{-2x}{-2} > \dfrac{-6}{-2}$ **Divide both sides by -2; the direction is reversed!**

$x < 3$

Therefore, from Example 2, any number less than 3 is a solution to the inequality $3(x - 4) > 5(x - 2) - 8$. The solution set can be represented on a number line by shading all points to the left of 3:

The open circle at 3 indicates that 3 is *not* a solution. A closed circle indicates that the number beneath the closed circle *is* a solution. The arrow shows that the solutions extend indefinitely to the left.

  7. Solve the inequality $0.192x + 70.0 > 0.101x + 77.5$ algebraically to determine the birth years in which men will be expected to live longer than women. How does your solution compare to the solutions determined numerically and graphically in Problems 3 and 4c?

Compound Inequality

Your aerobics instructor recommends that to achieve the most cardiovascular benefit from your workout, you should maintain your pulse rate between a lower and upper range of values. These values depend on your age.

8. If the variable a represents your age, then the lower and upper values for your pulse rate are determined by the following:

$$\text{lower value: } 0.72(220 - a)$$

$$\text{upper value: } 0.87(220 - a)$$

    a. Determine your lower value.

    b. Determine your upper value.

For the most cardiovascular benefit, a 20-year-old's pulse rate should be between 144 and 174. The phrase "between 144 and 174" means the pulse rate should be greater than 144 *and* less than 174. Symbolically, this combination or **compound inequality** is written as:

$$144 < \text{pulse rate } and \text{ pulse rate} < 174$$

This statement is written more compactly as: $144 < \text{pulse rate} < 174$.

The numbers that satisfy this compound inequality can be represented on a number line as:

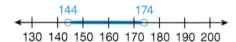

Other commonly used phrases that indicate compound inequalities involving the word "and" are given in the following example.

Example 3

STATEMENT, WHERE x REPRESENTS A REAL NUMBER	TRANSLATION TO A COMPOUND INEQUALITY
x is greater than or equal to 10 and less than 20	$10 \le x < 20$
x is greater than 10 and less than or equal to 20	$10 < x \le 20$
x is from 10 to 20 inclusive	$10 \le x \le 20$

9. Recall that the life expectancy for men is given by the expression $0.192x + 70.0$, where x represents the number of years since 1980. Use this expression to write a compound inequality that can be used to determine in what birth years men will be expected to live into their 80s.

The following example demonstrates how to solve a compound linear inequality algebraically and graphically.

Example 4   **a.** *Solve* $-4 < 3x + 5 \le 11$ *using an algebraic approach.*

SOLUTION

Note that the compound inequality has three parts: left: -4, middle: $3x + 5$, and right: 11. To solve this inequality, isolate the variable in the middle part.

$$-4 < 3x + 5 \le 11$$
$$\underline{-5 \qquad\quad -5 \quad -5}$$ Subtract 5 from each part.
$$-9 < 3x \le 6$$
$$-\frac{9}{3} < \frac{3x}{3} \le \frac{6}{3}$$ Divide each part by 3.
$$-3 < x \le 2$$

The solution can be represented on a number line as follows:

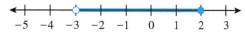

$$-5 \quad -4 \quad -3 \quad -2 \quad -1 \quad 0 \quad 1 \quad 2 \quad 3$$

b. *Solve* $-4 < 3x + 5 \le 11$ *using a graphical approach.*

SOLUTION

First, graph each part of the inequality on the same coordinate axes.

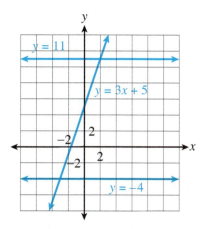

You need to determine values of x for which the graph of $y_2 = 3x + 5$ is above the graph of $y_1 = -4$ and on or below the graph of $y_3 = 11$. Using the graphs, verify that the solution is $-3 < x \le 2$.

10. a. Solve the compound inequality $80 \le 0.192x + 70.0 < 90$ from Problem 9 to determine in what birth years men will be expected to live into their 80s.

   b. Verify your results in part a by solving the compound inequality graphically. Use the window from Problem 4a.

Interval Notation

Interval notation is an alternative method to represent a set of real numbers described by an inequality. The **closed interval** $[-3, 4]$ represents all real numbers x for which $-3 \le x \le 4$. The square brackets [and] indicate that the endpoints of the interval are included. The **open** interval $(-3, 4)$ represents all real numbers x for which $-3 < x < 4$. Note that the parentheses (and) indicate that the endpoints are not included. The interval $(-3, 4]$ is said to be **half-open** or **half-closed.** The interval is open at -3 (endpoint not included) and closed at 4 (endpoint included).

Suppose you want to represent the set of real numbers x for which x is greater than 3. The symbol $+\infty$ (positive infinity) is used to indicate **unboundedness** in the positive direction. Therefore, the interval $(3, +\infty)$ represents all real numbers x for which $x > 3$. Note that $+\infty$ is always open.

The symbol $-\infty$ (negative infinity) is used to represent unboundedness in the negative direction. Therefore, the interval $(-\infty, 5]$ represents all real numbers x for which $x \leq 5$.

11. In parts a–d, express each inequality in interval notation.

a. $-5 \leq x \leq 10$ **b.** $4 \leq x < 8.5$ **c.** $x > -2$ **d.** $x \leq 3.75$

In parts e–h, express each of the following using inequalities.

e. $(-6, 4]$ **f.** $(-\infty, 1.5]$ **g.** $(-2, 2)$ **h.** $(-3, +\infty)$

SUMMARY: ACTIVITY 2.14

1. The **solution set** of an inequality is the set of all values of the variable that satisfy the inequality.

2. The direction of an inequality is not changed when

   i. the same quantity is added to or subtracted from both sides of the inequality.

      Stated algebraically,

      if $a < b$, then $a + c < b + c$ and $a - c < b - c$.

   ii. both sides of an inequality are multiplied or divided by the same positive number.

      If $a < b$, then $ac < bc$, where $c > 0$ and $\dfrac{a}{c} < \dfrac{b}{c}$, where $c > 0$.

3. The direction of an inequality is reversed if both sides of an inequality are multiplied by or divided by the same negative number. These properties can be written symbolically as:

   i. If $a < b$, then $ac > bc$, where $c < 0$.

   ii. If $a < b$, then $\dfrac{a}{c} > \dfrac{b}{c}$, where $c < 0$.

The two properties of inequalities above will still be true if $a < b$ is replaced by $a \leq b, a > b$, or $a \geq b$.

4. Inequalities such as $f(x) < g(x)$ can be solved using three different methods:

   i. a **numerical approach**, in which a table of x-y pairs is used to determine values of x for which $f(x) < g(x)$.

   ii. a **graphical approach**, in which values of x are located so that the graph of f is below the graph of g.

   iii. an **algebraic approach**, in which the properties of inequalities are used to isolate the variable.

Similar statements can be made for solving inequalities of the form $f(x) \leq g(x), f(x) > g(x)$, and $f(x) \geq g(x)$.

EXERCISES: ACTIVITY 2.14

In Exercises 1–6, translate the given statement into an algebraic inequality or compound inequality.

1. To avoid an additional charge, the sum of the length, l, width, w, and depth, d, of a piece of luggage to be checked on a commercial airline can be at most 61 inches.

2. A PG-13 movie rating means that your age, a, must be at least 13 years for you to view the movie.

3. The cost, $C(A)$, of renting a car from company A is less expensive than the cost $C(B)$ of renting from company B.

4. The label on a bottle of film developer states that the temperature, t, of the contents must be kept between 68° and 77°F.

5. You are in a certain tax bracket if your taxable income, i, is over \$24,650, but not over \$59,750.

6. The range of temperature, t, on the surface of Mars is from -153°C to 20°C.

Solve Exercises 7–14 graphically and algebraically.

7. $3x > -6$

8. $3 - 2x \le 5$

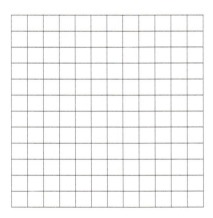
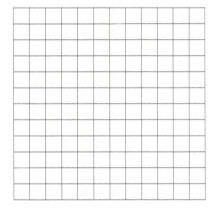

9. $x + 2 > 3x - 8$

10. $5x - 1 < 2x + 11$

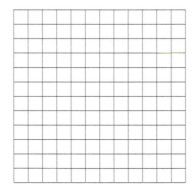

11. $8 - x \geq 5(8 - x)$

12. $5 - x < 2(x - 3) + 5$

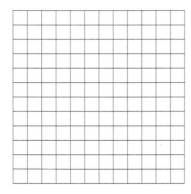

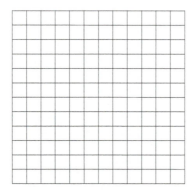

13. $\dfrac{x}{2} + 1 \le 3x + 2$

14. $0.5x + 3 \ge 2x - 1.5$

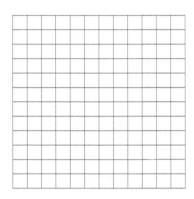

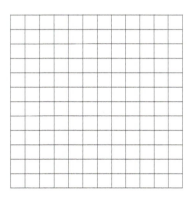

Solve Exercises 15–16 graphically and algebraically.

15. $1 < 3x - 2 < 4$

16. $-2 < \dfrac{x}{3} + 1 < 5$

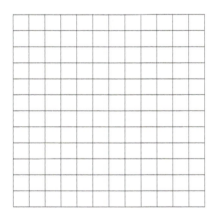

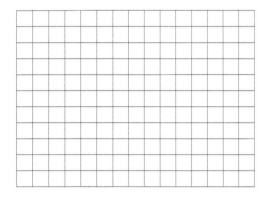

17. The consumption of cigarettes is declining. If t represents the number of years since 1990, then the consumption, C, is modeled by

$$C = -14.25t + 598.69,$$

where C represents the number of billions of cigarettes smoked per year.

   a. Write an inequality that can be used to determine the first year in which cigarette consumption is less than 200 billion cigarettes per year.

b. Solve the inequality in part a using an algebraic as well as a graphical approach.

18. You contact two local rental companies and obtain the following information for the one-day cost of renting a truck:

Company 1: $60 per day plus $0.75 per mile

Company 2: $30 per day plus $1.00 per mile

Let n represent the total number of miles driven in one day.

a. Write an expression to determine the total cost of renting a truck from company 1 and driving n miles.

b. Write an expression to determine the total cost of renting a truck from company 2 and driving n miles.

c. Use the expressions in parts a and b to write an inequality that can be used to determine for what number of miles it is less expensive to rent the truck from company 2.

d. Solve the inequality.

19. The sign on the elevator in a seven-story building states that the maximum weight it can carry is 1200 pounds. You need to move a large shipment of books to the sixth floor. Each box weighs 60 pounds.

a. Let n represent the number of boxes placed in the elevator. If you weigh 150 pounds, write an expression that represents the total weight in the elevator. Assume that only you and the boxes are in the elevator.

b. Using the expression in part a, write an inequality that can be used to determine the maximum number of boxes that you can place in the elevator at one time.

c. Solve the inequality.

20. The following equation is used in meteorology to determine the temperature humidity index:

$$T = \frac{2}{5}(w + 80) + 15,$$

where T represents the temperature and w represents the wet-bulb thermometer reading. For what values of w would T range from 70 to 75?

21. The temperature readings in the United States have ranged from a record low of $-79.8°F$ (Alaska, January 23, 1971) to a record high of $134°F$ (California, July 10, 1913).

    a. If F represents the Fahrenheit temperature, write a compound inequality that represents the interval of temperatures (in °F) in the United States.

    b. Recall that Fahrenheit and Celsius degrees are related by the formula

$$F = 1.8C + 32.$$

    Rewrite the compound inequality in part a to determine the temperature range in degrees Celsius.

    c. Solve the compound inequality.

22. You are enrolled in a half-year wellness course. You achieved grades of 70, 77, 81, and 83 on the first four exams. The final exam counts the same as any of the four exams already given.

    a. If x represents the grade on the final exam, write an expression that represents your course average.

    b. If your average is greater than or equal to 80 and less than 90, you will earn a B in the course. Using the expression from part a for your course average, write a compound inequality that must be satisfied to earn a B. The final exam is worth 100 points.

    c. Solve the inequality.

23. The local credit union is offering a special student checking account. The monthly cost of the account is $15. The first 10 checks are free, and each additional check costs $0.75. You search the Internet and find a bank that offers a student checking account with no monthly charge. The first 10 checks are free, but each additional check costs $2.50.

    a. Assume you will be writing more than 10 checks a month. Let n represent the number of checks written in a month. Write a function rule for the cost c of each account in terms of n.

b. Write an inequality to determine for what number of checks in the bank account would be more expensive than the credit union account.

c. Solve the inequality in part b.

d. Which student checking account would you choose? Explain.

24. Let x represent the number of years since 1990. The male population of the United States can be modeled by the linear function defined by

$$m(x) = 1.6x + 121,$$

where $m(x)$ represents the male population in millions. The female population of the United States can be modeled by the linear function defined by

$$f(x) = 1.54x + 127,$$

where $f(x)$ represents the female population in millions.

a. Determine the slope and vertical intercept of each linear model. Interpret the slope and intercept of each line. Interpret the slope and intercept in the context of this situation.

b. Will the male population exceed the female population? Explain.

c. Write an inequality in terms of x to determine after which year the male population will be greater than the female population.

d. Solve this inequality. What does the result mean in this situation? Is it realistic?

Activity 2.15

Will Trees Grow?

Objectives

1. Graph a linear inequality in two variables.

2. Solve a system of linear inequalities in two variables graphically.

3. Determine the corner points of the solution set of a system of linear inequalities.

While researching a term paper on global climate change, you discover a mathematical model that gives the relationship between temperature and amount of precipitation that is necessary for trees to grow. If t represents the average annual temperature in °F, and p represents the annual precipitation in inches, then

$$t \geq 35$$
$$5t - 7p < 70.$$

Source: Miller and Thompson, *Elements of Meteorology*.

These inequalities form a **system of inequalities in two variables**. The solution of the system is the set of all ordered pairs of the form (t, p) that make each of the inequalities in the system a true statement.

1. a. Will trees grow in a region in which the average annual temperature is 22°F and the annual precipitation is 30 inches?

b. Will trees grow if the average annual temperature is 55°F and the annual precipitation is 20 inches?

c. In Sydney, Australia, the average annual temperature is 64°F and the annual precipitation is 48 inches. Will trees grow in Sydney?

Actually, there are an infinite number of pairs of values for temperature t and amount of precipitation p for which trees will grow. Although it is not possible to list all the possible combinations, you can visualize all the solution pairs by graphing the solution set of the system. You will return to the Will Trees Grow? system later in this activity.

Solving a Linear Inequality in Two Variables by Graphing

To determine the solution set for a system of linear inequalities in two variables, you begin by graphing the solutions of each of the inequalities in the system. Problem 2 guides you through the process of graphing a linear equality.

2. a. To obtain the graph of $4x - 2y \geq 8$, first replace the inequality symbol $\geq$ with an equal sign and sketch a graph of the resulting equation.

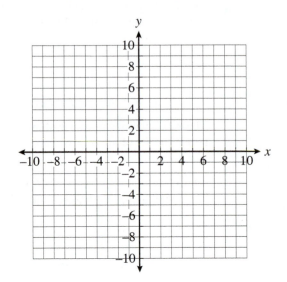

Note that the line $4x - 2y = 8$ divides the coordinate plane into three sets of points:

1. Points *on* the line that satisfy the equation $4x - 2y = 8$
2. Points *below* the line that satisfy the inequality $4x - 2y > 8$
3. Points *above* the line that satisfy the inequality $4x - 2y < 8$

A **half-plane** is a set of points on one side of a line. Therefore, the set of points above the line $4x - 2y = 8$ is a half-plane. The set of points below the line is also a half-plane. The solution set to a linear inequality will include all the points in one half-plane or the other.

b. Select a point in the half-plane below the line $4x - 2y = 8$. Use this test point to determine if the points in the half-plane are in the solution set of the inequality $4x - 2y \geq 8$.

c. Determine if the points in the half-plane above the line $4x - 2y = 8$ are solutions to the inequality $4x - 2y \geq 8$.

d. Shade the half-plane in the graph in part a that contains the test point that makes the inequality $4x - 2y \geq 8$ a true statement.

Therefore, the graph of the solution of the inequality $4x - 2y \geq 8$ is the set of all points on the line $4x - 2y = 8$ and in the half-plane below the line $4x - 2y = 8$. The graph gives you a visual representation of the solution set.

3. Graph the solution to the linear inequality $2x + 3y < 6$.

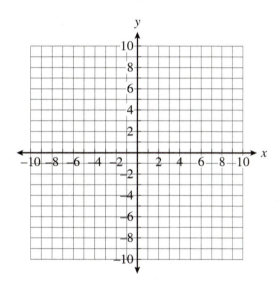

You may have discovered several patterns while graphing the inequalities in Problems 2 and 3. First, if the inequality symbol is $\leq$ or $\geq$, the points along the line are solutions to the inequality and the line is drawn solid. If the inequality symbol is $<$ or $>$, the corresponding line is drawn dashed, indicating that points along the line are not solutions to the inequality.

When the inequality **is written in slope-intercept** form (solved for y), another pattern is observed.

i. $y < mx + b$ or $y \leq mx + b$ means the half-plane **below** the line $y = mx + b$ is shaded.

ii. $y > mx + b$ or $y \geq mx + b$ means the half-plane **above** the line $y = mx + b$ is shaded.

4. Redo Problem 2 using the given patterns. Compare your results.

5. You can use the TI-83/84 Plus to graph a linear inequality such as $4x - 2y \geq 8$.

   a. Replace the inequality symbol with an equal sign and solve the corresponding equation $4x - 2y = 8$ for y.

   b. Enter the equation in part a into the calculator and show the graph in the standard window. Your screen should appear as follows:

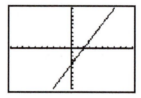

   c. To shade the half-plane below the line, press $y =$ and move the cursor to the extreme left. Next press ENTER key as many times as necessary to change the icon to ◣. Then graph to obtain the following screen:

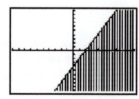

6. Use the graphing calculator to verify the results in Problem 3.

Solving a System of Linear Inequalities by Graphing

Recall that the solution to a system of linear **equations** is the point of intersection of the graphs of each line in the system. The (x, y) coordinates of the point of intersection makes each equation in the system a true statement. The solution to a system of linear **inequalities** such as

$$4x - 2y \geq 8$$
$$2x + 3y < 6$$

is the intersection of the solutions for each of the inequalities in the system. Therefore, to determine the solution set for a system of linear inequalities, you begin by graphing the solutions of each of the inequalities in the system.

7. a. Use the results from Problems 2 and 3 to graph the solution set of the inequalities $4x - 2y \geq 8$ and $2x + 3y < 6$ on the following coordinate system:

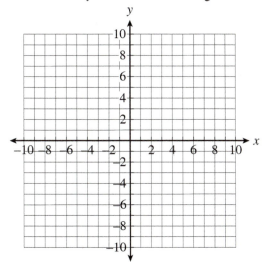

The solution of the system inequalities

$$4x - 2y \geq 8$$
$$2x + 3y < 6$$

is the intersection (if any) of all the solution regions of the inequalities in the system.

b. Graph the solution set of the system

$$4x - 2y \geq 8$$
$$2x + 3y < 6$$

on the coordinate system in part a.

c. Verify the results in part b using the graphing calculator. After solving each inequality for y, enter the corresponding equations in Y_1 and Y_2. Determine the intersection of the boundary lines using CALC feature. Shade the regions determined by each inequality. The solution set is that portion of the graph that is shaded twice. Your screen should appear as follows:

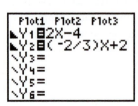 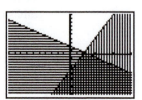

Note: The calculator does not draw dashed lines.

8. Solve each of the following systems of linear inequalities using a graphing approach. Verify using a graphing calculator.

a. $3x + 2y \geq 16$
     $x - 2y \leq 0$

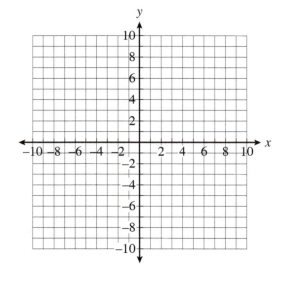

b. $y \leq -3x + 9$
$y \geq 2x - 3$
$x \geq 0$
$y \geq 0$

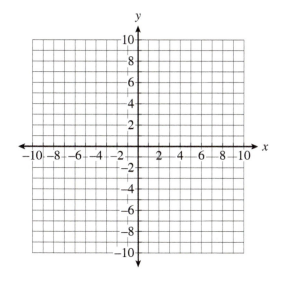

c. $y > 2x + 4$
$y \leq 2x - 6$

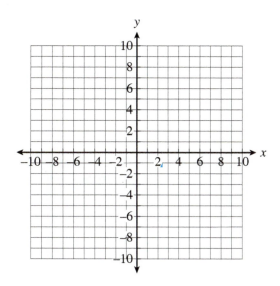

Corner Points

The graph of the system of inequalities in Problem 8a is

$$3x + 2y \geq 16$$
$$x - 2y \leq 0$$

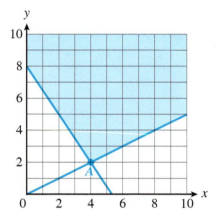

The point A of the intersection of the boundary lines of the shaded region is called a **corner point.**

9. a. Determine the coordinates of the corner point A in Problem 8a.

b. Is this corner point a solution of the given system of inequalities in Problem 8a? Explain.

10. By observing the boundary lines, determine if the corner point in Problem 7 is a solution to the system of inequalities.

11. Determine the corner points of the system in Problem 8b.

12. Return to the system of linear inequalities encountered in the Will the Trees Grow? situation,

$$t \geq 35$$
$$5t - 7p < 70,$$

where t represents the average annual temperature in °F, and p represents the annual precipitation in inches.

a. Graph the given system of inequalities. Determine ordered pairs of the form (t, p).

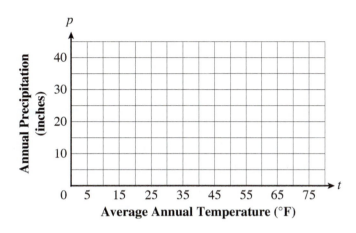

b. Determine the corner point. What is the significance of the point in this situation?

c. Determine whether or not trees will grow in each of the following regions.

REGION	AVERAGE TEMPERATURE (°F)	ANNUAL PRECIPITATION (inches)	(t, p)	CONCLUSION
Baghdad, Iraq	73	6		
Fairbanks, Alaska	29	14		
Lima, Peru	66	1		

Now, go to additional regions (such as Aden, Yemen) and confirm whether trees will grow there (without artificial support) by inspecting the graph of the system and by direct substitution.

SUMMARY: ACTIVITY 2.15

1. An inequality of the form $Ax + By < C$, where A and B cannot both equal zero, is called a **linear inequality in two variables**. The symbol $<$ can be replaced by $>$, $\leq$ or $\geq$.

2. The solution set of a linear inequality in two variables x and y is the collection of all the ordered pairs (x, y) whose coordinates satisfy the given inequality.

3. A **half-plane** is the set of all points on one side of a line.

4. To graph a linear inequality in two variables,

   i. Replace the inequality symbol with an equal sign and graph the resulting line. Draw a solid line if the inequality symbol is $\leq$ or $\geq$. If the inequality symbol is $<$ or $>$, draw a dashed line.

   ii. Select a test point not on the line and substitute its coordinates into the inequality. If the resulting statement is true, shade in the area on the same side of the line as the test point. If the resulting statement is false, shade in the area on the opposite side of the line as the test point

5. Two or more linear inequalities in two variables make up a **system of linear inequalities.**

6. The solution set of a system of linear inequalities in two variables is the collection of all ordered pairs whose coordinates satisfy each linear inequality in the system.

7. To graph a system of linear inequalities in two variables,

   i. Graph each linear inequality in the system.

   ii. The graph of the system is the intersection (overlap) of the shaded areas representing the solutions of the linear inequalities in the system, and any solid line common to all the inequalities in the system.

8. The points determined by the intersection of the boundary lines of the graph of a system of linear inequalities are called **corner points.**

EXERCISES: ACTIVITY 2.15

In Exercises 1–7, solve each of the systems of linear inequalities using a graphing approach. Verify using a graphing calculator.

1.
$$x \geq 0$$
$$y \geq 0$$
$$3x + 2y \leq 6$$

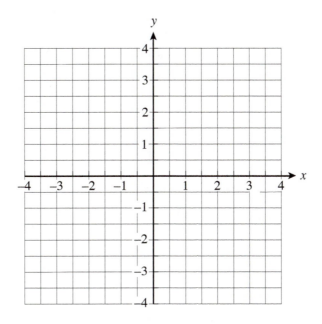

2. $x + y \leq 5$
$2x + y \geq 4$

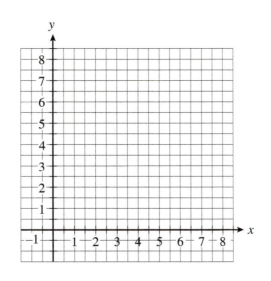

3. $y < x + 3$

$y \geq -2x + 6$

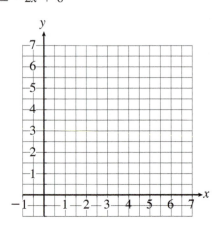

4. $-x + 2y \leq 10$

$2x - y \geq -4$

$y \leq 8$

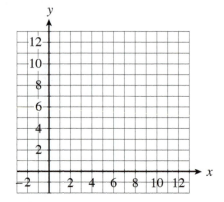

5. $2x - y \leq 0$

$2x + y \geq 4$

$-2x + 3y \leq 12$

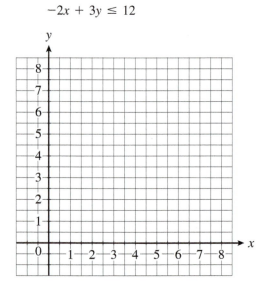

6. $x - 3y \leq 6$

$x - 3y > -6$

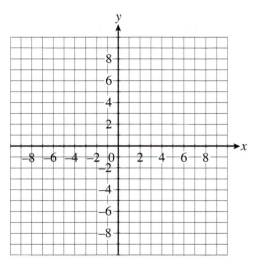

7. $x - 3y \geq 6$

$x - 3y < -6$

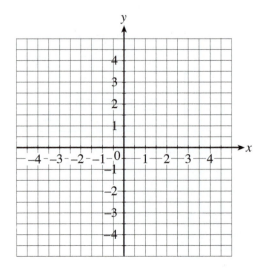

In Exercises 8–10, graph the given system of linear inequalities. Then, determine the corner points of the graph of the solution set.

8. $x \geq 1$
$\quad y \geq 0$
$\quad 2x + y \geq 4$

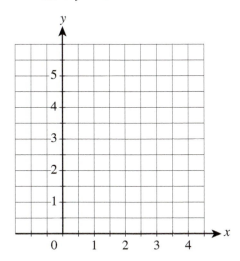

9. $y \geq 0$
$\quad y \leq 4$
$\quad x \geq 0$
$\quad x + y \leq 5$

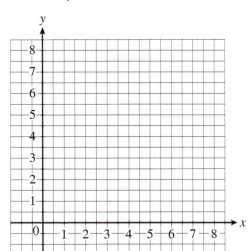

10. $3x - 5y \geq -10$
$\quad x + y \geq 6$
$\quad x \leq 5$

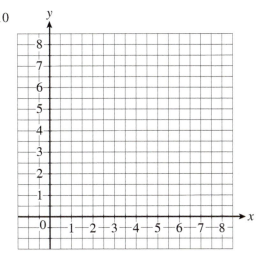

11. According to Miller and Thompson in their book *Elements of Meteorology,* grasslands will occur if the following system of inequalities is satisfied:

$$t \geq 35$$
$$5t - 7p \geq 70$$
$$3t - 35p \leq -140,$$

where t represents the average annual temperature in °F, and p represents the annual precipitation in inches.

a. Will grass grow in a region in which the average annual temperature is 40°F and the annual precipitation is 30 inches?

b. Graph the solution to the given system of linear inequalities.

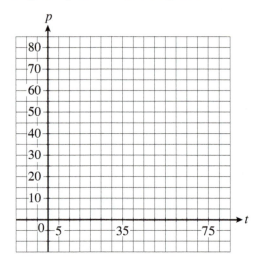

c. Determine the corner points.

12. The following mathematical model gives a relationship between a person's height and weight that is recommended for a healthy weight:

$$h > 0$$
$$w > 0$$
$$25h - 7w \leq 800$$
$$5h - w \geq 170,$$

where h represents height in inches, and w represents weight in pounds.

a. Your friend is 5 feet 7 inches tall and weighs 150 pounds. Is this a healthy weight for your friend? Explain.

b. According to the model, is 175 pounds a healthy weight for someone who is 6 feet tall? Explain.

c. Graph the solution set to the given system of linear inequalities. Use ordered pairs of the form (w, h).

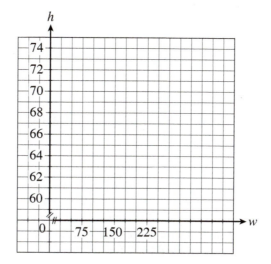

13. The local university requires that students applying for admission must have a combined SAT verbal and math score of at least 1150. The college of arts and sciences has an additional requirement that the verbal score must be at least 550.

a. Write a system of linear inequalities that gives the requirements (constraints) to be admitted to the college of arts and sciences.

b. The maximum score on the verbal part of the SAT is 800. The maximum score on the math portion of the SAT is also 800. Write two linear inequalities that express the constraints on the SAT scores.

c. The total system that describes the SAT requirements for admission to the college of arts and sciences is the combination of inequalities in parts a and b. Solve this system by graphing.

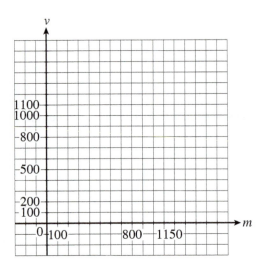

d. Determine two different combinations of verbal and math scores that satisfy the admission requirements to the college of arts and sciences.

14. A course in graphic design is being offered in the evening at the local high school. Because of limited space in the lab, the course is limited to 16 students. In order for the high school to cover expenses to run the course, at least $2800 must be received in student fees. The high school charges $280 per student if you are a resident student (live in the town where the high school is located). Students who live outside the town are charged $400 for the course.

a. Could a combination of 10 in-town students and 8 out-of-town students enroll in the course? Explain.

b. Could a combination of 6 in-town students and 2 out-of-town students enroll in the class? Explain.

c. Let x represent the number of in-town students enrolled in the class and y represent the number of out-of-town students enrolled. Write a system of linear inequalities that describes the constraints (limitations or restrictions) placed on x and y in this situation.

d. Solve the system in part c by graphing.

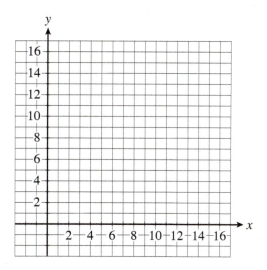

e. Give one combination of students (number of in-town and number of out-of-town) that could enroll in the class so that the constraints on the numbers and what they paid are satisfied.

f. Write an equation that gives the total amount of tuition, denoted by T, in terms of x and y.

Activities 2.11–2.15   **What Have I Learned?**

1. What is meant by a *solution to a system of linear equations?* How is a solution represented algebraically? Graphically.

2. Typically, a linear system of equations has one unique solution. Under what conditions is this not the case?

3. In this section, you solved 2 × 2 linear systems using a variety of methods. Briefly describe each method.

4. Describe a procedure that will combine the following two linear equations in three variables into a single linear equation in two variables.

$$2x + 3y - 5z = 10$$
$$3x - 2y + 2z = 4$$

5. Explain when the addition method would be more efficient to use than the substitution method as you solve a system of linear equations algebraically.

6. In solving an inequality, explain when you would change the direction of the inequality symbol.

7. a. Graph the solution set of the following system of linear inequalities.

$3x + y \geq 4$

$3x + y < 1$

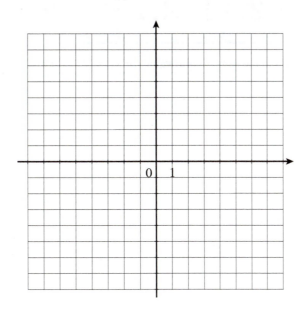

b. Does the system in part a have a solution? Explain why or why not.

Activities 2.11–2.15   How Can I Practice?

1. What input value results in the same output value for $y_1 = 2x - 3$ and $y_2 = 5x + 3$?

2. What point on the line given by $4x - 5y = 20$ is also on the line $y = x + 5$?

3. Solve the following systems both graphically and algebraically.

   a. $x + y = -3$
      $y = x - 5$

   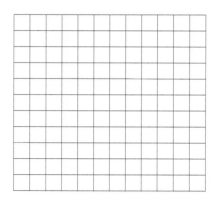

   b. $x - 2y = -1$
      $4x - 3y = 6$

   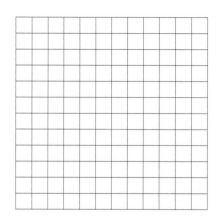

4. Rewrite the systems in Exercise 3 in the form

$$y = ax + b$$
$$y = cx + d$$

and check your solutions numerically using the table feature of your graphing calculator.

5. Solve the following inequalities algebraically. Check your solutions graphically.

a. $2.5x + 9.8 \geq 14.3$ **b.** $-3x + 14 < 32$ **c.** $-5 \leq 3x - 8 < 7$

6. You are going to help your grandmother plant a garden of tulips and daffodils. She has space for approximately 80 bulbs. The florist tells you that tulips cost \$0.50 per bulb and daffodils cost \$0.75 per bulb. How many of each can you purchase if her budget is \$52?

    a. Write the system of equations.

    b. Solve the system algebraically.

    c. Check your solution graphically using your graphing calculator.

7. You need some repair work done on your truck. Towne Truck charges \$80 just to examine the truck and \$40 per hour for labor costs. World Transport charges \$50 for the initial exam and \$60 per hour for the labor.

    a. Write a cost equation for each company. Use y to represent the total cost of doing the work and x to represent the number of hours of labor.

b. Complete the table of values for the cost functions.

x (NUMBER OF HOURS)	v. TOWNE TRUCK COST	v. WORLD TRANSPORT COST
2		
4		
6		

c. Graph the functions.

d. From the graph, estimate after how many hours the costs will be equal. What will be the total cost?

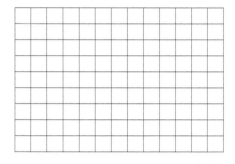

e. Check your solution in part d by solving the system algebraically.

f. You think that you have a transmission problem that will take approximately 4 hours to fix. Determine from the graph which company you will hire for this job. Explain.

8. Translate each of the following into an inequality statement.

a. x is greater than -5 and at most 6.

b. x is less than -5 or x is at least 3.

c. x is greater than or equal to -3 and less than 4.

9. You own a hot dog cart in Daytona Beach. Your monthly profit is determined from the expression $1.50x - 50$, where x represents the number of hot dogs sold each month. The number 1.50 in the expression is the profit for each hot dog. The cost of leasing the hot dog stand is the number 50 in the expression $1.50x - 50$.

a. To ensure a profit of at least $2000 per month, approximately how many hot dogs do you have to sell? Write the inequality and solve.

b. Your profit has been fluctuating between $1500 and $2200 per month. Determine approximately between what two values your hot dog sales have to be to realize this range of profit. Write the inequality and solve.

10. The formula $A = P + Prt$ represents the value, A, of an investment of P dollars at a yearly simple interest rate, r, for t years.

a. Write an equation to model the value, A, of an investment of $100 at 8% for t years.

b. Write an equation to model the value A of an investment of $120 at 5% for t years.

c. Assuming A has the same value, the equations in parts a and b form a system of two linear equations. Solve this system using an algebraic approach.

d. Interpret your answer in part c.

11. Assuming equal costs, when warehouse workers use hand trucks to load a boxcar, it costs the management $40 for labor for each boxcar. After management purchases a forklift for $2000, it costs only $15 for labor to load each boxcar.

a. Write an equation that models the cost, $C(n)$, of loading n boxcars with a hand truck.

b. Write an equation that models the total cost, $C(n)$, including the purchase price of the forklift, of loading n boxcars with the forklift.

c. The equations in parts a and b form a system of linear equations. Solve the system using an algebraic approach.

d. Interpret your answer in part c in the context of the boxcar situation.

12. The sign on the elevator in a 10-story building states that the maximum weight the elevator can carry is 1500 pounds. As part of your work-study program, you need to move a large shipment of books to the sixth floor. Each box weighs 80 pounds.

a. Let n represent the number of boxes that you place in the elevator. If you weigh 150 pounds, write an expression that represents the total weight, w, in the elevator. Assume that only you and the boxes are in the elevator.

b. Using the expression in part a, write an inequality that can be used to determine the maximum number of boxes that you can place in the elevator at one time.

c. Solve the inequality in part b.

13. Solve the following systems of linear inequalities.

a. $y \geq \dfrac{1}{3}x - 2$

$y < -x + 3$

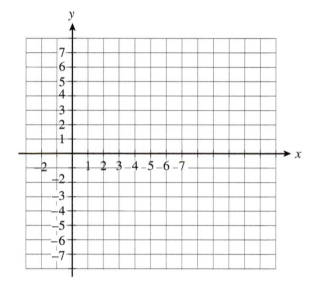

b. $x - y \leq 0$

$y \geq -3x + 9$

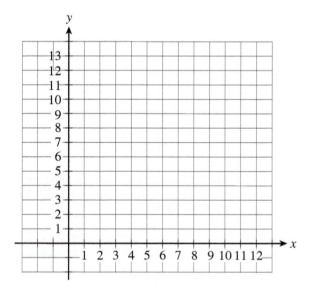

14. Determine the corner points of the solution set of the following system of linear inequalities.

$$x \geq 0$$

$$x \leq 4$$

$$y \geq 0$$

$$x + 2y \leq 10$$

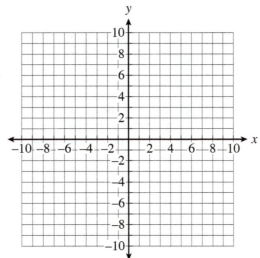

Chapter 2   Summary

The bracketed numbers following each concept indicate the activity in which the concept is discussed.

CONCEPT/SKILL	DESCRIPTION	EXAMPLE
Delta notation for change [2.1]	Let y_1 and y_2 represent the values corresponding to x_1 and x_2, respectively. As the variable x changes in value from x_1 to x_2, the change is represented by $\Delta x = x_2 - x_1$ and the change in y is represented by $\Delta y = y_2 - y_1$.	Given the x-y-pairs $(4, 14)$ and $(10, 32)$, $\Delta x = 10 - 4 = 6$ and $\Delta y = 32 - 14 = 18$.

Average rate of change over an interval [2.1]

The quotient

$$\frac{\Delta y}{\Delta x} = \frac{y_2 - y_1}{x_2 - x_1}$$

is called the average rate of change of y with respect to x over the x-interval from x_1 to x_2.

x	−3	4	7	10
y	0	14	27	32

The average rate of change over the interval from $x = 4$ to $x = 10$ is

$$\frac{\Delta y}{\Delta x} = \frac{32 - 14}{10 - 4} = \frac{18}{6} = 3.$$

Linear function [2.2]

A linear function is one whose average rate of change of y with respect to x from any one data point to any other data point is always the same (constant) value.

x	1	2	3	4
y	10	15	20	25

The average rate of change between any two of these points is 5.

Graph of a linear function [2.2]

The graph of every linear function is a straight line.

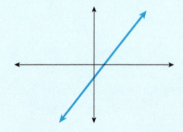

Slope of a line [2.2]

The slope of the line that contains the two points (x_1, y_1) and (x_2, y_2) is denoted by m; $m = \dfrac{\Delta y}{\Delta x} = \dfrac{y_2 - y_1}{x_2 - x_1}, x_2 \neq x_1$.

The slope of the line containing the two points $(2, 7)$ and $(5, 11)$ is

$$\frac{11 - 7}{5 - 2} = \frac{4}{3}.$$

Positive slope [2.2]

The graph of every linear function with positive slope is a line rising to the right.
A linear function is increasing if its slope is positive.

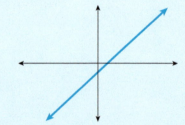

CONCEPT/SKILL	DESCRIPTION	EXAMPLE
Negative slope [2.2]	The graph of every linear function with negative slope is a line falling to the right. A linear function is decreasing if its slope is negative.	
Slope-intercept form of the equation of a line [2.3]	Represent the independent variable by x, the dependent variable by y. Denote the slope of the line by m, the y-intercept by $(0, b)$. Then the coordinate pair, (x, y), of *every* point on the line satisfies the equation $y = mx + b$.	The line with equation $y = 3x + 4$ has a slope of 3 and y-intercept $(0, 4)$. The point $(2, 10)$ is on the line because its coordinates satisfy the equation: $10 = 3(2) + 4$
x-intercept of a graph [2.3]	The x-intercept is the point at which the graph crosses the x-axis. Its ordered-pair notation is $(a, 0)$; that is, the y-value is equal to zero.	A line with x-intercept $(-3, 0)$ crosses the x-axis 3 units to the left of the origin.
y-intercept of a graph [2.3]	The y-intercept is the point at which the graph crosses the y-axis. Its ordered-pair notation is $(0, b)$; that is, the x-value is equal to zero.	A line with y-intercept $(0, 5)$ crosses the y-axis 5 units above the origin.
Zero slope [2.4]	The graph of every linear function with zero slope is a horizontal line. Every point on a horizontal line has the same y-value.	
Undefined slope [2.4]	A line whose slope is not defined (because its denominator is zero) is a vertical line. A vertical line is the only line that does not represent a function. Every point on a vertical line has the same input value.	
Equation of a horizontal line [2.4]	The slope, m, of a horizontal line is 0, and the y-value of each of its points is the same constant value, c. Its equation is $y = c$.	An equation of the horizontal line through the point $(-2, 3)$ is $y = 3$.
Equation of a vertical line [2.4]	The graph of $x = a$ is a vertical line.	The graph of $x = 2$ is a vertical line 2 units to the right of the y-axis.

CONCEPT/SKILL	DESCRIPTION	EXAMPLE
Rewriting the equation of a line in slope-intercept form [2.4]	A nonvertical line whose equation is not in slope-intercept form can be rewritten in slope-intercept form by solving the equation for y.	The equation of the line $5x - 2y = 6$ can be rewritten by solving for y: $$-2y = 6 - 5x \quad \text{Subtract } 5x.$$
Identifying the slope and y-intercept [2.4]	The slope can now be identified as the coefficient of the x-term. The y-intercept can be identified as the constant term.	$$\frac{-2y}{-2} = \frac{6}{-2} - \frac{5x}{-2} \quad \text{Divide by } -2.$$ $$y = -3 + \frac{5}{2}x \text{ or } y = \frac{5}{2}x - 3$$ The slope is $\frac{5}{2}$. The y-intercept is $(0, -3)$.
General form of a linear equation [2.4]	A linear function whose equation is in the form $Ax + By = C$, where A, B, and C are constants, is said to be written in general form.	$2x + 3y = 6$ is an equation of a linear function written in general form.
Reflection across the x-axis [2.5]	If the graph of $y = f(x)$ is reflected across the x-axis, then the equation of the resulting graph is $y = -1f(x)$. The graphs of $y = f(x)$ and $y = -1f(x)$ are mirror images across the x-axis. The y-values of each point in the original graph changes sign. The x-values remain the same.	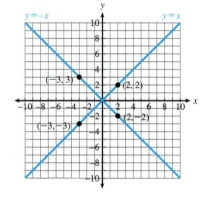 The graphs of $y = -1x$ is a reflection of $y = x$ across the x-axis.
Vertical stretch and shrink [2.5]	If the graph of $y = f(x)$ is vertically stretched (shrunk) by a factor of a, then the equation of the stretched (shrunk) graph is $y = af(x)$. The number a is called the stretch (shrink) factor. The y-value of each point in the graph of $y = f(x)$ is multiplied by a factor of a. The x-values remain the same.	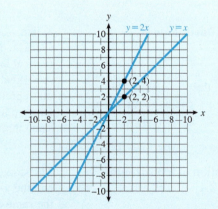 The graphs of $y = 2x$ is a vertical stretch of the graph of $y = x$. The stretch factor is 2. The y-value of each point of the graph of $y = x$ is doubled to obtain the graph of $y = 2x$.

CONCEPT/SKILL	DESCRIPTION	EXAMPLE
Graph of $y = c \cdot f(x)$, where c is a positive constant [2.5]	If c is a positive constant, then the graph of $y = c \cdot f(x)$ is the graph of $y = f(x)$ **i.** vertically shrunk by a factor of c, if $0 < c < 1$. **ii.** vertically stretched by a factor of c, if $c > 1$.	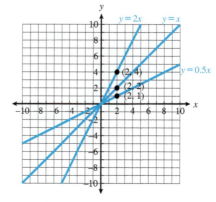 The graph of $y = 2x$ is a vertical stretch of the graph of $y = x$. The graph of $y = 0.5x$ is a vertical shrink of the graph of $y = x$.
Transformation [2. 5]	Vertical and horizontal shifts (also called *translations*), reflections across the x-axis, and vertical stretches and shrinks are all examples of transformations.	The graph of $y = -2x + 1$ can be interpreted as a combination of transformations, including vertical shift, stretch, and reflection.
Using the slope and y-intercept to write an equation of the line [2.6]	Given a line whose slope is m and y-intercept is $(0, b)$, an equation of the line is $$y = mx + b.$$	An equation of the line with slope $\dfrac{2}{3}$ and y-intercept $(0, -6)$ is $$y = \frac{2}{3}x - 6.$$
Determining the x-intercept of a line given its equation [2.6]	Because the x-intercept is the point whose y-coordinate is 0, set $y = 0$ in the equation and solve for x.	Given the line with equation $$y = 2x + 6,$$ set $y = 0$ to obtain equation $$0 = 2x + 6$$ $$-6 = 2x$$ $$-3 = x$$ The x-intercept is $(-3, 0)$.
Determining an equation of a line given the coordinates of two points on the line [2.7]	Given (x_1, y_1) and (x_2, y_2), calculate the slope, $m = \dfrac{\Delta y}{\Delta x} = \dfrac{y_2 - y_1}{x_2 - x_1}$. Substitute the coordinates of either given point for (h, k) and the value of the slope m into the equation $y - k = m(x - h)$. Then, solve for y.	See Example 1 in Activity 2.7
Linear regression equation [2.8, 2.9]	The linear regression equation is the linear equation that best fits a set of data.	See Problem 3 in Activity 2.9.
Interpolation [2.9]	Interpolation is the process of predicting a value of output for an input value that lies outside the range of the original data.	See Problem 6 in Activity 2.9.

CONCEPT/SKILL	DESCRIPTION	EXAMPLE
Extrapolation [2.9]	Extrapolation is the process of predicting a value of output for an input value that lies outside the range of the original data.	See Problem 6 in Activity 2.9.
System of equations [2.11]	Two equations that relate the same variables are called a system of equations. The solution of the system is the ordered pair(s) that satisfies the two equations.	For the system of equations $2x + 3y = 1$ $x - y = 3,$ the ordered pair $(2, -1)$ is a solution of the system because $x = 2, y = -1$ satisfy both equations.
Graphical method for solving a system of linear equations [2.11]	Graph both equations on the same coordinate system. If the two lines intersect, then the solution to the system is the coordinates of the point of intersection.	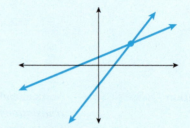
Substitution method for solving a system of linear equations [2.11]	• Solve one equation for either variable, say y as an expression in x. • In the other equation, replace y by the expression in x. (This equation should now contain only the variable x.) • Solve this equation for x. • Use this value of x to determine y.	For the system of equations $2x + 3y = 11$ $y = x - 3$ • In the second equation, y is already written as an expression in x. • Rewrite the first equation $2x + 3(y) = 11$ as $2x + 3(x - 3) = 11.$ • Solve this equation for x: $2x + 3x - 9 = 11$ $5x = 20$ $x = 4$ • Solve for y: $y = x - 3$, so $y = 4 - 3 = 1.$
Solution set of an inequality [2.14]	The solution set of an inequality is the set of all values of the variable that satisfy the inequality.	$x \leq 3$ is the solution set to the inequality $2x + 3 \leq 9.$

CONCEPT/SKILL	DESCRIPTION	EXAMPLE
Solving inequalities: direction of inequality sign [2.14]	**Property 1.** The direction of an inequality is not changed when **i.** the same quantity is added to or subtracted from both sides of the inequality. Stated algebraically, if $a < b$ then $a + c < b + c$ and $a - c < b - c$. **ii.** both sides of an inequality are multiplied or divided by the same positive number. If $a < b$, then $ac < bc$, where $c > 0$ and $\dfrac{a}{c} < \dfrac{b}{c}$, where $c > 0$. **Property 2.** The direction of an inequality is reversed if both sides of an inequality are multiplied by or divided by the same negative number. These properties can be written symbolically as: **i.** if $a < b$, then $ac > bc$, where $c < 0$. **ii.** if $a < b$, then $\dfrac{a}{c} > \dfrac{b}{c}$, where $c < 0$. Properties 1 and 2 will still be true if $a < b$ is replaced by $a \le b, a > b$, or $a \ge b$.	For example, because $-4 < 10$, then **i.** $-4 + 5 < 10 + 5$ or $1 < 15$ is true $-4 - 3 < 10 - 3$ or $-7 < 7$ is true. **ii.** $-4(6) < 10(6)$ or $-24 < 60$ is true; $\dfrac{-4}{2} < \dfrac{10}{2}$ or $-2 < 5$ is true. Because $-4 < 10$, then $-4(-5) > 10(-5)$ or $20 > -50$. Because $-4 < 10$, then $\dfrac{-4}{-2} > \dfrac{10}{-2}$ or $2 > -5$. These properties will be true if $a < b$ is replaced by $a \le b, a > b$, or $a \ge b$.
Solving inequalities of the form $f(x) < g(x)$ [2.14]	Inequalities of the form $f(x) < g(x)$ can be solved using three different methods. **Method 1** is a numerical approach, in which a table of input-output pairs is used to determine values of x for which $f(x) < g(x)$. **Method 2** is a graphical approach, in which values of x are located so that the graph of f is below the graph of g. **Method 3** is an algebraic approach, in which the two properties of inequalities are used to isolate the variable.	Solve $5x + 1 \ge 3x - 7$.  $5x + 1 \ge 3x - 7$ $2x \ge -8$ $x \ge -4$
A compound inequality [2.14]	A compound inequality is a statement that involves more than 1 inequality symbol $<, >, \le,$ or $\ge$.	$-3 < x + 7 \le 10$
Linear inequality in two variables [2.15]	An inequality of the form $Ax + By < C$, where A and B can not both equal zero, is a linear inequality in two variables. The symbol $<$ can be replaced by $>, \le,$ or $\ge$.	$2x + 3y \le 10$ is a linear inequality in two variables.

CONCEPT/SKILL	DESCRIPTION	EXAMPLE
Graph of a linear inequality in two variables [2.15]	The graph is the collection all ordered pairs (x, y) whose coordinates satisfy the given inequality.	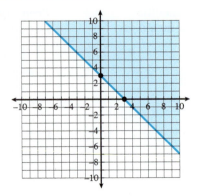 All points on the solid line $y = -x + 3$ and above the line $y = -x + 3$ represent the graph of $x + y \geq 3$.
Half-plane [2.15]	A half-plane is the set of all points on one side of a line.	All the points above the line $y = -x + 3$ represented in the shaded region in the graph above.
System of linear inequalities in two variables [2.15]	A system of linear inequalities consists of two or more linear inequalities.	$x + y \leq 5$ $y \leq 2x - 6$ $x \geq 0, y \geq 0$ represents a system of linear inequalities.
Graphing a system or linear inequalities in two variables [2.15]	Step 1. Graph each linear inequality in the system. Step 2. The graph of the system is the intersection of the shaded regions representing the solutions of the linear inequalities in the system, and any solid line common to all the inequalities in the system.	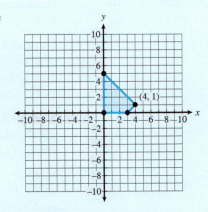
Corner points [2.15]	The points determined by the intersection of the boundary lines of the graph of a system of linear inequalities are called corner points.	The corner points in the region above arc $(0, 0)$, $(0, 5)$, $(3, 0)$, and $(4, 1)$.

1. For a certain yard, the fertilizer costs $20. You charge $8 per hour to do yard work. If x represents the number of hours worked on the yard and $f(x)$ represents the total cost, including fertilizer, complete the following table.

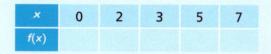

x	0	2	3	5	7
$f(x)$					

a. Is the total cost a function of the hours worked? Explain.

b. Which variable represents the input?

c. Which is the dependent variable?

d. Which value(s) of the domain would not be realistic for this situation? Explain.

e. What is the average rate of change from 0 to 3?

f. What is the average rate of change from 5 to 7?

g. What can you say about the rate of change between any two of the points?

h. What kind of relationship exists between the two variables?

i. Write this relationship in the form $f(x) = mx + b$.

j. What is the practical meaning of the slope in this situation?

k. What is the y-intercept? What is the practical meaning of this point?

l. Determine $f(4)$.

m. For what value(s) of x does $f(x) = 92$? Interpret your answer in the context of the situation.

2. Which of the following sets of data represent a linear function?

a.

x	0	2	4	6	8
$f(x)$	14	22	30	38	46

b.

x	5	10	15	20	25
y	4	2	0	−2	−4

c.

x	1	3	4	6	7
g(x)	10	20	30	40	50

d.

t	0	10	20	30	40
d	143	250	357	464	571

3. a. Determine the slope of the line through the points $(5, -3)$ and $(-4, 9)$.

b. From the equation $3x - 7y = 21$, determine the slope.

c. Determine the slope of the line from its graph.

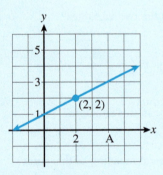

4. Write the equation of the line described in each of the following.

a. A slope of 0 and passing through the point $(2, 4)$.

b. A slope of 2 and a vertical intercept of $(0, 5)$.

c. A slope of -3 and passing through the point $(6, -14)$.

d. A slope of 2 and passing through the point $(7, -2)$.

e. A line with undefined slope passing through the point $(2, -3)$.

f. A slope of –5 and a horizontal intercept of $(4, 0)$.

g. A line passing through the points $(-3, -4)$ and $(2, 16)$.

h. Contains the point $(8, -3)$ and is parallel to $6x + 2y = 5$

5. Given the following graph of the linear function, determine the equation of the line.

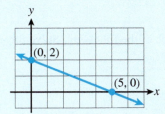

6. a. The building where the electronics store is located is 10 years old and has a current value of $200,000. When the building was 1-year-old, its value was $290,000. Assuming that the building's depreciation is linear, express the value of the building as a function, f, of its age, x, in years.

b. What is the slope of the line? What is the practical meaning of the slope in this situation?

c. What is the y-intercept? What is the practical meaning of the y-intercept in this situation?

d. What is the x-intercept? What is the practical meaning of the x-intercept in this situation?

7. Determine the y-intercept of the following functions. Solve for y if necessary.

a. $y = 2x - 3$

b. $y = -3$

c. $x - y = 3$

d. What relationship do the graphs of these functions have to one another?

e. Use your graphing calculator to graph the functions in parts a–c on the same coordinate axes. Compare your results with part d.

8. Determine the slopes and *y*-intercepts of each of the following. Solve for *y* if necessary.

    a. $y = -2x + 1$

    b. $2x + y = -1$

    c. $-4x - 2y = 6$

    d. What relationship do the graphs of these equations have to one another?

    e. Use your graphing calculator to graph the functions in parts a–c on the same coordinate axes. Compare your results with part d.

9. Determine the slopes and *y*-intercepts of each of the following. Solve for *y* if necessary.

    a. $y = -3x + 2$

    b. $3x + y = 2$

    c. $6x + 2y = 4$

    d. What relationship do the graphs of these equations have to one another?

    e. For two lines to be parallel to each other, what has to be the same?

    f. For two lines to lie on top of each other (coincide), what has to be the same?

    g. Use your graphing calculator to graph the functions in parts a–c on the same coordinate axes. Compare your results with part d.

10. a. Graph the function defined by $y = -2x + 150$. Determine the intercepts. Make sure to include some negative values of *x*.

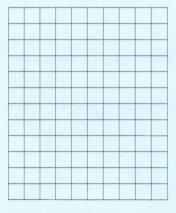

b. Using your graphing calculator, verify the graph you have drawn in part a.

c. Using the graph, determine the domain and range of the function.

d. Assume that a 150-pound person starts a diet and loses 2 pounds per week for 15 weeks. Write the equation modeling this situation.

e. Compare the equation you found in part d with the one given in part a.

f. What is the practical meaning of the x- and y-intercepts you found in part a?

g. What is the practical domain and range of this function for the situation given in part d?

11. a. You pay a flat fee of $35 per month for your trash to be picked up, and it doesn't matter how many bags of trash you have. Use x to represent the number of bags of trash, and write a function, f, in symbolic form to represent the total cost of your trash for the month.

b. Sketch the graph of this function.

c. What is the slope of the line?

12. During the years 2010–2014, the number of finishers in a large marathon increased. The following table gives the total number of finishers (to the nearest hundred) each year, where t represents the number of years after 2010.

Years After 2010, t	0	1	2	3	4
Number of Finishers, n	7900	9100	10,000	10,900	12,100

a. Enter the data from your table into your calculator. Determine the linear regression equation model and write the result.

b. What is the slope of the regression line? What is the practical meaning of the slope in this situation?

c. What is the n-intercept? What is the practical meaning of the n-intercept in this situation?

d. Use your graphing calculator to graph the regression line in the same screen as the scatterplot. How well do you think the line fits the data?

e. Use your regression model to predict the number of finishers in 2016.

f. Did you use interpolation or extrapolation to determine your result in part e? Explain.

g. Do you think that a prediction for the year 2029 will be as accurate as that in 2016? Explain.

13. Which input value results in the same output value for $y_1 = 2x - 3$ and $y_2 = 5x + 3$?

14. Which point on the line given by $4x - 5y = 20$ is also on the line $y = x + 5$?

15. You sell sets of hand-crafted jewelry online for $45 each. Your fixed costs are $500 per month, and each set of jewelry costs an average of $20 each to produce.

a. Let n represent the number of sets of jewelry you sell. Write an equation to determine the cost, C.

b. Write an equation to determine the revenue, R.

c. What is your break-even point (cost = revenue)?

16. Solve the system algebraically.

$y = -5x + 25$

$y = 5x + 30$

17. Use the addition method to solve the following system.

$4m - 3n = -7$

$2m + 3n = 37$

18. Solve the following system graphically and algebraically.

$y = 6x - 7$

$y = 6x + 4$

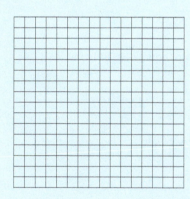

19. Consider the lines represented by the following pair of equations: $3x + 2y = 10$ and $x - 2y = -2$.

    a. Solve the system algebraically.

b. Determine the point of intersection of the lines by solving the system graphically.

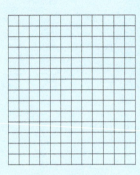

c. Determine the point of intersection of the lines by solving the system using the tables feature of your calculator, if available.

20. Use your graphing calculator to estimate the solution to the following system.

Note: Both equations first must be written in the form $y = mx + b$

$$342x - 167y = 418$$
$$-162x + 103y = -575$$

Use the window Xmin $= -10$, Xmax $= 0$,

Xscl $= 2$, Ymin $= -28$, Ymax $= 0$, Yscl $= 2$.

21. Solve the following systems of equations. Solve at least one system algebraically and at least one system graphically.

a. $3x - y = 10$

$5x + 2y = 13$

b. $4x + 2y = 8$

$x - 3y = -19$

22. The employees of a video arcade order lunch two days in a row from the corner deli. Lunch on the first day consists of five small pizzas and six cookies for a total of $27. On the second day, eight pizzas are ordered along with four cookies, totaling $39. To know how much money to pay, employees have to determine how much each pizza and each cookie cost. How much does the deli charge for each pizza and each cookie?

23. Solve the following inequalities algebraically.

    a. $2x + 3 \geq 5$ **b.** $3x - 1 \leq 4x + 5$

24. The weekly revenue, $R(x)$, and cost, $C(x)$, generated by a product can be modeled by the following equations:

$R(x) = 75x$

$C(x) = 50x + 2500$

Determine when $R(x) \geq C(x)$ numerically, algebraically, and graphically.

x	R(x)	C(x)
0		
50		
100		
150		
200		

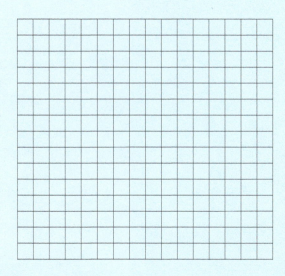

25. Graph the solution set for each of the following systems of linear inequalities:

a. $x + y \geq 10$
  $x + y \leq 12$
  $3y \leq x$
  $x \geq 0$
  $y \geq 0$

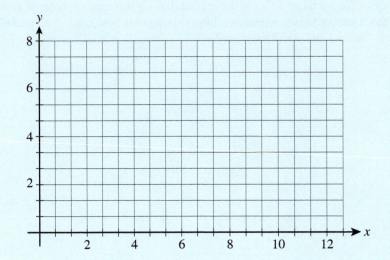

b. $x + y < 5$
  $2x + y > 4$
  $x > 0$
  $y > 0$

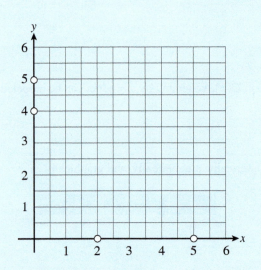

Chapter

3

Problem Solving with Quadratic and Variation Function Models

Activity 3.1

The Amazing Property of Gravity

Objectives

1. Evaluate functions of the form $y = ax^2, a \neq 0$.

2. Graph functions of the form $y = ax^2$.

3. Interpret the coordinates of points on the graph of $y = ax^2$ in context.

4. Solve an equation of the form $ax^2 = c$ graphically.

5. Solve an equation of the form $ax^2 = c$ algebraically by taking square roots.

Note: $a \neq 0$ in Objectives 1–5.

In the sixteenth century, scientists such as Galileo were experimenting with the physical laws of gravity. In a remarkable discovery, they learned that if the effects of air resistance are neglected, any two objects dropped from a height above earth will fall at exactly the same speed. That is, if you drop a feather and a brick down a tube whose air has been removed, the feather and brick will fall at the same speed. Surprisingly, the function that models the distance fallen by such an object in terms of elapsed time is a very simple one:

$$s = 16t^2,$$

where t represents the number of seconds elapsed and s represents distance (in feet) the object has fallen.

The function defined by $s = 16t^2$ indicates a sequence of two mathematical operations:

Start with a value for $t \rightarrow$ *square the value* $\rightarrow$ *multiply by 16* $\rightarrow$ to obtain values for s.

1. Use the given equation to complete the table.

t (sec)	s (ft)
0	
1	
2	
3	

2. **a.** How many feet does the object fall during the first second after it has been dropped?

   b. How many feet does the object fall during the first 2 seconds after it has been dropped?

3. **a.** Determine the average rate of change of distance fallen from time $t = 0$ to $t = 1$.

   b. What are the units of measurement of the average rate of change?

   c. Explain what the average rate of change indicates about the falling object.

353

4. Determine and interpret the average rate of change of distance fallen from time $t = 1$ to $t = 2$.

5. Is the function $s = 16t^2$ a linear function? Explain your answer.

6. a. If the object hits the ground after 5 seconds, determine the practical domain of the function.

   b. On the following grid, plot the points given in the table in Problem 1 and sketch a curve representing the distance function through the points.

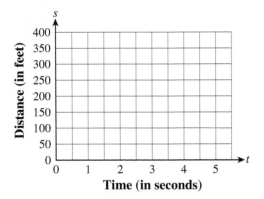

   c. Use your graphing calculator to verify the graph in part b in the window $0 \le x \le 5$, $0 \le y \le 400$. Your graph should resemble the one below.

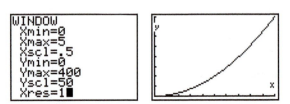

7. a. Confirm that the point $(2.5, 100)$ lies on the graph. What do the coordinates of this point indicate about the falling object?

b. Confirm that the point $(4.5, 324)$ lies on the graph. What do the coordinates of this point indicate about the falling object?

8. Use the graph to estimate the amount of time it takes the object to fall 256 feet.

Solving Equations of the Form $ax^2 = c$, $a \neq 0$

9. Use $s = 16t^2$ to write an equation to determine the amount of time it takes the object to fall 256 feet.

To solve the equation in Problem 9, you need to reverse the order of operations indicated by the function rule, replacing each operation by its inverse. Here, one of the operations is "square a number." The inverse of squaring is to take a square root, denoted by the symbol $\sqrt{}$, called a **radical sign**.

Start with a value for $s \rightarrow$ *divide by 16* $\rightarrow$ *take its square root* $\rightarrow$ to obtain t.

In particular, if $s = 256$ feet:

Start with $256 \rightarrow$ *divide by 16* $\rightarrow$ *take its square root* $\rightarrow$ to obtain t.

$$256 \div 16 = 16 \qquad \sqrt{16} = 4 \qquad t = 4$$

Therefore, you can conclude that it takes 4 seconds for the object to fall 256 feet.

10. Reverse the sequence of operations indicated by $s = 16t^2$ to determine the amount of time it takes an object to fall 1296 feet, approximately the height of a 100-story building.

Graph of a Parabola

Some interesting properties of the function defined by $s = 16t^2$ arise when you ignore the falling object context and consider just the algebraic rule itself.

Replace t with x and s with y in $s = 16t^2$, and consider the general equation $y = 16x^2$. First, by ignoring the context, you can allow x to take on a negative, positive, or zero value. For example, suppose $x = -5$. Then

$$y = 16(-5)^2 = 16 \cdot 25 = 400.$$

11. a. Use $y = 16x^2$ to complete the table.

x	−4	−3	−2.5	−2	−1.5	−1	−0.5	0	0.5	1	1.5	2	2.5	3	4
y								0		16		64		144	

      b. What pattern (symmetry) do you notice from the table?

12. a. Sketch the graph of $y = 16x^2$ by using the table in Problem 11. Plot the points and then draw a curve through them. Scale the axes appropriately.

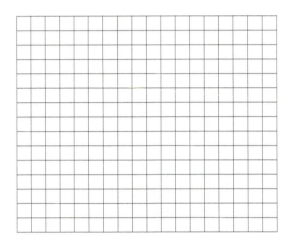

    b. Use a graphing calculator to produce a graph of this function in the window $-5 \leq x \leq 5, -100 \leq y \leq 400$. Your graph should resemble the one below.

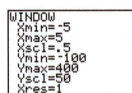
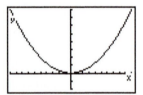

The U-shaped graph of $y = 16x^2$ is called a **parabola**.

13. a. The graph of the squaring function defined by $y = x^2$ is a parabola. How is the graph of $y = 16x^2$ related to the graph of $y = x^2$?

    b. The graphs of $y = 16x^2$ and $y = x^2$ are both U-shaped. Is the graph of $y = 16x^2$ wider or narrower than the graph of $y = x^2$? Explain.

    c. Is the graph of $y = 0.5x^2$ wider or narrower than the graph of $y = x^2$? Explain. Verify using a graphing calculator.

    d. If $0 < a < 1$, then is the graph of $y = ax^2$ wider or narrower than the graph of $y = x^2$?

    e. If $a > 1$, then is the graph of $y = ax^2$ wider or narrower than the graph of $y = x^2$?

Solving Equations of the Form $ax^2 = c, a \neq 0$: A Second Look

14. a. In the table in Problem 11a, how many points on the graph of $y = 16x^2$ lie 256 units above the x-axis?

b. Identify the points. What are their coordinates?

The x-values of the points on the graph of $y = 16x^2$ that lie 256 units above the x-axis can be determined algebraically by solving the equation $16x^2 = 256$.

Example 1   *Solve the equation* $16x^2 = 256$ *algebraically.*

SOLUTION

Step 1. Divide both sides by 16:

$$\frac{16x^2}{16} = \frac{256}{16} \text{ to obtain } x^2 = 16$$

Step 2. Calculate the square root:

When x is no longer restricted to positive (or nonnegative) values only, there are *two* square roots of 16, one denoted by $\sqrt{16}$ and the other denoted by $-\sqrt{16}$. These two square roots are often written in condensed form as $\pm\sqrt{16}$. Since $\sqrt{16} = 4$ and $-\sqrt{16} = -4$, the solutions to the equation $x^2 = 16$ are $x = \pm 4$.

The solution to $16x^2 = 256$ can be written in the following systematic manner.

$16x^2 = 256$

$\dfrac{16x^2}{16} = \dfrac{256}{16}$ Divide both sides by 16.

$x^2 = 16$ Simplify.

$x = \pm\sqrt{16}$ Take the positive and negative square roots.

or equivalently, $x = \pm 4$

15. a. How many points on the graph in Problem 12b lie 400 units above the x-axis?

b. Identify the points. What are their coordinates?

c. Set up the appropriate equation to determine the values of x for which $y = 400$, and solve it algebraically.

16. Solve the following equations algebraically:

    a. $x^2 = 36$ **b.** $2x^2 = 98$

    c. $3x^2 = 375$ **d.** $5x^2 = 50$

17. a. Refer to the graph in Problem 12 to determine how many points on the graph of $y = 16x^2$ lie 16 units *below* the x-axis.

    b. Set up an equation that corresponds to the question in part a.

    c. How many solutions does this equation have? Explain.

18. What does the graph of $y = 16x^2$ (Problem 12) indicate about the number of solutions to the following equations? (You do not need to solve these equations.)

    a. $16x^2 = 100$ **b.** $16x^2 = 0$ **c.** $16x^2 = -96$

19. Solve the following equations:

    a. $5x^2 = 20$ **b.** $4x^2 = 0$ **c.** $3x^2 = -12$

SUMMARY: ACTIVITY 3.I

1. The graph of a function of the form $y = ax^2, a \neq 0$, is a U-shaped curve and is called a **parabola**.

2. If $a > 0$, then the larger the value of a, the narrower the graph of $y = ax^2$.

3. An equation of the form $ax^2 = c, a \neq 0$, is solved algebraically by dividing both sides of the equation by a and then taking the positive and negative square roots of both sides.

4. Every positive number a has two square roots, one positive and one negative. The square roots are equal in magnitude. The positive, or **principal**, square root of a is denoted by $\sqrt{a}$. The negative square root of a is denoted by $-\sqrt{a}$. The symbol $\sqrt{}$ is called the **radical sign**.

5. The **square root** of a number a is the number that when squared produces a. For example, the square roots of 25 are

   $$\sqrt{25} = 5 \text{ because } 5^2 = 25$$

   and

   $$-\sqrt{25} = -5 \text{ because } (-5)^2 = 25.$$

   The square roots of 25 can be written in a condensed form as $\pm\sqrt{25} = \pm 5$.

EXERCISES: ACTIVITY 3.I

1. On the Earth's Moon, gravity is only one-sixth as strong as it is on Earth, so an object on the Moon will fall one-sixth the distance it would fall on Earth in the same time. This means that the gravity distance function for a falling object on the Moon is

   $$s = \frac{16}{6}t^2 \text{ or } s = \frac{8}{3}t^2,$$

   where t represents time since the object is released, in seconds, and s is the distance fallen, in ft.

   a. How far does an object on the Moon fall in 3 seconds?

   b. How long does it take an object on the Moon to fall 96 feet?

c. Graph the Moon's gravity function on a properly scaled and labeled coordinate axis or on a graphing calculator for $t = 0$ to $t = 5$.

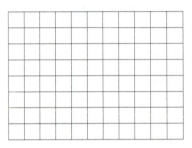

d. Use the graph to estimate how long it takes an object on the Moon to fall 35 feet.

e. Write an equation to determine the required time in part d. Solve the equation algebraically.

f. How does your answer in part e compare to your estimate in part d?

2. a. The graphs of $y = x^2$, $y = \frac{1}{3}x^2$, and $y = 5x^2$ appear in the following graphing calculator screen. Match the equation with the graph.

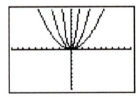

b. Is the graph of $y = 0.75x^2$ wider or narrower than the graph of $y = 10x^2$? Verify using a graphing calculator.

c. Is the graph of $y = 10x^2$ wider or narrower than the graph of $y = 5x^2$?

3. Solve the following equations:

   a. $5x^2 = 45$ **b.** $9x^2 = 0$ **c.** $-25x^2 = 100$

   d. $x^2 = 5$ **e.** $2x^2 = 20$ **f.** $\dfrac{x^2}{2} = 32$

4. Solve the following by first writing the equation in the form $x^2 = c$:

   a. $t^2 - 49 = 0$ **b.** $15 + c^2 = 96$ **c.** $3a^2 - 21 = 27$

In a right triangle, as shown in the following diagram, the side opposite the right angle is called the **hypotenuse**, and the other two sides are called **legs**. The **Pythagorean theorem** states that in any right triangle, the lengths of the three sides are related by the equation $c^2 = a^2 + b^2$.

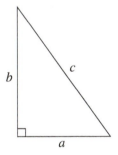

Use the Pythagorean theorem to answer Exercises 5 and 6.

5. Determine the length of the hypotenuse in a right triangle with legs 5 inches and 12 inches.

6. One leg of a right triangle measures 8 inches and the hypotenuse 17 inches. Determine the length of the other leg.

Activity 3.2

Baseball and the Willis Tower

Objectives

1. Identify functions of the form $f(x) = ax^2 + bx + c$, $a \neq 0$, as quadratic functions.

2. Explore the role of a as it relates to the graph of $f(x) = ax^2 + bx + c$.

3. Explore the role of b as it relates to the graph of $f(x) = ax^2 + bx + c$.

4. Explore the role of c as it relates to the graph of $f(x) = ax^2 + bx + c$.

Note: $a \neq 0$ in Objectives 1–4.

Imagine yourself standing on the roof of the 1450-foot-high Willis Tower (formerly called the Sears Tower) in Chicago. When you release and drop a baseball from the edge of the roof of the tower, *the ball's height above the ground, H* (in feet), can be described as a function of the time, t (in seconds), since it was dropped. This height function is defined by

$$H(t) = -16t^2 + 1450.$$

1. Sketch a diagram illustrating the Willis Tower and the path of the baseball as it falls to the ground.

2. **a.** Complete the following table.

TIME, t (sec.)	$H(t) = -16t^2 + 1450$
0	1450
1	
2	
3	
4	
5	
6	
7	
8	
9	
10	

b. How far does the baseball fall during the first second?

c. How far does it fall during the interval from 1 to 3 seconds?

3. Using the height function, $H(t) = -16t^2 + 1450$, determine the average rate of change of H with respect to t over the given interval. Remember:

$$\text{average rate of change} = \frac{\text{change in output}}{\text{change in input}}.$$

a. $0 \leq t \leq 1$ **b.** $1 \leq t \leq 3$

c. Based on the results of parts a and b, do you believe that $H(t) = -16t^2 + 1450$ is a linear function? Explain.

4. a. What is the value of H when the baseball strikes the ground? Use the table in Problem 2a to estimate the time when the ball is at ground level.

b. What is the practical domain of the height function?

c. Determine the practical range of the height function.

d. On the following grid, plot the points in Problem 2a that satisfy part b (practical domain) and sketch a curve representing the height function.

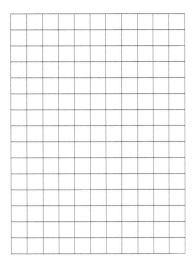

e. Is the graph of the height function in part d the actual path of the object (see Problem 1)? Explain.

Some interesting properties of the function defined by $H = -16t^2 + 1450$ arise when you ignore the falling object context. Replace H with y and t with x and consider the general function defined by $y = -16x^2 + 1450$.

5. The graph of $y = -16x^2 + 1450$ can be obtained by performing three transformations on the graph of $y = x^2$. In parts a–c,

    i. Use a graphing calculator to sketch a graph resulting from the given transformation. Set the window parameters at Xmin $= -10$, Xmax $= 10$, Ymin $= -1000$, and Ymax $= 1500$.

    ii. Write the equation of the graph.

    a. A vertical stretch of the graph of $y = x^2$ by a factor of 16.

   b. A reflection of the graph in part a over the x-axis.

   c. A vertical shift of the graph in part b upward 1450 units.

   d. The graph of $y = x^2$ is U-shaped and is said to open upward. The graph of $y = -16x^2 + 1450$ is also U-shaped. Does the graph open upward or open downward?

6. a. Graph the function defined by $y = -16x^2 + 1450$, setting the window parameters at Xmin $= -10$ and Xmax $= 10$ for the input and Ymin $= -50$ and Ymax $= 1500$ for the output. Your graph should appear as follows.

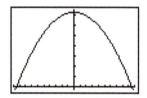

   b. Describe the important features of the graph of $y = -16x^2 + 1450$. Discuss the shape, symmetry, and intercepts.

Quadratic Functions

The graph of the function defined by $y = -16x^2 + 1450$ is a parabola. The graph of a **parabola** is a ∪-shaped figure that opens upward, ∪, or downward, ∩. Parabolas are graphs of a special category of functions called **quadratic functions**.

Definition

Any function defined by an equation of the form $y = ax^2 + bx + c$ or $f(x) = ax^2 + bx + c$, where a, b, and c represent real numbers and $a \neq 0$, is called a **quadratic function**. The output variable y is defined by an expression having three terms: the **quadratic term**, ax^2, the **linear term**, bx, and the **constant term**, c. The numerical factors of the quadratic and linear terms, a and b, are called the **coefficients** of the terms.

Example 1   *$H(t) = -16t^2 + 1450$ defines a quadratic function. The quadratic term is $-16t^2$. The linear term is $0t$, although it is not written as part of the expression defining $H(t)$. The constant term is 1450. The numbers -16 and 0 are the coefficients of the quadratic and linear terms, respectively. Therefore, $a = -16, b = 0,$ and $c = 1450$.*

7. For each of the following quadratic functions, identify the value of a, b, and c.

QUADRATIC FUNCTION	a	b	c
$y = 3x^2$			
$y = -2x^2 + 3$			
$y = x^2 + 2x - 1$			
$y = -x^2 + 4x$			

The Constant Term c: A Closer Look

Consider once again the height function $H(t) = -16t^2 + 1450$ from the beginning of the activity.

8. a. What is the H-intercept of the graph? Explain how you obtained the results.

b. What is the practical meaning of the H-intercept in this situation?

c. Predict what the graph of $H(t) = -16t^2 + 1450$ would look like if the constant term 1450 were changed to 800. That is, the baseball is dropped from a height of 800 feet rather than 1450 feet. Verify your prediction by graphing $H(t) = -16t^2 + 800$. What does the constant term tell you about the graph of the parabola?

The constant term c of a quadratic function $f(x) = ax^2 + bx + c$ always indicates the vertical (y-) intercept of the parabola. The vertical intercept of any quadratic function is $(0, c)$ since $f(0) = a \cdot 0^2 + b \cdot 0 + c = c$.

9. Graph the parabolas defined by the following quadratic equations. Note the similarities and differences among the graphs, especially the y-intercepts. Be careful in your choice of a window.

a. $f(x) = 1.5x^2$

b. $g(x) = 1.5x^2 + 7$

c. $q(x) = 1.5x^2 + 4$

d. $s(x) = 1.5x^2 - 4$

The Effects of the Coefficient a on the Graph of $y = ax^2 + bx + c$

10. a. Using transformations, describe how to obtain the graph of $y = 16x^2 + 1450$ from the graph of $y = x^2$.

b. Compare the transformations used to graph $y = 16x^2 + 1450$ to the transformations used to graph $y = -16x^2 + 1450$ (see Problem 5). What is different?

c. Graph the quadratic function defined by $y_1 = 16x^2 + 1450$ on the same screen as $y_2 = -16x^2 + 1450$. Use the window settings Xmin $= -10$, Xmax $= 10$, Ymin $= -50$, and Ymax $= 3000$.

d. What effect does the sign of the coefficient of x^2 appear to have on the graph of the parabola?

11. Graph the functions $y_3 = -16x^2 + 100$, $y_4 = -6x^2 + 100$, $y_5 = -40x^2 + 100$ in the same window. What effect does the magnitude of the coefficients of x^2 (namely, $|-16| = 16$, $|-6| = 6$, and $|-40| = 40$) appear to have on the graph of that particular parabola? Use window settings Xmin $= -10$, Xmax $= 10$, Ymin $= -50$, and Ymax $= 120$.

The results from Problems 10 and 11 regarding the effects of the coefficient a can be summarized as follows.

The graph of a quadratic function defined by $f(x) = ax^2 + bx + c$ is called a **parabola**.

- If $a > 0$, the parabola opens upward.

- If $a < 0$, the parabola opens downward.

- The magnitude of a affects the width of the parabola. The larger the absolute value of a, the narrower the parabola.

12. a. Is the graph of $h(x) = 0.3x^2$ wider or narrower than the graph of $f(x) = x^2$?

b. How do the output values of h and the output values of f compare for the same input value?

c. Is the graph of $g(x) = 3x^2$ wider or narrower than the graph of $f(x) = x^2$?

d. How do the output values of g and f compare for the same input value?

e. Describe the effect of the magnitude of the coefficient a on the width of the graph of the parabola.

f. Describe the effect of the magnitude of the coefficient of a on the output value.

The Effects of the Coefficient b on the Turning Point

Assume for the time being that you are back on the roof of the 1450-foot Willis Tower. Instead of merely releasing the ball, suppose you *throw it down* with an initial velocity of 40 feet per second. Then the function describing its height above ground as a function of time is modeled by

$$H_{\text{down}}(t) = -16t^2 - 40t + 1450.$$

If you tossed the ball straight up with an initial velocity of 40 feet per second, then the function describing its height above ground as a function of time is modeled by

$$H_{\text{up}}(t) = -16t^2 + 40t + 1450.$$

13. Predict what features of the graphs of H_{down} and H_{up} have in common with

$$H(t) = -16t^2 + 1450.$$

14. a. Graph the three functions $H(t)$, $H_{\text{down}}(t)$, and $H_{\text{up}}(t)$ using the same window settings given in Problem 6a.

b. What effect do the $-40t$ and $40t$ terms seem to have upon the turning point of the graphs?

If $b = 0$, the turning point of the parabola is located on the vertical axis. If $b \neq 0$, the turning point will not be on the vertical axis.

15. For each of the following quadratic functions, identify the value of b and then, without graphing, determine whether or not the turning point is on the y-axis. Verify your conclusion by graphing the given function. Set the window of your calculator to Xmin $= -8$, Xmax $= 8$, Ymin $= -20$, and Ymax $= 20$.

a. $y = x^2$ **b.** $y = x^2 - 4x$

c. $y = x^2 + 4$

d. $y = x^2 + x$

e. $y = x^2 - 3$

16. Match each function with its corresponding graph below, and then verify using your graphing calculator.

a. $f(x) = x^2 + 4x + 4$

b. $g(x) = 0.2x^2 + 4$

c. $h(x) = -x^2 + 3x$

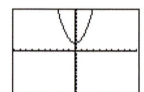

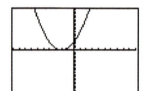

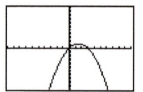

SUMMARY: ACTIVITY 3.2

I. The equation of a **quadratic function** with x as the input variable and y as the output variable has the standard form

$$y = ax^2 + bx + c,$$

where a, b, and c are real numbers and $a \neq 0$.

2. The graph of a quadratic function is called a **parabola**.

3. For the quadratic function defined by $f(x) = ax^2 + bx + c$:

  • If $a > 0$, the parabola opens upward.
  • If $a < 0$, the parabola opens downward.

The magnitude of a affects the width of the parabola. The larger the absolute value of a, the narrower the parabola.

4. If $b = 0$, the turning point of the parabola is located on the vertical axis. If $b \neq 0$, the turning point will not be on the vertical axis.

5. The constant term, c, of a quadratic function $f(x) = ax^2 + bx + c$ always indicates the vertical (y-) intercept of the parabola. The vertical (y-) intercept of any quadratic function is $(0, c)$.

EXERCISES: ACTIVITY 3.2

1. a. Complete the following table for $y = x^2$.

x	−3	−2	−1	0	1	2	3
$y = x^2$							

b. Use the results of part a to sketch a graph $y = x^2$. Verify using a graphing calculator.

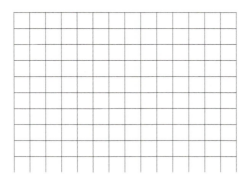

c. What is the coefficient of the term x^2?

d. From the graph, determine the domain and range of the function.

e. Create a table similar to the one in Exercise 1a to show the output for $g(x) = -x^2$.

x	−3	−2	−1	0	1	2	3
$g(x) = -x^2$							

f. Sketch the graph of $g(x) = -x^2$ on the same coordinate axis in part a. Verify using a graphing calculator.

g. What is the coefficient of the term $-x^2$?

h. How can the graph of $y = -x^2$ be obtained from the graph of $y = x^2$?

2. In each of the following functions defined by an equation of the form $y = ax^2 + bx + c$, identify the value of a, b, and c.

a. $y = -2x^2$

b. $y = \dfrac{2}{5}x^2 + 3$

c. $y = -x^2 + 5x$

d. $y = 5x^2 + 2x - 1$

3. Predict what the graph of each of the following quadratic functions will look like. Use your graphing calculator to verify your prediction.

a. $y = 3x^2 + 5$ **b.** $y = -2x^2 + 1$

c. $y = 0.5x^2 - 3$

4. Graph the following pairs of functions, and describe any similarities as well as any differences that you observe in the graphs.

a. $f(x) = 3x^2, g(x) = -3x^2$ **b.** $h(x) = \frac{1}{2}x^2, f(x) = 2x^2$ **c.** $g(x) = 5x^2, h(x) = 5x^2 + 2$

d. $f(x) = 4x^2 - 3, g(x) = 4x^2 + 3$ **e.** $f(x) = 6x^2 + 1, h(x) = -6x^2 - 1$

5. Use your graphing calculator to graph the two functions $y_1 = 3x^2$ and $y_2 = 3x^2 + 2x - 2$.

a. What is the y-intercept of the graph of each function?

b. Compare the two graphs to determine the effect of the linear term $2x$ and the constant term -2 on the graph of $y_1 = 3x^2$.

For Exercises 6–10, determine

a. whether the parabola opens upward or downward and
b. the y-intercept.

6. $f(x) = -5x^2 + 2x - 4$ **7.** $g(t) = \frac{1}{2}t^2 + t$ **8.** $h(v) = 2v^2 + v + 3$

9. $r(t) = 3t^2 + 10$ **10.** $f(x) = -x^2 + 6x - 7$

11. Does the graph of $y = -2x^2 + 3x - 4$ have any x-intercepts? Explain.

12. Put the following in order from narrowest to widest.

    a. $y = 0.5x^2$ **b.** $y = 8x^2$ **c.** $y = -2.3x^2$

13. Use transformations of the graph of $y = x^2$ to obtain a graph of each of the following. Verify using a graphing calculator.

    a. $y = -0.5x^2$

    b. $y = 2x^2 + 3$

    c. $y = -5x^2 - 2$

    d. $y = x^2 + 2x + 1$ *Note:* $x^2 + 2x + 1 = (x + 1)^2$

Activity 3.3

The Shot Put

Objectives

1. Determine the vertex or turning point of a parabola.

2. Determine the axis of symmetry of a parabola.

3. Identify the domain and range.

4. Determine the y-intercept of a parabola.

5. Determine the x-intercept(s) of a parabola graphically.

6. Interpret the practical meaning of the vertex and intercepts in a given problem.

7. Identify the vertex of a parabola having equation written in the form $y = a(x - h)^2 + k$, $a \neq 0$.

Parabolas are good models for a variety of situations that you encounter in everyday life, including situations involving motion. Examples include the path of a golf ball after it is struck, the arch (cable system) of a bridge, the path of a baseball thrown from the outfield to home plate, the stream of water from a drinking fountain, and the path of a cliff diver.

At the London Olympics in 2012, Poland's Tomasz Majewski became the first repeat shot put champion since 1956. His winning throw traveled 71.80 feet. The path of the throw can be modeled by the quadratic function defined by

$$H(x) = -0.015091x^2 + x + 6$$

where x is the horizontal distance from the point of the throw and $H(x)$ is the vertical height in feet of the shot above the ground.

1. **a.** After inspecting the equation for the path of the winning throw, which way do you expect the parabola to open? Explain.

   b. What is the H-intercept of the graph of the parabola? What practical meaning does this intercept have in this situation?

2. Use your graphing calculator to produce a plot of the path of the winning throw. Be sure to adjust your window settings so that all of the important features of the parabola (including x-intercepts) appear on the screen. Your graph should resemble the following.

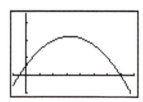

3. **a.** Use the graph to estimate the practical domain of the function.

   b. What does the practical domain mean in the shot put situation?

   c. Use the graph to estimate the practical range of the function.

   d. What does the practical range mean in the shot put situation?

4. Use the table feature of your graphing calculator to complete the following table:

x	10	20	30	40	50
$H(x)$					

Vertex of a Parabola

An important feature of the graph of any quadratic function defined by $f(x) = ax^2 + bx + c$ is its **turning point**, also called the **vertex**. As discussed in Activity 3.2, the turning point of a parabola that opens downward or upward is the point at which the parabola changes direction from increasing to decreasing or decreasing to increasing.

5. a. Use the results of Problem 4 to estimate the coordinates of the vertex of the shot put function.

   b. Use the Trace feature of your graphing calculator to approximate the vertex of the shot put function.

6. The vertex is often very important in a situation. What is the significance of the coordinates of the turning point in this problem?

The coordinates of the vertex of a parabola having equation $y = ax^2 + bx + c$ can be determined from the values of a, b, and c in the equation.

> **Definition**
>
> The **vertex** or turning point of a parabola having equation $y = f(x) = ax^2 + bx + c$ has coordinates
> $$\left(-\frac{b}{2a}, -\frac{b^2 - 4ac}{4a}\right) \text{ or } \left(\frac{-b}{2a}, f\left(-\frac{b}{2a}\right)\right),$$
> where a is the coefficient of the x^2 term and b is the coefficient of the x term.

Note that the y-coordinate of the vertex can be determined by substituting the x-coordinate of the vertex into the equation of the parabola and evaluating the resulting expression.

Example 1   *Determine the vertex of the parabola defined by the equation* $y = -3x^2 + 12x + 5.$

SOLUTION

Step 1. Determine the x-coordinate of the vertex by substituting the values of a and b into the formula $x = \dfrac{-b}{2a}$.

Because $a = -3$ and $b = 12$, you have

$$x = \frac{-(12)}{2(-3)} = \frac{-12}{-6} = 2.$$

Step 2. The y-coordinate of the vertex can be determined in two ways.

i. Substitute $a = -3$, $b = 12$, and $c = 5$ into the expression $-\dfrac{b^2 - 4ac}{4a}$ and evaluate as follows.

$$-\frac{12^2 - 4(-3)(5)}{4(-3)} = -\frac{144 + 60}{-12} = \frac{-204}{-12} = 17$$

ii. The y-value of the vertex is the corresponding output value for $x = 2$. Substituting 2 for x in the equation, you have

$$y = f(2) = -3(2)^2 + 12(2) + 5 = 17.$$

Therefore, the vertex is $(2, 17)$.

Because the parabola in Example 1 opens downward ($a = -3 < 0$), the vertex is the high point (maximum) of the parabola as demonstrated by the following graph of the parabola.

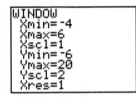

7. Determine the vertex of the parabola defined by $H = -0.015091x^2 + x + 6$ (the shot put function).

Rather than using the Trace feature to approximate the vertex of the parabola in Example 1, you can determine the vertex by selecting the maximum option in the Calc menu of your graphing calculator. Follow the prompts to obtain the coordinates of the maximum point (vertex).

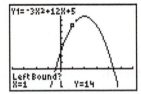
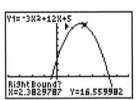

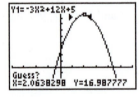
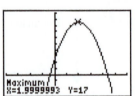

For further help with the TI-83/84 Plus, see Appendix A.

8. a. Use your graphing calculator to determine the vertex of the parabola having equation $H = -0.015091x^2 + x + 6$ (the shot put function).

b. How do the coordinates you determined in Problem 7 compare with your results in part a?

c. What is the practical meaning of the coordinates of the vertex in this situation?

Axis of Symmetry of a Parabola

> **Definition**
>
> The **axis of symmetry** is a vertical line that divides the parabola into two symmetrical parts that are mirror images in the line.

Example 2   *Consider the parabola from Example 1 having equation $y = -3x^2 + 12x + 5$. The axis of symmetry of the parabola is $x = 2$. Note that the line of symmetry passes through the vertex of the parabola.*

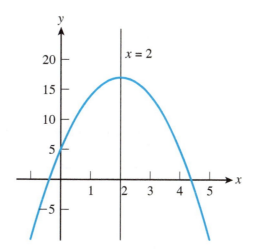

Because the vertex (turning point) of a parabola lies on the axis of symmetry, the equation of the axis of symmetry is

$$x = \frac{-b}{2a}.$$

9. What is the axis of symmetry of the shot put function?

Intercepts of the Graph of a Parabola

The y-intercept of the graph of the parabola defined by $y = -3x^2 + 12x + 5$ (see Example 1) can be determined directly from the equation. If $x = 0$, then

$$y = -3(0)^2 + 12(0) + 5 = 5$$

and the y-intercept is $(0, 5)$.

> In general, the y-intercept of the parabola defined by $y = ax^2 + bx + c$ is $(0, c)$.

Because the vertex $(2, 17)$ of the parabola having equation $y = -3x^2 + 12x + 5$ is a point above the x-axis (the y-coordinate is positive) and the parabola opens downward, the parabola must intersect the x-axis in two places. This is verified by the following graph:

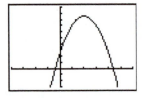

The x-intercepts can be determined using the zero option in the Calc menu of your graphing calculator. Follow the prompts to obtain one x-intercept at a time. The screens should appear as follows:

 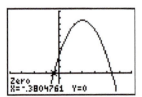 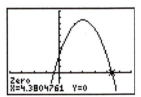

10. a. Use your graphing calculator to determine the x-intercept(s) for the shot put function having equation $H = -0.015091x^2 + x + 6$. The right-most intercept appears in the following screen:

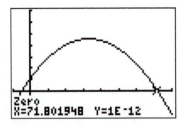

 b. Is either x-intercept determined in part a significant to the problem situation? Explain.

11. a. Use your result from Problem 10 to determine the practical domain of the shot put function. How does this compare with your answer in Problem 3a?

b. Sketch the path of the winning throw of the shot put. Be sure to label all key points, including the vertex and intercepts.

c. From the graph of the winning throw, over what horizontal distance (*x*-interval) is the height of the shot put increasing?

d. Determine the *x*-interval over which the height of the shot put is decreasing.

e. What is the practical range?

The graph of $H = -0.015091x^2 + x + 6$ has two *x*-intercepts. Does the graph of every parabola have *x*-intercepts? Problems 12 and 13 will help answer this question.

12. a. Use the values of *a*, *b*, and *c* to determine the coordinates of the vertex of the graph of $y = x^2 + 6x + 12$.

b. Use the *y*-coordinate of the vertex to determine if the vertex is above or below the *x*-axis.

c. Use the value of *a* in $y = x^2 + 6x + 12$ to determine if the parabola opens upward or downward.

d. Use the results from parts b and c to determine if the parabola has *x*-intercepts.

e. Use your graphing calculator to verify your answer to part d.

13. a. Use the values of *a*, *b*, and *c* to determine the coordinates of the vertex of the graph of $y = -x^2 + 8x - 21$.

b. Use the *y*-coordinate of the vertex to determine if the vertex is above or below the *x*-axis.

c. Use the value of a in $y = -x^2 + 8x - 21$ to determine if the parabola opens upward or downward.

d. Use the results from parts b and c to determine if the parabola has x-intercepts.

e. Use your graphing calculator to verify your answer to part d.

If a parabola opens upward and the vertex is above the x-axis there are no x-intercepts. If a parabola opens downward and the vertex is below the x-axis, there are no x-intercepts.

General Form of an Equation of a Quadratic Function

14. a. Sketch a graph of the quadratic function defined by $y = 3x^2$.

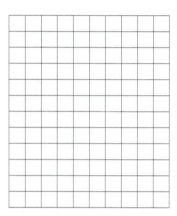

b. The graph of $y = 3x^2$ is shifted horizontally 2 units to the right. Sketch a graph of the new function. Write the equation of the transformed graph.

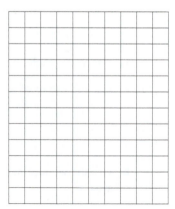

c. On the same coordinate axis, shift the graph in part b vertically 5 units upward. Write the equation of this new graph.

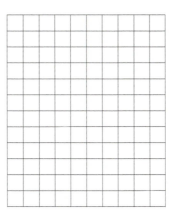

d. What is the vertex of the original parabola having equation $y = 3x^2$?

e. What is the vertex of the transformed parabola in part c?

The shift of the graph of $y = 3x^2$ horizontally 2 units to the right, followed by a vertical shift 5 units upward, moved the vertex of the parabola from $(0, 0)$ to $(2, 5)$. The resulting equation of the new parabola, $y = 3(x - 2)^2 + 5$, represents another way of writing the equation of a quadratic function.

> If (h, k) represents the vertex of a parabola, then $y = f(x) = a(x - h)^2 + k, a \neq 0$, is another form of the equation of a quadratic function. This form is especially convenient because the vertex (h, k) of the parabola is easily identified.

15. Determine the vertex of the graph of each of the following parabolas.

a. $y = -2(x - 3)^2 + 4$

b. $y = 1.25(x + 4)^2 + 3$

c. $y = -3(x + 2)^2 - 4.5$

The sign of a in $y = f(x) = a(x - h)^2 + k$, as in $y = g(x) = ax^2$, determines the direction in which the parabola opens. If $a > 0$, then the parabola opens upward. If $a < 0$, the parabola opens downward.

16. a. Graph the quadratic function defined by $y = f(x) = 3(x - 1)^2 + 2$. Be sure to identify the vertex and y-intercept.

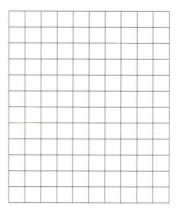

b. Graph the quadratic function defined by $y = g(x) = 3x^2 - 6x + 5$.

c. Compare the graphs in parts a and b.

d. Write the equation $y = 3(x - 1)^2 + 2$ in the form of $y = ax^2 + bx + c$. How does this equation compare to the equation of the parabola in part b?

SUMMARY: ACTIVITY 3.3

The following characteristics are commonly used in analyzing the quadratic function defined by $f(x) = ax^2 + bx + c, a \neq 0$, and its graph.

1. The **axis of symmetry** is a vertical line that separates the parabola into two mirror images. The equation of the vertical axis of symmetry is given by $x = \dfrac{-b}{2a}$.

2. The **vertex** (turning point) always falls on the axis of symmetry. The x-coordinate of the vertex is given by $\dfrac{-b}{2a}$. Its y-coordinate is determined by evaluating the function at this value.

In other words, the y-coordinate of the vertex is given by $f\left(-\dfrac{b}{2a}\right)$.

3. The **y-intercept**, the point where the parabola crosses the y-axis (that is, where its x-coordinate is zero), is always given by $(0, c)$.

4. The **x-intercept** is the point or points (if any) where the parabola crosses the x-axis (that is, where its y-coordinate is zero).

5. If a parabola opens upward and the vertex is above the x-axis there are no x-intercepts. If a parabola opens downward and the vertex is below the x-axis, there are no x-intercepts.

6. The **domain** of the general quadratic function is the set of all real numbers.

7. If the parabola opens upward, the **range** is all real numbers greater than or equal to the y-value of the vertex. If the parabola opens downward, the *range* is all real numbers less than or equal to the y-value of the vertex.

8. The equation of a quadratic function can be written in the general form $y = f(x) = a(x - h)^2 + k$, where (h, k) is the vertex and $a \neq 0$.

EXERCISES: ACTIVITY 3.3

For Exercises 1–8, determine the following characteristics of each quadratic function:

  a. the direction in which the graph opens

  b. the axis of symmetry

  c. the turning point (vertex)

  d. the y-intercept

1. $f(x) = x^2 - 3$

2. $g(x) = x^2 + 2x - 8$

3. $y = x^2 + 4x - 3$

4. $f(x) = 3x^2 - 2x$

5. $h(x) = x^2 + 3x + 4$

6. $g(x) = -x^2 + 7x - 6$

7. $y = 2x^2 - x - 3$

8. $f(x) = x^2 + x + 3$

For Exercises 9–16, use your graphing calculator to sketch the graphs of the functions and then determine each of the following:

    a. the coordinates of the x-intercepts for each function, if they exist

    b. the domain and range for each function

    c. the x-interval over which each function is increasing

    d. the x-interval over which each function is decreasing

9. $g(x) = -x^2 + 7x - 6$

10. $h(x) = 3x^2 + 6x + 4$

11. $y = x^2 - 12$

12. $f(x) = x^2 + 4x - 5$

13. $g(x) = -x^2 + 2x + 3$

14. $h(x) = x^2 + 2x - 8$

15. $y = -5x^2 + 6x - 1$

16. $f(x) = 3x^2 - 2x + 1$

17. You shoot an arrow straight into the air from a height of 5 feet with an initial velocity of 96 feet per second. The height, h, in feet above the ground, at any time, t (in seconds), is modeled by

$$h = 5 + 96t - 16t^2.$$

    a. Determine the maximum height the arrow will attain.

    b. Approximately when will the arrow reach the ground?

    c. What is the significance of the h-intercept?

d. What are the practical domain and practical range in this situation?

e. Use your graphing calculator to determine the *t*-intercepts. Determine the practical meaning of these intercepts in this situation.

18. As part of a recreational waterfront grant, the city council plans to enclose a rectangular area along the waterfront of the Gulf of Mexico and create a park and swimming area. The budget calls for the purchase of 3000 feet of fencing. *Note:* There is no fencing along the water.

a. Draw a picture of the planned recreational area. Let *x* represent the length of one of the two equal sides that are perpendicular to the water.

b. Write an expression that represents the width (side opposite the water) in terms of *x*. *Note:* You have 3000 feet of fencing.

c. Write an equation that expresses the area *A* of this rectangular site as a quadratic function of *x*.

d. Determine the value of *x* for which *A* is a maximum.

e. What is the maximum area that can be enclosed?

f. What are the dimensions of the enclosed area?

g. Use your graphing calculator to graph the area function. What point on the graph represents the maximum area?

h. What is the *A*-intercept? Does this point have any practical meaning in this situation?

i. From the graph, determine the *x*-intercepts. Do they have any practical meaning in this situation? Explain.

19. The cost to produce metal statues for local parks is given by

$$C = 2x^2 - 120x + 2000,$$

where x represents the number of statues produced and C is the cost of producing them.

a. Use your graphing calculator to graph the cost function and determine the coordinates of the turning point.

b. Determine the vertex algebraically.

c. How do your answers in parts a and b compare?

d. Is the vertex a minimum or maximum point?

e. What is the practical meaning of the vertex in this situation?

f. What is the C-intercept? What is the practical meaning of this intercept?

20. You are manufacturing ceramic lawn ornaments to sell online. After several months, your accountant tells you that your profit P can be modeled by

$$P = -0.002n^2 + 5.5n - 1200,$$

where n is the number of ornaments sold each month.

a. Use your graphing calculator to produce a graph of this function. Use the table feature set at TblStart $= 0$ and ΔTbl $= 500$ to help you set your window. Include the x-intercepts and the vertex.

b. Determine the x-intercepts of the graph of the profit function.

c. Determine the practical domain of the profit function.

d. Determine the practical range of the profit function.

e. How many ornaments must be sold to maximize the profit?

f. Write the equation that must be solved to determine the number of ornaments that must be sold to produce a profit of $2300.

g. Solve the equation in part f graphically.

Activity 3.4

Per Capita Personal Income

Objectives

1. Solve quadratic equations numerically.

2. Solve quadratic equations graphically.

According to statistics from the U.S. Department of Commerce, the per capita personal income (or average annual income) of each resident of the United States from 1960 to 2012 can be modeled by the equation

$$P(t) = 10.4t^2 + 263.0t + 1347.8,$$

where t represents the number of years since 1960.

1. What is the practical domain for the model represented by the function P?

2. Use your graphing calculator to complete the following table of values for t, the number of years since 1960, and $P(t)$, the per capita income. Round your output to the nearest dollar.

Year	1960	1970	1980	1990	2000	2010	2012
t	0	10					
$P(t)$							

3. Sketch a graph of the function using your graphing calculator using the window Xmin $= -5$, Xmax $= 55$, Ymin $= -2000$, and Ymax $= 45{,}000$. The graph should appear as follows:

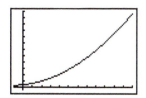

4. Estimate the per capita personal income in the year 1989 ($t = 29$).

5. You want to determine in which year the per capita personal income reached \$20,500. Write an equation to determine the value of t when $P = 20{,}500$.

The equation in Problem 5 is called a **quadratic equation**. The standard form of a quadratic equation is $ax^2 + bx + c = 0$, $a \neq 0$. Examples of quadratic equations include $x^2 + 3x - 1 = 9$, $2x^2 - 4x + 1 = 0$, and $6x^2 = 18$.

One method of approximating the solution to the equation in Problem 5 is numerical, using a table of appropriate data points. Example 1 demonstrates this approach.

Example 1 *Solve the quadratic equation $x^2 + 3x - 1 = 9$ numerically (using tables of data.)*

SOLUTION

Create a table in which x is the input and $y = x^2 + 3x - 1$ is the output. The solution is the x-value corresponding to a y-value of 9. Using the graphing calculator, the solution is $x = 2$.

```
Plot1 Plot2 Plot3
\Y1■X²+3X-1
\Y2■9■
\Y3=
\Y4=
\Y5=
\Y6=
\Y7=
```

X	Y1	Y2
1	3	9
2	9	9
3	17	9
4	27	9
5	39	9
6	53	9
7	69	9

X=1

A second solution is $x = -5$; try it yourself.

6. Determine the solution to $20{,}500 = 10.4x^2 + 263.0x + 1347.8$ numerically using a table of appropriate data points. What is your approximation using this approach?

Solving Quadratic Equations Graphically

A second method of solving the quadratic equation in Problem 5 is a graphical approach using your graphing calculator. Recall from Chapter 1 that you can solve the equation $10.4t^2 + 263.0t + 1347.8 = 20{,}500$ by solving the following system of equations graphically:

$$y_1 = 10.4t^2 + 263.0t + 1347.8$$

$$y_2 = 20{,}500$$

The expression for y_1 gives the per capita personal income in any given year. The value y_2 is the specific per capita personal income in which you are interested. The solution to the equation is the x-value for which $y_1 = y_2$. To do this, determine the point of intersection of these two graphs. If you use the intersect option under the Calc menu, the graph should appear as follows:

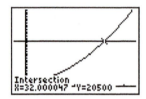

Another graphical method for solving the problem is to rearrange the quadratic equation

$$10.4t^2 + 263.0t + 1347.8 = 20{,}500 \tag{1}$$

so the right-hand side is equal to zero. Subtracting 20,500 from each side you have

$$10.4t^2 + 263.0t - 19{,}152.2 = 0$$

If you let $y = 10.4t^2 + 263.0t - 19152.2$, then the solutions to equation (1) are the t-values for which $y = 0$, if they exist. These are the t-values of the t-intercepts of the graph, also called the **zeroes** of the function.

7. a. Use your graphing calculator to sketch a graph of

$$y = 10.4t^2 + 263.0t - 19152.2.$$

The screen should appear as follows:

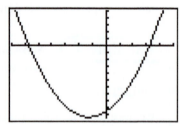

b. What are the t-intercepts of the new function defined by $y = 10.4t^2 + 263.0t - 19{,}152.2$?

c. Using the results from part b, determine the solutions to the equation $20{,}500 = 10.4t^2 + 263.0t + 1347.8$. Are both of the values relevant to our problem? Explain.

8. Describe two different ways to solve the equation $2x^2 - 4x + 3 = 2$ using a graphing approach. Solve the equation using each graphing method. How do your answers compare?

SUMMARY: ACTIVITY 3.4

1. A **quadratic equation** is an equation having standard form $y = ax^2 + bx + c = 0, a \neq 0$.

2. To solve $f(x) = n$ **numerically**, construct a table for $y = f(x)$, and determine the x-values that produce n as their y-value.

3. To solve $f(x) = n$ **graphically**:

   a. Graph $y_1 = f(x)$, graph $y_2 = n$, and determine the x-values of the points of intersection.

   b. Or graph $y = f(x) - n$, and determine the x-intercepts.

EXERCISES: ACTIVITY 3.4

In Exercises 1–4, solve the quadratic equation numerically (using tables of x- and y-values). Verify your solutions graphically.

1. $-4x = -x^2 + 12$

2. $x^2 + 9x + 18 = 0$

3. $2x^2 = 8x + 90$

4. $x^2 - x - 3 = 0$

In Exercises 5–8, solve the quadratic equation graphically using at least two different approaches. When necessary, give your solutions to the nearest hundredth.

5. $x^2 + 12x + 11 = 0$

6. $2x^2 - 3 = 2x$

7. $16x^2 - 400 = 0$

8. $4x^2 + 12x = -4$

In Exercises 9–12, solve the equation by using either a numeric or a graphing approach.

9. $x^2 + 2x - 3 = 0$

10. $x^2 + 11x + 24 = 0$

11. $x^2 - 2x - 8 = x + 20$

12. $x^2 - 10x + 6 = 5x - 50$

13. The stopping distance, d (in feet), for a car moving at a velocity (speed) v miles per hour is modeled by the equation

$$d = 0.04v^2 + 1.1v.$$

a. What is the stopping distance for a velocity of 55 miles per hour?

b. What is the speed of the car if it takes 200 feet to stop?

14. An international rule for determining the number, n, of board feet (usable finished lumber) in a 16-foot log is modeled by the equation

$$n = 0.22d^2 - 0.71d,$$

where d is the diameter of the log in inches.

a. How many board feet can be obtained from a 16-foot log with a 14-inch diameter?

b. Sketch a graph of this function. What is the practical domain of this function?

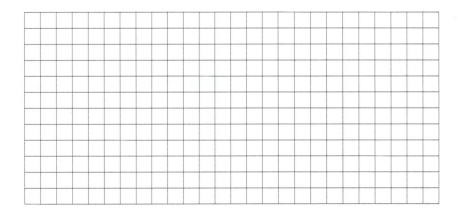

c. Use the graph to approximate the d-intercept(s). What is the practical meaning in this situation?

d. What is the diameter of a 16-foot log that has 200 board feet?

Activity 3.5

Sir Isaac Newton

Objectives

1. Factor expressions by removing the greatest common factor.

2. Factor trinomials using trial and error.

3. Use the zero-product principle to solve equations.

4. Solve quadratic equations by factoring.

Sir Isaac Newton XIV, a descendant of the famous physicist and mathematician, takes you to the top of a building to demonstrate a physics property discovered by his famous ancestor. He throws a softball straight up into the air. The ball's distance, s, above the ground as a function of time, x, is modeled by

$$s = -16x^2 + 16x + 32.$$

1. When the ball strikes the ground, what is the value of s?

2. Write the equation that you must solve to determine when the ball strikes the ground.

The quadratic equation in Problem 2 can be solved by using a numerical or a graphical approach. However, an algebraic technique is efficient in this case and will give an exact answer. The algorithm is based on the algebraic principle known as the **zero-product principle.**

Zero-Product Principle

If a and b are any numbers and $a \cdot b = 0$, then either a or b, or both, must be equal to zero.

Example 1 *Solve the equation $x(x + 5) = 0$.*

SOLUTION

The two factors in this equation are x and $x + 5$. The zero-product principle says one of these factors must equal zero. That is,

$$x = 0 \quad \text{or} \quad x + 5 = 0.$$

The first equation tells you that $x = 0$ is a solution. To determine a second solution, solve $x + 5 = 0$.

$$\begin{array}{r} x + 5 = 0 \\ \underline{-5 \quad -5} \\ x = -5 \end{array}$$

There are two solutions, $x = 0$ and $x = -5$.

Your graphing calculator verifies the solutions as follows.

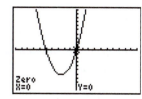 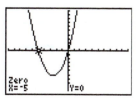

3. Solve each of the following equations using the zero-product principle.

   a. $3x(x - 2) = 0$　　　　　　　　　**b.** $(2x - 3)(x + 2) = 0$

   c. $(x + 2)(x + 3) = 0$

For the zero-product principle to be applied, one side of the equation must be zero. Therefore, at first glance, the zero-product principle can be used to solve the quadratic equation $3x^2 - 6x = 0$. However, a second condition must be satisfied. The nonzero side of the equation must be written as a product.

The process of writing an expression such as $3x^2 - 6x$ as a product is called *factoring*.

Definition

Rewriting an expression as a product is called **factoring.**

Factoring Common Factors

A **common factor** is a number or an expression that is a factor of each term of the entire expression. Whenever you wish to factor a polynomial, look first for a common factor.

Procedure

Removing a Common Factor from a Polynomial: First, identify the common factor, and then apply the distributive property in reverse.

Example 2   *Given the binomial $3x + 6$, 3 is a common factor because 3 is a factor of both terms $3x$ and 6. Applying the distributive property in reverse, you write*

$$3x + 6 \text{ as } 3(x + 2).$$

You may always check the factored binomial by multiplying:

$$3(x + 2) = 3(x) + 3(2) = 3x + 6$$

When you look for a common factor, determine the largest or **greatest common factor** (or GCF). You can see that 3 is a common factor of $6x + 24$ because 3 is a factor of both 6 and 24. However, there is a larger common factor, 6. Therefore,

$$6x + 24 = 6(x + 4).$$

Example 3   *Given $6x^2 + 14x - 30$, you can see that 2 is a common factor. Is 2 the greatest common factor? Yes, because no larger number is a factor of every term.*

If you divide each term by 2, you obtain $3x^2 + 7x - 15$. The expression $6x^2 + 14x - 30$ can now be written in factored form as $2(3x^2 + 7x - 15)$. Check the factored trinomial by multiplying.

Example 4   *Factor $4x^3 - 8x^2 + 28x$.*

SOLUTION

Four is a factor of each term, but x is as well. Therefore, the greatest common factor is $4x$. You remove the GCF by dividing each term by $4x$. This leads to the factored form $4x(x^2 - 2x + 7)$.

You can check your factoring by applying the distributive property.

4. Factor the following polynomials by removing the greatest common factor.

     a. $9a^6 + 18a^2$ **b.** $21xy^3 + 7xy$

     c. $3x^2 - 21x + 33$ **d.** $4x^3 - 16x^2 - 24x$

Factoring Trinomials

With patience, you can factor trinomials of the form $ax^2 + bx + c$ by trial and error, using the FOIL method in reverse.

Procedure

Factoring Trinomials by Trial and Error

1. Remove the greatest common factor, GCF.

2. To factor the resulting trinomial into the product of two binomials, try combinations of factors for the first and last terms in two binomials.

3. Check the outer and inner products to match the middle term of the original trinomial.

     a. If the constant term, c, is positive, both of its factors are positive or both are negative.

     b. If the constant term is negative, one factor is positive and one is negative.

4. If the check fails, repeat steps 2 and 3.

Example 5   *Factor $6x^2 - 7x - 3$.*

SOLUTION

Step 1.   There is no common factor, so go to step 2.

Step 2.   You could factor the first term, $6x^2$, as $6x(x)$ or as $2x(3x)$. The last term, -3, has factors $3(-1)$ or $-3(1)$. Try $(2x + 1)(3x - 3)$.

Step 3.   The outer product is $-6x$. The inner product is $3x$. The sum is $-3x$, not $7x$. The check fails.

Step 4.   Try $(2x - 3)(3x + 1)$. The outer product is $2x$. The inner product is $-9x$. The sum is $-7x$. It checks.

5. Factor the following trinomials.

   a. $x^2 - 7x + 12$ **b.** $x^2 - 8x - 9$

   c. $x^2 + 14x + 49$ **d.** $25 + 10w + w^2$

Solving Quadratic Equations by Factoring

The following example demonstrates the procedure for solving quadratic equations written in standard form, $ax^2 + bx + c = 0$, by factoring.

Example 6   *Solve the equation $3x^2 - 2 = -x$ by factoring.*

Step 1. Rewrite the equation in the form $ax^2 + bx + c = 0$ (called *standard form*).

$$3x^2 - 2 = -x$$
$$\underline{+ \; x \qquad\qquad +x}$$
$$3x^2 + x - 2 = 0$$

Step 2. Factor the expression on the nonzero side of the equation.

$$(x + 1)(3x - 2) = 0$$

Step 3. Use the zero-product principle to set each factor equal to zero, and then solve each equation.

$$(x + 1)(3x - 2) = 0$$

$x + 1 = 0$	$3x - 2 = 0$
$x = -1$	$3x = 2$
	$x = \dfrac{2}{3}$

Therefore, the solutions are $x = -1$ and $x = \dfrac{2}{3}$.

These solutions can be verified graphically as follows.

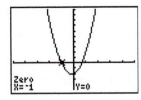 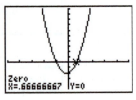

6. a. Returning to the ball problem from the beginning of this activity, solve the equation from Problem 2 by factoring.

   b. Are both solutions to the equation ($x = 2$ and $x = -1$) also solutions to the question, "At what time does the ball strike the ground"? Explain.

7. a. You want to know at what time the ball is 32 feet above the ground. Write a quadratic equation that represents this situation.

b. Solve the quadratic equation in part a by factoring.

8. Solve each of the following quadratic equations by factoring.

a. $2x^2 - x - 6 = 0$ **b.** $3x^2 - 6x = 0$

c. $x^2 + 4x = -x - 6$

9. Determine the zeros of the function defined by $f(x) = 2x^2 - 3x - 2$.

SUMMARY: ACTIVITY 3.5

I. To remove a **common factor** from a polynomial, first

a. identify the common factor, and then
b. apply the distributive property in reverse.

2. The **zero-product principle** says that if $ab = 0$ is a true statement, then either $a = 0$ or $b = 0$.

3. To factor trinomials of the form $ax^2 + bx + c$ by **trial and error**,

a. remove the greatest common factor.
b. try combinations of factors for the first and last terms in two binomials.
c. check the outer and inner products to match the middle term of the original trinomial.

  • If the constant term, c, is positive, both factors of c are positive or both are negative.
  • If the constant term is negative, one factor is positive and one is negative.

d. If the check fails, repeat steps 3b and 3c.

4. To solve equations by **factoring**,

    a. use the addition principle to remove all terms from one side of the equation; this results in a polynomial being set equal to zero.

    b. combine like terms, and then factor the nonzero side of the equation.

    c. use the zero-product principle to set each factor containing a variable equal to zero, and then solve the equations.

    d. check your solutions in the original equation.

EXERCISES: ACTIVITY 3.5

In Exercises 1–4, factor the polynomials by removing the GCF (greatest common factor).

1. $12x^5 - 18x^8$

2. $14x^6y^3 - 6x^2y^4$

3. $2x^3 - 14x^2 + 26x$

4. $5x^3 - 20x^2 - 35x$

In Exercises 5–13, completely factor the polynomials. Remember to look first for the GCF.

5. $x^2 + x - 6$

6. $p^2 - 16p + 48$

7. $x^2 + 7xy + 10y^2$

8. $x^2 - 4x - 32$

9. $12 + 8x + x^2$

10. $2x^2 + 7x - 15$

11. $3x^2 + 19x - 14$

12. $8x^4 - 47x^3 - 6x^2$

13. $20b^4 - 65b^3 - 60b^2$

In Exercises 14–21, solve each quadratic equation by factoring.

14. $x^2 - 5x + 6 = 0$

15. $x^2 + 2x - 3 = 0$

16. $x^2 - x = 6$

17. $x^2 - 5x = 14$

18. $3x^2 + 11x - 4 = 0$

19. $3x^2 - 12x = 0$

20. $x^2 - 7x = 18$

21. $3x(x - 6) - 5(x - 6) = 0$

22. Your neighbors have just finished installing a new swimming pool at their home. The pool measures 15 feet by 20 feet. They would like to plant a strip of grass of uniform width around three sides of the pool, the two short sides and one of the longer sides.

    a. Sketch a diagram of the pool and the strip of lawn, using x to represent the width of the uniform strip.

    b. Write an equation for the area, A, in terms of x, that represents the lawn area around the pool.

    c. They have enough seed for 168 square feet of lawn. Write an equation that relates the quantity of seed to the area of the uniform strip of lawn.

    d. Solve the equation in part c to determine the width of the uniform strip that can be seeded.

Ups and Downs

Objectives

1. Use the quadratic formula to solve quadratic equations.

2. Identify the solutions of a quadratic equation with points on the corresponding graph.

Suppose a soccer goalie punted the ball in such a way as to kick the ball as far as possible down the field. The height of the ball above the field as a function of time can be approximated by

$$y = -0.017x^2 + 0.98x + 0.33,$$

where y represents the height of the ball (in yards) and x represents the horizontal distance (in yards) down the field from where the goalie kicked the ball.

In this situation, the graph of $y = -0.017x^2 + 0.98x + 0.33$ is the actual path of the flight of the soccer ball. The graph of this function appears below:

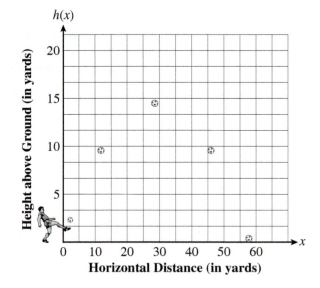

1. Use the graph to estimate how far downfield from the point of contact the soccer ball is 10 yards above the ground. How often during its flight does this occur?

2. **a.** Use $y = -0.017x^2 + 0.98x + 0.33$ to write the quadratic equation that arises in determining when the ball is 10 yards above the ground.

   b. Rewrite the equation in standard form and simplify if possible.

   c. What does the graph tell you about the number of solutions to this equation?

By some straightforward (but somewhat messy) algebraic manipulations, a quadratic equation can be turned into a formula that produces the solutions. This formula is called the **quadratic formula**.

The Quadratic Formula

For a quadratic equation in standard form, $ax^2 + bx + c = 0, a \neq 0$, the solutions are

$$x = \frac{-b \pm \sqrt{b^2 - 4 \cdot a \cdot c}}{2 \cdot a}.$$

The $\pm$ in the formula indicates that there are two solutions, one in which the terms in the numerator are added, and another in which the terms are subtracted.

When you use this formula, the quadratic equation *must* be in standard form so that the values (and signs) of the three coefficients a, b, and c are correct. You also want to be careful when you key the expression into your calculator. The square root and quotient computations need to be done carefully; you often must include appropriate sets of parentheses.

Example 1 *Use the quadratic formula to solve the quadratic equation*
$6x^2 - x = 2.$

SOLUTION

$6x^2 - x - 2 = 0$ Write equation in standard form, $ax^2 + bx + c = 0$

$a = 6, b = -1, c = -2$ Identify a, b, and c.

$x = \dfrac{-(-1) \pm \sqrt{(-1)^2 - 4(6)(-2)}}{2(6)}$ Substitute for a, b, and c in $x = \dfrac{-b \pm \sqrt{b^2 - 4ac}}{2a}$.

$x = \dfrac{1 \pm \sqrt{1 + 48}}{12} = \dfrac{1 \pm 7}{12}$

$x = \dfrac{1 + 7}{12} = \dfrac{2}{3}, x = \dfrac{1 - 7}{12} = \dfrac{-6}{12} = -\dfrac{1}{2}$

3. Use the quadratic formula to solve the quadratic equation in Problem 2b.

4. Solve $3x^2 + 20x + 7 = 0$ using the quadratic formula.

SUMMARY: ACTIVITY 3.6

The solutions of a quadratic equation in standard form, $ax^2 + bx + c = 0, a \neq 0$, are

$$x = \frac{-b \pm \sqrt{b^2 - 4 \cdot a \cdot c}}{2 \cdot a}.$$

The $\pm$ symbol in the formula indicates that there are two solutions, one in which the terms in the numerator are added and one in which the terms are subtracted.

EXERCISES: ACTIVITY 3.6

Use the quadratic formula to solve the equations in Exercises 1–5.

1. $x^2 + 2x - 15 = 0$

2. $4x^2 + 32x + 15 = 0$

3. $-2x^2 + x + 1 = 0$

4. $2x^2 + 7x = 0$

5. $-x^2 + 10x + 9 = 0$

6. a. Determine the x-intercepts, if any, of the graph of $y = x^2 - 2x + 5$.

b. Sketch the graph of $y = x^2 - 2x + 5$ using your graphing calculator and verify your answer to part a.

7. The following data from the National Health and Nutrition Examination Survey indicates that the number of American adults who are overweight or obese is increasing.

Years Since 1960, t	1	12	18	31	39	44	46	50
Percentage of American Adults Who Are Overweight or Obese, $P(t)$	45	47	47	56	64.5	66.3	66.9	69.2

This data can be modeled by the equation $P(t) = 0.0076t^2 + 0.1549t + 44.015$

a. Determine the percentage of overweight or obese Americans in the year 2020.

b. Using the quadratic formula, determine the year when the model predicts that the percentage of overweight Americans will first reach 75%.

8. The height of a bridge arch located in the Thousand Islands is modeled by the function $y = -0.04x^2 + 28$, where x is the distance, in feet, from the center of the arch and y is the height of the arch.

   a. Sketch a picture of this arch on a grid using the vertical axis as the center of the arch.

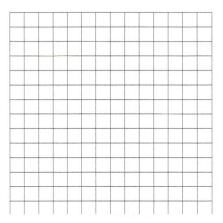

   b. Determine the y-intercept. What is the practical meaning of this intercept in this situation?

   c. Determine the x-intercepts algebraically using the quadratic formula.

   d. Graph the function on your graphing calculator and check the accuracy of the intercepts you found in part c.

   e. If the arch straddles the river exactly, how wide is the river?

   f. A sailboat is approaching the bridge. The top of the mast measures 30 feet. Will the boat clear the bridge? Explain.

   g. You want to install a flagpole on the bridge at an arch height of 20 feet. Write the equation that you must solve to determine how far to the right of center the arch height is 20 feet.

   h. Solve the equation in part g using the quadratic formula. Use your graphing calculator to check your result.

9. The number n (in millions) of cell phone subscribers in the United States from 2000 to 2012 is given in the following table.

Year	2000	2002	2004	2006	2007	2011	2012
Number of Subscribers (millions)	109.5	149.8	182.1	233.0	255.4	316.0	326.4

This data can be approximated by the quadratic model

$$n(t) = -0.3553t^2 + 22.84t + 106.3$$

where $t = 0$ corresponds to the year 2000.

a. Use your graphing calculator to sketch a graph of the function.

b. Use the graph to estimate the year in which there were 300 million cell phone subscribers in the United States.

c. Use the quadratic formula to answer part b. How does your answer compare to the estimate you obtained using the graphical approach?

d. How confident are you in the solutions in part c? Explain your answer.

Activity 3.7

Heat Index

Objectives

1. Determine quadratic regression models using a graphing calculator.

2. Solve problems using quadratic regression models.

On very hot and humid summer days, it is common for meteorologists at the National Weather Service to issue warnings due to a very high heat index. The heat index is a measurement that combines air temperature and relative humidity to determine a relative temperature. The heat index is the temperature you perceive (how it feels) on a hot and humid day. Heat index is similar to windchill, the temperature you perceive on a cold and windy day. Heat indices are not usually calculated until the air temperature reaches 80 degrees Fahrenheit.

The following table gives the heat index for various temperatures when the relative humidity is 90%. Both air temperature and the heat index (perceived temperature) are measured in degrees Fahrenheit.

Temperature (Degrees Fahrenheit) t	80	82	84	86	88	90	92
Heat Index (°F), h	86	91	98	105	113	122	131

Note: Relative humidity is 90%.

1. Sketch a scatterplot of the data. Let the air temperature, t, represent the input variable, and the heat index, h, represent the output variable. Could the data be modeled by a quadratic function? Explain.

2.

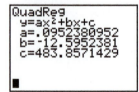

2. Use the regression feature of your graphing calculator to determine and plot a quadratic function that best fits this data. Your graph should appear as follows.

3. **a.** How does the plot of the quadratic regression equation compare with your scatterplot of the data?

   b. Do you believe the quadratic regression model is a good model for the heat index when the relative humidity is 90%?

4. What is the practical domain of this function?

5. a. When the humidity is 90%, use the quadratic regression equation to estimate the heat index for each of the following temperatures.

    i. 45 degrees   **ii.** 85 degrees   **iii.** 100 degrees

  b. Which, if any, of these estimates do you think is most reliable? Explain.

6. a. If the relative humidity is 90%, estimate the temperatures for which the heat index is equal to 100° using each of the following methods.

    i. The given table (numerical)

    ii. The graph of the quadratic regression equation (graphical)

  b. Use the quadratic formula (algebraic) to estimate the temperature for which the heat index is 100 degrees when the relative humidity is 90%.

  c. How do the results in parts a and b compare?

7. a. In Problem 6b, the stated relative humidity was 90%, and you wrote an equation to determine the temperature for which the heat index (perceived temperature) was equal to 100 degrees. Write an equation to determine the temperature for which the air temperature and the heat index are equal.

  b. Solve the equation in part a graphically to estimate the temperatures for which the air temperature and the heat index are equal.

8. a. The National Weather Service does not calculate heat index unless the air temperature is at least 80 degrees Fahrenheit and the relative humidity is at least 40%. Based on your results in Problems 5, 6, and 7, does this policy seem reasonable? Explain.

    b. According to this policy, will the heat index reported by the National Weather Service ever be equal to the actual air temperature if the relative humidity is 90%? Explain.

9. The following data from the National Health and Nutrition Examination Survey appeared in Problem 7 of Activity 3.6. The data indicates that the number of American adults who are overweight or obese is increasing.

Years Since 1960, t	1	12	18	31	39	44	46	50
Percentage of Americans Who Are Overweight or Obese, $P(t)$	45	47	47	56	64.5	66.3	66.9	69.2

Use the regression feature of your graphing calculator to verify that this data can be modeled by the equation $P(t) = 0.0076t^2 + 0.1549t + 44.015$.

SUMMARY: ACTIVITY 3.7

Parabolic data can be modeled by a **quadratic regression equation**.

EXERCISES: ACTIVITY 3.7

1. During one game, the Buffalo Bills punter was called upon to punt the ball eight times. On one of these punts, the punter struck the ball at his own 30-yard line. The height, h, of the ball above the field in feet as a function of time, t, in seconds can be partially modeled by the following table:

t	0	0.6	1.2	1.8	2.4	3.0
$h(t)$	2.50	28.56	43.10	46.12	37.12	17.60

a. Sketch a scatterplot of the data using your graphing calculator.

b. Use your graphing calculator to obtain a quadratic regression function for these data. Round the values of a, b, and c to four decimal places.

c. Graph the equation from part b on the same coordinate axes as the data points. Does the curve appear to be a good fit for the data? Explain.

d. In this model, what is the practical domain of the quadratic regression function?

e. Estimate the practical range of this model.

f. How long after the ball was struck did the ball reach 35 feet above the field? Explain.

g. How many results did you obtain for part f? Do you think you have all of the solutions? Explain.

2. Use the following data set to perform the tasks in parts a–e.

x	0	3	6	9	12
y	5	28	86	180	310

a. Determine an appropriate scale, and plot these points.

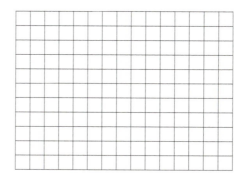

b. Use your graphing calculator to determine the quadratic regression equation for this data set.

c. Graph the regression equation on the same coordinate axes as the data points in part a.

d. Compare the predicted *y*-values with the *y*-values given in the table.

e. Predict the *y*-value for $x = 7$ and for $x = 15$.

3. The following table shows the stopping distance for a car at various speeds on dry pavement:

Speed (mph)	25	35	45	55	65	75
Distance (ft)	65	108	167	245	340	450

a. Use your graphing calculator to determine a quadratic regression equation that represents this data.

b. Use the regression equation to predict the stopping distance at 90 mph.

c. What speed would produce a stopping distance of 280 feet? (Round to the nearest tenth.) Explain how you arrived at your conclusion.

4. According to the U.S. Energy Information Administration, the net imports of crude oil and petroleum products to the United States is decreasing. The following table indicates the net imports of crude oil and petroleum products, I, (thousands of barrels per day) to the United States from 2002 to 2012.

Year Since 2002	0	1	2	3	4	5	6	7	8	9	10
Imports (thousands of barrels per day)	10.5	11.2	12.1	12.5	12.4	12.0	11.1	9.7	9.4	8.5	7.4

a. Create a scatterplot of the data using your graphing calculator.

b. Use your graphing calculator to obtain a quadratic regression function for these data. Round the values of a, b, and c to three decimal places.

c. Graph the equation from part b on the same coordinate axes as the data points. Does the curve appear to be a good fit for the data? Explain your answer.

d. What does the regression equation predict for net import of oil and petroleum products in 2027? Does this seem reasonable? Explain your answer.

e. In what year does the model predict that we stop importing oil products?

5. Since 1988, the number of voters who identify themselves as Hispanic has been increasing. The following table shows the number of Hispanic voters for each election since 1988.

Year	1988	1992	1996	2000	2004	2008	2012
Number of Hispanic Voters in the Presidential Election, V (millions)	3.7	4.1	5.0	5.9	7.6	9.7	11.2

a. Sketch the scatterplot of the data. Let *t* represent the number of years since 1988. Therefore, *t* = 0 corresponds to the year 1988. Does the data appear to be quadratic?

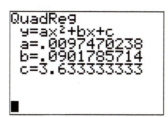

b. Use the quadratic regression feature of your graphing calculator to determine and plot a quadratic function that fits these data. Round your results to the first three digits after the leading 0's. Your graph should appear as follows:

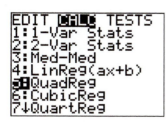 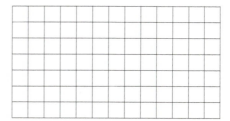 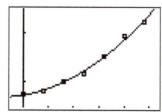

c. Identify the *V*-intercept from the regression equation. What does it indicate? How does it compare to the data?

d. Use the regression equation from part b to predict the number of Hispanic voters in 2024.

e. Use the graph of the regression equation to predict when the number of Hispanic voters in presidential elections will first exceed 25 million. Your screen should appear as shown.

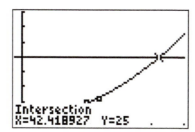

Collecting and Analyzing Data

6. Designing and building bridges is a marvelous engineering feat. Often, the shape of the main support cable in a suspension bridge can be modeled by a parabola.

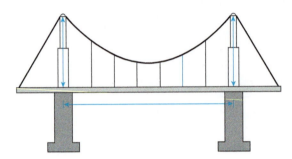

Use resources from the library or the Internet to determine specific dimensions, such as the distance between the main support columns, for one of the following bridges: Golden Gate Bridge, George Washington Bridge, or Verrazano Bridge.

Use the dimensions to develop points that lie on the graph of the main support cable of the bridge you selected. Draw a sketch of the support cable on graph paper. The orientation of the coordinate system, especially the origin, is very important. You will need at least three points that lie on the graph of the support cable in your drawing.

Use the points to determine a quadratic regression equation to represent the main support cable of the bridge. Use the model to determine the minimum distance from the cable to the highway.

Be prepared to give a presentation of your findings.

Activities 3.1–3.7 | **What Have I Learned?**

1. a. In order for the graph of the equation $y = ax^2 + bx + c$ to be a parabola, the value of the coefficient of x^2 cannot be zero. Explain.

b. What is the vertex of the parabola having an equation of the form $y = ax^2$?

c. Describe the relationship between the vertex and the y-intercept of the graph of $y = ax^2 + c$.

2. Determine if the vertex is a minimum point or a maximum point of $y = ax^2 + bx + c$ in each of the following situations.

a. $a < 0$ **b.** $a > 0$

3. a. What are the possibilities for the number of y-intercepts of a quadratic function?

b. What are the possibilities for the number of x-intercepts of a parabola?

4. What is the relationship between the vertex and the x-intercept of the graph of $y = x^2 - 4x + 4$?

5. a. The vertex of a parabola is $(3, 1)$. Using this information, complete the following table:

x	1	2	3	4	5
y	5	2			

b. If the vertex of a parabola is $(2, 4)$, complete the following table:

x	−2	0	2	4	6
y	0	3			

6. The vertex of the parabola having equation $g(x) = x^2 + 2x - 8$ is $(-1, -9)$.

a. The graph of g is shifted vertically upward 5 units. Write the equation of the resulting graph.

b. Determine the vertex of the graph in part a.

c. Compare the y-values of the vertex of the original parabola and the translated graph.

7. a. Given the following graph, explain why choices i, ii, and iii do not fit the curve:

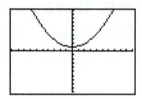

    i. $f(x) = ax^2 + bx$ with $a > 0, b < 0$

    ii. $g(x) = ax^2 + c$ with $a < 0, c > 0$

    iii. $h(x) = ax^2 + bx + c$ with $a < 0, b > 0, c < 0$

  b. What restrictions on a, b, and c are necessary to fit $y = ax^2 + bx + c$ to this graph?

Recognizing the correspondence among situations, equations, and graphs is an important skill. Exercises 8–12 contain a description of a free-falling object where s is the distance an object is above the ground at time t. Match the description to the quadratic models. Then match each description (and equation) to the graph that represents the situation, selecting from graphs i–vii.

8. A rock is thrown straight up with a velocity of 20 meters per second from a building that is 50 meters high.

9. A model rocket is shot straight up into the air from a launch pad at ground level with a velocity of 70 feet per second.

10. A telephone repairperson drops her vice grips from a 50-foot tower.

11. A ball is hit directly upward with a velocity of 100 feet per second from a height of 6 feet above the ground.

12. A ball is thrown straight down with a velocity of 50 feet per second from the roof of a building that is 50 feet high.

  a. $s(t) = -16t^2 + 50$

  b. $s(t) = -4.9t^2 - 20t + 50$

  c. $s(t) = -16t^2 + 100t + 6$

  d. $s(t) = -16t^2 + 70t$

  e. $s(t) = -16t^2 - 50t + 50$

  f. $s(t) = -4.9t^2 + 20t + 50$

  g. $s(t) = -16t^2 + 50t$

i.

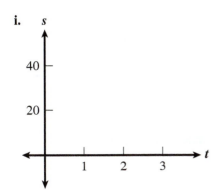

ii.

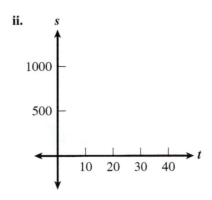

iii.

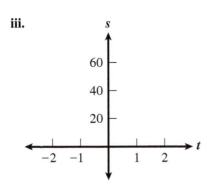

iv.

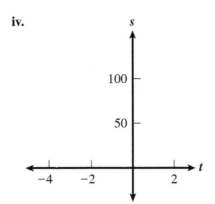

v.

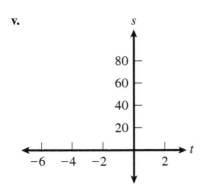

vi.

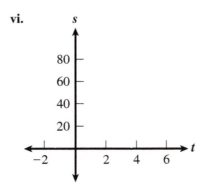

vii.

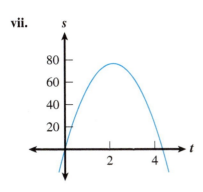

Activities 3.1–3.7   How Can I Practice?

1. Complete the following table:

EQUATION OF THE FORM $y = ax^2 + bx + c$	VALUE OF a	VALUE OF b	VALUE OF c
$y = 5x^2$			
$y = \dfrac{1}{3}x^2 + 3x - 1$			
$y = -2x^2 + x$			

For Exercises 2–7, determine the following characteristics for each graph:

    a. the direction in which the parabola opens

    b. the equation of the axis of symmetry

    c. the vertex

    d. the y-intercept

2. $y = -2x^2 + 4$ **3.** $y = \dfrac{2}{3}x^2$ **4.** $f(x) = -3x^2 + 6x + 7$

5. $f(x) = 4x^2 - 4x$ **6.** $y = x^2 + 6x + 9$ **7.** $y = x^2 - x + 1$

For Exercises 8–11, use your graphing calculator to sketch the graph of each quadratic function, and then determine the following for each function.

    a. the coordinates of the x-intercepts (if they exist)

    b. the domain and range

    c. the x-interval over which the function is increasing

    d. the x-interval over which the function is decreasing

8. $y = -x^2 + 4$ **9.** $y = x^2 - 5x + 6$

10. $y = -3x^2 - 6x + 8$ **11.** $y = 0.22x^2 - 0.71x + 2$

12. Use your graphing calculator to approximate the vertex of the graph of the parabola defined by the equation

$$y = -2x^2 + 3x + 25.$$

13. Determine one solution of the following quadratic equations numerically.

   a. $5x^2 = 7$ **b.** $x^2 - 7x + 10 = 5$ **c.** $3x^2 - 5x = 2$

a.

x	1	1.1	1.2	1.3	1.4	1.5
y						

b.

x	0.5	0.6	0.7	0.8	0.9	1
y						

c.

x	0	1	2	3	4	5
y						

14. Solve each of the equations from Exercise 13 using the quadratic formula. When necessary, round your solutions to the nearest tenth. Check your solutions by graphing.

15. Solve each of the following equations by factoring.

   a. $4x^2 - 8x = 0$ **b.** $x^2 - 6 = 7x + 12$

   c. $2x(x - 4) = -6$ **d.** $x^2 - 8x + 16 = 0$

e. $x^2 - 2x - 24 = 0$ **f.** $y^2 - 2y - 35 = -20$

g. $a^2 + 2a + 1 = 3a + 7$ **h.** $4x^2 + 4x - 3 = -3x - 1$

16. A fastball is hit straight up over home plate. The ball's height, h (in feet), from the ground is modeled by

$$h = -16t^2 + 80t + 5,$$

where t is measured in seconds.

a. What is the maximum height of the ball above the ground?

b. Write an equation to determine how long will it take for the ball to reach the ground. Solve the equation using the quadratic formula. Check your solution by graphing.

c. Write the equation you would need to determine when the ball is 101 feet above the ground.

d. Solve the equation you determined in part c algebraically to determine the time it will take for the ball to reach a height of 101 feet. Verify your results graphically.

17. The distance between the support towers of a suspension bridge (shown in the accompanying figure) is 100 meters. The bridge is supported by cables attached to the tops of towers, which are 35 meters above the roadway. The cables hang from the towers approximately in the shape of a parabola. The height, h (in meters), of the cables above the surface of the roadway is modeled by

$$h = 0.01x^2 - x + 35,$$

where x is the horizontal distance measured from the point where the tower and roadway meet.

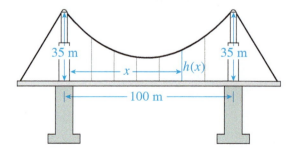

a. Use your graphing calculator to examine the height function. What is the practical domain of this function?

b. What is the minimum distance of the center cables from the roadway?

18. Use the following data set to perform the tasks in parts a–f:

x	0	1	3	5	7	8
y	10	4	−18	−54	−107	−145

a. Determine an appropriate scale and plot these points.

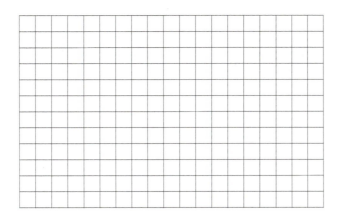

b. Use your graphing calculator to determine the quadratic regression equation for this data set.

c. Graph the regression equation on the same coordinate axes as the data points.

d. Compare the predicted outputs with the given outputs in the table.

e. What is the predicted output for $x = 4$ and for $x = 9$?

f. For what value of x is $y = -40$? Use the quadratic formula.

19. According to the Centers for Disease Control and Prevention the consumption of cigarettes by adults in the United States is decreasing. The following table represents the year consumption of cigarettes, C, (in billions of cigarettes) for various years between 2000 and 2011.

Year	2000	2002	2004	2006	2008	2010	2011
Consumption of Cigarettes by U.S. Adults (billions), C	435.6	415.7	397.7	380.6	346.4	300.5	292.8

Data Source: Centers for Disease Control and Prevention

a. Let $t = 0$ correspond to the year 2000. Sketch a scatterplot of the data.

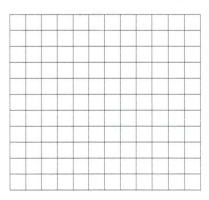

b. Use your graphing calculator to determine a linear regression function for these data. Round the values of a and b to two decimal places.

c. Determine the quadratic model for the data. Round the values of a, b, and c to three decimal places.

d. Which model best represents the data? Explain your answer.

e. Use each model to predict the consumption of cigarettes in 2020.

f. How confident are you in your predictions? Explain your answer.

20. Use transformations to help describe the differences and the similarities in the graphs of the following quadratic functions. Verify using a graphing calculator.

a. $y = 3x^2, y = 3x^2 + 5$

b. $y = 2x^2, y = -2x^2$

c. $y = 2x^2 + 1, y = 2x^2 - 4$

d. $y = 4x^2, y = 4(x - 1)^2$

Activity 3.8

A Thunderstorm

Objectives

1. Recognize the equivalent forms of a direct variation statement.

2. Determine the constant of proportionality in a direct variation problem.

3. Solve direct variation problems.

One of nature's more spectacular events is a thunderstorm. The skies light up, delighting your eyes, and seconds later your ears are bombarded with the boom of thunder. Because light travels faster than sound, during a thunderstorm you see the lightning before you hear the thunder. The formula

$$d = 1080t$$

describes the distance, d in feet, you are from the storm's center if it takes t seconds for you to hear the thunder.

1. Complete the following table for the model $d = 1080t$.

t, in Seconds	1	2	3	4
d, in Feet				

2. What does the ordered pair $(3, 3240)$ from the above table mean in a practical sense?

3. As the value of t increases, what happens to the value of d?

The relationship between time t and distance d in this situation is an example of **direct variation**. As t increases, d also increases.

Definition

Two variables, x and y, are said to **vary directly** if $y = kx$ for some constant k. In this situation, whenever the magnitude of x increases by a scale factor, the magnitude of y increases by the same factor.

4. Graph the distance, d, as a function of time, t, using the values in Problem 1.

5. If the time it takes for you to hear the thunder after you see the lightning decreases, what is happening to the distance between you and the center of the storm?

The following statements are equivalent. The number represented by k is called the **constant of proportionality** or **constant of variation**.

a. y varies directly as x

b. y is directly proportional to x

c. $y = kx$ for some constant k.

6. The amount of garbage, G, varies directly with the population, P. The population of Grand Prairie, Texas, is 0.13 million and creates 2.6 million pounds of garbage each week. Determine the amount of garbage produced by Houston with a population of 2 million. Assume both locations have the same constant of variation.

   a. Write an equation relating G and P and the constant of variation k.

   b. Determine the value of k.

   c. Rewrite the equation in part a using the value of k from part b.

   d. Use the equation in part c and the population of Houston to determine the weekly amount of garbage produced.

7. Use Problem 6 to outline a procedure to solve direct variation problems.

Note that direct variations are not always increasing. For example, $y = -1.2x$ is a direct variation. From the graph it is clear that the function is decreasing.

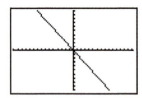

8. Does this contradict the definition of a direct variation? Explain.

9. A worker's gross wages, w, vary directly as the number of hours the worker works, h. The following table shows the relationship between the wages and the hours worked.

Hours Worked, h	15	20	25	30	35
Wages, w	$172.50	$230.00	$287.50	$345.00	$402.50

a. Graph the gross wages, w, as a function of hours worked, h, using the values in the preceding table

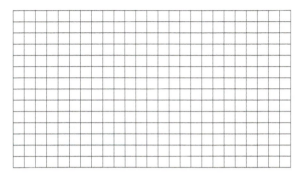

b. The relationship is defined by $w = kh$. Pick one ordered pair from the table and use it to determine the value of k.

c. What does k represent in this situation?

d. Use the formula to determine the gross wages for a worker who works 40 hours.

SUMMARY: ACTIVITY 3.8

1. Two variables are said to **vary directly** if as the magnitude of one increases, the magnitude of the other does as well.

2. The following statements are equivalent. The number represented by k is called the **constant of proportionality** or **constant of variation**.

   a. y varies directly as x

   b. y is directly proportional to x

   c. $y = kx$ for some constant x

EXERCISES: ACTIVITY 3.8

1. The amount of sales tax, s, on any item is directly proportional to the list price of an item, p. The sales tax in your area is 8%.

    a. Write a formula that relates the amount of sales tax, s, to the list price, p, in this situation.

    b. Use the formula in part a to complete the following table:

LIST PRICE, p	SALES TAX, s
$10	
$20	
$30	
$50	
$100	

    c. Graph the sales tax, s, as a function of list price, p, using the values in the preceding table

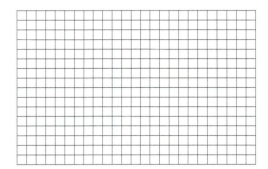

    d. Determine the list price of an item for which the sales tax is $3.60.

2. The amount you tip a waiter, t, is directly proportional to the amount of the check, c.

    a. Assuming you tip your waiter 15% of the amount of the check, write a formula that relates the amount of the tip, t, to the amount of the check, c.

    b. Use the formula in part a to complete the following table.

AMOUNT OF CHECK, c	AMOUNT OF THE TIP, t
$15	
$25	
$35	
$50	
$100	

c. Graph the amount of the tip, t, as a function of the amount of the check, c, using the values in the preceding table.

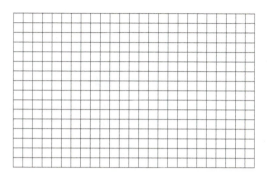

d. A company has a policy that no tip can exceed $12.00 and still be reimbursed by the company. What is the maximum cost of a meal for which all of the tip will be reimbursed if the tip is 15% of the check?

3. y varies directly as x, $y = 25$ when $x = 5$. Determine y when $x = 13$.

4. y varies directly as x, $y = 7$ when $x = 21$. Determine x when $y = 5$.

5. Given that y varies directly as x, consider the following table:

x	2	4	7		12
y	4	8		20	

a. Determine a formula that relates x and y.

b. Use the formula in part a to complete the table above.

Activity 3.9

The Power of Power Functions

Objectives

1. Identify a direct variation function.

2. Determine the constant of variation.

3. Identify the properties of graphs of power functions defined by $y = kx^n$, where n is a positive integer and $k \neq 0$.

4. Graph transformations of power functions.

Direct variations can involve higher powers of x, like x^2, x^3, or in general, x^n. For example, in Activity 3.1, you investigated the formula $s = 16t^2$ used to model the distance s, in feet, an object had fallen t seconds since it had been released. The equation $s = 16t^2$ is in the form

$$s = kt^n, \text{ where } k = 16 \text{ and } n = 2.$$

In general, the equation

$$y = kx^n,$$

where $k \neq 0$ and n is a positive integer, defines a **direct variation** function in which y varies directly as x^n. The constant, k, is called the **constant of variation** or the **constant of proportionality**.

> **Example 1** The constant of variation, k, in the free-falling object situation defined by $s = 16t^2$ is 16.

1. What is the constant of variation for the direct variation function defined by $V = \dfrac{4}{3}\pi r^3$?

Suppose you know only that the distance, s, varies directly as the square of t and one data pair. You are able to determine the constant of variation k using the same process discussed in Activity 3.8. Example 2 demonstrates this process.

> **Example 2** Let s vary directly as the square of t. If $s = 64$ when $t = 2$, determine the direct variation equation.

SOLUTION

Because s varies directly as the square of t, you have

$$s = kt^2,$$

where k is the constant of variation. Substituting 64 for s and 2 for t, you have

$$64 = k(2)^2 \quad \text{or} \quad 64 = 4k \quad \text{or} \quad k = 16.$$

Therefore, the direct variation equation is

$$s = 16t^2.$$

2. For each table, determine the pattern and complete the table. Determine the constant k of variation and then write a direct variation equation for each table.

   a. y varies directly as x.

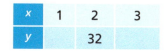

x	1	2	4	8	12
y		12			

   b. y varies directly as x^3.

x	1	2	3
y		32	

3. The length, L, of skid distance left by a car varies directly as the square of the initial velocity, v (in miles per hour), of the car.

   a. Write a general equation for L as a function of v. Let k represent the constant of variation.

   b. Suppose a car traveling at 40 miles per hour leaves skid distance of 60 feet. Use this information to determine the value of k.

   c. Use the function to determine the length of the skid distance left by the car traveling at 60 miles per hour.

Power Functions

The direct variation functions that have equations of the form $y = kx^n$, where n is a positive integer and $k \neq 0$, are also called **power functions**. The graphs of this family of functions are very interesting and are useful in problem solving.

4. Sketch a graph of each of the following power functions. Use a graphing calculator to verify the graph.

a. $y = x$

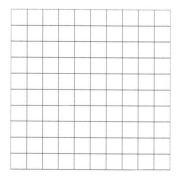

b. $y = x^2$

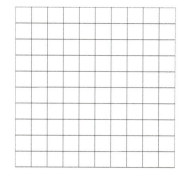

c. $y = x^3$

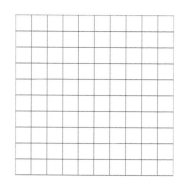

d. $y = x^4$

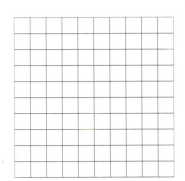

e. $y = x^5$

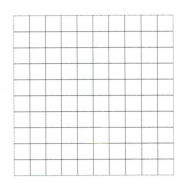

f. $y = x^6$

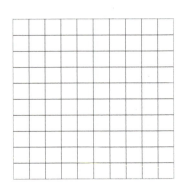

5. Each graph in Problem 4 has an equation of the form $y = x^n$, where n is a positive integer.

    a. What is the basic shape of the graph if

        i. n is even?   **ii.** n is odd?

    b. If n is even, what happens to the graph as n gets larger in value?

    c. If n is odd, is the function increasing or decreasing?

6. Use the patterns from Problem 5 in combination with graphing techniques you have learned previously to sketch a graph of each of the following without using a graphing calculator.

a. $y = x^2 + 1$

b. $y = -2x^4$

c. $y = 3x^8 + 1$

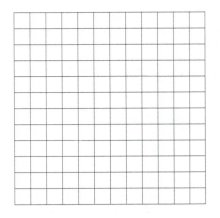

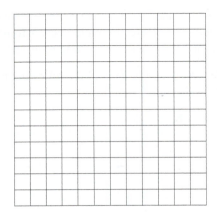

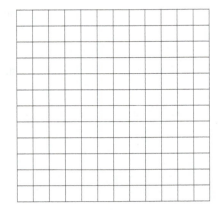

d. $y = -2x^5$ **e.** $y = x^{10}$ **f.** $y = 5x^3 + 2$

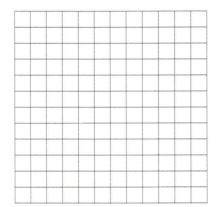 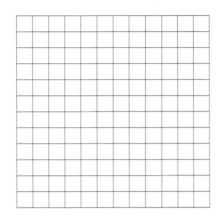 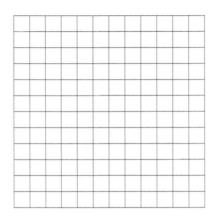

SUMMARY: ACTIVITY 3.9

I. The equation $y = kx^n$, where $k \neq 0$ and n is a positive integer, defines a **direct variation function**. The constant, k, is called the **constant of variation**.

2. The direct variation functions that have equations of the form $y = kx^n$, where n is a positive integer, are also called **power functions**.

    a. Power functions, in which n is even, resemble parabolas. As n increases in value, the graph flattens near the vertex.

    b. Power functions, in which n is odd, resemble the graph of $y = kx^3$. If k is positive, the graph is increasing. If k is negative, the graph is decreasing.

EXERCISES: ACTIVITY 3.9

I. For each table, determine the pattern and complete the table. Then write a direct variation equation for each table.

    a. y varies directly as x.

x	$\frac{1}{4}$	1	4	8
y		8		

    b. y varies directly as x^3.

x	$\frac{1}{2}$	1	3	6
y		1		

2. The area, A, of a circle is given by the function $A = \pi r^2$, where r is the radius of the circle.

    a. Does the area vary directly as the radius? Explain.

    b. What is the constant of variation k?

3. Assume that y varies directly as the square of x, and that when $x = 2$, $y = 12$. Determine y when $x = 8$.

4. The distance, d, that you drive at a constant speed varies directly as the time, x, that you drive. If you can drive 150 miles in 3 hours, how far can you drive in 6 hours?

5. The number of meters, d, that a skydiver falls before her parachute opens varies directly as the square of the time, t, that she is in the air. A skydiver falls 20 meters in 2 seconds. How far will she fall in 2.5 seconds?

In Exercises 6–10, sketch a graph of the given power function. Verify your graphs using a graphing calculator.

6. $y = -3x^2$

7. $y = x^4 + 1$

8. $y = -2x^5$

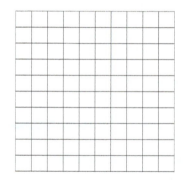

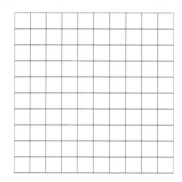

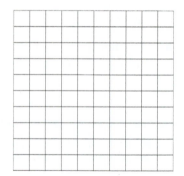

9. $f(x) = x^6$

10. $g(x) = 3x^3 - 3$

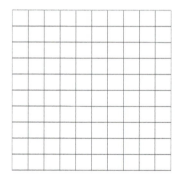

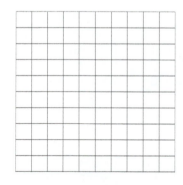

11. Determine the x-interval over which the function $f(x) = \dfrac{1}{2}x^4$ is increasing.

12. Does the function $g(x) = -\dfrac{1}{2}x^6$ have a maximum or a minimum point? Explain.

13. For $x > 1$, is the graph of $y = x^2$ rising faster or slower than the graph of $y = x^3$? Explain.

14. Is the graph of $y = \dfrac{3}{2}x^4$ wider or narrower than the graph of $y = x^4$?

15. How are the graphs of $y = -2x^3$ and $y = 2x^3 + 1$ different? How are the graphs similar?

Activity 3.10

Diving Under Pressure, or Don't Hold Your Breath

Objectives

1. Recognize functions of the form $y = \dfrac{k}{x}, x \neq 0$, as nonlinear.

2. Recognize equations of the form $xy = k$ as inverse variations.

3. Graph an inverse variation relationship.

4. Solve equations of the form $\dfrac{a}{x} = b, x \neq 0$.

Do you know why you shouldn't hold your breath when scuba diving? Safe diving depends on a fundamental law of physics that states that the volume of a given mass of gas (air in your lungs) will increase as the pressure decreases (when the temperature remains constant). This law is known as **Boyle's Law**, named after its discoverer, the seventeenth-century scientist Sir Robert Boyle.

In this activity, you will discover the answer to the opening question.

1. You have a balloon filled with air that has a volume of 10 liters at sea level. The balloon is under a pressure of 1 atmosphere (atm) at sea level. Measurements show 1 atm to be equal to 14.7 pounds of pressure per square inch.

   a. For every 33-foot increase in depth, the pressure will increase by 1 atm. Therefore, if you take the balloon underwater to a depth of 33 feet below sea level, it will be under 2 atm of pressure. At a depth of 66 feet, the balloon will be under 3 atm of pressure. Complete column 2 (pressure) in the table below part b.

   b. At sea level, the balloon is under 1 atm of pressure. At a depth of 33 feet, the pressure is 2 atm. Since the pressure is now twice as much as it was at sea level, Boyle's Law states that the volume of the balloon decreases to one-half of its original volume, or 5 liters. Taking the balloon down to 66 feet, the pressure compresses the balloon to one-third of its original volume, or 3.33 liters; at 99 feet down the volume is one-fourth the original, or 2.5 liters, and so on. Complete the third column (volume) in the following table.

DEPTH OF WATER (ft) BELOW SEA LEVEL	PRESSURE (ATM)	VOLUME OF AIR IN BALLOON (L)	PRODUCT OF PRESSURE AND VOLUME
Sea level	1	10	$1 \cdot 10 = 10$
33	2	5	
66	3	$3.3\overline{3}$	
99		2.5	
132			
297			

   c. Note, in column 4 of the table, that the product of the volume and pressure at sea level is 10. Compute the product of the volume and pressure for the other depths in the table, and record them in the last column.

   d. What is the result of your computations in part c?

2. Let p represent the pressure of a gas and v the volume at a given (constant) temperature. Use what you observed in Problem 1 to write an equation for Boyle's Law for this experiment.

Boyle's Law is an example of **inverse variation** between two variables. In the case of Boyle's Law, when one variable increases by a multiplicative factor p, the other decreases by the multiplicative factor $\dfrac{1}{p}$, so that their product is always the same positive constant.

> In general an inverse variation is represented by an equation of the form $xy = k$, where x and y represent the variables and k represents the constant.

Sometimes when you are investigating an inverse variation, you may want to consider one variable as a function of the other variable. In that case, you can solve the equation $xy = k$ for y as a function of x. By dividing each side of the equation by x, you obtain $y = \dfrac{k}{x}$.

3. Write an equation for Boyle's Law where the volume, v, is expressed as a function of the pressure, p.

4. a. Graph the volume, v, as a function of pressure, p, using the values in the preceding table.

   b. If you decrease the pressure (by rising to the surface of the water), what will happen to the volume? (Use the graph in part a to help answer this question.)

5. a. Suppose you are 99 feet below sea level, where the pressure is 4 atm and where the volume is one-fourth of the sea level volume. You use a scuba tank to fill the balloon back up to a volume of 10 liters. What is the product of the pressure and volume in this case?

   b. Now suppose you take the balloon up to 66 feet where the pressure is 3 atm. What is the volume now?

   c. You continue to take the balloon up to 33 feet where the pressure is 2 atm. What is the volume?

   d. Finally you are at sea level. What is the volume of the balloon?

   e. Suppose the balloon can expand only to 30 liters. What will happen before you reach the surface?

6. Explain why the previous problem shows that, when scuba diving, you should not hold your breath if you use a scuba tank to fill your lungs.

7. a. Use the graph in Problem 4 to determine the volume of the balloon when the pressure is 12 atmospheres.

   b. From the graph, what do you estimate the pressure to be when the volume is 11 liters?

   c. Use the equation $v = \dfrac{10}{p}$ to determine the volume when the pressure is 12 atm. How does your answer compare to the result in part a?

   d. Use the equation $v = \dfrac{10}{p}$ to determine the pressure when the volume is 11 liters. How does your answer compare to the result in part b?

8. Solve each of the following:

   a. $\dfrac{20}{x} = 10$ **b.** $\dfrac{150}{x} = 3$

9. Suppose another balloon contains 12 liters of air at sea level. The inverse variation between pressure and volume is modeled by the equation

$$pv = 12.$$

   a. Use the equation to write pressure as a function of the volume for this balloon.

b. Choose 5 values for *v* between 0 and 12. For each value, determine the corresponding value of *p*. Record your results in the following table:

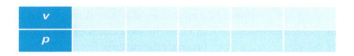

v					
p					

c. On the following grid, use the points you obtained in part b to graph the pressure as a function of the volume.

SUMMARY: ACTIVITY 3.10

1. For a function defined by $y = \dfrac{k}{x}$, where $k > 0$, as the independent variable x increases, the dependent variable y decreases.

2. The function rule $y = \dfrac{k}{x}$, $k > 0$ can also be expressed in the form $xy = k$. This form shows explicitly that as one variable increases by a multiplicative factor p the other decreases by a multiplicative factor $\dfrac{1}{p}$ so that the product of the two variables is always the constant value, k.

The relationship between the variables defined by this function is known as **inverse variation**.

3. The value $x = 0$ is not in the domain of the function defined by $y = \dfrac{k}{x}$.

EXERCISES: ACTIVITY 3.10

1. A guitar string has a number of frequencies at which it will naturally vibrate when it is plucked. These natural frequencies are known as the harmonics of the guitar string and are related to the length of the guitar string in an inverse relationship. For the first harmonic (also known as the fundamental frequency) the relationship can be expressed by the formula $K = 2f \cdot L$, where f represents the frequency in cycles per second (Hertz) and L represents the length of the guitar string in meters. The quantity K is about 420 m/s.

a. Determine the fundamental frequency in hertz (first harmonic) of the string if its length is 70 centimeters.

b. To observe the effect that changing the length of the string has on the first harmonic, write the frequency, f, as a function of L in the formula $420 = 2f \cdot L$.

c. Use the formula you wrote in part b to determine the frequencies for the string lengths given in the following table.

LENGTH OF GUITAR STRING (m)	FREQUENCY (Hz)
0.2	
0.3	
0.4	
0.5	
0.6	

d. Graph the points in the table in part c on the following grid.

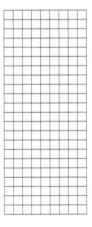

e. From your graph, estimate the frequency for a string length of 35 centimeters.

f. What string length would produce a frequency of 840 hertz?

You may notice that all the strings in a guitar are the same length. Their difference in pitch is due to their weight and tension as well as their length. String length is only one element of the guitar's design.

2. A bicyclist, runner, jogger, and walker each completed a journey of 30 miles. Their results are recorded in the table below.

	CYCLIST	RUNNER	JOGGER	WALKER
v, Average Speed (mph)	30	15	10	5
t, Time Taken (hr)	1	2	3	6
d = vt				

a. Compute the product *vt* for each person, and record the products in the last row of the table.

b. Write the time to complete the journey as a function of average speed.

c. On the following grid, graph the function you wrote in part b for a journey of 30 miles.

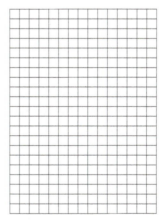

d. From the graph, determine how long the cyclist would take if she decreased her average speed to 20 miles per hour. Verify your answer from the graph by using the equation you wrote in part b.

e. From the graph, determine how long it would take the runner to complete her run if she slowed to 12 miles per hour. Verify your answer from the graph by using your equation from part b.

3. According to the architect's blueprint, the floor area of the stage in the new auditorium at your high school must be rectangular and equal to 1200 square feet. The width of the stage is key to all of the theater productions. Therefore, in this situation, the stage's depth is a function of its width.

   a. Let d represent the depth (in feet) and w represent the width (in feet). Write an equation that expresses d as a function of w.

   b. Complete the following table using the equation from part a.

w (feet)	30	35	40	50	60
d (feet)					

   c. What happens to the depth as the width increases?

   d. What happens if the width is 100 feet? Is this realistic? Explain.

   e. Can the width be zero? Explain.

   f. What do you think is the practical domain for this function?

   g. What type of a function do you have in this situation?

4. According to a physics principle, the weight, w, of a body varies inversely as the square of its distance, d, from the center of the Earth. The equation $wd^2 = 3.2 \times 10^9$ represents this relationship for a 200-pound person, where w is in pounds and d is in miles.

   a. Rewrite the equation to express the weight of a body, w, as a function of the distance, d, from the center of the Earth.

   b. How much would a 200-pound man weigh if he were 1000 miles above the surface of the Earth? The radius of the Earth is approximately 4000 miles. Explain how you solve this problem.

5. **a.** Solve $xy = 120$ for y. Is y a function of x? What is its domain?

b. Solve $xy = 120$ for x. Is x a function of y? What is its domain?

6. Solve the following equations:

   a. $30 = \dfrac{120}{x}$ **b.** $2 = \dfrac{250}{x}$ **c.** $-8 = \dfrac{44}{x}$

Activity 3.11

Loudness of a Sound

Objectives

1. Graph an inverse variation function defined by an equation of the form $y = \dfrac{k}{x^n}$, where n is any positive integer $x \neq 0$.

2. Describe the properties of graphs having equation $y = \dfrac{k}{x^n}$.

3. Determine the constant of proportionality (also called the constant of variation).

The loudness (or intensity) of any sound is a function of the listener's distance from the source of the sound. In general, the relationship between the intensity I and the distance d can be modeled by an equation of the form

$$I = \frac{k}{d^2},$$

where I is measured in microwatts per square meter, d is measured in meters, and k is a constant determined by the source of the sound and nature of the surroundings.

1. The intensity, I, of a typical iPod® at maximum setting can be given by the formula $I = \dfrac{64}{d^2}$. Complete the following table.

d (m)	0.1	0.5	1	2	5	10	20	30
I (μW/m²)								

2. a. What is the practical domain of the function?

   b. Sketch a graph that shows the relationship between intensity of sound and distance from the source of the sound. Use the table in Problem 1 to help determine an appropriate scale.

3. As you move closer to the person speaking, what happens to the intensity of the sound?

4. As you move away from the person speaking, what happens to the intensity of the sound?

Functions Defined by $y = \dfrac{k}{x^2}$, Where k Is a Nonzero Constant

The function defined by $I = \dfrac{64}{d^2}$ belongs to a family of functions having an equation of the form $y = \dfrac{k}{x^2}$, where k represents some nonzero constant. Other examples of this type of function are $f(x) = \dfrac{1}{x^2}$ and $g(x) = \dfrac{10}{x^2}$.

5. a. What is the domain of functions f and g defined above?

b. Complete the following table:

x	−20	−10	−5	−1	−0.5	−0.1	0	0.1	0.5	1	5	10	20
f(x)													
g(x)													

c. Sketch the graphs of f and g on the same coordinate system. Verify your sketch using your graphing calculator.

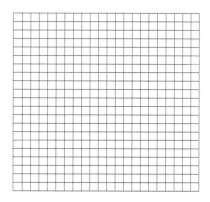

d. Explain why no part of each graph appears below the x-axis.

e. What happens to the y-values as the x-values increase infinitely in the positive direction or decreases infinitely in the negative direction?

f. What happens to the y-values as the positive x-values get closer to zero?

g. What happens to the y-values as the negative x-values get closer to zero?

h. Do the functions have a maximum y-value or a minimum y-value?

6. Describe how the graphs of $y = \dfrac{1}{x}$ and $y = \dfrac{1}{x^2}$ are similar and how they are different.

7. a. Complete the following table:

x	-20	-10	-5	-1	-0.5	-0.1	0	0.1	0.5	1	5	10	20
$g(x) = \dfrac{10}{x^2}$													
$h(x) = \dfrac{-10}{x^2}$													

b. Sketch graphs of functions g and h on the same coordinate system. Use labels or different colors to differentiate the graphs. Verify your sketch using your graphing calculator.

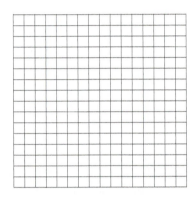

c. Describe how to obtain the graph of function h using the graph of function g.

8. Sketch a graph of $h(x) = \dfrac{1}{x^3}$. Is the graph similar to the graph of $f(x) = \dfrac{1}{x}$ or $g(x) = \dfrac{1}{x^2}$?

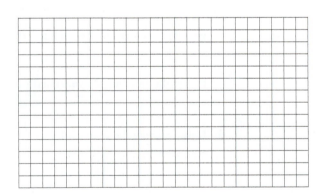

9. Sketch a graph $R(x) = \dfrac{1}{x^4}$. Is the graph similar to the graph of $f(x) = \dfrac{1}{x}$ or $g(x) = \dfrac{1}{x^2}$?

Inverse Variation Functions

The function defined by $I = \dfrac{64}{d^2}$ is another type of inverse variation function. In general, any function defined by an equation of the form $y = \dfrac{k}{x^n}$, where k is a nonzero constant and n is a positive integer, belongs to the family of **inverse variation functions**.

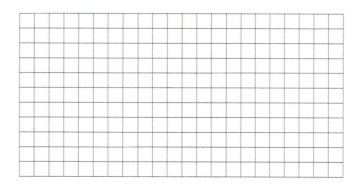

Example 1   *For the inverse variation function given by $y = \dfrac{4}{x^2}$, y varies inversely as x^2, or y is inversely proportional to x^2. The number 4 is called the constant of variation or the constant of proportionality. The following table demonstrates that as x doubles in value, the corresponding y-values are reduced by one-quarter.*

x	2	4	8	16
$y = \dfrac{4}{x^2}$	1	$\dfrac{1}{4}$	$\dfrac{1}{16}$	$\dfrac{1}{64}$

10. For the function defined by $I = \dfrac{64}{d^2}$ (see Problem 1), answer the following questions:

a. I varies inversely as what quantity?

b. What is the constant of proportionality?

c. If d is doubled, what is the effect on I?

11. Using the patterns of inverse variation, complete the following table if y is inversely proportional to the cube of x:

x	0.5	1	2	6
y		8		

12. In this activity, the relationship between the intensity, I, of an iPod® and the distance, d, from the individual was given by $I = \dfrac{64}{d^2}$, where I is measured in microwatts per square meter, d is measured in meters, and 64 is the constant of proportionality. The constant of proportionality depends on the source of the sound and the surroundings. If the source of the sound changes, the value of the constant of proportionality will also change.

a. The intensity of the sound made by a heavy truck 20 meters away is 1000 microwatts per square meter. Determine the constant of proportionality.

b. Write a formula for the intensity, I, of the sound made by a truck when it is d meters away.

c. Use the formula from part b to determine the intensity of the sound made by the truck when it is 100 meters away.

SUMMARY: ACTIVITY 3.11

- Functions defined by equations of the form $y = \dfrac{k}{x^n}$, where k is a positive integer, have the following properties.

   1. The domain consists of all real numbers except zero.

   2. The graph of f has the following general shape:

   a. Where $k > 0$, and n is an even integer.

   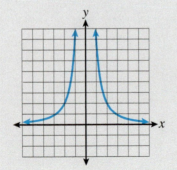

   b. Where $k > 0$, and n is an odd integer.

   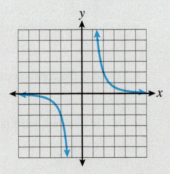

c. Where $k < 0$, and n
is an even integer.

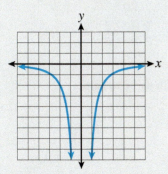

d. Where $k < 0$, and n
is an odd integer.

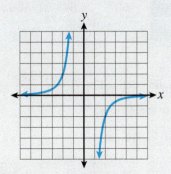

3. There are no x- or y-intercepts.

4. There is no maximum or minimum y-value.

• Functions defined by $y = \dfrac{k}{x^n}$ are called **inverse variation functions** in which

1. y is said to vary inversely as the nth power of x.

2. k is called the constant of variation or constant of proportionality.

EXERCISES: ACTIVITY 3.11

1. Doctors sometimes use a patient's body-mass index to determine whether or not the patient
should lose weight. The formula for the body-mass index, B, is

$$B = \frac{705w}{h^2},$$

where w is the weight in pounds and h is height in inches.

a. What is your body-mass index?

b. Suppose your friend weighs 170 pounds. Substitute this value into the body-mass index formula to obtain an equation for B in terms of height.

c. B varies inversely as what quantity? What is the constant of proportionality?

d. What is the practical domain of the body-mass index function in part b?

e. Complete the following table using the formula for the body-mass index of a 170-pound person:

h, Height in inches	60	64	68	72	76	80
B						

f. Sketch a graph of the function defined by $B = \dfrac{119{,}850}{h^2}$. Use the data values in part e to help determine an appropriately scaled axis.

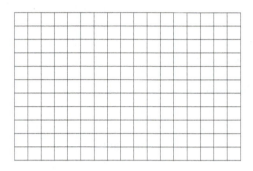

g. What happens to the body-mass index as height increases? Does this make sense in the context of the situation? Explain why or why not.

h. A body-mass index of less than 10 or greater than 50 is extremely rare. It is recommended that a person's body-mass index be between 19 and 25. Use the graph and the trace key on your calculator to approximate the values of h for which $19 < B < 25$.

2. Sketch a graph of the functions $f(x) = \dfrac{3}{x^2}$ and $g(x) = \dfrac{-3}{x^2}$ on the same coordinate system.

Describe how the graph of g is related to the graph of f.

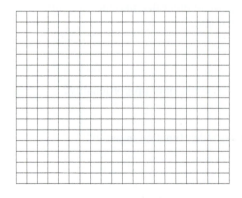

3. Match the following functions with the accompanying graphs.

   i. $f(x) = \dfrac{10}{x^4}$

   ii. $g(x) = \dfrac{100}{x^5}$

   iii. $h(x) = \dfrac{-10}{x^3}$

   iv. $F(x) = \dfrac{-1}{x^2}$

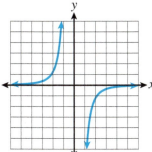

a.

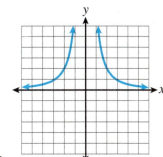

b.

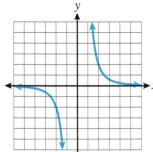

c.

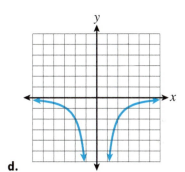

d.

4. Describe how the graphs of $y = \dfrac{1}{x^2}$ and $y = \dfrac{1}{x^3}$ are similar and how they are different.

5. Consider the family of functions of the form $y = \dfrac{k}{x^n}$, where k is a nonzero constant and n is a positive integer.

   a. What is the domain of the function?

b. Use several different values of k and n, where $k > 0$ and n is an odd positive integer, to determine the general shape of the graph of f.

c. Use several different values of k and n, where $k > 0$ and n is an even positive integer, to determine the general shape of the graph of f.

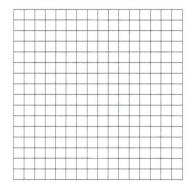

6. How will the general shapes of the graphs in Exercise 5 change if $k < 0$?

7. If y varies inversely as the cube of x, determine the constant of proportionality if $y = 16$ when $x = 2$.

8. For each inverse variation below, at least one ordered pair (x, y) is given in the table. Use this ordered pair to determine the constant of proportionality. Then use the equation to complete the table.

a. y varies inversely as x.

x	y
$\frac{1}{2}$	
1	2
2	
6	

b. y varies inversely as x^3.

x	y
$\frac{1}{2}$	
1	8
2	1
6	

9. The amount of current, I, in a circuit varies inversely as the resistance R. A circuit containing a resistance of 10 ohms has a current of 12 amperes. Determine the current in a circuit containing a resistance of 15 ohms.

10. The intensity, I, of light varies inversely as the square of the distance, d, between the source of light and the object being illuminated. A light meter reads 0.25 unit at a distance of 2 meters from a light source. What will the meter read at a distance of 3 meters from the source?

11. You are investigating the relationship between the volume, V, and pressure, P, of a gas. In a laboratory, you conduct the following experiment: While holding the temperature of a gas constant, you vary the pressure and measure the corresponding volume. The data that you collect appears in the following table.

P (psi)	20	30	40	50	60	70	80
V (ft³)	82	54	41	32	27	23	20

a. Sketch a graph of the data.

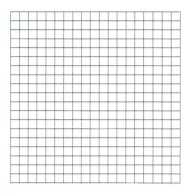

b. One possible model for the data is that V varies inversely as the square of P. Does the data fit the model $V = \dfrac{k}{P^2}$? Explain.

c. Another possible model for the data is that V varies inversely as P. Does $V = \dfrac{k}{P}$ model the data? Explain.

d. Predict the volume of the gas if the pressure is 65 pounds per square inch.

Activities 3.8–3.11 **What Have I Learned?**

1. In a direct variation situation, as the magnitude of the independent variable, x, increases, what happens to the magnitude of the dependent variable, y?

2. In an inverse variation situation, as the magnitude of independent variable, x, decreases, what happens to the magnitude of the dependent variable, y?

3. Does an equation of the form $xy = k$ indicate a direct variation or an inverse variation situation? Explain.

4. Which of the following statements is not equivalent to the other two?

   a. y varies directly as x

   b. $y = kx$ for some constant k

   c. $y = \dfrac{k}{x}$ for some constant k

5. For what values of k will the graph of $y = \dfrac{k}{x}$ be in the second and fourth quadrants?

6. Is the graph of $y = 3x^4$ narrower or wider than the graph of $y = x^2$? Explain.

Activities 3.8–3.11   How Can I Practice?

1. y varies directly as x and $y = 21$ when $x = 6$. Determine the value of y when $x = 14$.

2. y varies inversely as x and $y = 11$ when $x = 8$. Determine the value of y when $x = 16$.

3. y varies directly as x^2. When $x = 3$, $y = 45$. Determine y when $x = 6$.

4. Recall if d represents the distance (in feet) of the lightning from the observer, then d varies directly as the time, t (in seconds), it takes to hear the thunder. The relationship is modeled by

$$d = 1080t.$$

  a. As the time t doubles (say from 3 to 6), the corresponding d-values _____

  b. What is the value of k, the constant of variation, in this situation? What significance does k have in this problem?

5. The velocity, v, of a falling object varies directly to the time, t, of the fall. After 3 seconds, the velocity of the object is 96 feet per second. What will be its velocity after 4 seconds?

6. A propane gas bill varies directly as the amount of gas used. The bill for 56 gallons of propane was $184.80. What is the bill for 70 gallons of propane?

7. The power P generated by a wind turbine varies directly as the square of the wind speed w. The turbine generates 750 watts of power in a 25 mph wind. Determine the power it generates in a 45 mph wind.

8. The time it takes to drive a certain distance varies inversely as the rate of travel. If it takes 6 hours at 48 miles per hour to drive the distance, how long would it take at 72 miles per hour?

9. The sound intensity of a speaker varies inversely as the square of the distance from the speaker. The intensity is 20 microwatts per square meter when you are 10 feet from the speaker. Determine the intensity when you are 5 feet from the speakers.

10. Describe the relationship between the graphs of $f(x) = \dfrac{1}{x}$ and $g(x) = \dfrac{-1}{x}$.

11. Describe the relationship between the graphs of $f(x) = \dfrac{1}{x}$ and $g(x) = \dfrac{1}{x^3}$.

12. a. Suppose you are taking a trip of 145 miles. Assume that you drive the entire distance at a constant speed. Express your time to take this trip as a function of your speed.

b. What is the practical domain of this function?

c. Using the equation for the function, determine the domain.

13. The intensity of a light source varies inversely as the square of the distance from the light source. For a particular light the intensity measures 24 foot-candles at a distance of 12 feet from the source.

a. Determine the equation that relates the intensity, I, to the distance from the light source, s, in this case.

b. Use the equation in part a to complete the following table:

Distance from the Light Source, s	2	3	4	6	8	12
Intensity of the Light, I						

c. Explain how this table indicates that the relationship between s and I is an inverse variation.

d. Graph the relationship between *I* and *s* using the last five pairs from the table in part b.

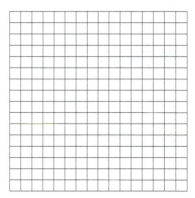

e. If the intensity from the light source in this case measures 6 foot-candles, how far are you from the light source?

The bracketed numbers following each concept indicate the activity in which the concept is discussed.

CONCEPT/SKILL	DESCRIPTION	EXAMPLE
Parabola [3.1]	A parabola is the U-shaped graph described by a function of the form $y = ax^2 + bx + c$.	
Solving an equation of the form $ax^2 = c, a \neq 0$ [3.1]	Solve an equation of this form algebraically by dividing both sides of the equation by a and then taking the square root of both sides.	$2x^2 = 8$ $$\frac{2x^2}{2} = \frac{8}{2}$$ $x^2 = 4, x = \pm 2$
Quadratic function [3.2]	The quadratic function with x as the variable has the standard form $y = ax^2 + bx + c$, where a, b, and c represent real numbers and $a \neq 0$.	$y = 2x^2 - 3x - 2$
Graph a quadratic function (a parabola) [3.2]	For the quadratic function defined by $y = ax^2 + bx + c$; if $a > 0$, the parabola opens upward; if $a < 0$, the parabola opens downward.	The graph of $y = 2x^2 - 3x - 2$ is a parabola that opens upward.
y-intercept of the graph of a quadratic function [3.2]	The constant term c of a quadratic function $y = ax^2 + bx + c$ always indicates the y-intercept of the parabola. The y-intercept of any quadratic function is $(0, c)$.	The y-intercept of the graph of $y = 2x^2 - 3x - 2$ is $(0, -2)$.
Axis of symmetry [3.3]	The axis of symmetry of a parabola is a vertical line that separates the parabola into two mirror images. The equation of the vertical axis of symmetry is given by $x = \dfrac{-b}{2a}$.	See Example 2 in Activity 3.3.
Vertex (turning point) [3.3]	The vertex of a parabola defined by $y = ax^2 + bx + c$ is the point where the graph changes direction. It is given by $\left(\dfrac{-b}{2a}, f\left(-\dfrac{b}{2a} \right) \right)$.	See Example 1 in Activity 3.3.
x-intercept(s) [3.3]	An x-intercept is the point or points (if any) where the parabola crosses the x-axis (that is, where its y-coordinate is zero).	The x-intercepts of the parabola defined by $y = 2x^2 - 3x - 2$ are $(2, 0)$ and $(-0.5, 0)$.

CONCEPT/SKILL	DESCRIPTION	EXAMPLE
Domain of the quadratic function [3.3]	The domain of all quadratic functions is all real numbers.	The domain of $y = 2x^2 - 3x - 2$ is all real numbers.
Range of the quadratic function [3.3]	If the parabola opens upward, the range is $y \geq y$-value of vertex. If the parabola opens downward, the range is $y \leq y$-value of vertex.	The range of the parabola defined by $y = 2x^2 - 3x - 2$ is $y \geq \dfrac{-25}{8}$.
Solving $f(x) = c$ graphically [3.4]	Graph $y = f(x)$, graph $y = c$, and determine the x-values of the points of intersection. Or graph $y = f(x) - c$ and determine the x-intercepts.	See Example 1 in Activity 3.4.
Greatest common factor (or GCF) [3.5]	The GCF is the largest factor common to all terms in an expression.	The GCF of $3x^4 - 6x^3 + 18x^2$ is $3x^2$.
Zero-product principle [3.5]	If a and b are any numbers and $a \cdot b = 0$, then either a or b, or both, must be equal to zero.	Example 1, Activity 3.5
Factoring trinomials by trial and error [3.5]	To factor trinomials by trial and error, 1. Remove the GCF. 2. Try combinations of factors for the first and last terms in two binomials. 3. Check the outer and inner products to match middle term of the original trinomial. 4. If the check fails, repeat steps 2 and 3.	Example 5, Activity 3.5
Solving quadratic equations by factoring [3.5]	To solve a quadratic equation by factoring, 1. Use the addition principle to remove all terms from one side of the equation. This results in a quadratic polynomial being set equal to zero. 2. Combine like terms and then factor the nonzero side of the equation. 3. Use the zero-product principle to set each factor containing a variable equal to zero and then solve the equations. 4. Check your solutions in the original equation.	Example 6, Activity 3.5
Quadratic formula [3.6]	$x = \dfrac{-b \pm \sqrt{b^2 - 4ac}}{2a}$	
Solving a quadratic equation of the form $ax^2 + bx + c = 0$, $a \neq 0$, using the quadratic formula [3.6]	To solve a quadratic equation of the form $ax^2 + bx + c = 0$, $a \neq 0$ using the quadratic formula $$x = \frac{-b \pm \sqrt{b^2 - 4ac}}{2a},$$ 1. Set the quadratic equation equal to zero. 2. Identify the coefficients a and b and the constant term c. 3. Substitute these values into the formula, and simplify. 4. Check your solutions.	Example 1, Activity 3.6.

CONCEPT/SKILL	DESCRIPTION	EXAMPLE
Direct variation [3.8]	Two variables, x and y, are said to vary directly if $y = kx$ for some constant k. In this situation, whenever the magnitude of x increases by a scale factor, the magnitude of y increases by the same factor.	In the function $y = 2x$, y varies directly as x. As x increases in magnitude so does y.
The constant of proportionality or constant of variation, k [3.8]	The following statements are equivalent: **a.** y varies directly as x **b.** y is directly proportional to x **c.** $y = kx$ for some constant k	In the direct variation function $y = 2x$, 2 is the constant of variation.
Direct variation [3.8], [3.9]	Two variables are said to vary directly if as the magnitude of one increases, the magnitude of the other does as well.	In the function $y = 2x$, y varies directly as x. As x increases in magnitude so does y.
The constant of proportionality or constant of variation, k [3.8], [3.9]	The following statements arc equivalent: **a.** y varies directly as x **b.** y is directly proportional to x **c.** $y = kx$ for some constant k	In the direct variation function $y = 2x$, 2 is the constant of variation.
Power functions [3.9]	The direct variation function having an equation of the form $y = kx^n$, where n is a positive integer, is also called a power function.	$y = 4x^3$ is a third-power function.
Inverse variation function [3.10]	An inverse variation function is a function defined by the equation $xy = k\left(\text{or } y = \dfrac{k}{x}\right)$. When $k > 0$ and x increases by a constant multiplicative factor p, the y decreases by the multiplicative factor $\dfrac{1}{p}$. The value $x = 0$ is not in the domain of this function.	$y = \dfrac{360}{x}$ When $x = 4$, $y = \dfrac{360}{4} = 90$. If x doubles to 8, y is halved and becomes 45.
Domain of $y = \dfrac{k}{x^n}$ [3.11]	The domain consists of all real numbers except zero.	For $y = \dfrac{4}{x^3}$, the domain is all real numbers except 0.
Graph of $y = \dfrac{k}{x^n}$ [3.11]	The graphs will vary depending on the values of k and n.	See the summary at the end of Activity 3.11.
Inverse variation functions [3.11]	Functions defined by $y = \dfrac{k}{x^n}$ are called inverse variation functions in which y is said to vary inversely as the nth power of x; k is called the constant of variation.	For the function $y = \dfrac{4}{x^3}$, y varies inversely as the cube of x, and 4 is the constant of variation.

Chapter 3   Gateway Review

In Exercises 1–8, determine the following characteristics of each quadratic function by inspecting its equation.

    a. the direction in which the graph opens

    b. the equation of the axis of symmetry

    c. the vertex

    d. the y-intercept

1. $y = x^2 + 2$

2. $y = -3x^2$

3. $y = -3x^2 + 4$

4. $y = 2x^2 - x$

5. $y = x^2 + 5x + 6$

6. $y = x^2 - 3x + 4$

7. $y = x^2 - 2x + 1$

8. $y = -x^2 + 5x - 6$

In Exercises 9–15, sketch the graph of each quadratic function using your graphing calculator. Then determine each of the following using the graph.

    a. the coordinates of the x-intercepts; if they exist

    b. the domain and the range of the function

    c. the x-interval in which the function is increasing

    d. the x-interval in which the function is decreasing

9. $y = x^2 + 4x + 3$

10. $y = x^2 + 2x - 3$

11. $y = x^2 - 3x + 1$

12. $y = 2x^2 + 8x + 5$

13. $y = -2x^2 + 8$

$(2, 0), (-2, 0)$

14. $y = -3x^2 + 4x - 1$

15. $y = 4x^2 + 5$

In Exercises 16–19, solve the quadratic equation numerically (using tables). Verify your solutions graphically.

16. $x^2 + 4x + 4 = 0$

17. $x^2 - 5x + 6 = 0$

18. $3x^2 = 18x + 10$

19. $-x^2 = 3x - 10$

In Exercises 20 and 21, solve the equation. Round your answer to the nearest tenth when necessary.

20. $8x^2 = 10$

21. $5x^2 + 25x = -5$

22. Completely factor the following polynomials.

   a. $9a^5 - 27a^2$

   b. $24x^3 - 6x^2$

   c. $4x^3 - 16x^2 - 20x$

   d. $5x^2 - 16x + 6$

   e. $x^2 - 5x - 24$

   f. $t^2 + 10t + 25$

In Exercises 23–27, solve each equation by factoring. Verify your answer graphically or by substitution of the solutions in the equations.

23. $x^2 - 9 = 0$

24. $-x^2 + 36 = 0$

25. $x^2 - 7x + 12 = 0$

26. $x^2 - 6x = 27$

27. $x^2 = -x$

In Exercises 28–32, write each of the equations in the form $ax^2 + bx + c = 0$. Then identify a, b, and c, and solve the equation using the quadratic formula. Verify your solutions by substitution.

28. $x^2 + 5x + 3 = 0$

29. $2x^2 - x = 3$

30. $x^2 = 81$

31. $3x^2 + 5x = 12$

32. $2x^2 = 3x + 5$

33. For the quadratic function $f(x) = 2x^2 - 8x + 3$, determine the zeros of the graph, if they exist. First, approximate the zeros using your graphing calculator. Second, write and solve the appropriate equation using the quadratic formula. Approximate your answers to the nearest hundredth.

34. The height, h (in feet), of a golf ball is a function of the time, t (in seconds), it has been in flight. The approximate height of the ball above the ground is modeled by

$$h = -16t^2 + 80t.$$

a. Sketch a graph of the function. What is the practical domain in this situation?

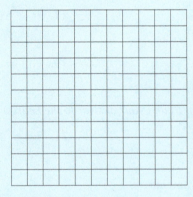

b. Determine the vertex of the parabola. What is the practical meaning of this point?

c. What is the h-intercept, and what is its practical meaning in this situation?

d. Determine the t-intercepts. What is the significance of these intercepts?

e. What assumption are you making in this situation about the elevation of the spot where the ball is struck and the point where the ball lands?

35. To use the regression feature of your calculator to determine the equation of a parabola, you need at least three distinct points. The stream of water flowing out of a water fountain is in the shape of a parabola. Suppose you let the origin of a coordinate system correspond to the point where the water begins to flow out of the nozzle (see figure).

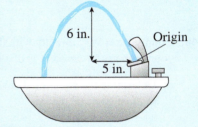

The maximum height of water stream occurs approximately 5 inches measured horizontally from the nozzle. The maximum height of the stream of water is measured to be approximately 6 inches.

a. What is the vertex of the parabola?

b. You already have two points that lie on the parabola. What are they? Use symmetry to obtain a third point.

c. Using these three points and the regression feature of your graphing calculator, determine the equation of the stream of water.

36. A model rocket is shot straight up from five feet above the ground level. The height, h (in feet), of the rocket from the ground is modeled by

$$h(t) = -16t^2 + 80t + 5,$$

where t is measured in seconds.

a. What is the maximum height of the rocket above the ground?

b. How long will it take for the rocket to reach the ground?

37. Safe automobile spacing, S (in feet), is modeled by

$$S(v) = 0.03125v^2 + v + 18,$$

where v is average velocity in feet per second.

a. Suppose a car is traveling at 44 feet per second. To be safe, how far should it be from the car in front of it?

b. If the car is following 50 feet behind a van, what is a safe speed for the car to be traveling? How fast is this in miles per hour (60 miles per hour $\approx$ 88 feet per second)?

38. y varies directly as x and $y = 88$ when $x = 16$. Determine the value of x when $y = 77$.

39. The sales tax on an item varies directly as the cost of the item. The sales tax in your area is 7%. The sales tax on your item was $10.50. What was the cost of the item before the sales tax was applied?

40. County taxes vary directly with the assessed value of your property. Your tax bill indicated that you owed $1245 in taxes on your property assessed at $105,000.

    a. Determine the equation that relates the county tax due, t, to the assessed value of your property, v, in this case. Round k to three decimal places.

    b. Use the equation in part a to complete the following table:

Assessed Value of Your Property, v	50,000	75,000	100,000	125,000	200,000	450,000
Amount of County Taxes, t						

    c. Explain how this table indicates that the relationship between v and t is a direct variation.

    d. Graph the relationship between v and t using the first five pairs from the table in part b.

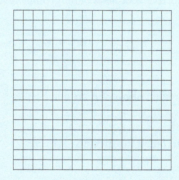

    e. If the taxes on your property are $1164, what is the assessed value of your property.

41. a. If y varies inversely as x, and $x = 10$ when $y = 12$, then determine the value of y when $x = 30$.

    b. The sound intensity of a stereo varies inversely to the square of the distance from the speaker to the person listening. If the sound intensity is 1000 microwatts per square meter at a distance of 4 meters, then what is the intensity when the listener is 10 meters from the speaker?

    c. When the volume of a circular cylinder is constant, the height varies inversely as the square of the radius. If the radius is 2 inches when the height is 8 inches, determine the height when the radius is 5 inches.

Chapter 4

Modeling with Exponential and Logarithmic Functions

Activity 4.1

Going Shopping

Objectives

1. Define growth factor.

2. Determine growth factors from percent increases.

3. Apply growth factors to problems involving percent increases.

4. Define decay factor.

5. Determine decay factors from percent decreases.

6. Apply decay factors to problems involving percent decreases.

To obtain revenue (income), many state and local governments require merchants to collect sales tax on the items they sell. In several localities, the sales tax is assessed at as much as 8% of the selling price and is passed on directly to the purchaser. The Texas state sales tax rate is currently 6.25%. The combined rate of local sales and use taxes cannot exceed 2%. Therefore, the highest possible combined sales tax rate is 8.25%.

1. Determine the total cost (including the 8% sales tax) to the customer of the following items. Include each step of your calculation in your answer.

    a. A greeting card selling for $3.50

    b. A Blu-ray player priced at $120.

Many people correctly determine the total costs for Problem 1 in two steps:

- They first compute the sales tax on the item.
- Then they add the sales tax to the selling price to obtain the total cost.

It is also possible to compute the total cost in one step by using the idea of a **growth factor**.

Growth Factors

2. **a.** If a quantity increases by 50%, how does its new value compare to its original value? That is, what is the ratio of the new value to the original value? The first row of the following table is completed for you. Complete the remaining rows to confirm or discover your answer.

A Matter of Values

ORIGINAL VALUE	NEW VALUE (INCREASED BY 50%)	RATIO OF NEW VALUE TO ORIGINAL VALUE		
		FRACTIONAL FORM	DECIMAL FORM	PERCENT FORM
20	10 + 20 = 30	$\frac{30}{20} = \frac{3}{2} = 1\frac{1}{2}$	1.50	150%
50				
100				
Your choice of an original value				

   b. Use the results from the preceding table to answer the question, "What is the ratio of the new value to the original value of any quantity that increases by 50%?" Express this ratio as a reduced fraction, a mixed number, a decimal and a percent.

Definition

For any specified percent *increase*, no matter what the original value may be, the ratio of the new value to the original value is always the same. This ratio $\dfrac{\text{new value}}{\text{original value}}$ is called the **growth factor** associated with the specified percent increase.

The growth factor is most often written in decimal form. As you saw previously, the growth factor associated with a 50% increase is 1.50.

Procedure

The growth factor is formed by adding the specified percent increase to 100% and then changing this percent into its decimal form.

Example 1

The growth factor corresponding to a 50% increase is

$$100\% + 50\% = 150\%.$$

If you change 150% to a decimal, the growth factor is **1.50.**

   3. Determine the growth factor of any quantity that increases by the given percent.

   a. 30%

   b. 75%

   c. 15%

   d. 7.5%

4. A growth factor can always be written in decimal form. Explain why this decimal will always be greater than 1.

Procedure

When a quantity increases by a specified percent, its new value can be obtained by multiplying the original value by the corresponding growth factor. That is,

$$\text{original value} \cdot \text{growth factor} = \text{new value.}$$

---

5. a. Use the growth factor 1.50 to determine the new value of a stock portfolio that increased 50% over its original value of $400.

b. Use the growth factor 1.50 to determine the population of a town that has grown 50% over its previous size of 120,000 residents.

6. a. Determine the growth factor for any quantity that increases 20%.

b. Use the growth factor to determine this year's budget, which has increased 20% over last year's budget of $75,000.

7. a. Determine the growth factor of any quantity that increases 8%.

b. Use the growth factor to determine the total cost of each item in Problem 1.

In the previous problems, you were given an original (earlier) value and a percent increase and were asked to determine the new value. Suppose you are asked a "reverse" question: Sales in your small retail business have increased 20% over last year. This year you had gross sales receipts of $75,000. How much did you gross last year?

Using the growth factor 1.20, you can write

$$\text{original value} \cdot 1.20 = 75{,}000.$$

To determine last year's sales receipts, you divide both sides of the equation by 1.20.

$$\text{original value} = 75{,}000 \div 1.20.$$

Therefore, last year's sales receipts totaled $62,500.

Procedure

When a quantity has *already* increased by a specified percent, its original value is obtained by dividing the new value by the corresponding growth factor. That is,

$$\text{new value} \div \text{growth factor} = \text{original value.}$$

8. You determined a growth factor for an 8% increase in Problem 7. Use it to answer the following questions.

    a. The cash register receipt for your new tablet, which includes the 8% sales tax, totals $182.52. What was the ticketed price of the tablet?

    b. The credit card receipt for a new sofa for your apartment, which includes the 8% sales tax, totals $1620. What was the ticketed price of the sofa?

Decay Factors

It is the sale you have been waiting for all season. Everything in the store is marked 40% off the original ticket price.

9. Determine the discounted price of a pair of sunglasses originally selling for $25.

Many people correctly determine the discounted prices for Problem 9 in two steps:

- They first compute the actual dollar discount on the item.
- Then they subtract the dollar discount from the original price to obtain the sale price.

It is also possible to compute the sale price in one step by using the idea of a **decay factor**.

10. a. If a quantity decreases 20%, how does its new value compare to its original value? That is, what is the ratio of the new value to the original value? Complete the rows in the following table. The first row has been done for you. Notice that the new values are less than the originals, so the ratios of new values to original values will be less than 1.

Up and Down

| | NEW VALUE | RATIO OF NEW VALUE TO ORIGINAL VALUE | | |
ORIGINAL VALUE	(DECREASED BY 20%)	FRACTIONAL FORM	DECIMAL FORM	PERCENT FORM
20	16	$\dfrac{16}{20} = \dfrac{4}{5}$	0.80	80%
50				
100				
Choose any value				

    b. Use the results from the preceding table to answer the question, "What is the ratio of the new value to the original value of any quantity that decreases 20%?" Express this ratio in reduced-fraction, decimal, and percent forms.

Definition

For a specified percent *decrease*, no matter what the original value may be, the ratio of the new value to the original value is always the same. This ratio $\dfrac{\text{new value}}{\text{original value}}$ is called the **decay factor** associated with the specified percent decrease.

As you have seen, the decay factor associated with a 20% decrease is 80%. For calculation purposes, a decay factor is usually written in decimal form, so the decay factor 80% is expressed as 0.80.

Procedure

Note that a percent decrease describes the percent that has been *removed* (such as the discount taken off an original price), but the corresponding decay factor always represents the percent remaining (the portion that you still must pay). Therefore, a decay factor is formed by *subtracting* the specified percent decrease from 100% and then changing this percent into its decimal form.

Example 2

The decay factor corresponding to a 10% decrease is

$$100\% - 10\% = 90\%.$$

If you change 90% to a decimal, the decay factor is 0.90.

---

11. Determine the decay factor of any quantity that decreases by the following given percents:

a. 30%

b. 75%

c. 15%

d. 7.5%

Remember that multiplying the original value by the decay factor will always result in the amount remaining, not the amount that has been removed.

12. Explain why every decay factor will have a decimal format that is less than 1.

Procedure

When a quantity *decreases* by a specified percent, its new value is obtained by multiplying the original value by the corresponding decay factor. That is,

$$\text{original value} \cdot \text{decay factor} = \text{new value}.$$

13. Use the decay factor 0.80 to determine the new value of a stock portfolio that has lost 20% of its original value of $400.

In the previous questions, you were given an original value and a percent decrease and were asked to determine the new (*smaller*) value. Suppose you are asked a "reverse" question: "Voter turnout in the local school board elections declined 40% from last year. This year only 3600 eligible voters went to the polls. How many people voted in last year's school board elections?"

Using the decay factor 0.60 you can write

$$\text{original value} \cdot 0.60 = 3600.$$

To determine last year's voter turnout, divide both sides of the equation by 0.60.

$$\text{original value} = 3600 \div 0.60$$

Therefore, last year 6000 people voted in the school board elections.

> ### Procedure
>
> When a quantity has already *decreased* by a specified percent, its original value is obtained by dividing the new value by the corresponding decay factor. That is,
>
> $$\text{new value} \div \text{decay factor} = \text{original value}.$$

14. a. You purchased a smartphone on sale for $150. The discount was 40%. What was the original price?

b. Determine the original price of a treadmill that is on sale for $1140. Assume that the store has reduced all merchandise by 25%.

SUMMARY: ACTIVITY 4.1

1. When a quantity increases by a specified percent, the ratio $\dfrac{\text{new value}}{\text{original value}}$ is called the **growth factor**. The growth factor is formed by adding the specified percent increase to 100% and then changing this percent into its decimal form.

2. a. When a quantity increases by a specified percent, its new value can be obtained by multiplying the original value by the corresponding growth factor. So,

$$\text{original value} \cdot \text{growth factor} = \text{new value}.$$

b. When a quantity has already increased by a specified percent, its original value can be obtained by dividing the new value by the corresponding growth factor. So,

$$\text{new value} \div \text{growth factor} = \text{original value}.$$

3. When a quantity decreases by a specified percent, the ratio $\dfrac{\text{new value}}{\text{original value}}$ is called the **decay factor** associated with the specified percent decrease. The decay factor is formed by subtracting the specified percent decrease from 100%, and then changing this percent into its decimal form.

4. a. When a quantity decreases by a specified percent, its new value can be obtained by multiplying the original value by the corresponding decay factor. So,

$$\text{original value} \cdot \text{decay factor} = \text{new value}.$$

b. When a quantity has already decreased by a specified percent, its original value can be obtained by dividing the new value by the corresponding decay factor. So,

$$\text{new value} \div \text{decay factor} = \text{original value}.$$

EXERCISES: ACTIVITY 4.1

1. Complete the following table:

Percent Increase	5%		15%		100%	300%	
Growth Factor		1.35		1.045			11.00

2. You are purchasing a new car. The price you negotiated with a car dealer is $17,944, excluding sales tax. The local sales tax rate is 8.25%.

a. What growth factor is associated with the sales tax rate?

b. Use the growth factor to determine the total cost of the car.

3. A college in Michigan reported that its enrollment increased from 11,952 students in spring 2013 to 12,144 students in spring 2014.

a. What is the ratio of the number of students in spring 2014 to the number of students in spring 2013?

b. From your result in part a, determine the growth factor. Write it in decimal form to the nearest hundredth.

c. By what percent did the enrollment increase?

4. In 1967, your grandparents bought a house for $26,000. The U.S. Bureau of Labor Statistics reports that, due to inflation, the cost of the same house in 2014 would be $184,984.

a. What is the inflation growth factor of housing from 1967 to 2014?

b. What is the inflation rate (percent increase to the nearest whole number percent) for housing from 1967 to 2014?

c. In 2014, your grandparents sold their house for $245,000. What profit did they make in terms of 2014 dollars?

5. Your friend plans to move from Newark, New Jersey, to Anchorage, Alaska. She earns $50,000 a year in Newark. How much must she earn in Anchorage to maintain the same standard of living if the cost of living in Anchorage is 35% higher?

6. You decide to invest $3000 in a 1-year certificate of deposit (CD) that earns an annual percentage yield (APY) of 2.4%. How much will your investment be worth in a year?

7. Your company is moving you from New York City to one of the other cities listed in the following table. For each city, the table lists the cost of a basket of goods and services that would cost $100 in New York City.

In this problem, you will compare the cost of the same basket of goods in each of the cities listed to the $100 cost in New York City. The comparison ratio in each case will be a growth factor relative to the cost in New York City.

Living Abroad

CITY	COST OF BASKET	GROWTH FACTOR IN COST OF LIVING	PERCENT INCREASE IN SALARY
New York, NY United States	100		
Brussels, Belgium	103		
Tokyo, Japan	126		
Sydney, Australia	107		
Paris, France	112		
Oslo, Norway	124		
Zurich, Switzerland	116		

a. Determine the growth factor for each city in the table and list it in the third column.

b. Determine the percent increase in your salary that would allow you the same cost of living as you have in New York City. List the percent increase in the fourth column of the table.

c. You earn a salary of $45,000 in New York City. What would your salary have to be in Tokyo, Japan, to maintain the same standard of living?

8. Since 2007, the International Data Corporation (IDC) has been keeping track of the amount of digital information created and replicated on computers and the Internet. IDC calls this set of information the digital universe. In 2012, IDC reported that the size of the digital universe was 2.8 trillion gigabytes. In 2011, that number was 1.8 trillion gigabytes.

a. What was the growth factor in the amount of digital data from 2011 to 2012? Round your answer to the nearest hundredth.

b. Use the growth factor from part a to predict the size of the digital universe in 2013, assuming that the growth factor remains the same. Round your answer to the nearest trillion.

9. Complete the following table:

Percent Decrease	5%		15%		12.5%
Decay Factor		0.45		0.94	

10. You wrote an eight-page article that will be published in a journal. The editor has asked you to revise the article and reduce the number of pages to six. By what percent must you reduce the length of your article?

11. A car dealer will sell you a car you want for $18,194, which is just $200 over the dealer invoice price (the price the dealer pays the manufacturer for the car). You tell him that you will think about it. The dealer is anxious to meet his monthly quota of sales, so he calls the next day to offer you the car for $17,994 if you agree to buy it tomorrow. You decide to accept the deal.

a. What is the decay factor associated with the decrease in the price to you? Write your answer in decimal form to the nearest thousandths.

b. What is the percent decrease of the price, to the nearest tenth of a percent?

c. The sales tax is 6.5%. How much did you save on sales tax by taking the dealer's second offer?

12. Your company is moving you from New York City to one of the other cities listed in the following table. For each city, the table lists the cost of a basket of goods and services that would cost $100 in New York City.

In this problem, you will compare the cost of the same basket of goods in each of the cities listed to the $100 cost in New York City. The comparison ratio in each case will be a decay factor relative to the cost in New York City.

Over There

CITY	COST OF BASKET	DECAY FACTOR IN COST OF LIVING	PERCENT DECREASE IN COST OF LIVING
New York, NY United States	100		
Lisbon, Portugal	93		
Berlin, Germany	97		
Jakarta, Indonesia	82		
Melbourne, Australia	96		
Mumbai, India	76		
Montreal, Canada	84		

a. Determine the decay factor for each cost in the table and list it in the third column.

b. Determine the percent decrease in the cost of living for each city compared to living in New York City. List the percent decrease in the fourth column of the table.

c. You earn a salary of $45,000 in New York City. If you move to Jakarta, Indonesia, how much of your salary will you need for the cost of living there?

d. How much of your salary can you save if you move to Lisbon, Portugal?

13. You need a fax machine but plan to wait for a sale. Yesterday, the model you want was reduced by 30%. It originally cost $129.95. Use the decay factor to determine the sale price of the fax machine.

14. Your doctor advises you that losing 10% of your body weight will significantly improve your health.

a. Before you start dieting, you weigh 175 pounds. Determine and use the decay factor to calculate your goal weight to the nearest pound.

b. After six months, you reach your goal weight. However, you still have more weight to lose to reach your ideal weight range, which is 125–150 pounds. If you lose 10% of your new weight, will you be in your ideal weight range? Explain.

15. Aspirin is typically absorbed into the bloodstream from the duodenum (the beginning portion of the small intestine). In general, 75% of a dose of aspirin is eliminated from the bloodstream in an hour. A patient is given a dose of 650 milligrams. How much aspirin remains in the patient after one hour?

16. Airlines often encourage their customers to book online by offering a 5% discount on tickets. You are traveling from Kansas City to Denver. A fully refundable fare is $558.60. A restricted, nonrefundable fare is $273.60. Determine and use the decay factor to calculate the cost of each fare if you book online.

**Take an Additional
20% Off**

Objectives

1. Define consecutive growth and decay factors.

2. Determine a consecutive growth or decay factor from two or more consecutive percent changes.

3. Apply consecutive growth and/or decay factors to solve problems involving percent changes.

Cumulative Discounts

Your friend arrives at your house. Today's newspaper contains a 20%-off coupon at Old Navy. The $100 jacket she had been eyeing all season was already reduced by 40%. She clipped the coupon, drove to the store, selected her jacket, and walked up to the register. The cashier brought up a price of $48; your friend insisted that the price should have been only $40. The store manager arrived and re-entered the transaction, and, again, the register displayed $48. Your friend left without purchasing the jacket and drove to your house to tell you her story.

1. How do you think your friend calculated a price of $40?

2. You grab a pencil and start your own calculation. First, you determine the sale price that reflects the 40% reduction at which the Old Navy is selling the jacket? Explain how you calculated this price.

3. To what price does the 20%-off coupon apply?

4. Apply the 20% discount to determine the final price of the jacket.

You are now curious: could you justify a better price by applying the discounts in the reverse order? That is, applying the 20% off coupon, followed by a 40% reduction. You start a new set of calculations.

5. Starting with the list price, determine the sale price after taking the 20% reduction.

6. Apply the 40% discount to the intermediate sale price.

7. Which sequence of discounts gives a better sale price?

The important point to remember here is that when multiple discounts are given, they are always applied sequentially, one after the other—never all at once.

In the following example, you will see how to use decay factors to simplify calculations such as the ones above.

> **Example 1**   *A stunning $2000 gold and diamond necklace you saw was far too expensive to even consider. However, over several weeks you tracked the following successive discounts: 20% off list; 30% off marked price; and an additional 40% off every item. Determine the selling price after each of the discounts is taken.*

SOLUTION

To calculate the first reduced price, apply the decay factor corresponding to a 20% discount. Recall that you form a decay factor by subtracting the 20% discount from 100% to obtain 80%, or 0.80 as a decimal. Apply the 20% reduction by multiplying the original price by the decay factor.

The first sale price $= \$2000 \cdot 0.80$; the selling price is now $1600.

In a similar manner, determine the decay factor corresponding to a 30% discount by subtracting 30% from 100% to obtain 70%, or 0.70 as a decimal. Apply the 30% reduction by multiplying the already discounted price by the decay factor.

The second sale price $= \$1600 \cdot 0.70$; the selling price is now $1120.

Finally, determine the decay factor corresponding to a 40% discount by subtracting 40% from 100% to obtain 60%, or 0.60 as a decimal. Apply the 40% reduction by multiplying the most current discounted price by the decay factor.

$\$1120 \cdot 0.60$; the final selling price is $672.

You may have noticed that it is possible to calculate the final sale price from the original price using a single chain of multiplications:

$\$2000 \cdot 0.80 \cdot 0.70 \cdot 0.60 = \$672$

The final sale price is $672.

8. Use the chain of multiplications in Example 1 to determine the final sale price of the necklace if the discounts are taken in the reverse order (40%, 30%, and 20%).

You can form a single decay factor that represents the cumulative effect of applying the three consecutive percent decreases; the single decay factor is the *product* of the three decay factors.

In Example 1, the effective decay factor is given by the product $0.80 \cdot 0.70 \cdot 0.6$, which equals 0.336 (or 33.6%). The effective discount is calculated by subtracting the decay factor (in percent form) from 100%, to obtain 66.4%. Therefore, the effect of applying 20%, 30%, and 40% consecutive discounts is identical to a single discount of 66.4%.

9. a. Determine a single decay factor that represents the cumulative effect of consecutively applying Old Navy's 40% and 20% discounts.

   b. Use this decay factor to determine the effective discount on your friend's jacket.

10. a. Determine a single decay factor that represents the cumulative effect of consecutively applying discounts of 40% and 50%.

   b. Use this decay factor to determine the effective discount.

Cumulative Increases

Your computer repair business is growing faster than you had ever imagined. Last year you had 100 employees statewide. This year you opened several additional locations and increased the number of workers by 30%. With demand so high, next year you will open new stores nationwide and plan to increase your employee roll by an additional 50%.

To determine the projected number of employees next year, you can use growth factors to simplify the calculations.

11. a. Determine the growth factor corresponding to a 30% increase.

    b. Apply this growth factor to calculate your current workforce.

12. a. Determine the growth factor corresponding to a 50% increase.

    b. Apply this growth factor to your current workforce to determine the projected number of employees next year.

13. Starting from last year's workforce of 100, write a single chain of multiplications to calculate the projected number of employees.

> You can form a single growth factor that represents the cumulative effect of applying the two consecutive percent increases; the single growth factor is the *product* of the two growth factors.

In Problem 13, the effective growth factor is given by the product $1.30 \cdot 1.50$, which equals 1.95 (an increase of 95%), nearly double the number of employees last year.

What Goes Up Often Comes Down

You purchased $1000 of a recommended stock last year and watched gleefully as it rose quickly by 30%. Unfortunately, the economy turned downward, and your stock recently fell 30% from last year's high. Have you made or lost money on your investment? The answer might surprise you. Find out by solving the following sequence of problems.

14. a. Determine the growth factor corresponding to a 30% increase.

    b. Determine the decay factor corresponding to a 30% decrease.

You can form a single factor that represents the cumulative effect of applying the consecutive percent increase and decrease—the single factor is the *product* of the growth and decay factors. In this example, the effective factor is given by the product $1.30 \cdot 0.70$, which equals 0.91.

15. Does 0.91 represent a growth factor or a decay factor? How can you tell?

16. a. What is the current value of your stock?

    b. What is the cumulative effect (as a percent change) of applying a 30% increase followed by a 30% decrease?

17. What is the cumulative effect if the 30% decrease had been applied first, followed by the 30% increase?

SUMMARY: ACTIVITY 4.2

1. The cumulative effect of a sequence of percent changes is the *product* of the associated growth or decay factors. For example,

   a. To calculate the effect of consecutively applying 20% and 50% increases, form the respective growth factors and multiply. The effective growth factor is

$$1.20 \cdot 1.50 = 1.80,$$

   which represents an effective increase of 80%.

   b. To calculate the effect of applying a 25% *increase* followed by a 20% *decrease*, form the respective growth and decay factors, and then multiply them. The effective factor is $1.25 \cdot 0.80 = 1.00$, which indicates neither growth nor decay. That is, the quantity has returned to its original value.

2. The cumulative effect of a sequence of percent changes is the same, regardless of the order in which the changes are applied. For example, the cumulative effect of applying a 20% *increase* followed by a 50% *decrease* is equivalent to having first applied the 50% *decrease*, followed by the 20% *increase*.

EXERCISES: ACTIVITY 4.2

1. A $300 suit is on sale for 30% off. You present a coupon at the cash register for an additional 20% off.

   a. Determine the decay factor corresponding to each percent decrease.

   b. Use these decay factors to determine the price you paid for the suit.

2. Your union has just negotiated a three-year contract containing annual raises of 3%, 4%, and 5% during the term of contract. Your current salary is $42,000. What will you be earning in three years?

3. You anticipated a large demand for a popular toy and increased your inventory of 1600 by 25%. You sold 75% of your inventory. How many toys remain?

4. You deposit $2000 in a five-year certificate of deposit that pays 1.85% interest compounded annually. Determine to the nearest dollar, your account balance when your certificate comes due.

5. Budget cuts have severely crippled your department over the last few years. Your operating budget of $600,000 has decreased 5% in each of the last three years. What is your current operating budget?

6. When you became a manager, your $60,000 annual salary increased 25%. You found the new job too stressful and requested a return to your original job. You resumed your former duties at a 20% reduction in salary. How much more are you making at your old job after the transfer and return?

7. A coat with an original price tag of $400 was marked down by 40%. You have a coupon good for an additional 25% off.

   a. What is its final cost?

   b. What is the effective percent discount?

8. A digital camera with an original price tag of $500 was marked down by 40%. You have a coupon good for an additional 30% off.

   a. What is the decay factor for the first discount?

   b. What is the decay factor for the additional discount?

   c. What is the effective decay factor?

   d. What is the final cost of the camera?

   e. What is the effective percent discount?

9. Your friend took a job 3 years ago that started at $30,000 a year. Last year, she got a 5% raise. She stayed at the job for another year but decided to make a career change and took another job that paid 5% less than her current salary. Was the starting salary at the new job more, or less, or the same as her starting salary at her previous job?

10. You wait for the price to drop on a diamond-studded watch at Macy's. Originally, it cost $2500. The first discount was 20% and the second discount is 50%.

   a. What are the decay factors for each of the discounts?

   b. What is the effective decay factor for the two discounts?

   c. Use the effective decay factor in part b to determine how much you will pay for the watch.

Activity 4.3

Inflation

Objectives

1. Recognize an exponential function as a rule for applying a growth factor or a decay factor.

2. Graph exponential functions from numerical data.

3. Recognize exponential functions from equations.

4. Graph exponential functions from equations.

Exponential Growth

Inflation means that a current dollar will buy less in the future. According to the U.S. Consumer Price Index, the inflation rate for the 12 months from July 2012 to July 2013 was 2%. This means that a 1-pound loaf of white bread that cost a dollar in July of 2012 cost $1.02 in July of 2013. The change in price is usually expressed as an annual percentage rate, known as the **inflation rate**.

1. a. At the current inflation rate of 2%, how much will an $80 pair of athletic shoes cost next year?

b. Assume that the rate of inflation remains at 2% next year. How much will the shoes cost in the year following next year?

2. If you assume that the inflation rate increases somewhat and remains at 4% per year for the next decade, you can calculate the cost of a currently priced $12 pizza for each of the next ten years. Complete the following table. Round to the nearest cent.

Years from Now, t	0	1	2	3	4	5	6	7	8	9	10
Cost of Pizza, $c(t)$, $											

3. Plot the data from Problem 2 on appropriately scaled and labeled axes. Use the variable t, years from now, as the independent variable and $c(t)$, the cost of pizza, as the dependent variable.

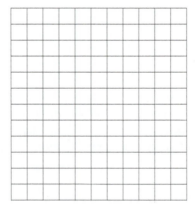

4. **a.** Determine the average rate of change of the cost of the pizza in the first year (from $t = 0$ to $t = 1$), the fifth year (from $t = 4$ to $t = 5$), and the tenth year (from $t = 9$ to $t = 10$). Explain what these results mean for the cost of pizza over the next ten years.

b. Is the function linear? Explain.

To complete the table in Problem 2, you could calculate each cost value by multiplying the previous output by 1.04, the inflation's growth factor. Thus, to obtain the cost after ten years, you would multiply the original cost, $12, by 1.04, a total of ten times. Symbolically, you write $12(1.04)^{10}$. Therefore, you can model the cost of pizza algebraically as

$$c(t) = 12(1.04)^t,$$

where $c(t)$ represents the cost and t represents the number of years from now.

> **Definition**
>
> When the independent variable of a function appears as an exponent of the growth factor, the function, in this case $c(t) = 12(1.04)^t$, is called an **exponential function**. The result is an increasing exponential function.

5. **a.** Under a constant annual inflation rate of 4%, what will a pizza cost after 20 years?

b. Use the cost function to calculate the cost of a pizza after five years.

6. a.

6. **a.** A graph of the function $c(t) = 12(1.04)^t$ is shown to the left in a window for t between -20 and 20 and $c(t)$ between 0 and 25. Use the same window to graph this function on a graphing calculator.

b. Use the trace or table feature of the calculator to examine the coordinates of some points on the graph. Does your plot of the numerical data in Problem 3 agree with the graph shown in part a?

X	Y1
0	12
1	12.48
2	12.979
3	13.498
4	14.038
5	14.6
6	15.184
X=0	

X	Y1
7	15.791
8	16.423
9	17.08
10	17.763
11	18.473
12	19.212
13	19.981
X=7	

c. Is the entire graph in part a really relevant to the original problem? Explain.

d. Resize your window to include only the first quadrant from $x = 0$ to $x = 20$ and $y = 0$ to $y = 25$. Regraph the function. Your screens should appear as follows:

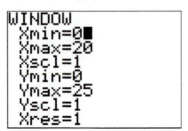 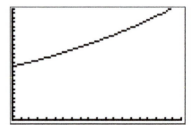

7. a. How many years will it take for the price of a pizza to double? Explain how you determined your answer.

b. How many years will it take for the price of a pizza to triple? Explain how you find your answer.

Exponential Decay

When you take medicine, your body metabolizes and eliminates the medication until there is none left in your body. The half-life of a medication is the time it takes for your body to eliminate one-half of the amount present. For many people the half-life of the medicine Prozac is one day.

8. a. What fraction of a dose is left in your body after one day?

b. What fraction of the dose is left in your body after two days? (This is one-half of the result from part a.)

c. Complete the following table. Let t represent the number of days since a dose of Prozac is taken and let Q represent the fraction of the dose of Prozac remaining in your body.

t, Days	0	1	2	3	4
Q	1				

9. a. The values of the fraction of the dosage, Q, can be written as powers of $\frac{1}{2}$. For example, $1 = \left(\frac{1}{2}\right)^0$, $\frac{1}{2} = \left(\frac{1}{2}\right)^1$, and so on. Complete the following table by writing each value of Q in the table in Problem 8c as a power of $\frac{1}{2}$. The values for 1 and $\frac{1}{2}$ have already been entered.

t, Days	0	1	2	3	4
Q	$\left(\frac{1}{2}\right)^0$	$\left(\frac{1}{2}\right)^1$			

b. Use the result of part a to write an equation for Q in terms of t.

c. What is the practical domain of the half-life function?

d. Sketch a scatterplot of the data in part a on an appropriately scaled and labeled coordinate axis.

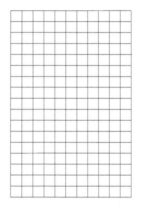

e. Is this function discrete or continuous?

Notice that the equation $Q = \left(\frac{1}{2}\right)^t$ fits the equation form of an exponential function in which the independent variable appears as an exponent. However, the value of the base for this exponential function is $\frac{1}{2}$, which is not greater than 1. Therefore, the base $\frac{1}{2}$ is a decay factor.

> When the base, b, of an exponential function is between 0 and 1, the base is a **decay factor**. The result is a **decreasing exponential function**.

SUMMARY: ACTIVITY 4.3

An **exponential function** is a function in which the independent variable appears as an exponent of the growth factor (for example, $c = 12(1.04)^t$) or a decay factor $\left(\text{for example, } Q = \left(\frac{1}{2}\right)^t\right)$.

EXERCISES: ACTIVITY 4.3

1. a. Complete the following table, in which $f(x) = \left(\dfrac{1}{2}\right)^x$ and $g(x) = 2^x$.

x	−2	−1	0	1	2
f(x)					
g(x)					

b. Use the points in part a to sketch a graph of the given functions f and g on the same coordinate axes. Compare the graphs. List the similarities and differences.

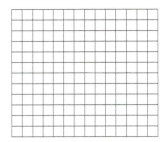

c. Use a graphing calculator to graph the functions $f(x) = \left(\dfrac{1}{2}\right)^x$ and $g(x) = 2^x$ given in part a in the window $\text{Xmin} = -2$, $\text{Xmax} = 2$, $\text{Ymin} = -2$, and $\text{Ymax} = 5$. Compare to the graphs you obtained in part b.

2. Suppose the inflation rate is 7% per year and remains the same for the next seven years.

a. Complete the following table for the cost of a pair of sneakers that costs $45 now. Round to the nearest cent:

t, Years from Now	0	1	2	3	4	5	6	7
c(t), Cost of Sneakers ($)	45							

b. Determine the growth factor for a 7% inflation rate.

c. If the yearly inflation rate remains at 7%, what exponential function would you use to determine the cost of $45 sneakers after t years?

d. What would be the cost of the sneakers in ten years if the inflation rate stayed at 7%?

3. An exponential function may be increasing or decreasing. Determine which is the case for each of the following functions. Explain how you determined each answer.

a. $y = 5^x$

b. $y = \left(\dfrac{1}{2}\right)^x$

c. $y = 1.5^t$

d. $y = 0.2^P$

4. a. Evaluate the functions in the following table for the input values, x:

Input x	0	1	2	3	4	5
$g(x) = 3x$						
$f(x) = 3^x$						

b. Compare the functions $f(x) = 3^x$ and $g(x) = 3x$ from $x = 0$ to $x = 5$ by comparing the output values in the table.

c. Compare the graphs of the functions f and g that are shown in the following window. Approximate the interval in which the exponential function f grows slower than the linear function g and the interval where it grows faster.

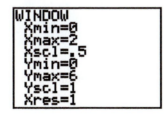
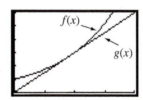

d. Compare the rate of increase of the functions $f(x) = 3^x$ and $g(x) = 3x$ from $x = 0$ to $x = 5$ by calculating the average rate of change for each function from $x = 0$ to $x = 5$. Determine which function grows faster on average in the given interval.

5. a. For investment purposes, you recently bought a house for $100,000 in a neighborhood where the price of houses has been rising $10,000 a year. If the rise in housing prices continues in the same manner, how much will your investment be worth in a year and in two years?

   b. You decided to buy a second investment house, again for $100,000 but in another area. Here the prices have been rising at a rate of 10% a year. If this rate of increase continues, how much will your investment be worth in a year and in two years?

   c. Which investment will give you the better return?

6. As a radiology specialist you use the radioactive substance iodine-131 to diagnose conditions of the thyroid gland. Iodine-131 decays (loses radiation energy and changes to a nonradioactive form of iodine) at the rate of 8.3% per day. Your hospital currently has a 20-gram supply. In parts a–f, create representations for this situation, such as formulas, tables, and graphs, to communicate your mathematical ideas.

   a. What is the decay factor for the decay of the iodine?

   b. Use the decay factor from part a to determine the number of grams remaining for the days listed in the following table. Record your results in the table to the nearest hundredth.

t (Number of Days Starting from a 20-Gram Supply of Iodine-131)	0	4	8	12	16	20	24
N, Number of Grams of Iodine-131 Remaining from a 20-Gram Supply	20.00						

   c. Write an exponential decay formula for N, the number of grams of iodine-131 remaining, in terms of t, the number of days from the current supply of 20 grams.

   d. Determine the number of grams of iodine-131 remaining from a 20-gram supply after two months (60 days).

e. Graph the decay formula for iodine-131, $N = 20(0.917)^t$ as a function of the time t (days). Use appropriate scales and labels on the following grid or an appropriate window on a graphing calculator.

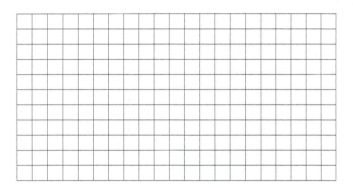

f. How long will it take for iodine-131 to decrease to half its original value? Explain how you determined your answer.

The Summer Job

Objectives

1. Determine the growth or decay factor of an exponential function.

2. Identify the properties of the graph of an exponential function defined by $y = b^x$, where $b > 0$ and $b \neq 1$.

Your brother will be attending college in the fall, majoring in mathematics. On July 1, he goes to your neighbor's house looking for summer work to help pay for college expenses. Your neighbor is interested since he needs some odd jobs done. Your brother can start right away and will work all day July 1 for 2 cents. This gets your neighbor's attention, but you wonder if there is a catch. Your brother says that he will work July 2 for 4 cents, July 3 for 8 cents, July 4 for 16 cents, and so on for *every* day of the month of July.

For Problems 1–2, assume that the neighbor hires your brother.

1. a. Complete the following table.

Making Cents of It All								
Day in July (Input)	1	2	3	4	5	6	7	8
Pay in Cents (Output)								

b. Do you notice a pattern in the output values? Describe how you can obtain the pay on a given day knowing the pay on the previous day.

c. Use what you discovered in part b to determine the pay on July 9.

2. a. The pay on any given day can be written as a power of 2. Write each pay entry in the output row of the table in Problem 1 as a power of 2. For example, $2 = 2^1, 4 = 2^2$.

b. Let n represent the number of days worked. Write an equation for the daily pay, $P(n)$, (in cents) as a function of n, the number of days worked. Note that the number of days worked is the same as the July date.

c. Use the equation from part b to determine how much your brother will earn on July 20. That is, determine the value of $P(n)$ when $n = 20$. What are the units of measurement of your answer?

d. How much will he earn on July 31? Be sure to indicate the units of your answer.

Therefore, it was not a good idea for the neighbor to hire your brother! This situation demonstrates the growth power of exponential functions.

3. a. Determine the average rate of change of $P(n)$ as n increases from $n = 3$ to $n = 4$. What are the units of measurement of your answer?

b. Determine the average rate of change of $P(n)$ as n increases from $n = 7$ to $n = 8$. Include units in your answer.

c. Is the function linear? Explain.

4. a. What is the practical domain of the function defined by $P(n) = 2^n$?

b. Sketch a scatterplot of ordered pairs of the form $(n, P(n))$ from July 1 to July 10 on appropriately scaled and labeled axes.

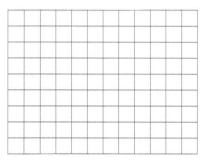

The function defined by $P(n) = 2^n$ gives the relationship between the pay $P(n)$ (in cents) and the given July date, n, worked. This function belongs to a family of **exponential functions**.

If x represents the independent variable and y represents the dependent variable, then some **exponential functions** can be defined by equations of the form $y = b^x$, where the base b is a constant such that b is a positive number not equal to 1 ($b > 0$ and $b \neq 1$). As discussed in Activity 4.3, the base b is a growth or decay factor.

Example 1   *Some examples of exponential functions are*

$g(x) = 10x$, where $b = 10$, $h(x) = (1.08)^x$, where $b = 1.08$,

$V(x) = \left(\dfrac{1}{2}\right)^x$, where $b = \dfrac{1}{2}$, and $T(x) = (0.75)^x$, where $b = 0.75$.

Graphs of Increasing Exponential Functions

Because n in $P(n) = 2^n$ (the summer job situation) represents a given day in July, the practical domain (whole numbers from 1 to 31) limits the investigation of the exponential function.

5. a. Consider the general function defined by $f(x) = 2^x$. Use a graphing calculator to sketch a graph of this function. Use the window Xmin $= -10$, Xmax $= 10$, Ymin $= -2$, and Ymax $= 10$. Your screens should appear as follows:

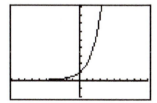

b. Because the graph of the general function $f(x) = 2^x$ is continuous (it has no holes or breaks), what appears to be the domain of the function f? What is the range of the function f?

c. Determine the y-intercept of the graph of f by substituting 0 for x in the equation $y = 2^x$ and solving for y.

d. Is the function f increasing or decreasing?

Definition

If the base b of an exponential function defined by $y = b^x$ is greater than 1, then b is the **growth factor**. The graph of $y = b^x$ is increasing if $b > 1$. For each increase of 1 of the value of x, y increases by a factor of b.

Example 2   *The base 2 of $f(x) = 2^x$ is the growth factor because each time the input, x, is increased by 1, the output is multiplied by 2.*

6. Identify the growth factor, if any, for the given function.

a. $y = 1.08^x$

b. $h(x) = 0.8^x$

c. $y = 8x$

d. $g(x) = 10^x$

7. Return to the graph of $f(x) = 2^x$.

a. Does the graph of $f(x) = 2^x$ appear to have an x-intercept?

b. Use your calculator to complete the following table.

x	-1	-2	-4	-6	-8	-10
$f(x) = 2^x$						

Note: By definition, $b^{-n} = \dfrac{1}{b^n}$. Therefore,

$$2^{-1} = \frac{1}{2^1} = 0.5, \ 2^{-4} = \frac{1}{2^4} = 0.0625, \text{ and } 2^{-10} = \frac{1}{2^{10}} \approx 0.000977.$$

c. As the values of the input variable x decrease, what happens to the output values?

d. Use the trace feature of your graphing calculator to trace the graph of $f(x) = 2^x$ for $x < 0$. What appears to be the relationship between the graph of $y = 2^x$ and the x-axis when x becomes more negative?

— **Definition** —

A horizontal axis having equation $y = 0$ is called a **horizontal asymptote** of the graph of a function defined by $y = b^x$, where $b > 0$ and $b \neq 1$. The graph of the function gets closer and closer to the x-axis ($y = 0$) as the input gets farther from the origin, in the negative direction.

Example 3   *The x-axis is the horizontal asymptote of $y = 3^x$ and $y = 7^x$ because, as x gets more negative, the graph gets closer and closer to the x-axis. See the graph that follows.*

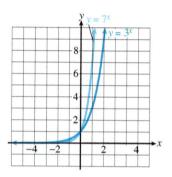

8. a. Complete the following table.

x	−3	−2	−1	0	1	2	3	4	5
$f(x) = 2^x$									
$g(x) = 10^x$									

b. Sketch the graph of the functions f and g on your graphing calculator. Use the window Xmin = −5, Xmax = 5, Ymin = −2, and Ymax = 9.

c. Use the results from parts a and b to describe how the graphs of $f(x) = 2^x$ and $g(x) = 10^x$ are similar and how they are different. Be sure to include domain, growth factor, x- and y-intercepts, and horizontal asymptotes. Also discuss whether the graph of g increases faster or slower than the graph of f.

9. In parts a–d, sketch a graph of the given transformation of the graph of $y = 2^x$. Indicate any intercepts. Then, write the equation of the transformed graph. Verify the graph using a graphing calculator.

a. reflection of the graph across the *x*-axis.

b. vertical stretch of the graph by a factor of 5.

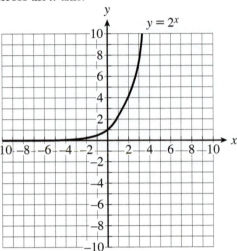

c. horizontal shift of the graph 3 units to the right.

d. vertical shift of the graph 3 units upward.

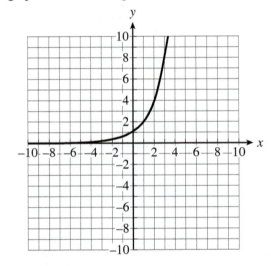

10. Describe in words how the graph of $y = 2^x$ is transformed by each equation.

   a. $y = 0.5 \cdot 2^x$

   b. $y = 2^{x+1}$

   c. $y = 2^x + 1$

   d. $y = -3 \cdot 2^x$

Graphs of Decreasing Exponential Functions

11. a. Complete the following table.

x	−3	−2	−1	0	1	2	3	4	5
$y = \left(\dfrac{1}{2}\right)^x$									

b. Describe how you can obtain the y-value for $x = 6$, using the y-value for $x = 5$.

c. Sketch the graph of $y = \left(\dfrac{1}{2}\right)^x$. Verify your sketch using your graphing calculator.

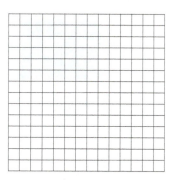

d. What are the domain and range of the exponential function?

e. Determine the y-intercept of the graph.

f. Is the function defined by $y = \left(\dfrac{1}{2}\right)^x$ increasing or decreasing?

Definition

If the base b of an exponential function $y = b^x$ is between 0 and 1, then b is the **decay factor**. The graph of $y = b^x$ is decreasing if $0 < b < 1$. For each increase of 1 of the value of the input, the output decreases by a factor of b.

Example 4 *The base $\dfrac{1}{2}$ in the function $y = \left(\dfrac{1}{2}\right)^x$ is the decay factor because each time x is increased by 1, the output value is multiplied by $\dfrac{1}{2}$.*

12. Identify the decay factor, if any, for the function defined by the given equation.

a. $y = 0.98^x$

b. $h(x) = 1.8^x$

c. $y = 0.8x$

d. $g(x) = \left(\dfrac{2}{7}\right)^x$

13. Return to the graph of $y = \left(\dfrac{1}{2}\right)^x$.

 a. Does the graph of $y = \left(\dfrac{1}{2}\right)^x$ have an x-intercept?

 b. Complete the following table.

x	1	3	5	7	10
$y = \left(\dfrac{1}{2}\right)^x$					

 c. As the values of the input variable x get larger, what happens to the y-values?

 d. Does the graph of $y = \left(\dfrac{1}{2}\right)^x$ have a horizontal asymptote? Explain.

14. In parts a–c, describe in words how the graph of $y = \left(\dfrac{1}{2}\right)^x$ is transformed. Sketch a graph of the given equation and verify using a graphing calculator.

 a. $y = \left(\dfrac{1}{2}\right)^x - 2$

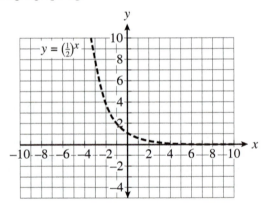

 b. $y = \left(\dfrac{1}{2}\right)^{x+1}$

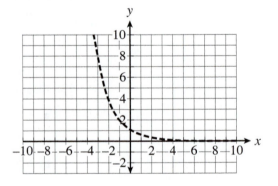

c. $y = -2\left(\dfrac{1}{2}\right)^x$

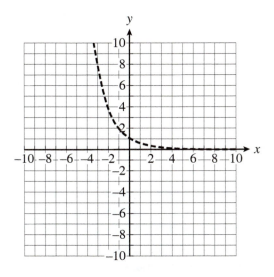

15. a. For each of the following exponential functions, identify the base, b, and determine whether the base is a growth or decay factor. Graph each function on your graphing calculator, and complete the table below.

FUNCTION	BASE, b	GROWTH OR DECAY FACTOR	x-INTERCEPT	y-INTERCEPT	HORIZONTAL ASYMPTOTE	INCREASING OR DECREASING
$h(x) = (1.08)^x$						
$T(x) = (0.75)^x$						
$f(x) = (3.2)^x$						
$r(x) = \left(\dfrac{1}{4}\right)^x$						

b. Without graphing, how might you determine which of the functions in part a increase and which decrease? Explain.

16. Examine the output pattern to determine which of the following data sets is linear and which is exponential. For the linear set, determine the slope. For the exponential set, determine the growth or decay factor.

a.

x	−2	−1	0	1	2	3	4
y	−8	−4	0	4	8	12	16

b.

	−2	−1	0	1	2	3	4
y	$\dfrac{1}{16}$	$\dfrac{1}{4}$	1	4	16	64	256

17. Determine the decay factor of the function represented by the data and complete the table.

x	−2	−1	0	1	2
f(x)	16	4			

SUMMARY: ACTIVITY 4.4

Functions defined by equations of the form $y = b^x$, where $b > 0$ and $b \neq 1$, are called **exponential functions** and have the following properties.

1. The domain is all real numbers.

2. The range is $y > 0$.

3. If $0 < b < 1$, the function is decreasing and has the following general shape.

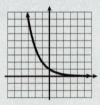

  In this case, b is called the **decay factor**.

4. If $b > 1$, the function is increasing and has the following general shape.

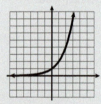

  In this case, b is called the **growth factor**.

5. The y-intercept is $(0, 1)$.

6. The graph does not intersect the horizontal axis. There is no x-intercept.

7. The line $y = 0$ (the x-axis) is a **horizontal asymptote**.

8. The function is continuous.

EXERCISES: ACTIVITY 4.4

I. a. Complete the following tables.

x	−3	−2	−1	0	1	2	3
$h(x) = 5^x$							

x	−3	−2	−1	0	1	2	3
$g(x) = \left(\dfrac{1}{5}\right)^x$							

b. Sketch graphs of h and g on the following grid.

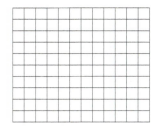

c. Use the tables and graphs in parts a and b to complete the following table.

FUNCTION	BASE, b	GROWTH OR DECAY FACTOR	x-INTERCEPT	y-INTERCEPT	HORIZONTAL ASYMPTOTE	INCREASING OR DECREASING
$h(x) = 5^x$						
$g(x) = \left(\dfrac{1}{5}\right)^x$						

2. a. Complete the following table.

x	−3	−2	−1	0	1	2	3
$f(x) = 3^x$							
$g(x) = x^3$							
$h(x) = 3x$							

b. Sketch a graph of each of the given functions f, g, and h.

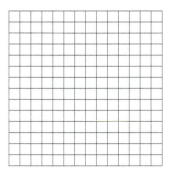

c. Describe any similarities or differences that you observe in the graphs.

3. Using your graphing calculator, investigate the graphs of the following families (groups) of functions. Describe any relationships within each family, including domain and range, growth or decay factors, vertical and horizontal intercepts, and asymptotes. Identify the functions as increasing or decreasing.

a. $f(x) = \left(\dfrac{3}{4}\right)^x$, $g(x) = \left(\dfrac{4}{3}\right)^x$

b. $f(x) = 10^x$, $g(x) = -10^x$

c. $f(x) = 3^x$, $g(x) = \left(\dfrac{1}{3}\right)^x$

4. Determine which of the following data sets are linear and which are exponential. For the linear sets, determine the slope. For the exponential sets, determine the growth factor or the decay factor.

a.

x	−2	−1	0	1	2	3	4
y	$\frac{1}{9}$	$\frac{1}{3}$	1	3	9	27	81

b.

x	−2	−1	0	1	2	3	4
y	2	2.5	3	3.5	4	4.5	5

c.

x	−2	−1	0	1	2	3	4
y	0.75	1.5	3	6	12	24	48

d.

x	−2	−1	0	1	2	3	4
y	6.25	2.5	1	0.4	0.16	0.064	0.0256

5. Assume that y is an exponential function of x.

a. If the growth factor is 1.08, then complete the following table.

x	0	1	2	3
y	23.1			

b. If the decay factor is 0.75, then complete the following table.

x	0	1	2	3
y	10			

6. Would you expect $f(x) = 3^x$ to increase faster or slower than $g(x) = 2.5^x$ for $x > 0$? Explain. (*Hint*: You may want to use your graphing calculator for help.)

7. Match each graph with either $f(x) = 2^x$ or $g(x) = \left(\dfrac{1}{2}\right)^x$.

i.

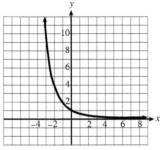

ii.

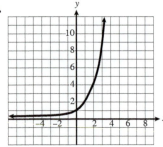

8. Take a piece of paper from your notebook. Let x represent the number of times you fold the paper in half and $f(x)$ represent the number of sections the paper is divided into after the folding.

a. Complete the table of values.

x	0	1	2	3	4	5
f(x)						

b. If you could fold the paper eight times, how many individual sections will there be on the paper?

c. Does this data represent an exponential function? Explain.

d. What is the practical domain and range in this situation?

In Exercises 9–13, match each graph with its equation.

9. $y = 3^x$

10. $y = 3^{x+2}$

11. $y = 2 \cdot 3^x$

12. $y = -3^x$

13. $y = 3^{-x}$

a.

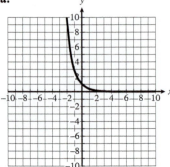

b.

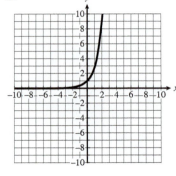

c.

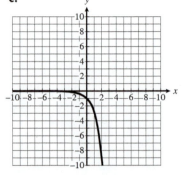

d.

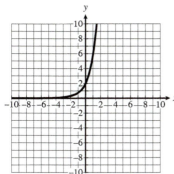

e.

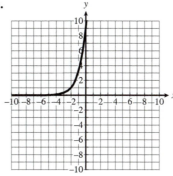

14. Identify if the exponential function is increasing or decreasing. Verify using a graphing calculator.

a. $y = 5^x$ b. $y = \left(\dfrac{4}{5}\right)^x$ c. $y = (2.5)^x$ d. $y = (0.65)^x$

15. In parts a–f, graph the function defined by the given equation. Using transformations, describe how the graph is related to the graph of $y = 4^x$. Verify the graph using a graphing calculator.

a. $y = 4^{x+2}$ b. $y = -4^x$

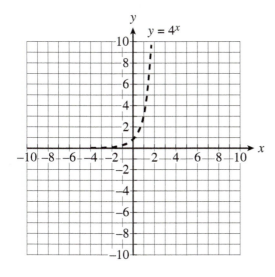

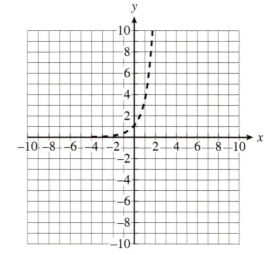

c. $y = 4^{-x}$

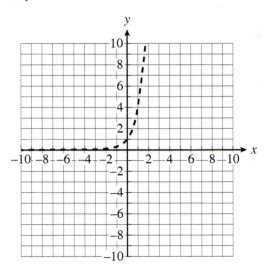

d. $y = 3 \cdot 4^x$

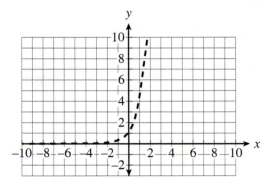

e. $y = \dfrac{1}{3} \cdot 4^x$

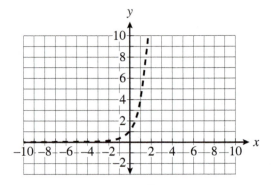

f. $y = 4^x + 2$

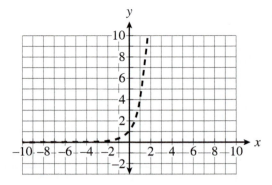

Activity 4.5

National Debt

Objectives

1. Determine the growth and decay factor of an exponential function represented by a table of values or an equation.

2. Graph exponential functions defined by $y = ab^x$, where $a \neq 0$, $b > 0$ and $b \neq 1$.

3. Determine the doubling and halving time.

The amount of money owed by the federal government to individuals and countries around the world is not hard to track. The U.S. Treasury Department publishes this data daily.

The following table shows how the National Debt grew from 2006 to 2012.

NATIONAL DEBT	
YEAR	TOTAL DEBT AS OF SEPT 30 (trillions of dollars)
2006	8.5
2007	9.0
2008	10.0
2009	11.9
2010	13.6
2011	14.8
2012	16.1

Data Source: U.S. Treasury Department

1. Is this a linear function? How do you know?

2. a. Use the data in the table above to evaluate the following ratios to complete the following table.

Debt in 2007 / Debt in 2006	Debt in 2008 / Debt in 2007	Debt in 2009 / Debt in 2008	Debt in 2010 / Debt in 2009	Debt in 2011 / Debt in 2010	Debt in 2012 / Debt in 2011

b. What do you notice about all of the values in the table?

In an exponential function with base b, equally spaced x-values yield y-values whose successive ratios are constant. If the x-values increase by increments of 1, the common ratio is the base b. If $b > 1$, b is the growth factor; if $0 < b < 1$, b is the decay factor.

3. a. Can the data in the table preceding Problem 1 be modeled by an exponential function? Explain.

b. What is the growth factor?

c. As a consequence of the result found in part b, you can start with the national debt in 2006 and approximate the national debt in 2007 by multiplying by the growth factor, b. You can then approximate the national debt in 2008 by multiplying the national debt in 2007 by b, and so on. Verify this with your calculator. Note that because the exponential function is a mathematical model, the results vary slightly from the actual values given in the table preceding Problem 1.

Once you know the growth factor (b), you can determine the equation that models the national debt in trillions of dollars as a function of t, the number of years since 2006. Note that $t = 0$ corresponds to 2006, $t = 1$ to 2007, and so on.

4. a. Using a growth factor of 1.12, complete the following table.

t	CALCULATION OF THE NATIONAL DEBT	EXPONENTIAL FORM	NATIONAL DEBT (in trillions of dollars)
0	8.50	$8.50(1.12)^0$	8.50
1	(8.50)1.12		
2			
3			

     b. Use the pattern in the preceding table to help you write the equation of the form $D(t) = a \cdot b^t$, where $D(t)$ represents the national debt in trillions of dollars and t represents the number of years since 2006.

     c. What is the practical domain of this function?

     d. Graph the function D on your graphing calculator, and then sketch the result on an appropriately scaled and labeled axis.

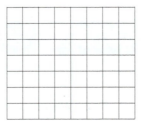

     e. Determine the D-intercept of the graph of D by substituting 0 for the value t. What is the practical meaning of the vertical intercept in this situation?

     f. What does the value of a represent in part b?

Definition

Many exponential functions can be represented symbolically by $y = a \cdot b^x$, where a is the value of y when $x = 0$ and b is the growth or decay factor. If the x-variable of $y = a \cdot b^x$ represents time, then the coefficient a is called the **initial value**.

Example 1

The exponential function defined by $y = 5 \cdot 2^x$ has y-intercept $(0, 5)$ and growth factor $b = 2$. The exponential function defined by $y = \dfrac{1}{2}(0.75)^x$ has y-intercept $\left(0, \dfrac{1}{2}\right)$ and decay factor $b = 0.75$.

5. Use the function defined by $D = 8.50 \cdot 1.12^t$ to estimate the national debt in 2016. Do you think this is a good estimate? Explain.

6. a. Use the graph of the exponential function $D = 8.50 \cdot 1.12^t$ on your graphing calculator to estimate how long it will take for the national debt to double from 8.50 trillion dollars to 17.00 trillion dollars.

   b. Estimate the time necessary for the national debt to double from 17.00 trillion dollars to 34.00 trillion dollars. Verify your estimate using your calculator.

   c. At the 2006 rate, how long will it take the national debt to double at any given point in time?

Definition

The **doubling time** of an exponential function is the time it takes for an output to double. The doubling time is determined by the growth factor and remains the same for all output values.

Example 2

The balance B(t), in dollars, of an investment account is defined by $B(t) = 5500(1.12)^t$, where t is the number of years. The initial value for this function is \$5500. Determine the value of t when the balance is doubled or equal to \$11,000.

SOLUTION

If you use the table feature of your calculator, the doubling time is estimated at 6.1 years (see the following calculator graphic). The intersect feature on the graphing calculator shows the doubling time to be 6.12 years to the nearest hundredth.

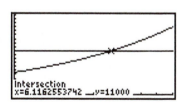

Decreasing Exponential Functions

You have just purchased a new car for $26,000. Much to your dismay, you have just learned that you should expect the value of your car to depreciate by 30% per year! The following table shows the book value of the car for the next several years, where V is the value in thousands of dollars:

DEPRECIATION: TAKING ITS TOLL

t (year)	0	1	2	3	4
V (in thousands of dollars)	26	18.2	12.7	8.9	6.2

The values of the independent variable, t, are incremented by 1, and a value of the V is obtained by multiplying the previous value by a constant factor. This is another example of an exponential function. However, because the value of the car is decreasing, the constant factor is a *decay factor* and its value will be between 0 and 1.

7. a. Using the information in the table, calculate the decay factor, b, in this situation.

b. You can now start with 26, the initial value of the car (in thousands of dollars), and obtain the value after one year by multiplying by the decay factor, $b = 0.70$. The value of the car after two years is the value of the car after one year times the decay factor. Verify this on your calculator, and compare your results with the entries in the graphic above.

8. a. Complete the following table:

T	CALCULATION OF THE VALUE OF THE CAR	EXPONENTIAL FORM	VALUE, $V = a \cdot b^t$ (in thousands of dollars)
0	26	$26(0.70)^0$	26
1	$26 \cdot 0.70$	$26(0.70)^1$	
2			
3			

b. Use the pattern in the preceding table to help you write an equation of the form $V = a \cdot b^t$, where V represents the value of the car as a function of time, t.

c. What is the practical domain of this function?

d. Graph this function on your graphing calculator, and sketch the result below on an appropriate scaled and labeled set of axes.

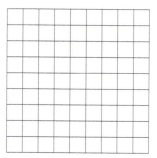

e. Determine the *V*-intercept of the graph by substituting 0 for the input *t*. What is the practical meaning of the intercept in this situation?

9. a. Use your graphing calculator to estimate the number of years it takes to halve the value of the automobile from $26,000 to $13,000.

b. Estimate the time necessary to halve the value from $13,000 to $6500. Verify your answer using your graphing calculator.

c. How long will it take for any specific value of the car to halve?

Definition

The **half-life** of an exponential function is the time it takes for a *y*-value to decay by one-half. The half-life is determined by the decay factor and remains the same for all *y*-values.

Example 3

The population of Detroit, Michigan, can be modeled by the equation $D(t) = 1461 \cdot (0.985)^t$, with $t = 0$ representing the year 1970 and D(t) representing the population of Detroit in thousands. If the population of Detroit continues to decline at the same rate, determine the number of years it will take for the population of Detroit to be one-half of the 1970 population.

SOLUTION

The equation indicates that the population of Detroit, Michigan in 1970 ($t = 0$) is $D(0) = 1461 \cdot (0.985)^0 = 1461$ thousand people. Therefore, half of that is 730.5 thousand people. Use the table of your graphing calculator; the halving time is estimated at 46 years.

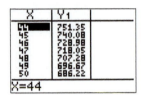

Use the intersect feature of your graphing calculator and you will determine the halving time to be 46 years to the nearest year.

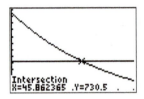

10. Homemade chocolate chip cookies lose their freshness over time. Let the taste quality be one when the cookies are fresh. The taste quality decreases according to the function:

$$y = 0.8^x,$$

where x is the number of days since the cookies were baked.

Determine when the taste quality will be one-half of its value. Use the intersect feature of your calculator to determine when y is $\frac{1}{2}$ of 1 or 0.5.

SUMMARY: ACTIVITY 4.5

1. For **exponential functions** defined by $y = ab^x$, a is the value of y when $x = 0$ (sometimes called the initial value), and b is the growth or decay factor.

2. The y-intercept of these functions is $(0, a)$.

3. In an exponential function, equally spaced x-values yield y-values whose successive ratios are constant. If the x-values increase by one unit, then
   a. the constant ratio is the **growth factor** if the y-values are increasing.
   b. the constant ratio is the **decay factor** if the y-values are decreasing.

4. The **doubling time** of an increasing exponential function is the time it takes for y-value to double. The doubling time is determined by the growth factor and remains the same for all y-values.

5. The **half-life** of a decreasing exponential function is the time it takes for y-values to decay by one-half. The half-life is determined by the decay factor and remains the same for all y-values.

EXERCISES: ACTIVITY 4.5

1. The population (in millions) of Russia in selected years is given in the following table.

Year	1995	1996	2000	2006	2011
Population (in millions)	148.0	147.6	146.0	142.0	140.2

a. Let 1995 correspond to $t = 0$. Let b be the ratio between the population of Russia in 1996 and 1995. Determine an exponential function of the form $y = a \cdot b^t$ to represent the population P of Russia symbolically. Round to three decimal places.

b. Does the function in part a give an accurate value of the population of Russia in 2000? Explain.

c. Use the model in part a to predict the population of Russia in 2014.

2. Without using your graphing calculator, match each graph with its equation. Then check your answer using your graphing calculator.

a. $f(x) = 0.5(0.73)^x$ **b.** $g(x) = 3(1.73)^x$ **c.** $h(x) = -2(1.73)^x$

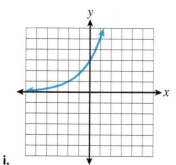

i.

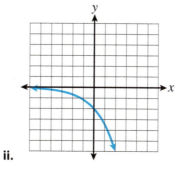

ii.

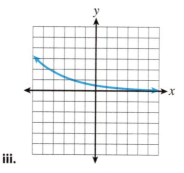

iii.

3. Which of the following tables represent exponential functions? Indicate the growth or decay factor for the data that is exponential.

a.

x	0	1	2	3	4
y	0	2	16	54	128

b.

x	0	1	2	3	4
y	1	4	16	64	256

c.

x	1	2	3	4	5
y	1750	858	420	206	101

4. a. Sketch a graph of $f(x) = 2^x$ and $g(x) = 3 \cdot 2^x$ on the same coordinate axis.

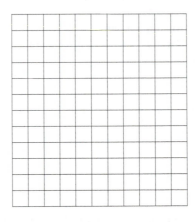

    b. Describe how the graphs of f and g are similar and how they are different.

5. If $f(x) = 3 \cdot 4^x$, determine the exact value of each of the following, when possible. Otherwise, use your calculator to approximate the value to the nearest hundredth.

    a. $f(-2)$ **b.** $f\left(\dfrac{1}{2}\right)$

    c. $f(2)$ **d.** $f(1.3)$

6. According to industry reports, the total global sales of smartphones can be modeled by the equation $P(t) = 132.2 \cdot 1.49^t$, where t represents the number of years since 2008 and $P(t)$ represents global smartphone sales measured in millions of units.

    a. Complete the following table

t, Number of Years Since 2008	0	1	2	3	4	5
P(t), Global Smartphone Sales (millions)						

    b. Determine the growth factor of the smartphone sales.

c. Sketch a graph of this exponential function. Use $0 \le t \le 5$ and $0 \le P(t) \le 1000$.

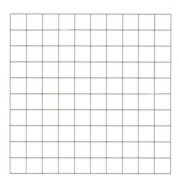

d. Use the equation to determine the projected smartphone sales in 2020. Do you believe this is reasonable?

e. Use your graphing calculator to approximate the year in which smartphone sales first exceed 1.5 billion units.

7. Chlorine is used to disinfect swimming pools. The chlorine concentration should be between 1.5 and 2.5 parts per million (ppm). On sunny, hot days, 30% of the chlorine dissipates into the air or combines with other chemicals. The chlorine concentration, $A(x)$, (in parts per million) in a pool after x sunny days can be modeled by

$$A(x) = 2.5(0.7)^x.$$

a. What is the initial concentration of chlorine in the pool?

b. Complete the following table:

x	0	1	2	3	4	5
A(x)						

c. Sketch a graph of the chlorine function.

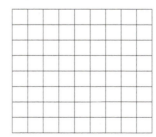

d. What is the chlorine concentration in the pool after three days?

e. Approximate graphically and numerically the number of days before chlorine should be added.

8. The population of Clarksville, Tennessee, from 2005 to 2012 is approximated in the following table.

Year	2005	2006	2007	2008	2009	2010	2011	2012
Population (in thousands)	113.4	114.1	119.6	120.3	124.6	133.7	136.2	142.5

a. Can the relationship in the table be reasonably modeled by an exponential function? Explain.

b. What is the growth factor?

c. Determine the equation that models the population N in thousands, of Clarksville as a function of t, the number of years since 2005. Note that $t = 0$ corresponds to 2005, and so on.

d. Graph the function.

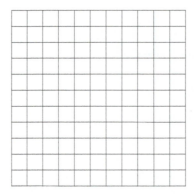

e. What is the N-intercept? What is the practical meaning of this intercept in this situation?

f. Use the equation to estimate the population of Clarksville, TN, in the year 2025. Do you think this is a good estimate? Explain.

g. Use the graph of the exponential function and the graph or table feature of your graphing calculator to estimate the number of years it takes for the population of Clarksville to double from 113.4 thousand to 226.8 thousand people.

9. As a radiology specialist, you use the radioactive substance iodine-131 to diagnose conditions of the thyroid gland. Your hospital currently has a 20-gram supply of iodine-131. The following table gives the number of grams remaining after a specified number of days.

t (Number of Days Starting from a 20-Gram Supply of Iodine-131)	0	1	2	3	4	5	6
N, Number of Grams of Iodine-131 Remaining from a 20-Gram Supply	20.00	18.34	16.82	15.42	14.14	12.97	11.89

a. Does the relationship represent an exponential function? Explain.

b. What is the decay factor?

c. Write an exponential decay formula for N, the number of grams of iodine-131 remaining, in terms of t, the number of days from the current supply of 20 grams.

d. Determine the number of grams of iodine-131 remaining from a 10-gram supply after one month (30 days).

e. Graph the decay formula for iodine-131, $N = 20(0.917)^t$, as a function of the time t (days). Use appropriate scales and labels on the following grid or an appropriate window on a graphing calculator.

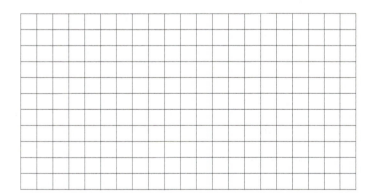

f. How long will it take for iodine-131 to decrease to half its original value? Explain how you determined your answer.

Population Growth

Objectives

1. Determine annual growth or decay rate of an exponential function represented by a table of values or an equation.

2. Graph an exponential function having equation $y = a(1 + r)^x, a \neq 0$.

According to the U.S. Census Bureau, in 2013 the city of Greenville, North Carolina, was one of the fastest growing cities in the nation. The population of Greenville, NC, in 2013 was 87,834.

1. **a.** Assuming that the population increases at a constant percent rate of 1.5% determine the population of Greenville (in thousands) in 2014.

   b. Determine the population of Greenville (in thousands) in 2015.

   c. Divide the population in 2014 by the population in 2013 and record this ratio.

   d. Divide the estimated population in 2015 by the estimated population in 2014 and record this ratio.

   e. What do you notice about the ratios in parts c and d? What do these ratios represent?

   Linear functions represent quantities that change at a constant rate (slope). Exponential functions represent quantities that change at a constant rate, expressed as a percent.

   Example 1 *Population growth, sales and advertising trends, compound interest, spread of disease, and concentration of a drug in the blood are examples of quantities that increase or decrease at a constant rate expressed as a percent.*

2. Let t represent the number of years since 2013 ($t = 0$ corresponds to 2013). Use the results from Problem 1 to complete the following table:

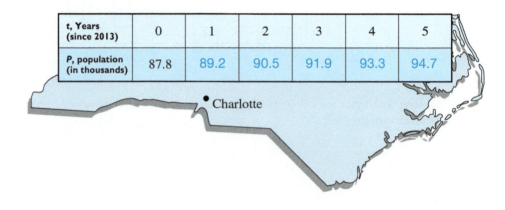

t, Years (since 2013)	0	1	2	3	4	5
P, population (in thousands)	87.8	89.2	90.5	91.9	93.3	94.7

Charlotte

Once you know the growth factor ($b = 1.015$), you can determine the exponential model that describes the population of Greenville, NC, as a function of t, where $t = 0$ corresponds to the year 2013.

3. a. Complete the following table:

t	CALCULATION FOR POPULATION (in thousands)	EXPONENTIAL FORM	P, POPULATION (in thousands)
0	87.8	$87.8(1.015)^0$	87.8
1	87.8(1.015)	$87.8(1.015)^1$	
2	87.8(1.015)(1.015)		
3			

Note the values in the last column of the table do not quite match the values in the table in Problem 2 due to round off error.

b. Use the pattern in the table in part a to help you write an equation for P, the population of Greenville, NC (in thousands), using t, the number of years since 2013, as the independent variable.

The equation $P = 87.8(1.015)^t$ has the general form $P = P_0(1 + r)^t$, where r is the annual **growth rate**, $(1 + r)$ is the **growth factor** or the base, b, of the exponential function, t is the time in years, and P_0 is the initial value, the population when $t = 0$.

Example 2

a. Determine the growth factor and the growth rate of the function defined by
$y = 250(1.7)^x.$

SOLUTION

The growth factor $1 + r$ is the base 1.7. To determine the growth rate, solve the equation $1 + r = 1.7$ for r.

$$r = 0.7 \text{ or } 70\%$$

b. If the growth rate of a function is 5%, determine the growth factor.

SOLUTION

If $r = 5\%$ or 0.05, the growth factor is $1 + r = 1 + 0.05 = 1.05$.

4. a. Determine the growth factor in the Greenville population function $P = 87.8(1.015)^t$.

b. Determine the growth rate. Express your answer as a percent.

5. a. Using the function defined by $P = 87.8(1.015)^t$, determine the population of Greenville, NC, in 2020. That is determine P when $t = 7$.

   b. Graph the population function with your graphing calculator, adjusting the window to show the population between 2010 and 2025. Use the window Xmin $= -3$, Xmax $= 16$, Ymin $= 80$, and Ymax $= 115$. What type of function does the graph resemble?

   c. Reset the window to Xmin $= -50$, Xmax $= 200$, Ymin $= -20$, Ymax $= 1000$ and display the graph. Now what type of function does the graph resemble?

   d. Determine P when $t = 0$. What is the graphical and practical meaning of this number?

6. a. Use your model to predict Greenville's population in 2023.

   b. Verify your prediction on the graph from Problem 5c.

7. a. Use the graph to estimate when Greenville's population will reach 105 thousand, assuming it continues to grow at the same rate. Remember P is the population in thousands.

   b. Evaluate P when $t = 20$ and describe what it means.

8. Use the model to estimate the population of Greenville in 2015 and 2025. In which prediction are you more confident? Why?

9. a. Assuming that the growth rate remains constant, how long will it take for Greenville to double its 2013 population?

   b. Explain how you reached your conclusion in part a.

Wastewater Treatment Facility

You are a chemical engineer working at a wastewater treatment facility. You are presently treating water contaminated with 18 micrograms of pollutant per liter. Your process is designed to remove 20% of the pollutant during each treatment. Your goal is to reduce the pollutant to less than 3 micrograms per liter.

10. a. What percent of pollutant present at the start of a treatment remains at the end of the treatment?

b. The concentration of pollutant is 18 micrograms per liter at the start of the first treatment. Use the result of part a to determine the concentration of pollutant at the end of the first treatment.

c. Complete the following table. Round the results to the nearest tenth.

n, Number of Treatments	0	1	2	3	4	5
C, Concentration of Pollutant, in $\mu g/l$, at the End of the nth Treatment	18	14.4				

d. Write an equation for the concentration, C, of the pollutant as a function of the number of treatments, n.

The equation $C = 18(0.80)^n$ has the general form $C = C_0(1 - r)^n$, where r is the **decay rate**, $(1 - r)$ is the **decay factor** or the base of the exponential function, n is the number of treatments, and C_0 is the initial value, the concentration when $n = 0$.

Example 3 **a.** *Determine the decay factor and the decay rate of the function defined by $y = 123(0.43)^x$.*

SOLUTION

The decay factor $1 - r$ is the base, 0.43. To determine the decay rate, solve the equation $1 - r = 0.43$ for r.

$$r = 0.57 \text{ or } 57\%$$

b. *If the decay rate of a function is 5%, determine the decay factor.*

SOLUTION

If $r = 5\%$ or 0.05, the decay factor is $1 - r = 1 - 0.05 = 0.95$.

11. a. If the decay rate is 2.5%, what is the decay factor?

b. If the decay factor is 0.76, what is the decay rate?

12. a. Use the function defined by $C = 18(0.8)^n$ to predict the concentration of contaminants at the wastewater treatment facility after seven treatments.

b. Sketch a graph of the concentration function on your graphing calculator. Use the table in Problem 10c to set a window. Does the graph look like you expected it would? Explain.

c. What is the C-intercept? What is the practical meaning of the intercept in this situation?

d. Reset the window of your graphing calculator to Xmin $= -5$, Xmax $= 15$, Ymin $= -10$, and Ymax $= 50$. Does the graph have a horizontal asymptote? Explain what this means in this situation.

13. Use the table or trace feature of your graphing calculator to estimate the number of treatments necessary to bring the concentration of pollutant below 3 micrograms per liter.

SUMMARY: ACTIVITY 4.6

1. Exponential functions are used to describe phenomena that grow or decay by a constant percent rate over time.

2. If r represents the **annual growth rate**, the exponential function that models the quantity, P, can be written as

$$P = P_0(1 + r)^t,$$

where P_0 is the initial amount, t represents the number of elapsed years, and $1 + r$ is the growth factor.

3. If r represents the **annual decay rate**, the exponential function that models the amount remaining can be written as

$$P = P_0(1 - r)^t,$$

where $1 - r$ is the decay factor.

EXERCISES: ACTIVITY 4.6

1. Determine the growth and decay factors and growth and decay rates in the following table.

GROWTH FACTOR	GROWTH RATE		DECAY FACTOR	DECAY RATE
1.02			0.77	
	2.9%			68%
2.23			0.953	
	34%			19.7%
1.0002			0.9948	

2. In 2005, the U.S. Census Bureau estimated the population of Boston, MA, as 609.7 thousand people and the population of Detroit, MI, as 921.1 thousand people. Since 2005, Boston's population has been increasing at approximately 0.62% per year. Detroit's population has been decreasing at approximately 3.82% per year. Assume the growth and decay rates remain constant.

a. Let $P(t)$ represent the population t years after 2005. Determine the exponential functions that model the population of both cities.

b. Use your models to predict the population of both cities in the year 2015.

c. Estimate the number of years for the population of Detroit to halve.

d. Using the table and/or graphs of these functions, predict when the populations of these cities will be equal.

3. You have just taken over as the city manager of a small city. The personnel expenses were $8,500,000 in 2014. Over the past five years, the personnel expenses have increased at a rate of 3.2% annually.

a. Assuming that this rate continues, write an equation describing personnel costs, C, in millions of dollars, where $t = 0$ corresponds to 2014.

b. Sketch a graph of this function up to the year 2030 ($t = 16$).

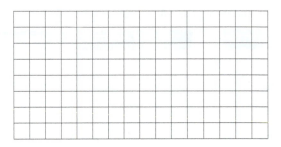

c. What are your projected personnel costs in the year 2019?

d. What is the C-intercept? What is the practical meaning of the intercept in this situation?

e. In what year will the personnel expenses be double the 2014 personnel expenses?

4. According to the U.S. Census Bureau, the population of the United States (in millions) can be modeled by $P(t) = 123.3 \cdot 1.0118^t$, where t represents the number of years since 1930.

a. Use your graphing calculator to sketch a graph of the U.S. population model.

b. Determine the annual growth rate and the growth factor from the equation.

c. Use the model to determine the population (in millions) of the United States in 2010. How does your answer compare to the actual population of 309.3 million?

5. You have recently purchased a new truck for $20,000, by arranging financing for the next five years. You are curious to know what your new truck will be worth when the loan is completely paid off.

a. Assuming that the value depreciates at a constant rate of 15%, write an equation that represents the value V of the truck t years from now.

b. What is the decay rate in this situation?

c. What is the decay factor in this situation?

d. Use the equation from part a to estimate the value of your truck five years from now.

e. Use the trace and table features of your graphing calculator to check your results in part d.

f. Use the trace or table features of your graphing calculator to determine when your truck will be worth $10,000.

6. Suppose the inflation rate is 5% per year and remains the same for the next seven years.

a. Complete the following table for a pair of sneakers that cost $65 now. Round to the nearest cent.

t, Years from Now	0	1	2	3	4	5	6	7
c(t), Cost of Sneakers ($)	65							

b. Determine the growth factor for a 5% inflation rate.

c. If the yearly inflation rate remains at 5%, what exponential function would you use to determine the cost of $65 sneakers after t years.

d. Use the equation in part c to determine the cost of the sneakers in ten years. What assumption are you making regarding the inflation rate?

7. Beginning in 1988, infestations of zebra mussels started spreading through North American waters. These mussels spread at an alarming rate and threatened entire ecosystems. Zoologists have approximated the growth rate of an area of an infestation of zebra mussels to be 350% per year.

a. In one study, ten zebra mussels were discovered in a small lake. Assuming the 350% growth rate, determine a function to represent the expected number N of zebra mussels t years after the study began.

b. Use the function from part a to complete the following table.

Years Since Study Began, t	1	2	3	4	5
Expected Number of Zebra Mussels, N					

c. Use your graphing calculator to determine how many years it will take for the ten zebra mussels to become 1,000,000 zebra mussels.

Activity 4.7

Bird Flu

Objectives

1. Determine the equation of an exponential function that best fits the given data.

2. Make predictions using an exponential regression equation.

3. Determine whether a linear or exponential model best fits the data.

Scientists, including biologists, medical researchers, and epidemiologists, study the causes and spread of infectious diseases. In 2005, the avian flu, also known as bird flu, received international attention. Although 13 died in Cambodia on July 2, 2013, there are very few documented cases of the avian flu infecting humans worldwide. World health organizations, including the Centers for Disease Control in Atlanta, express concern that a mutant strain of the bird flu virus capable of infecting humans might develop and produce a worldwide pandemic.

The infection rate (the number of people that any single infected person will infect) and the incubation period (the time between exposure and the development of symptoms) of this flu cannot be known precisely, but they can be approximated by studying the infection rates and incubation periods of existing strains of the virus.

A very conservative infection rate would be 1.5 and a reasonable incubation period would be about 15 days or roughly half of a month. This means that the first infected person could be expected to infect 1.5 people in roughly half of a month. After the half of a month, that person cannot infect anyone else. This assumes that the spread of the virus is not checked by inoculation or vaccination.

So the total number of infected people 0.5 months after the first person was infected would be 2.5, the sum of the original infected person and the 1.5 newly infected people. During the second half-month the 1.5 newly infected people would infect $1.5 \times 1.5 = 2.25$ new people. This means you have 2.25 people to add to the 2.5 people previously infected, or approximately 5 people infected with bird flu at the end of the first month.

The following table represents the total number of people who could be infected with a mutant strain of bird flu over a period of five months.

Months Since the First Person was Infected	0	0.5	1	1.5	2	2.5	3	3.5	4	4.5	5
Number of Newly Infected		1.5	2.25	3.375	5						
Total Number of People Infected	1	2.5	5	8	13						

1. Complete the table above. Round each value to the nearest whole person.

2. Let t represent the number of months since the first person was infected and N represent the total number of people infected with the bird flu virus. Create a scatterplot of the data below.

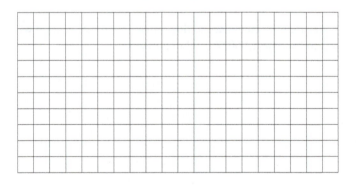

3. Does the scatterplot indicate a linear relationship between t and N? Explain.

4. a. Use your calculator to create a scatterplot of the bird flu data. Then use option 0: ExpReg in the STAT CALC menu to determine an exponential function that best fits the given data. Record that model below. Round a and b to the nearest 0.001.

  b. Sketch a graph of the exponential model using your calculator and add it to the scatterplot in part a.

  c. What is the practical domain of this function?

  d. What is the N-intercept of the graph? How does it compare to the actual initial value $(t = 0)$ from the table?

5. a. Use the exponential model on your graphing calculator to predict the total number of infected people one year after the initial infection $(t = 12)$ provided the virus is unchecked. Round your result to the nearest whole person.

  b. Use the exponential function to write an equation that can be used to determine when the virus will first infect 2,000,000 people.

  c. Solve the equation in part b using a graphing approach. Use the intersect feature of your calculator; the screen containing the solution should resemble the following:

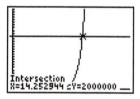

  d. Interpret the meaning of your solution in part c.

Increasing Exponential Model

According to the U.S. Department of Education, the number of students receiving Bachelor's degrees has increased significantly since 1970. The following table gives the number of Bachelor's degrees (in thousands) awarded from 1970 through 2011.

College Bound

YEAR	NUMBER OF BACHELOR'S DEGREES (thousands)
1970	840
1980	935
1990	1094
2000	1238
2010	1602
2011	1716

Data Source: U.S. Department of Education

6.

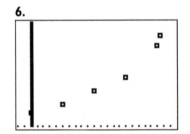

6. Let t represent the number of years since 1970 ($t = 0$ corresponds to 1970, $t = 10$ to 1980, etc.). Let N represent the number of college graduates (in thousands) at time t. Sketch a scatterplot of the given data on your graphing calculator. Your scatterplot should appear as seen to the left.

7. a. Use your graphing calculator to determine the equation of an exponential function that best fits the given data.

7. b.

   b. Sketch a graph of the exponential model using your graphing calculator. Your graph should appear as seen to the left.

   c. What is the practical domain of this exponential function?

   d. What is the N-intercept of the graph? How does it compare to the actual initial value ($t = 0$) from the table?

8. a. What is the base of the exponential model? Is the base a growth or decay factor? How do you know?

   b. What is the annual growth rate?

9. a. Use the exponential model to determine the number of Bachelor's degrees that will be awarded in 2018. ($t = 48$)

   b. Use the exponential model to write an equation that can be used to determine the year in which there will be 2 million Bachelor's degrees awarded. Remember that the number of Bachelor's degrees is measured in thousands.

c.

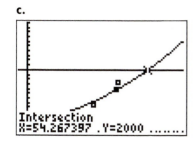

Intersection
X=54.267397 .Y=2000 ....

c. Solve the equation in part b using a graphing approach. Use the intersect feature of your graphing calculator: the screen should appear as seen to the left.

10. What is the doubling time for your exponential model? That is, approximately how many years will it take for the number of Bachelor's degrees awarded in any given year to double?

Decreasing Exponential Model

Students in U.S. public schools have had much greater access to computers in recent years. The following table shows the number of students per computer in a large school district in selected years:

Year	1997	1998	1999	2001	2003	2006	2009	2013
Number of Students Per Computer	125	75	50	32	22	16	10	5.7

11. a. Use your graphing calculator to determine the equation of an exponential function that models the given data. Let the independent variable, t, represent the number of years since 1997.

b.

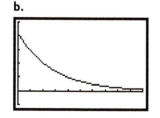

b. Sketch a graph of the exponential model using your graphing calculator. Your graph should appear as seen to the left.

c. What is the base of the exponential model? Is the base a growth or decay factor? How do you know?

d. What is the annual decay rate?

EXERCISES: ACTIVITY 4.7

1. The total amount of money spent on health care in the United States is increasing at an alarming rate. The following table gives the total national health care expenditures in billions of dollars for selected years from 1980 through 2011.

Year	1980	1990	2000	2005	2010	2011
Total Spent (billions of dollars)	256	724	1377	2163	2600	2701

Data Source: Centers for Medicare and Medicaid Services

a. Would the data in the preceding be better modeled by a linear model, $y = mx + b$, or an exponential model, $y = a \cdot b^x$? Explain.

b. Sketch a scatterplot of this data.

c. Does the graph reinforce your conclusion in part a? Explain.

d. Use your graphing calculator to determine the exponential regression equation that best fits the health care data in the preceding table. Let the independent variable, *t*, represent the number of years since 1980.

e. Using the regression equation from part d, determine the predicted total health care expenditures for the year 2000?

f. According to the exponential model, what is the growth factor for the total health care costs per year?

g. What is the growth rate?

h. According to the exponential model, in what year did the total health care costs first exceed $1 trillion?

i. What is the doubling time for the exponential model?

2. a. Consider the following data set for the variables *x* and *y*:

x	5	8	11	15	20
y	70.2	50.7	35.1	22.6	9.5

Plot these points on the following grid:

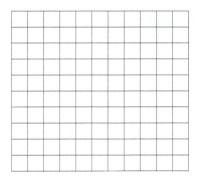

b. Use your graphing calculator to determine both a linear regression and an exponential regression model of the data. Record the equations for these models here.

c. Which model appears to fit the data better? Explain.

d. Use the better model to determine y when $x = 13$ and y when $x = 25$.

e. For the exponential model, what is the decay factor?

f. What does it mean that the decay factor is between 0 and 1?

g. What is the half-life for the exponential model?

3. Use the graph of $y = 5 \cdot 2^x$ as a check, and summarize the properties of the exponential function $y = a \cdot b^x$, where $a > 0$ and $b \neq 1$.

a. What is the domain?

b. What is the range?

c. When is $y = a \cdot b^x$ positive?

d. When is $y = a \cdot b^x$ negative?

e. What is the y-intercept of the graph of $y = a \cdot b^x$?

4. The number of transistors that can be placed on a single chip has grown significantly between 1970 and 2012. The following table gives the number of transistors (in millions) that can be placed on a specific chip in a given year.

YEAR	x, NUMBER OF YEARS SINCE 1970	CHIP	TRANSISTORS (in millions)
1971	1	4004	0.0023
1986	16	386DX	0.275
1989	19	486DX	1.2
1993	23	Pentium	3.3
1997	27	Pentium II	7.5
2000	30	Pentium IV	42
2006	36	Core 2 Duo	291
2010	40	16-Core	1000
2012	42	62-Core	5000

Data Source: Intel

a. Use a graphing calculator to determine an exponential regression model of the data. Let the independent variable x represent the number of years since 1970.

b. Assuming the rate of growth continues, approximate the number of transistors that can be placed on a chip in 2017?

Collecting and Analyzing Data

5. You need to find some real-world data in science that appears to be increasing exponentially. Newspapers, magazines, scientific journals, almanacs, and the Internet are good resources. Once the data has been obtained, model the data by an exponential function defined by $y = ab^x$. Explain why an exponential function would best represent the data. Be sure to describe the meaning of a and b in the exponential model in terms of the situation. Make a prediction about the dependent variable y for a specific value of the independent variable x. Describe the reliability of this prediction.

6. The combined populations of China and India currently represent over 38% of the world's population. Search the Internet to obtain the population of each country for every five years from 1960 to 2010.

a. Beginning with 1960, make a scatterplot of the data from each country's population. Let the independent variable represent the number of years since 1960 and the dependent variable represent the population of the country in billions. Describe any patterns you observe.

b. Determine whether a linear or exponential function best fits each set of data. Explain.

c. Determine a regression equation for each set of population data. Which country has the larger rate of increase in population?

d. Sketch the graph of each regression equation on the appropriate scatterplot in part a.

e. Predict the population of each country in 2015.

f. Determine the year in which the population of each country should reach 1.5 billion.

g. Use the intersect feature of a graphing calculator to estimate the year when the populations of China and India will be equal.

Activities 4.1–4.7   What Have I Learned?

I. a. Think of an example, or find one in a newspaper or magazine or on the Internet, and use it to show how to determine the growth factor associated with a percent increase. For instance, you might think of or find growth factors associated with yearly interest rates for a saving account.

  b. Show how you would use the growth factor to apply a percent increase twice, then three times.

2. Current health research shows that losing just 10% of body weight produces significant health benefits, including a reduced risk for a heart attack. Suppose a relative who weighs 199 pounds begins a diet to reach a goal weight of 145 pounds.

  a. Use the idea of a decay factor to show your relative how much he will weigh after losing the first 10% of his body weight.

  b. Explain to your relative how he can use the decay factor to determine how many times he has to lose 10% of his body weight to reach his goal weight of 145 pounds.

3. Consider a linear function defined by $g(x) = mx + b$, $m \neq 0$, and an exponential function defined by $f(x) = a \cdot b^x$, $a > 0$. Explain how you can determine from the equation whether the function is increasing or decreasing.

4. Suppose you have an exponential function of the form $f(x) = a \cdot b^x$, where $a > 0$ and $b > 0$ and $b \neq 1$. By inspecting the graph of f, can you determine if $b > 1$ or if $0 < b < 1$? Explain.

5. You are given a function defined by a table, and the x-values are in increments of 1. By looking at the table, can you determine whether or not the function can be approximated by an exponential model? Explain.

6. An exponential function $y = a \cdot b^x$ passes through the point (0, 2.6). What can you conclude about the values of a and b?

7. Explain the difference between growth rate and growth factor.

8. Explain why the base in an exponential function cannot equal 1.

Activities 4.1–4.7 | How Can I Practice?

1. A suit with an original price of $300 was marked down 30%. You have a coupon good for an additional 20% off.

   a. What is the suit's final cost?

   b. What is the equivalent percent discount?

2. You are planning to purchase a new motorcycle. You know that new prices are projected to increase at a rate of 4% per year for the next few years.

   a. Write an equation that represents the projected cost, C, of the motorcycle t years in the future, given that it costs $8490.

   b. Identify the growth rate and the growth factor.

   c. Use your equation in part a to project the cost of your motorcycle three years from now.

   d. Use your graphing calculator to approximate how long it will take the motorcycle to cost $12,000 if the price continues to increase at 4% per year.

3. Without using your graphing calculator, match the graph with its equation.

   a. $g(x) = 2.5(0.47)^x$ **b.** $h(x) = 1.5(1.47)^x$

   i. 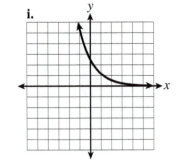   **ii.**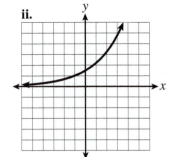

4. Explain the reasons for your choices in Problem 3.

5. Complete the following tables representing exponential functions. Round calculations to the nearest hundredth whenever necessary.

a.

x	0	1	2	3	4
y	2.00	5.10			

b.

x	0	1	2	3	4
y	3.50	2.10			

c.

x	0	1	2	3	4
y	$\frac{1}{6}$	6			

6. Write the equation of the exponential function that represents the data in each table in Problem 5.

a. **b.** **c.**

7. Without graphing, classify each of the following functions as increasing or decreasing and determine $f(0)$. (Use your graphing calculator to verify.)

a. $f(x) = 1.3(0.75)^x$ **b.** $f(x) = 0.6(1.03)^x$

c. $f(x) = 3\left(\frac{1}{5}\right)^x$

8. a. Given the following table, do you believe that it can be approximately modeled by an exponential function?

x	0	1	2	3	4	5	6
y	2	5	12.5	31.3	78.1	195.3	488.3

b. If you answered yes to part a, what is the constant ratio of successive y-values?

c. Determine an exponential equation that models this data.

9. a. Complete the following tables.

x	−3	−2	−1	0	1	2	3
$h(x) = 4^x$							

x	−3	−2	−1	0	1	2	3
$g(x) = \left(\frac{1}{4}\right)^x$							

b. Sketch graphs of h and g on the following grid.

c. Use the tables and graphs in part a and b to complete this table.

FUNCTION	BASE, b	GROWTH OR DECAY FACTOR	x-INTERCEPT	y-INTERCEPT	HORIZONTAL ASYMPTOTE	INCREASING OR DECREASING
$h(x) = 4^x$						
$g(x) = \left(\dfrac{1}{4}\right)^x$						

10. The starting salary at your new job is $22,000 per year. You are offered two options for salary increases:

    Plan 1: an annual increase of $1000 per year or
    Plan 2: an annual percentage increase of 4% of your salary.

Your salary is a function of the number of years of employment at your job.

a. Write an equation to determine the salary S after x years on the job using plan 1; using plan 2;

b. Complete the following table using the equations from part a:

x	0	1	3	5	10	15
S, Plan 1						
S, Plan 2						

c. Which plan would you choose? Explain.

11. You have just obtained a special line of credit at the local electronics store. You immediately purchase a stereo system for $415. Your credit limit is $500. Let's assume that you make no payments and purchase nothing more and there are no other fees. The monthly interest rate is 1.18%.

a. What is your initial credit balance?

b. What is the growth rate of your credit balance?

c. What is the growth factor of your credit balance?

d. Write an exponential function to determine how much you will owe (represented by y) after x months with no more purchases or payments.

e. Use your graphing calculator to graph this function. What is the y-intercept?

f. What is the practical meaning of this intercept in this situation?

g. How much will you owe after ten months? Use the table feature on your graphing calculator to determine the solution.

h. When you reach your credit limit of $500, the company will expect a payment. How long do you have before you will have to start paying the money back? Use the trace feature on your graphing calculator to approximate the solution.

12. You are working part-time for a computer company while going to high school. The following table shows the hourly wage, w, in dollars, that you earn as a function of time, t. Time is measured in years since the beginning of 2014 when you started working.

Time, t, Years, Since 2014	0	1	2	3	4	5
Hourly Wage, w, ($)	12.50	12.75	13.01	13.27	13.53	13.81

a. Calculate the ratios of the w-values to determine if the data in the table is exponential. Round each ratio to the nearest hundredth.

b. What is the growth factor?

c. Write an exponential equation that models the data in the table.

d. What percent raise did you receive each year?

e. For approximately how many years will you have to work for the company in order for your hourly wage to double? (Assume you will receive the same percentage increase each year.)

13. The number of licensed dairy farms in the United States has been declining since 1992. The following data shows the number of licensed dairy farms in the United States for various years since 1992.

Year	1992	1995	1998	2001	2004	2007	2010	2011
Number of Dairy Farms (thousands)	131.5	111.8	91.5	76.9	66.8	59.1	53.1	51.5

a. Make a scatterplot of this data.

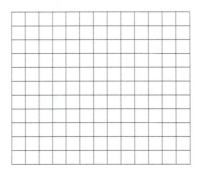

b. Does the scatterplot show that the data would be better modeled by a linear model or by an exponential model? Explain.

c. Let x represent the number of years since 1992. Let N represent the number of dairy farms in the United States in thousands. Use your graphing calculator to determine the exponential regression model that best fits the data.

d. Use the regression equation to predict the total number of dairy farms in the United States in 2020.

e. According to the exponential model, what is the decay factor for the total number of dairy farms in the United States?

f. What is the decay rate?

g. Use your graphing calculator to determine the halving time for your exponential model.

Collecting and Analyzing Data

14. The world's population has been increasing, but is it increasing steadily? Go to the United Nations' Population Database Web site http://esa.un.org/unpd/wpp/unpp/panel_population.htm to obtain the world population every five years from 1960 to 2010.

    a. Beginning with 1960, make a scatterplot of the world population. Let the independent variable represent the number of years since 1960 and the dependent variable represent the world population in billions. Describe any patterns you observe.

    b. Population growth is usually well described by an exponential fit. Determine a regression equation of an exponential function that models the world population.

    c. Sketch the graph of the regression equation on the scatterplot in part a.

    d. The United Nations' Population Database projects the world population. Make a second scatterplot of the world population for every five years from 2010 to 2050. Let the independent variable represent the number of years from 2010.

    e. Determine a regression equation of an exponential function that models the second set of world population data from 2010 to 2050.

    f. Use both models to estimate the population in 2014. Compare these estimates with each other and with the actual population in 2014.

    g. Compare the growth rates from the exponential models. Are the growth rates before and after 2010 similar or different? Explain.

    h. What does the difference in growth rates mean?

    i. Use both models from parts b and e to predict the 2100 population. Which model would be more trustworthy? Explain.

Activity 4.8

The Diameter of Spheres

Objectives

1. Define logarithm.

2. Write an exponential statement in logarithmic form.

3. Write a logarithmic statement in exponential form.

4. Determine log and ln values using a calculator.

Spheres are all around you (pardon the pun). You play sports with spheres such as baseballs, basketballs, and golf balls. You live on a sphere. Earth is a big ball in space, as are the other planets, the Sun, and the Moon. All spheres have properties in common. For example, the formula for the volume, V, of any sphere is $V = \frac{4}{3}\pi r^3$, and the formula for the surface area, S, of any sphere is $S = 4\pi r^2$, where r represents the radius of the sphere.

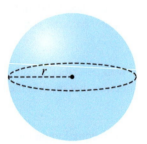

However, not all spheres are the same size. The following table gives the diameter, d, of some spheres you know. Recall that the diameter, d, of a sphere is twice the radius, r.

SPHERE	DIAMETER, d, IN METERS
Golf ball	0.043
Baseball	0.075
Basketball	0.239
Moon	3,476,000
Earth	12,756,000
Jupiter	142,984,000

If you want to determine either the volume or surface area of any of the spheres in the preceding table, the diameter of the given sphere would be the input value and would be referenced on the horizontal axis. But how would you scale this axis?

1. a. Plot the values in the first three rows of the table. Scale the axis starting at 0 and incrementing by 0.02 meter.

b. Can you plot the values in the last three rows of the table on the same axis? Explain.

2. a. Plot the values in the last three rows of the table on a different axis. Scale the axis starting at 0 and incrementing by 10,000,000 meters.

b. Can you plot the values in the first three rows of the table on the axis in part a? Explain.

Logarithmic Scale

3. There is a way to scale the axis so that you can plot all the values in the table on the same axis.

   a. Starting with the leftmost tick mark, give the first tick mark a value of 0.01 meter. Write 0.01 as a power of 10 as follows: $0.01 = \dfrac{1}{100} = \dfrac{1}{10^2} = 10^{-2}$ meters. Give the next tick mark a value of 0.1, written as 10^{-1} meters. Continue in this way by giving each consecutive tick mark a value that is one power of 10 greater than the preceding tick mark.

$$10^{-2} \quad 10^{-1}$$

   b. Complete the following table by writing all of the diameters from the preceding table in scientific notation.

SPHERE	DIAMETER, d, IN METERS	d, IN SCIENTIFIC NOTATION
Golf ball	0.043	
Baseball	0.075	
Basketball	0.239	
Moon	3,476,000	
Earth	12,756,000	
Jupiter	142,984,000	

   c. To plot the diameter of a golf ball, notice that 0.043 meter is between $10^{-2} = 0.01$ meter and $10^{-1} = 0.1$ meter. Now using the axis in part a, plot 0.043 meter between the tick mark labeled 10^{-2} and 10^{-1} meters, closer to the tick mark labeled 10^{-2} meters.

   d. To plot the diameter of Earth, notice that 12,756,000 meters is between $10^7 = 10,000,000$ meters and $10^8 = 100,000,000$ meters. Now plot 12,756,000 meters between the tick marks labeled 10^7 and 10^8 meters, closer to the tick mark labeled 10^7.

   e. Plot the remaining data in the same way by first determining between which two powers of 10 the number lies.

The scale you used to plot the diameter values is a *logarithmic*, or *log scale*. The tick marks on a logarithmic scale are usually labeled with just the exponent of the powers of 10.

4. **a.** Rewrite the axis from Problem 3a by labeling the tick marks with just the exponents of the powers of 10.

b. The axis looks like a standard axis with tick marks labeled -1, -2, 0, 1, and so on. However, it is quite different. Describe the difference between this log scale and a standard axis labeled in the same way. Focus on the values between consecutive tick marks.

Definition

The exponents used to label the tick marks of the preceding axis are **logarithms** or simply **logs**. Since these are exponents of powers of 10, the exponents are logs **base 10**, known as **common logarithms** or common logs.

Example 1

a. *The common logarithm of 10^3 is the exponent to which 10 must be raised to obtain a result of 10^3. Therefore, the common log of 10^3 is 3.*
b. *The common log of 10^{-2} is -2.*
c. *The common log of $100 = 10^2$ is 2.*

5. Determine the common log of each of the following.

a. 10^{-1} **b.** 10^4 **c.** 1000

d. 100,000   **e.** 0.0001

Logarithmic Notation

Remember that **a logarithm is an exponent.** The common log of x is an exponent, y, to which the base, 10, must be raised to get result x. That is, in the equation $10^y = x$, y is the logarithm. Using log notation, $\log_{10} x = y$. Therefore, $\log_{10} 10{,}000 = \log_{10} 10^4 = 4$.

Example 2

x, THE NUMBER	y, THE EXPONENT (LOGARITHM) TO WHICH THE BASE, 10, MUST BE RAISED TO GET x	LOG NOTATION $\log_{10} x = y$
10^3	3	$\log_{10} 10^3 = 3$
10^{-2}	-2	$\log_{10} 10^{-2} = -2$
100	2	$\log_{10} 100 = 2$

When using logs base 10, the notation $\log_{10}$ is shortened by dropping the 10. Therefore,

$$\log_{10} 10^3 = \log 10^3 = 3; \log_{10} 100 = \log 100 = 2.$$

6. Determine each of the following. Compare your result with those from Problem 5.

    a. $\log 10^{-1}$ **b.** $\log 10^4$ **c.** $\log 1000$

    d. $\log 100{,}000$ **e.** $\log 0.0001$

Bases for Logarithms

The logarithmic scale for the diameter of spheres situation was labeled with the exponents of powers of 10. Using 10 as the base for logarithms is common since the number 10 is the base of our number system. However, other numbers could be used as the base for logs. For example, you could use exponents of powers of 5 or exponents of powers of 2.

Example 3   *Base-5 logarithms: The log base 5 of a number, x, is the exponent to which the base, 5, must be raised to obtain x. For example,*

    a. $\log_5 5^4 = 4$, or in words, log base 5 of 5 to the fourth power equals 4.

    b. $\log_5 125 = \log_5 5^3 = 3$.

    c. $\log_5 \dfrac{1}{25} = \log_5 5^{-2} = -2$.

Base-2 logarithms: The log base 2 of a number, x, is the exponent to which 2 must be raised to obtain x. For example,

    a. $\log_2 2^5 = 5$, or log base 2 of 2 to the fifth power equals 5.

    b. $\log_2 16 = \log_2 2^4 = 4$.

    c. $\log_2 \dfrac{1}{8} = \log_2 2^{-3} = -3$.

In general, a statement in logarithmic form is $\log_b x = y$, where b is the base of the logarithm, x is a power of b, and y is the exponent. The base b for a logarithm can be any positive number except 1.

7. Determine each of the following.

    a. $\log_4 64$ **b.** $\log_2 \dfrac{1}{16}$ **c.** $\log_3 9$ **d.** $\log_3 \dfrac{1}{27}$

The examples and problems so far in this activity demonstrate the following property of logarithms.

Property of Logarithms

In general, $\log_b b^n = n$, where $b > 0$ and $b \neq 1$.

8. Determine each of the following:

    a. $\log 1$ **b.** $\log_5 1$ **c.** $\log_{\frac{1}{2}} 1$

    d. $\log 10$ **e.** $\log_5 5$ **f.** $\log_{1/2}\left(\dfrac{1}{2}\right)$

9. a. Referring to Problems 8a–c, write a general rule for $\log_b 1$.

    b. Referring to Problems 8d–f, write a general rule for $\log_b b$.

> **Property of Logarithms**
>
> In general, $\log_b 1 = 0$ and $\log_b b = 1$, where $b > 0, b \neq 1$.

Natural Logarithms

Because the base of a log can be any positive number except 1, the base can be the number e. The number e is a very important number in mathematics. This number is irrational, so its decimal representation never ends and never repeats.

10. Locate the number e on your calculator, and write its decimal approximation below.

> Many applications involve the use of log base e. Log base e is called the **natural** log and has the following special notation.
>
> $$\log_e x \text{ is written as } \ln x, \text{ read simply as el-n-x.}$$

Example 4   **a.** $\ln e^2 = \log_e e^2 = 2$ **b.** $\ln \dfrac{1}{e^4} = \ln e^{-4} = -4$

11. Evaluate the following.

    a. $\ln e^7$ **b.** $\ln\left(\dfrac{1}{e^3}\right)$ **c.** $\ln 1$

    d. $\ln e$ **e.** $\ln \sqrt{e}$

Logarithmic and Exponential Forms

Because logarithms are exponents, logarithmic statements can be written as exponential statements and exponential statements can be written as logarithmic statements.

For example, in the statement $3 = \log_5 125$, the base is 5, the exponent (logarithm) is 3, and the result is 125. This relationship can also be written as the equation $5^3 = 125$.

> In general, the logarithmic equation $y = \log_b x$ is equivalent to the exponential equation $b^y = x$.

Example 5   *Rewrite the exponential equation* $e^{0.5} = x$ *as an equivalent logarithmic equation.*

SOLUTION

In the equation $e^{0.5} = x$, the base is e, the result is x, and the exponent (logarithm) is 0.5. Therefore, the equivalent logarithmic equation is $0.5 = \log_e x$, or $0.5 = \ln x$.

12. Rewrite each exponential equation as a logarithmic equation and each log equation as an exponential equation.

a. $3 = \log_2 8$

b. $\ln e^3 = 3$

c. $\log_2 \dfrac{1}{16} = -4$

d. $6^3 = 216$

e. $e^1 = e$

f. $3^{-2} = \dfrac{1}{9}$

Logarithms and the Calculator

The numbers whose logarithms you have been working with have been exact powers of the base. However, in many situations, you must evaluate a logarithm where the number is not an exact power of the base. For example, what is log 20 or ln 15? Fortunately, the common log (base 10) and the natural log (base e) are functions on your calculator.

13. Use your calculator to evaluate the following.

a. $\log 20$

b. $\ln 15$

c. $\ln \dfrac{1}{2}$

d. $\log 0.02$

e. Use your calculator to check your answers to Problems 6 and 11.

14. a. Use your calculator to complete the following table.

SPHERE	DIAMETER, d, IN METERS	d, IN SCIENTIFIC NOTATION	log(d)
Golf ball	0.043	4.3×10^{-2}	
Baseball	0.075	7.5×10^{-2}	
Basketball	0.239	2.39×10^{-1}	
Moon	3,476,000	3.476×10^{6}	
Earth	12,756,000	1.2756×10^{7}	
Jupiter	142,984,000	1.4298×10^{8}	

b. Plot the values from the log column in the preceding table on the following axis.

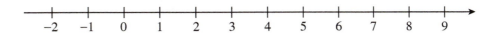

      c. Compare the preceding plot with the plot on the log-scaled axis in Problem 3a and comment.

SUMMARY: ACTIVITY 4.8

1. The notation for logarithms is $\log_b x = y$, where b is the base of the log, x is the resulting power of b, and y is the exponent. The base, b, can be any positive number except 1; x can be any positive number. The range of y values includes all real numbers.

2. The notation for the **common logarithm**, or base-10 logarithm, is $\log_{10} x = \log x$.

3. The notation for the **natural logarithm**, or base e logarithm, is $\log_e x = \ln x$.

4. The **logarithmic equation** $y = \log_b x$ is equivalent to the **exponential equation** $b^y = x$.

5. If $b > 0$ and $b \neq 1$,

    a. $\log_b 1 = 0$

    b. $\log_b b = 1$

    c. $\log_b b^n = n$

EXERCISES: ACTIVITY 4.8

1. Use the definition of logarithm to determine the exact value of each of the following.

    a. $\log_2 32$ **b.** $\log_3 27$ **c.** $\log 0.1$

    d. $\log_2\left(\dfrac{1}{64}\right)$ **e.** $\log_5 1$ **f.** $\log_{1/2}\left(\dfrac{1}{4}\right)$

    g. $\log_7 \sqrt{7}$
       (Hint: $\sqrt{7} = 7^{1/2}$)   **h.** $\log_{100} 10$ **i.** $\log 1$

    j. $\log_2 1$ **k.** $\ln e^5$ **l.** $\ln\left(\dfrac{1}{e^2}\right)$ **m.** $\ln 1$

2. Evaluate each common logarithm without the use of a calculator.

    a. $\log\left(\dfrac{1}{1000}\right) =$ _____   **b.** $\log\left(\dfrac{1}{100}\right) =$ _____

    c. $\log\left(\dfrac{1}{10}\right) =$ _____   **d.** $\log 1 =$ _____

    e. $\log 10 =$ _____   **f.** $\log 100 =$ _____

    g. $\log 1000 =$ _____   **h.** $\log \sqrt{10} =$ _____

3. Rewrite the following equations in logarithmic form.

   a. $3^2 = 9$

   b. $\sqrt{121} = 11$ (Hint: First rewrite $\sqrt{121}$ in exponential form.)

   c. $4^t = 27$

   d. $b^3 = 19$

4. Rewrite the following equations in exponential form.

   a. $\log_3 81 = 4$

   b. $\dfrac{1}{2} = \log_{100} 10$

   c. $\log_9 N = 12$

   d. $y = \log_7 x$

   e. $\ln \sqrt{e} = \dfrac{1}{2}$

   f. $\ln \left(\dfrac{1}{e^2} \right) = -2$

5. Estimate between what two integers the solutions for the following equations fall. Then solve each equation exactly by changing it to log form. Use your calculator to approximate your answer to three decimal places.

   a. $10^x = 3.25$

   b. $10^x = 590$

   c. $10^x = 0.0000045$

**Walking Speed
of Pedestrians**

Objectives

1. Determine the inverse of
   the exponential function.

2. Identify the properties of
   the graph of a logarithmic
   function.

3. Graph the natural
   logarithmic function.

On a recent visit to Boston, you notice that people seem rushed as they move about the city. Upon returning to college, you mention this observation to your psychology instructor. The instructor refers you to a psychology study that investigates the relationship between the average walking speed of pedestrians and the population of the city. The study cites statistics presented graphically as follows.

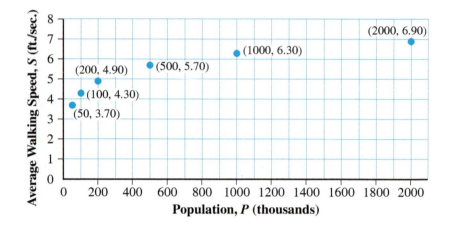

I. a. Does the data appear to be linear? Explain.

b. Does the data appear to be exponential? Explain.

This data is actually logarithmic. Situations that can be modeled by logarithmic functions will be the focus of this and the following activity.

Introduction to the Logarithmic Function

The logarithmic function base b is symbolized by $y = \log_b x$, where

i. b represents the base of the logarithmic function $(b > 0, b \neq 1)$.

ii. x is the input and represents a power of the base b (x is also called the argument) and y is the output and is the exponent needed on the base b to obtain x.

2. a. Evaluate $\log_{10}(-100)$ using your calculator. What do you observe? Does it seem reasonable? Explain.

b. Is it possible to determine $\log(0)$? Explain.

c. What is the domain for the function defined by $y = \log x$?

d. What is the range? Remember, the output y is an exponent.

3. The exponential function defined by $f(x) = 10^x$ has a special relationship with the corresponding logarithmic function defined by $g(x) = \log_{10} x = \log x$.

   a. Complete the following tables for $f(x) = 10^x$ and $g(x) = \log x$.

x	$f(x) = 10^x$
-2	
-1	
0	
1	
2	

x	$g(x) = \log x$
0.01	
0.1	
1	
10	
100	

   b. Compare the input and output values for functions f and g.

   c. Sketch the graphs of $Y1 = 10^x$ and $Y2 = \log_{10} x$ using your graphing calculator. Use the window $Xmin = -4$, $Xmax = 4$, $Ymin = -3$, and $Ymax = 3$. Your screen should appear as follows:

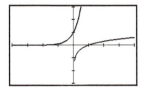

   d. Graph $y = x$ on the same coordinate axes as functions f and g. Describe in a sentence or two the symmetry you observe in the graphs of f and g.

Function f and g in Problem 3 are inverse functions. The inverse function interchanges the domain and range of the original function. Also, the graph of an inverse function is the reflection of the original function about the line $y = x$. Therefore, the results in Problem 3 demonstrate that $f(x) = 10^x$ and $g(x) = \log x$ are inverse functions.

You can determine the equation of the inverse function by solving the defining equation for the input (x-value) and then interchanging the input (x-value) and the output (y-value).

Example 1   *Determine the equation of the inverse of the function defined by $y = 5^x$.*

SOLUTION

Step 1. Solve the equation for x by writing the statement in logarithmic notation. $x = \log_5 y$

Step 2. Interchange the x and y variables. $y = \log_5 x$

   4. Use the algebraic approach demonstrated in Example 1 to verify that $y = \log x$ is the inverse of $y = 10^x$.

Problems 2, 3, and 4 illustrate the following properties of the common logarithmic function.

> **Properties of the Common Logarithmic Function Defined by $f(x) = \log x$**
>
> **1.** The domain of f is the set of all positive real numbers $(x > 0)$.
>
> **2.** The range of f is all real numbers.
>
> **3.** f is the inverse of the function defined by $g(x) = 10^x$.

The Graph of the Natural Logarithmic Function

5. a. Using your calculator, complete the following table. Round your answers to 3 decimal places.

x	0.1	0.5	1	5	10	20	50
$y = \ln x$							

b. Sketch a graph of $y = \ln x$.

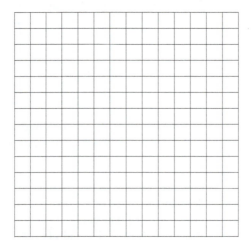

c. Verify your graph in part b using your graphing calculator. Using the window Xmin $= -1$, Xmax $= 4$, Ymin $= -2.5$, and Ymax $= 2.5$, your screen should appear as follows:

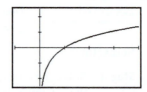

d. What are the domain and range of the function defined by $y = \ln x$?

e. Determine the intercepts of the graph.

f. Does the graph of $y = \ln x$ have a horizontal asymptote? Explain.

g. Complete the following table using your calculator. Round your answers to the nearest tenth.

x	1	0.5	0.25	0.1	0.01	0.001
$y = \ln x$						

h. As the input values take on values closer and closer to zero, what happens to the corresponding output values?

The y-axis (the line $x = 0$) is a vertical asymptote of the graph of $y = \ln x$.

Definition

A **vertical asymptote** is a vertical line, $x = a$, that the graph of a function becomes very close to but never touches. As the input values get closer to $x = a$, the output values get larger in magnitude. That is, the output values become very large positive or very large negative values.

Example 2   *The vertical asymptote of the graphs of $y = \log x$ and $y = \ln x$ is the vertical line $x = 0$ (the y-axis).*

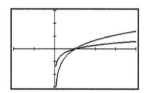

6. a. Graph $y = e^x$, $y = \ln x$, and $y = x$ on the same set of coordinate axes using the window Xmin $= -7.5$, Xmax $= 7.5$, Ymin $= -5$, and Ymax $= 5$. Describe the symmetry that you observe. Your graph should appear as follows.

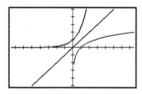

b. Use an algebraic approach to determine the inverse of the exponential function defined by $y = e^x$.

7. A horizontal shift of the graph of $y = \ln x$ changes the vertical asymptote.

a. What is the equation of the vertical asymptote of the graph of $y = \ln (x + 3)$?

b. What is the equation of the vertical asymptote of the graph of $y = \ln (x - 2)$?

SUMMARY: ACTIVITY 4.9

1. Properties of the logarithmic function defined by $y = \log_b x$, where $b > 1$.

  a. The domain of f is $x > 0$.

  b. The range of f is all real numbers.

  c. f is the inverse of the function defined by $g(x) = b^x$.

2. The graph of a logarithmic function defined by $y = \log_b x$, where $b > 1$,

  a. is increasing for all $x > 0$

  b. has an x-intercept of $(1, 0)$

  c. has no y-intercept

  d. has a vertical asymptote of $x = 0$, the y-axis

  e. resembles the following graph:

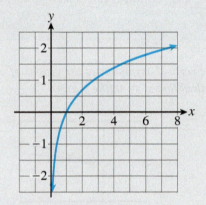

3. The **common logarithmic function** is defined by

$$y = \log x = \log_{10} x.$$

4. The **natural logarithmic function** is defined by

$$y = \ln x = \log_e x.$$

5. Because the base of the natural log function is $e > 1$, the graph of the natural logarithmic function $y = \ln x$

  a. is increasing for all $x > 0$

  b. has an x-intercept of $(1, 0)$

  c. has a vertical asymptote of $x = 0$, the y-axis

6. The logarithmic function defined by $y = \log_b x$, where $b > 1$, is continuous over its domain, $x > 0$.

7. You can determine the equation of the inverse of the function by solving the equation of the function for x and then interchanging the input (x-values) and the output (y-values) in the new equation.

EXERCISES: ACTIVITY 4.9

1. Using the graph of $y = \log x$ as a check, summarize the following properties of the common logarithmic function.

a. What is the domain?

b. What is the range?

c. For what values of x is $\log x$ positive?

d. For what values of x is $\log x$ negative?

e. For what values of x does $\log x = 0$?

f. For what values of x does $\log x = 1$?

2. a. Complete the following table using your calculator. Round your answers to the nearest tenth.

x	0.001	0.01	0.1	0.25	0.5	1
y = log x						

b. As the positive input values take on values closer to 0, what happens to the corresponding output values?

c. Determine the vertical asymptote of the graph of $y = \log x$.

3. The exponential function defined by $y = 2^x$ has an inverse. Determine the equation of the inverse function. Write your answer in logarithmic form.

4. Using the graph of $y = \ln x$ as a check, summarize the following properties of the natural logarithmic function.

a. What is the domain?

b. What is the range?

c. For what values of x is $\ln x$ positive?

d. For what values of x is $\ln x$ negative?

e. For what values of x does $\ln x = 0$?

f. For what values of x does $\ln x = 1$?

5. The life expectancy of a piece of equipment is the number, n, of years for the equipment to depreciate to a known salvage value, V. The life expectancy, n, is given by the formula

$$n = \frac{\log V - \log C}{\log (1 - r)},$$

where C is the initial cost of the piece of equipment and r is the annual rate of depreciation expressed as a decimal. If a backhoe costs \$45,000 and has a salvage value of \$2500, what is the life expectancy if the annual rate of depreciation is 40%?

6. The Pew Research Center's Internet & American Life Project Surveys gathered information that included the percent of Americans 18 years and older, who use the Internet. Some of that information is summarized in the following table.

Years since 1994, t	1	3	5	7	9	11	13	15	17
Percent of U.S. Adults Using the Internet, P	14	30	40	61	63	70	72	79	78

a. Use your graphing calculator to produce a scatterplot of the data. Do you believe that the data can be best modeled by a linear, exponential, or logarithmic model?

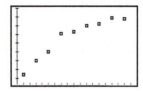

This data can be modeled by $P = 8.59 + 24.76 \ln (t)$

b. Use the model to predict the percent of Americans 18 years and older who will be using the Internet in 2020.

c. Use your model from part b to predict the percent of Americans 18 years old and older who will be using the Internet in 2040. Do you have confidence in this result? Explain.

Activity 4.10

Walking Speed of Pedestrians, continued

Objectives

1. Compare the average rate of change of increasing logarithmic, linear, and exponential functions.

2. Determine the regression equation of a natural logarithmic function having equation $y = a + b \ln x$ that best fits a set of data.

In Activity 4.9 you looked at a psychology study that investigated the relationship between the average walking speed of pedestrians and the population of the city. Graphically the data was presented as follows.

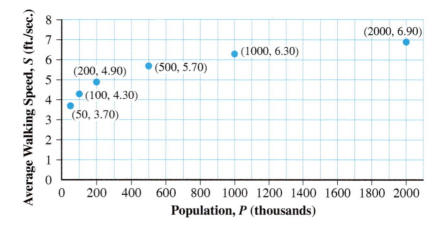

1. a. Does the data appear to be logarithmic? Explain.

b. Use the data in the graph to complete the following table.

Population, P (thousands)	50	100	200	500	1000	2000
Average Walking Speed, S						

The natural logarithmic function can be used to model a variety of scientific and natural phenomena. The natural logarithmic function is so prevalent that on most graphing calculators, it has its own built-in regression finder.

2. a. Use your graphing calculator and the table in Problem 1b to produce a scatterplot of the average walking speed data.

b. Use the regression feature of your calculator to produce a natural logarithmic curve that approximates the data in the table. Use option 9 from the STAT CALC menu.

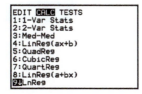

The LnReg option will generate a regression equation of the form $y = a + b \ln x$. Round a and b to the nearest thousandth, and record the function below.

c. Enter the function from part b into your graphing calculator. Verify visually that this function is a good model for your data.

d. What is the practical domain of this function?

e. Use the function from part b to predict the average walking speed in Boston, population 589,121. [**Note:** P is in thousands (589.121 thousands).]

f. Use the model to predict the average walking speed in New York City, population 8,008,278.

3. a. If the average walking speed in a certain city is 5.2 feet per second, write an equation that can be used to estimate the population P of the city.

b. Solve the equation using a graphical approach.

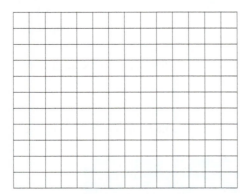

Comparing the Average Rate of Change of Logarithmic, Linear, and Exponential Functions

4. a. Complete the following table using the function defined by $S = 0.303 + 0.868 \ln P$.

P, Population (thousands)	10	20	150	250
S, Average Walking Speed (ft./sec.)				

b. Determine the average rate of change of S as the population increases from

i. 10 to 20 thousand

ii. 20 to 150 thousand

iii. 150 to 250 thousand

c. What can you say in general about the average rate of change in the walking speed as the population increases?

You should have discovered that the average rate of change in this situation is always positive. This means that the walking speed increases as the population increases. Nevertheless, in general, the increase gets smaller as the population increases. This is characteristic of logarithmic functions.

As the input of a logarithmic function with $b > 1$ increases, the output increases at a slower rate (the graph becomes less steep).

5. Complete the following statements by describing the rate at which the output values change.

   a. For an increasing linear function, as the input variable increases, the output

   _____

   b. For an increasing exponential function, as the input increases, the output

   _____

   c. For an increasing logarithmic function, as the input increases, the output

   _____

6. Consider the graphs of

   i. $f(x) = e^x$ **ii.** $h(x) = x$ **iii.** $g(x) = \ln x$

   using the window Xmin $= -7.5$, Xmax $= 7.5$, Ymin $= -5$, and Ymax $= 5$.

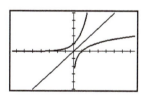

   a. Which of the functions are increasing?

   b. Which of the functions are decreasing?

   c. As the input values get larger, which of the functions grows fastest?

   d. As the input values get larger, which of the functions grows most slowly?

   e. Do any of these functions have a horizontal asymptote?

   f. Do any of these functions have a vertical asymptote?

    g. Compare the domains of these functions.

    h. Compare the ranges of these functions.

Problem 6 illustrates some of the relationships between $f(x) = b^x$, where $b > 1$; $g(x) = \log_b x$, where $b > 1$; and $y = mx + b$, where $m > 0$.

Application

  7. You are working on the development of an "elastic" ball for the IBF Toy Company. The question you are investigating is, If the ball is launched straight up, how far has it traveled vertically when it hits the ground for the tenth time?

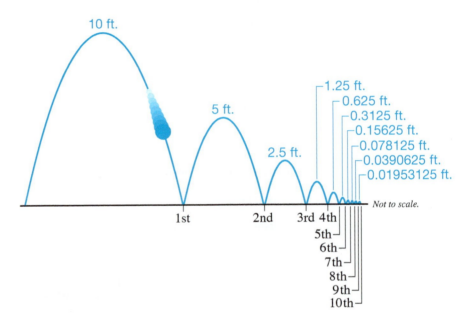

Your launcher will project the ball 10 feet into the air. This means that it will travel 20 feet (10 feet up and 10 feet down) before it hits the ground the first time. Assuming that the ball returns to 50% of its previous height, it will rebound 5 feet and travel 10 feet before it hits the ground again. The following table summarizes this situation.

N, Times the Ball Hits the Ground	1	2	3	4	5	6
Distance Traveled since Last Time (ft.)	20	10	5	2.5	1.25	0.625
T, Total Distance Traveled (ft.)	20	30	35	37.5	38.75	39.375

    a. Using the window Xmin = 0, Xmax = 7, Ymin = 0, and Ymax = 45, a plot of N versus T should resemble the following.

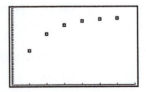

b. Do the table and scatterplot indicate that the data is linear, exponential, or logarithmic?

c. Use your graphing calculator to produce linear, exponential, and natural log regression equations for the given data.

d. Graph each equation, and visually determine which of the regression models best fits the data.

e. Use the equation of best fit to predict the total distance the ball traveled when it hits the ground for the tenth time.

SUMMARY: ACTIVITY 4.10

1. As the input of a logarithmic function increases, the output increases at a slower rate (the graph becomes less steep).

2. The relationships among the graphs of $f(x) = b^x$, where $b > 1$, $g(x) = \log_b x$, where $b > 1$, and $y = mx + b$, where $m > 0$, are identified in the following table.

FUNCTION	INCREASING OR DECREASING	GROWTH RATE	HORIZONTAL OR VERTICAL ASYMPTOTE	DOMAIN	RANGE
$f(x) = b^x$, $b > 1$	increasing	fastest	horizontal asymptote	all real numbers	$y > 0$
$g(x) = \log_b x$, $b > 1$	increasing	slowest	vertical asymptote	$x > 0$	all real numbers
$y = mx + b$, $m > 0$	increasing	constant	none	all real numbers	all real numbers

EXERCISES: ACTIVITY 4.10

1. The percent of American adults (age 18 and older) who own a cell phone continues to rise. The following table gives data on the percent of American adults who own a cell phone from 2004 to 2013.

Years, t, since 2000	4	6	8	9	10	11	13
Percent of American Adults Who Own a Cell Phone, P	65	73	82	83	85	89	91

a. Plot the points on the following grid

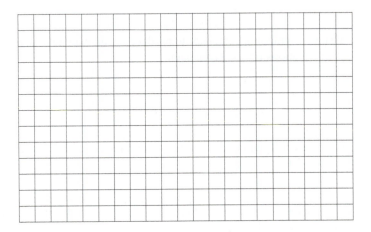

b. Does the scatterplot indicate that the data is logarithmic? Explain.

c. Determine the natural log regression equation. Record the regression equation below, and add a sketch of the regression curve to the scatterplot in part a.

d. Does this appear to be a good fit? Explain.

e. Use your model to predict the percent of American adults who own a cell phone in 2017.

2. a. Consider a data set for the variables x and y.

x	1	4	7	10	13
$f(x)$	3.0	4.5	5.0	5.2	5.8

Plot these points on the following grid.

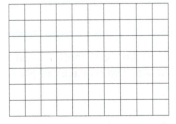

b. Does the scatterplot indicate that the data is more likely linear, exponential, or logarithmic? Explain.

c. Use your graphing calculator to determine a logarithmic regression model that represents this data.

d. Use your model to determine $f(11)$ and $f(20)$.

3. The barometric pressure, P, in inches of mercury at a distance x miles from the eye of a moderate hurricane can be modeled by

$$P = f(x) = 0.48 \ln (x + 1) + 27.$$

a. Determine $f(0)$. What is the practical meaning of the value in this situation?

b. Sketch a graph of this function.

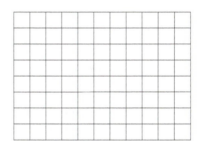

c. Describe how air pressure changes as one moves away from the eye of the hurricane.

4. The following data were collected during an experiment in science class. The table shows the yield y of a substance (in milligrams) after x minutes of a chemical reaction.

Minutes, x	1	2	3	4	5	6	7	8
Yield, y (milligrams)	1.5	7.4	10.2	13.4	15.8	16.3	18.2	18.3

a. Produce a scatterplot of the chemical reaction data.

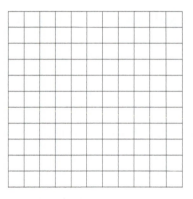

b. Use the regression feature of a calculator to obtain a natural logarithmic model and a linear model that approximates the data in the table. Graph each equation on the scatterplot in part a.

c. Determine which model best fits the data.

5. The formula $R = 80.4 - 11 \ln x$ is used to approximate the minimum required ventilation rate, R, as a function of the air space per child in a public school classroom. The rate R is measured in cubic feet per minute, and x is measured in cubic feet.

a. Sketch a graph of the rate function for $100 \le x \le 1500$.

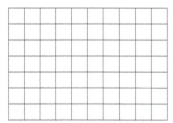

b. Determine the required ventilation rate if the air space per child is 300 cubic feet.

6. You have recently accepted a job working in the coroner's office of a large city. Because of the large numbers of homicides, it has been difficult for the coroners to complete all of their work. In part, your job is to assist them in the paperwork. On one particular day, you are working on a case in which you are attempting to establish the time of death.

The coroner tells you that to establish the time of death, he uses the formula

$$t = 4 \ln\left(\frac{98.6 - T_s}{T_b - T_s}\right),$$

where t is the number of hours the victim has been dead,

   T_b represents the temperature of the body when discovered, and

   T_s represents the temperature of his surroundings.

The coroner also tells you that the thermostat was set at 68°F in the apartment in which the body was found and that the victim's body temperature was 78°F.

a. Using the preceding formula, determine the number of hours the victim has been deceased. Use your calculator to approximate your answer to 1 decimal place.

b. If the body was discovered at 10:07 P.M., what do you estimate for the time of death?

7. The following formula can be used to determine the time, t, it takes for an investment to double or triple, and so forth in value.

$$t = \frac{\ln m}{n \ln\left(1 + \dfrac{r}{n}\right)},$$

where m represents the number of times the investment is to grow in value ($m = 2$ is double, $m = 3$ is triple, etc.),

   r is the annual interest rate expressed as a decimal,

   n is the number of corresponding compounding periods per year.

a. How many years will it take an investment to double if you are receiving an annual rate of 5.5% compounded quarterly ($n = 4$)?

b. How many years will it take the investment in part a to triple in value?

c. Suppose the interest on the investment in part a was compounded monthly ($n = 12$). How long will it take the value to double?

Collecting and Analyzing Data

8. The cost of a gallon of gas has fluctuated dramatically over the past few years. In the spring of 2008, it was speculated that the price of a gallon of regular unleaded gasoline would soon exceed $4 per gallon.

Go to the Internet to obtain the average price for 1 gallon of regular unleaded gasoline in the United States in each year from 2000 through 2014.

a. Beginning with 2000, make a scatterplot of the data collected. Let the independent variable represent the average price (in cents) of 1 gallon of regular unleaded gasoline and the dependent variable represent the number of years since 2000. Describe any patterns you observe.

b. Determine a logarithmic regression equation for the data.

c. How well does the logarithmic model in part b fit the data? Explain.

d. Predict the year in which the average price of 1 gallon of regular unleaded gasoline will reach $4 per gallon.

e. How reliable is this prediction? Explain.

Activity 4.11

The Elastic Ball

Objectives

1. Apply the log of a product property.

2. Apply the log of a quotient property.

3. Apply the log of a power property.

4. Apply the change of base formula.

You are continuing your work on the development of the elastic ball. You are still investigating the question, if the ball is launched straight up, how far has it traveled vertically when it hits the ground for the tenth time? However, your supervisor tells you that you cannot count the initial launch distance. You must calculate only the rebound distance.

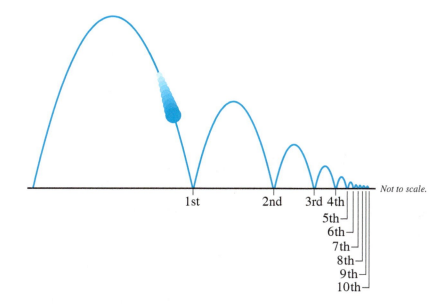

Using some physical properties, timers, and your calculator, you collect the following data.

N, Number of Times the Ball Hits the Ground	1	2	3	4	5	6
T, Total Rebound Distance (ft.)	0	9.0	13.5	16.3	18.7	21.0

1. Does the data seem reasonable? Explain.

2. Use your graphing calculator to construct a scatterplot of the data with N as the input and T as the output. Using a window of Xmin = 0, Xmax = 7, Ymin = 0, and Ymax = 25, your graph should resemble the following.

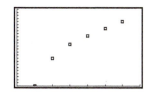

3. Do you believe that the data can be modeled by a logarithmic function? Explain.

4. This data can be modeled by $T = 26.75 \log N$. Use your graphing calculator to verify visually that this is a reasonable model for the given data.

5. a. Using the log model, complete the following table. Round values to the nearest hundredth.

N	2	5	10
T = 26.75 log N			

b. How are the T-values for $N = 2$ and $N = 5$ related to the T-value for $N = 10$?

c. Using the results from part b, how could you determine the total rebound distance after 10 bounces?

The results from Problem 5 can be written as follows.

$$\underbrace{26.75} = \underbrace{8.05} + \underbrace{18.70}$$

$$26.75 \log 10 = 26.75 \log 2 + 26.75 \log 5$$
$$26.75 \log (2 \cdot 5) = 26.75 \log 2 + 26.75 \log 5$$

Dividing both sides by 26.75, you have

$$\log (2 \cdot 5) = \log 2 + \log 5.$$

This result illustrates an important property of logarithms.

Property of the Logarithm of a Product

If $A > 0, B > 0$, then $\log_b (A \cdot B) = \log_b A + \log_b B$, where $b > 0, b \neq 1$. Expressed verbally, this property states that the logarithm of a product is the sum of the individual logarithms.

Example 1

a. $\log_2 32 = \log_2 (4 \cdot 8) = \log_2 4 + \log_2 8 = 2 + 3 = 5$

b. $\log (5st) = \log 5 + \log s + \log t$

c. $\ln (xy) = \ln x + \ln y$

6. Use the property of the logarithm of a product to write the following as the sum of two or more logarithms.

a. $\log_b (7 \cdot 13)$ **b.** $\log_3 (xyz)$

c. $\log 15$ **d.** $\ln (3xy)$

7. Write the following as the logarithm of a single expression.

a. $\ln a + \ln b + \ln c$ **b.** $\log_4 3 + \log_4 9$

Logarithm of a Quotient

Consider the following table from Problem 5.

N	2	5	10
$T = 26.75 \log N$	8.05	18.70	26.75

This table also indicates that the rebound distance after the ball has hit the floor twice (8.05 feet) is the total rebound distance when the ball has hit the ground 10 times (26.75 feet) minus the total rebound distance when the ball has hit the ground 5 times (18.70 feet).

This can be written as

$$\underbrace{8.05}\quad=\quad\underbrace{26.75}\quad-\quad\underbrace{18.70}$$

$$26.75 \log 2 = 26.75 \log 10 - 26.75 \log 5$$

$$\log 2 = \log 10 - \log 5.$$

Substituting $\log\left(\dfrac{10}{5}\right)$ for $\log 2$, you have

$$\log\left(\frac{10}{5}\right) = \log 10 - \log 5.$$

This suggests another important property of logarithms. The property is demonstrated further in Problem 8.

8. a. Complete the following table. Round your answers to the nearest thousandth.

x	$Y1 = \log\left(\dfrac{x}{4}\right)$	$Y2 = \log x - \log 4$
1		
5		
10		
23		

b. Is the expression $\log\left(\dfrac{x}{4}\right)$ equivalent to $\log x - \log 4$? Explain.

c. Sketch the graph of $y = \log\left(\dfrac{x}{4}\right)$ and $y = \log x - \log 4$ using your graphing calculator. What do the graphs suggest about the relationship between $\log\left(\dfrac{x}{4}\right)$ and $\log x - \log 4$?

> **Property of the Logarithm of a Quotient**
>
> If $A > 0, B > 0$, then $\log_b\left(\dfrac{A}{B}\right) = \log_b A - \log_b B$, where $b > 0, b \neq 1$. Expressed verbally, this property states that the logarithm of a quotient is the difference of the logarithm of the numerator and the logarithm of the denominator.

> **Example 2**
>
> **a.** $\log_3\left(\dfrac{81}{27}\right) = \log_3 81 - \log_3 27 = 4 - 3 = 1$
>
> Note that $\log_3\left(\dfrac{81}{27}\right) = \log_3 3 = 1$.
>
> **b.** $\log\left(\dfrac{2x}{y}\right) = \log 2x - \log y = \log 2 + \log x - \log y$
>
> **c.** $\ln\left(\dfrac{x^2}{5}\right) = \ln x^2 - \ln 5$

9. Use the properties of logarithms to write the following as the sum or difference of logarithms.

     a. $\log_6 \dfrac{17}{3}$ **b.** $\ln \dfrac{x}{23}$

     c. $\log_3 \dfrac{2x}{y}$ **d.** $\log \dfrac{3}{2z}$

10. Write the following expressions as the logarithm of a single expression.

     a. $\log x - \log 4 + \log z$ **b.** $\log x - (\log 4 + \log z)$

11. **a.** Use your graphing calculator to sketch the graphs of $y = \log x + \log 4$ and $y = \log (x + 4)$.

     b. How do these graphs compare?

     c. What do the graphs suggest about the relationship between $\log (A + B)$ and $\log A + \log B$?

Logarithm of a Power

Before calculators, logarithms were used to help in computing products and quotients of numbers. More important, logarithms were used to compute powers such as 734.21^3 and $\sqrt{0.0761} = (0.0761)^{1/2}$. In such a case, the first step was to take the logarithm of the power and rewrite the resulting expression. To determine how to rewrite $\log 734.21^3$, you can investigate the expression $\log x^3$.

12. **a.** Complete the following table. Round to the nearest thousandth.

x	Y1 = $\log x^3$	Y2 = $3 \log x$
2		
7		
15		

b. Sketch the graphs of $y = \log x^3$ and $y = 3 \log x$ using your graphing calculator.

c. What do the results of part a and part b demonstrate about the relationship between $\log x^3$ and $3 \log x$?

The results in Problem 12 illustrate another property of logarithms.

> **Property of the Logarithm of a Power**
>
> If $A > 0$ and p is any real number, then $\log_b A^p = p \cdot \log_b A$, where $b > 0, b \neq 1$. In words, the property states that the logarithm of a power is equivalent to the exponent times the logarithm of the base.

Example 3
  a. $\log_3 9^2 = 2 \log_3 9 = 2 \cdot 2 = 4$
  b. $\log_5 x^4 = 4 \log_5 x$

  c. $\ln (xy)^7 = 7 \ln (xy)$
  d. $\ln x^{1/4} = \dfrac{1}{4} \ln x$

  e. $\log \sqrt{63} = \log 63^{1/2} = \dfrac{1}{2} \log 63$

13. Use the properties of logarithms to write the given logarithms as the sum or difference of two or more logarithms or as the product of a real number and a logarithm. All variables represent positive numbers.

  a. $\log_3 x^{1/2}$
  b. $\log_5 x^3$

  c. $\ln t^2$
  d. $\log \sqrt[3]{50}$ (Hint: $\sqrt[3]{50} = 50^{1/3}$)

  e. $\log_5 \dfrac{x^2 y^3}{z}$
  f. $\log_3 \dfrac{3x^2}{y^3}$

14. Write each of the following as the logarithm of a single expression with coefficient 1.

  a. $2 \log_3 5 + 3 \log_3 2$
  b. $\dfrac{1}{2} \log x^4 - \dfrac{1}{2} \log y^5$

  c. $3 \log_b 10 - 4 \log_b 5 + 2 \log_b 3$
  d. $3 \ln 4 - (4 \ln 5 + 2 \ln 3)$

Using the properties of logarithms to solve exponential equations algebraically will be investigated in the next activity.

Change of Base Formula

Because most graphing calculators have only the log base 10 (log) and the log base e (ln) keys, you cannot graph a logarithmic function such as $y = \log_2 x$ directly. Consider the following argument to rewrite the expression $\log_2 x$ as an equivalent expression using log base 10.

By definition of logs, $y = \log_2 x$ is the same as $x = 2^y$. Taking the log base 10 of both sides of the second equation, $x = 2^y$, you have

$$\log x = \log 2^y.$$

Using the property of the log of a power, $\log x = y \log 2$. Solving for y, you have

$$y = \frac{\log x}{\log 2}.$$

Therefore, the equation $y = \log_2 x$ is equivalent to $y = \dfrac{\log x}{\log 2}$.

15. To graph $y = \log_2 x$, enter $\log(X)/\log(2)$ for Y1 in your calculator. Your graph should resemble the following.

 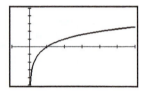

16. a. Write $y = \log_6 x$ as an equivalent equation using base 10.

b. Use the result from part a to graph $y = \log_6 x$.

c. What is the domain of the function?

d. What is the x-intercept of the graph?

> The formula you used in Problems 15 and 16 for graphing log functions of different bases is a special case of the formula
>
> $$\log_b x = \frac{\log_a x}{\log_a b}, \text{ where } a > 0, a \neq 1, b > 0, \text{ and } b \neq 1.$$
>
> This is often called the **change of base formula**.

The change of base formula is used to change from base b to base a. Because most calculators have log base 10 (log) and log base e (ln) keys, you usually convert to one of those bases. For those bases,

$$\log_b x = \frac{\log x}{\log b} \quad \text{or} \quad \log_b x = \frac{\ln x}{\ln b}.$$

Example 4   *Change the equation $y = \log_5 x$ to an equivalent equation in base 10 and/or base e.*

$$y = \log_5 x = \frac{\log x}{\log 5} \quad \text{or} \quad y = \log_5 x = \frac{\ln x}{\ln 5}$$

17. Use each of the change of base formulas to determine $\log_4 1024$.

    a. Using base 10:   **b.** Using base e:

    c. How do the results in parts a and b compare?

SUMMARY: ACTIVITY 4.11

Properties of the Logarithmic Function

If $A > 0, B > 0, b > 0$, and $b \neq 1$, then

1. $\log_b (A \cdot B) = \log_b A + \log_b B$

2. $\log_b \left(\dfrac{A}{B} \right) = \log_b A - \log_b B$

3. $\log_b (x + y) \neq \log_b x + \log_b y$

4. $\log_b A^p = p \log_b A$

5. You can use the calculator to change logarithms in base b to common or natural logarithms by

$$\log_b x = \frac{\log x}{\log b} \quad \text{or} \quad \log_b x = \frac{\ln x}{\ln b}.$$

EXERCISES: ACTIVITY 4.11

1. Use the properties of logarithms to write the following as a sum or difference of two or more logarithms.

    a. $\log_b (3 \cdot 7)$ **b.** $\log_3 (3 \cdot 13)$

    c. $\log_7 \dfrac{13}{17}$ **d.** $\log_3 \dfrac{xy}{3}$

2. Write the following expressions as the logarithm of a single number.

    a. $\log_3 5 + \log_3 3$ **b.** $\log 25 - \log 17$

    c. $\log_5 x - \log_5 5 + \log_5 7$ **d.** $\ln (x + 7) - \ln x$

3. a. Sketch the graphs of $y = \log(2x)$ and $y = \log x + \log 2$ on your graphing calculator.

   b. Are you surprised by the results? Explain.

4. a. Sketch the graphs of $y = \log\left(\dfrac{3}{x}\right)$ and $y = \log x - \log 3$ on your graphing calculator.

   b. Are you surprised by your results? Explain.

   c. If your graphs in part a are not identical, can you modify the second function to make the graphs identical? Explain.

5. You have been hired to handle the local newspaper advertising for a large used car dealership in your community. The owner tells you that your predecessor in this position used the formula

$$N(A) = 7.4 \log A$$

to decide how much to spend on newspaper advertising over a 2-week period. The owner admitted that he didn't know much about the formula except that $N(A)$ represented the number of cars that the owner could expect to sell and A was the amount of money that was spent on local newspaper advertising. He also indicated that the formula seemed to work well. You can purchase small ads in the local paper for $15 per day, larger ads for $50 per day, and giant ads for $750 per day.

   a. How many cars do you expect to sell if you purchase one small ad?

   b. To understand the relationship between the amount spent on advertising and the number of cars sold, you set up a table. Complete the following table.

AD COST, A	EXPECTED CAR SALES, N(A)
15	
50	
750	

   c. How do the expected car sales from one small ad and one larger ad compare with the expected car sales from just one giant ad?

   d. Are the results in the table in part b consistent with what you know about the properties of logarithms? Explain.

   e. What are you going to advise the owner regarding the purchase of a giant ad?

6. Use the properties of logarithms to write the given logarithms as the sum or difference of two or more logarithms or as the product of a real number and a logarithm. Simplify if possible. All variables represent positive numbers.

a. $\log_3 3^5$

b. $\log_2 2^x$

c. $\log_b \dfrac{x^3}{y^4}$

d. $\ln \dfrac{\sqrt[3]{x}\sqrt[4]{y}}{z^2}$

e. $\log_3 (2x + y)$

7. Write each of the following as the logarithm of a single expression with coefficient 1.

a. $2\log_2 7 + \log_2 5$

b. $\dfrac{1}{4}\log x^3 - \dfrac{1}{4}\log z^5$

c. $2\ln 10 - 3\ln 5 + 4\ln z$

d. $\log_5 (x + 2) + \log_5 (x + 1) - 2\log_5 (x + 3)$

8. Given that $\log_a x = 6$ and that $\log_a y = 25$, determine the numeric value of each of the following.

a. $\log_a \sqrt{y}$

b. $\log_a x^3$

c. $3 + \log_a x^2$

d. $\log_a \dfrac{x^2 y}{a}$

9. Use the change of base formula and your calculator to determine a decimal approximation of each of the following to the nearest ten thousandth.

a. $\log_7 5$

b. $\log_6 \sqrt{15}$

c. $\log_{13} 47$

d. $\log_5 \sqrt[3]{31}$

10. The formula

$$P = 95 - 30 \log_2 t$$

gives the percentage, P, of students who could recall the important content of a classroom presentation as a function of time, t, where t is the number of days that have passed since the presentation was given.

a. Sketch a graph of the function.

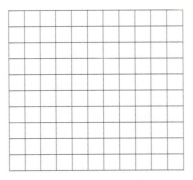

b. After three days, what percentage of the students will remember the important content of the presentation?

c. According to the model, after how many days do only half $(P = 50)$ of the students remember the important features of the presentation? Use a graphing approach.

tivity *4.12*

nging Demographics

jective

e exponential equations
graphically and
braically.

The number of people in the United States who claim Hispanic origins has been increasing rapidly since 1970. The following table gives U.S. Census Bureau data on the Hispanic population.

Year	1970	1980	1990	2000	2010
Hispanic Population (millions of people)	9.1	14.6	22.4	35.3	50.5

1. Do you believe that the data is better modeled by a linear or exponential function?

2. Let t represent the number of years since 1970. Use your graphing calculator to produce a scatterplot of the U.S. Hispanic population. Your screen should appear as follows:

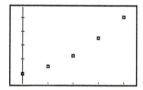

3. Use your graphing calculator to determine the regression equation of the exponential model that best represents the Hispanic population data. Remember that the input variable, t, is the number of years since 1970. In your regression equation, $P = a \cdot b^t$, round the value for a to two decimal places and the value of b to three decimal places. Record your model below.

4. Use your graphing calculator to visually check how well the equation in Problem 3 fits the data. Your graph should resemble the following.

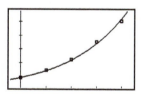

5. Use the exponential model from Problem 3 to determine what the projected Hispanic population will be in the United States in 2020.

6. a. Using your model from Problem 3, write an equation that can be used to determine the year in which the Hispanic population in the United States will first reach 75 million.

b. Solve this equation using a graphing approach. Your screen should resemble the following. What is the equation of the horizontal line in the graph?

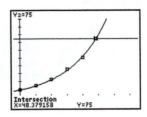

To solve the equation $9.34 \cdot (1.044)^t = 75$ for t using an algebraic approach, you need to remove t as an exponent. The following problem guides you through this process. As you will discover, logarithms are essential in this algebraic approach.

7. Solve $9.34(1.044)^t = 75$ for t using an algebraic approach.

a. Isolate the exponential factor $(1.044)^t$ on one side of the equation.

b. Take the log (or ln) of each side of the equation in part a.

c. Apply the appropriate property of logarithms on the left side of the equation to "remove" t as an exponent.

d. Solve the resulting equation in part c for t.

e. How does your solution in part d compare with the estimate you obtained graphically in Problem 6b?

8. As the Hispanic population of the country has grown, so has the number of Hispanic voters in presidential elections. The following table shows the number of Hispanic voters in presidential elections from 1988 to 2012

Year	1988	1992	1996	2000	2004	2008	2012
Number of Hispanic Voters (millions)	3.7	4.1	5.0	5.9	7.6	9.7	11.2

a. Determine an exponential model for the number of Hispanic voters, V, in presidential elections. Let the variable, t, represent the number of years since 1988. In the regression equation $V = a \cdot b^t$, round the value of a to 1 decimal place and the value of b to 3 decimal places.

b. Use the exponential model to predict the number of Hispanic voters in the 2016 presidential election.

c. Write an equation that can be used to determine the year in which the number of Hispanic voters in a presidential election will first exceed 20 million.

d. Solve the equation in part c using an algebraic approach. Keep in mind that presidential elections take place every four years.

Radioactive Decay

Radioactive substances such as uranium-235, strontium-90, iodine-131, and carbon-14 decay continuously with time. If P_0 represents the original amount of a radioactive substance, then the amount P present after a time t (usually measured in years) is modeled by

$$P = P_0 e^{-kt},$$

where k represents the rate of continuous decay.

Example 1 *One type of uranium decays at a rate of 0.35% per day. If 40 pounds of this uranium are available today, how much will be available after 90 days?*

SOLUTION

The uranium decays at a constant rate of $0.35\% = 0.0035$ per day. The initial amount, the amount available on the first day, is 40 pounds, so the equation for the amount available after t days is

$$P = 40e^{-0.0035t}.$$

To determine the amount available after 90 days, let $t = 90$. The amount available 90 days from now is

$$P = 40e^{-0.0035(90)} \approx 29.2 \text{ lb.}$$

9. Strontium-90 decays continuously at a constant rate of 2.4% per year. Therefore, the equation for the amount P of strontium-90 after t years is

$$P = P_0 e^{-0.024t}.$$

a. If 10 grams of strontium-90 are present initially, determine the number of grams present after 20 years.

b. How long will it take the given quantity to decay to 2 grams?

c. How long will it take the given amount of strontium-90 to decay to one-half of its original size (called its half-life)? Round to the nearest whole number.

d. Do you think that the half-life of strontium-90 is 29 years regardless of the initial amount? Answer part c using P_0 as the initial amount. (Hint: Find t when $P = \frac{1}{2}P_0$.)

SUMMARY: ACTIVITY 4.12

To solve exponential equations of the form $ab^x = c$, where $a > 0, b > 0, b \neq 1$, and $c > 0$,

Step 1. Isolate the exponential factor on one side of the equation

Step 2. Take the log (or ln) of each side of the equation

Step 3. Apply the property $\log b^x = x \log b$ to remove the variable x as an exponent

Step 4. Solve the resulting equation for the variable

EXERCISES: ACTIVITY 4.12

1. The amount of money the U.S. government owes to countries and individuals is known as the national debt. Since the year 2000, the national debt has been growing at historic rates. The following table shows the national debt, D, in trillions of dollars for various years since 2000.

Year	2000	2002	2004	2006	2008	2010	2012
National Debt, D (trillions of dollars)	5.7	6.2	7.4	8.5	10.0	13.6	16.1

a. Use your graphing calculator to determine the exponential regression model $D = a \cdot b^t$, where t represents the number of years since 2000. Round the value of a to 1 decimal place and b to 3 decimal places.

b. According to the model, what was the national debt in 2009?

c. According to the model, when will the national debt first reach $25 trillion?

2. The U.S. Department of Transportation recommended that states adopt a 0.08% blood-alcohol concentration as the legal measure of drunk driving. Medical research has shown that as the concentration of alcohol in the blood increases, the risk of having a car accident increases exponentially. The risk, R, expressed as a percentage, is modeled by

$$R(x) = 6e^{12.77x},$$

where x is the blood-alcohol concentration, expressed as a percent.

a. What is the risk of having a car accident if a driver's blood-alcohol concentration is 0.08% $(x = 0.08)$?

b. What blood-alcohol concentration has a corresponding 25% risk of a car accident?

3. In 1990, the International Panel on Climate Change projected the following future amounts of carbon dioxide (in parts per million or ppm) in the atmosphere.

Year	1990	2000	2075	2175	2275
Amount of Carbon Dioxide (ppm)	353	375	590	1090	2000

a. Use your graphing calculator to create a scatterplot of the data. Let t represent the number of years since 1990 and $A(t)$ represent the amount of carbon dioxide (in ppm) in the atmosphere. Do the carbon dioxide levels appear to be growing exponentially?

b. Use your graphing calculator to determine the regression equation of an exponential model that best fits the data.

c. Use the model in part b to determine in what year the 1990 carbon dioxide level is expected to double.

d. Verify your result in part c graphically.

In Exercises 4–9, solve each equation using an algebraic approach. Verify your answers graphically.

4. $2^x = 14$

5. $3^{2x} = 8$

6. $1000 = 500(1.04)^t$

7. $e^{0.05t} = 2$ (Hint: Take the natural log of both sides.)

8. $2^{3x+1} = 100$

9. $e^{-0.3t} = 2$

10. a. Iodine-131 disintegrates at a continuous constant rate of 8.6% per day. Determine its half-life. Use the model

$$P = P_0 e^{-0.086t},$$

where t is measured in days. Round your answer to the nearest whole number.

b. If dairy cows eat hay containing too much iodine-131, their milk will be unsafe to drink. Suppose that hay contains five times the safe level of iodine-131. How many days should the hay be stored before it can be fed to dairy cows?

(Hint: Find t when $P = \dfrac{1}{5} P_0$.)

11. a. In 1969, a report written by the National Academy of Sciences (U.S.) estimated that Earth could reasonably support a maximum world population of 10 billion. The world's population was approximately 3.6 billion and growing continuously at 2% per year. If this growth rate remained constant, in what year would the world population reach 10 billion, referred to as Earth's carrying capacity? Use the model

$$P = P_0 e^{kt},$$

where P is the population (in billions), $P_0 = 3.6$, $k = 0.02$, and t is the number of years since 1969.

b. According to your growth model, when would this 1969 population double?

c. The world population in 1995 was approximately 5.7 billion. How does this compare with the population predicted by your growth model in part a?

d. The growth rate in 1995 was 1.5%. Assuming this growth rate remains constant, determine when Earth's carrying capacity will be reached. Use the model $P = P_0 e^{kt}$.

Collecting and Analyzing Data

12. Texas and California are two of the most populous states in the United States, both experiencing tremendous growth over the past century. Use the Internet to obtain the population of each state every ten years between 1950 and 2010.

a. Beginning with 1950, make a scatterplot of the data for each state's population. Let the independent variable represent the number of years since 1950 and the dependent variable represent the population. Describe the pattern you observe.

b. Determine whether a linear or exponential function best fits each set of data. Explain.

c. Using the best models determined in part b, use the regression equations to graph each function on the appropriate scatterplot. Which state appears to be growing the fastest?

d. Predict the population of each state in 2020.

e. Assuming these growth patterns continue into the future, determine the year in which the population of each state will reach 50 million.

f. Use the intersect feature of your graphing calculator to estimate the year when the population of Texas and California will be equal.

Activity 4.13

Frequency and Pitch

Objective

Solve logarithmic equations both graphically and algebraically.

Raising a musical note one octave has the effect of doubling the pitch, or frequency, of the sound. However, you do not perceive the note to sound "twice as high," as you might predict. Perceived pitch is given by the function

$$P(f) = 2410 \log (0.0016f + 1),$$

where P is the perceived pitch in mels (units of pitch) and f is the frequency in hertz.

1. Let frequency (input) vary in value from 10 to 100,000 hertz, and let the perceived pitch (output) vary from 0 to 6000 mels. Graph this equation on your graphing calculator, using the following window: Xmin $= 0$, Xmax $= 100,000$, Ymin $= 0$, and Ymax $= 6000$.

2. What is the perceived pitch, P, for the input value 10,000 hertz?

3. a. Write an equation that can be used to determine what frequency, f, gives an output value of 2000 mels.

b. Solve the equation in part a using a graphing approach.

To determine the exact answer in Problem 3, you can use an algebraic approach. The following problem guides you through this process.

4. Solve the equation $2410 \log (0.0016f + 1) = 2000$ using an algebraic approach.

a. Solve the equation for $\log (0.0016f + 1)$. That is, isolate the log on one side of the equation.

b. The equation in part a is now in the form $\log_b N = E$, where $b = 10$,

$N = 0.0016x + 1$, and $E = \dfrac{2000}{2410}$. Write the equation from part a in exponential

form, $b^E = N$.

c. In exponential form, the equation in part a should be

$$0.0016f + 1 = 10^{2000/2410}.$$

Solve this equation for f. Of course, you will need to approximate a value of $10^{2000/2410}$ using your calculator.

d. How does your answer to part c compare to your answer to Problem 3b?

5. a. Use an algebraic approach to determine the frequency, f, that produces a perceived pitch of 3000 mels.

b. Verify your answer in part a using a graphing approach.

6. The formula $W = 0.35 \ln P + 2.74$ is a model for the average walking speed, W, in feet per second for a resident of a city with population P, measured in thousands.

    a. Determine the walking speed of a resident of a small city having a population of 500,000.

    b. If the average walking speed of a resident is 4.5 feet per second, what is the population of the city? Round your answer to the nearest thousand.

SUMMARY: ACTIVITY 4.13

To solve a logarithmic equation algebraically,

Step 1. Rewrite the equation in the form $\log_b (f(x)) = c$, where $b > 0, b \neq 1, c > 0$, and $f(x) > 0$.

Step 2. Rewrite the resulting equation from step 1 in exponential form, $f(x) = b^c$.

Step 3. Solve the resulting equation from step 2 algebraically.

Step 4. Check the solutions in the original equation.

EXERCISES: ACTIVITY 4.13

In Exercises 1–6, solve each equation using an algebraic approach. Then verify your answer using a graphical approach.

1. $\log_2 x = 5$

2. $\ln x = 10$

3. $3 \log_5 (x + 2) = 5$

4. $\log_5 (x - 4) = 2$

5. $20 = 3.5 \ln x$

6. $4 + 1.75 \ln x = 31$

7. Stars have been classified into magnitude according to their brightness. Stars in the first six magnitudes are visible to the naked eye; those of higher magnitudes are visible only through a telescope. The magnitude, m, of the faintest star that is visible with a telescope having lens diameter d, in inches, is modeled by

$$m = 8.8 + 5.1 \log d.$$

What is the highest magnitude of a star that is visible with the 200-inch telescope at Mount Palomar, California?

8. The consumption of ethanol fuel in the United States can be modeled by the equation

$$C(x) = 258.63 + 347.64 \ln(x)$$

where x is the number of years since 2005, and $C(x)$ is the amount of ethanol fuel consumed measured in thousands of barrels per day. According to the model, in what year will the consumption of ethanol fuel reach 1 million barrels per day?

9. The acidity or alkalinity of any solution is determined by the concentration of hydrogen ions, $[H^+]$, in the substance, measured in moles per liter (mol/l). Acidity (or alkalinity) is measured on a pH scale, using the model

$$pH = -\log\,[H^+].$$

The pH scale ranges from 0 to 14. Values below seven have progressively greater acidity; values greater than seven are progressively more alkaline. Normal unpolluted rain has a pH of about 5.6. The acidity of rain over the northeastern United States, caused primarily by sulfur dioxide emissions, has had very damaging effects. One of the most acidic rainfalls on record had a pH of 2.4. What was the concentration of hydrogen ions?

10. The Richter scale is a well-known method of measuring the magnitude of an earthquake in terms of the amplitude, A (height), of its shock waves. The magnitude of any given earthquake is given by

$$m = \log\left(\frac{A}{A_0}\right),$$

where A_0 is a constant representing the amplitude of an average earthquake.

a. The magnitude of the 1906 San Francisco earthquake was 8.3 on the Richter scale. Write an equation that gives the amplitude, A, of the San Francisco earthquake in terms of A_0.

b. An earthquake with a magnitude of 5.5 will begin to cause serious damage. Write an equation that gives the amplitude, A, of a serious-damage earthquake in terms of A_0.

c. Determine the ratio of the amplitude of the San Francisco earthquake to the amplitude of a serious-damage (magnitude 5.5) earthquake. What is the significance of this number?

Activities 4.8–4.13 **What Have I Learned?**

1. A logarithm is an exponent. Explain how this fact relates to the following properties of logarithms.

 a. $\log_b (x \cdot y) = \log_b x + \log_b y$

 b. $\log_b \dfrac{x}{y} = \log_b x - \log_b y$

 c. $\log_b x^n = n \cdot \log_b x$

2. You have $20,000 to invest. Your broker tells you that the value of shares of mutual fund A has been growing exponentially for the past two years and that shares of mutual fund B have been growing logarithmically over the same period. If you make your decision based solely on the past performances of the funds, in which fund would you choose to invest? Explain.

3. Study the following graphs, which show various types of functions you have encountered in this course.

a.
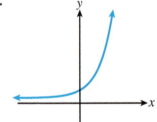

b.
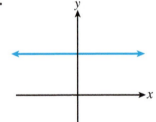

c.
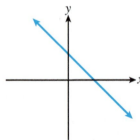

d.
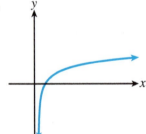

e.
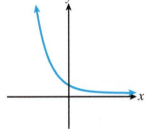

f.
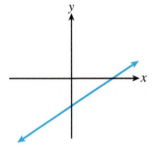

Complete the following table with respect to the preceding graphs.

DESCRIPTION	GRAPH LETTER	GENERAL EQUATION
Constant function	b	
Linearly decreasing function	c	
Logarithmically increasing function	d	
Exponentially decreasing function	e	
Exponentially increasing function	a	
Linearly increasing function	f	

4. The graph of $y = \log_b x$ will never be located in the second or third quadrants. Explain.

5. What function would you enter into Y1 on your graphing calculator to graph the function $y = \log_4 x$?

6. What values of x cannot be inputs in the function $y = \log_b (3x - 2)$?

7. What is the relationship between the functions $y = \log x$ and $y = 10^x$? How are the graphs related?

Activities 4.8–4.13   How Can I Practice?

1. Write each equation in logarithmic form.

   a. $4^2 = 16$ **b.** $0.0001 = 10^{-4}$ **c.** $3^{-4} = \dfrac{1}{81}$

2. Write each equation in exponential form.

   a. $\log_2 32 = 5$ **b.** $\log_5 1 = 0$

   c. $\log_{10} 0.001 = -3$ **d.** $\ln e = 1$

3. Solve each equation for the unknown variable.

   a. $\log_4 x = -3$ **b.** $\log_b 32 = 5$ **c.** $\log_5 125 = y$

4. a. Complete the table of values for the function $f(x) = \log_4 x$.

x	0.25	0.5	1	4	16	64
f(x)						

   b. Sketch a graph of the function f.

   c. Use your graphing calculator to check your result in parts a and b.

   d. Determine the x-intercept.

e. What is the domain of the function?

f. What is the range?

g. Is the function increasing or decreasing?

h. Does the graph have a vertical or horizontal asymptote?

i. Use your graphing calculator to determine $f(32)$.

j. Use your graphing calculator to determine x when $f(x) = 3.25$.

5. Write each of the following as a sum, difference, or multiple of logarithms. Assume that x, y, and z represent positive numbers.

a. $\log_b \dfrac{xy^2}{z}$

b. $\log_3 \dfrac{\sqrt{x^3 y}}{z}$

c. $\log_5 (x\sqrt{x^2 + 4})$

d. $\log_4 \sqrt[3]{\dfrac{xy^2}{z^2}}$

6. Rewrite the following as the logarithm of a single quantity.

a. $\log x + \dfrac{1}{3}\log y - \dfrac{1}{2}\log z$

b. $3 \log_3 (x + 3) + 2 \log_3 z$

c. $\dfrac{1}{3}\log_3 x - \dfrac{2}{3}\log_3 y - \dfrac{4}{3}\log_3 z$

7. Use the change of base formula and your calculator to approximate the following.

a. $\log_5 17$

b. $\log_{13} \sqrt[3]{41}$

8. Solve each of the following using an algebraic approach.

a. $25 + 3 \ln x = 10$

b. $1.5 \log_4 (x - 1) = 7$

9. Solve the following algebraically. Check your solutions using graphs or tables.

   a. $3^x = 17$ **b.** $42 = 3e^{1.7x}$

10. You have invested \$10,000 in a money market account that will pay you 4% interest compounded continuously.

   a. Write an equation that relates the current value of the account, V, to the number of years you have held the account, t.

   b. Use the equation in part a to determine algebraically the number of years it will take the value of the account to double.

   c. Verify your result in part b graphically.

11. Data collected from over 100 countries showed that the relationship between per capita health care expenditures, H, in dollars and average life expectancy, E, could be modeled by the formula $E = 0.035 + 9.669 \ln(H)$, where $0 < H \leq 4500$.

   a. Sketch a graph of the health care/life expectancy model using the domain $0 < H \leq 4500$.

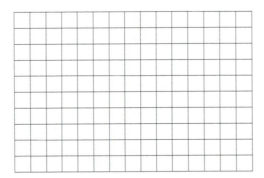

   b. Use this model to predict the average life expectancy in a country whose per capita health care expenditure is \$1500 per year.

   c. Use the model and your graphing calculator to predict per capita health care expenditures in a country whose average life expectancy is 77 years.

12. Sketch a graph of each of the following equations. Using transformations, describe how the graph is related to either the graph of $y = \ln x$ or the graph of $y = \log x$.

a. $y = 3 \ln x$

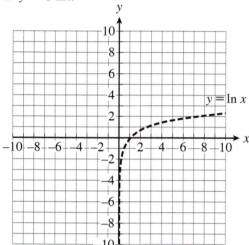

b. $y = 4 + 2 \log x$

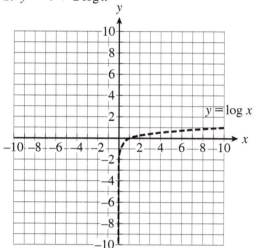

c. $y = -0.25 \ln x$

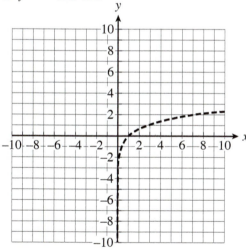

d. $y = 4.5 + 1.5 \ln x$

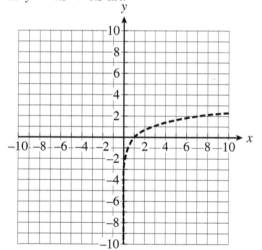

e. $y = 3 \log (x + 5)$

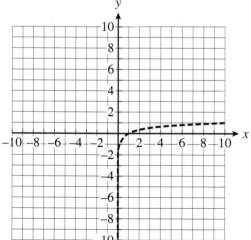

Chapter 4   Summary

The bracketed numbers following each concept indicate the activity in which the concept is discussed.

CONCEPT/SKILL	DESCRIPTION	EXAMPLE
Growth factor [4.1]	When a quantity increases by a specified percent, the ratio of its new value to the original value is called the growth factor associated with the specified percent increase. The growth factor is formed by adding the specified percent increase to 100% and then changing this percent into its decimal form.	To determine the growth factor associated with a 25% increase, add 25% to 100% to obtain 125%. Change 125% into a decimal to obtain the growth factor 1.25.
Applying a growth factor to an original value [4.1]	original value $\times$ growth factor = new value	A $120 item increases by 25%. Its new value is $120 \times 1.25 = $150.
Applying a growth factor to a new value [4.1]	new value $\div$ growth factor = original value	An item has already increased by 25% and is now worth $200. Its original value was $200 \div 1.25 = $160.
Decay factor [4.1]	When a quantity decreases by a specified percent, the ratio of its new value to the original value is called the *decay factor* associated with the specified percent decrease. The decay factor is formed by subtracting the specified percent decrease from 100%, and then changing this percent into its decimal form.	To determine the decay factor associated with a 25% decrease, subtract 25% from 100%, to obtain 75%. Change 75% into a decimal to obtain the decay factor 0.75.
Applying a decay factor to an original value [4.1]	original value $\cdot$ decay factor = new value	A $120 item decreases by 25%. Its new value is $120 \times 0.75 = $90.
Applying a decay factor to a new value [4.1]	new value $\div$ decay factor = original value	An item has already decreased by 25% and is now worth $180. Its original value was $180 \div 0.75 = $240.
Applying a sequence of percent changes [4.2]	The cumulative effect of a sequence of percent changes is the product of the associated growth or decay factors.	The effect of a 25% increase followed by a 25% decrease is $1.25 \times 0.75 = 0.9375$. That is, the item is now worth only 93.75% of its original value. The item's value has decreased by 6.25%.

CONCEPT/SKILL	DESCRIPTION	EXAMPLE
Exponential function [4.3]	An exponential function is a function whose y-values exhibit a constant percent growth (or decay). In an equation defining an exponential function, $y = ab^x$, $b > 0$, $b \neq 1$, the x-values variable always occurs as the exponent on the growth (decay) factor.	$y = 16 \cdot (2.1)^x$
Decay factor of an exponential function [4.3]	If $0 < b < 1$, the function is decreasing, and b is called the decay factor.	The exponential function $y = \left(\dfrac{1}{2}\right)^x$ has a decay factor of $\dfrac{1}{2}$.
Growth factor of an exponential function [4.3]	If $b > 1$, the function is increasing, and b is called the growth factor.	The exponential function $y = 3^x$ has a growth factor of 3.
y-intercept of an exponential function [4.4]	The y-intercept of an exponential function $y = a \cdot b^x$ is $(0, a)$.	The graph of $y = 5 \cdot 2^x$ passes through the point $(0, 5)$.
Horizontal asymptote of an exponential function [4.4]	The line $y = 0$ is a horizontal asymptote of an exponential function $y = b^x$.	As x gets smaller, the output values of $y = 3^x$ approach 0.
Doubling time [4.5]	The doubling time of an exponential function is the time it takes for an output to double. The doubling time is set by the growth factor and remains the same for all output values.	Example 2, Activity 4.5.
Half-life [4.5]	The half-life of an exponential function is the time it takes for an output to decay by one-half. The half-life is determined by the decay factor and remains the same for all output values.	Example 3, Activity 4.5.
Growth model [4.5, 4.6]	If r represents the annual growth rate, the exponential function that models the quantity P can be written as $P(t) = P_0(1 + r)^t$, where P_0 is the initial amount, t represents the number of elapsed years, and $1 + r$ is the growth factor.	
Decay model [4.5, 4.6]	If r represents the annual percent that decays, the exponential function that models the amount remaining can be written as $P(t) = P_0(1 - r)^t$, where $1 - r$ is the decay factor.	
Logarithm [4.8]	In the equation $y = b^x$, where $b > 0$ and $b \neq 1$, x is called a logarithm, or log.	For the equation $3^4 = 81$, 4 is the logarithm of 81, base 3.
Notation for logarithms [4.8]	The notation for logarithms is $\log_b x = y$, where b is the base of the log, x (a positive number) is the power of b, and y is the exponent.	In the equation $\log_2 16 = 4$, 2 is the base, 4 is the log or exponent, and 16 is the power of 2.

CONCEPT/SKILL	DESCRIPTION	EXAMPLE
Common logarithm [4.8]	A common logarithm is a base-10 logarithm. The notation is $\log_{10} x = \log x$.	$1000 = 10^3$. The common logarithm of 10^3 is 3 (i.e., $\log 1000 = 3$).
Natural logarithm [4.8]	A natural logarithm is a base-e logarithm. The notation is $\log_e x = \ln x$.	$\log_e e^3 = \ln e^3 = 3$
Logarithmic equation [4.8]	The logarithmic equation $y = \log_b x$ is equivalent to the exponential equation $b^y = x$.	The equations $6 = \log_4 x$ and $x = 4^6$ are equivalent.
Basic properties of logarithms [4.8]	If $b > 0$ and $b \neq 1$, $\log_b 1 = 0$, $\log_b b = 1$, and $\log_b b^n = n$.	$\log_4 1 = 0, \log_7 7 = 1,$ $\log_6 6^4 = 4$
Logarithmic function [4.9]	If $b > 0$ and $b \neq 1$, the logarithmic function is defined by $y = \log_b x$.	$y = \log_4 x$
Graph of the logarithmic function [4.9]	The graph of $y = \log_b x$, where $b > 1$ is increasing for all $x > 0$, has an x-intercept of $(1, 0)$ and has a vertical asymptote of $x = 0$, the y-axis.	
Comparison of the graphs of $f(x) = b^x$, where $b > 1$, and $g(x) = \log_b x$, where $b > 1$ [4.10]	Both graphs increase. The exponential function increases faster as x increases; the log function increases slower as x increases. The domain of the exponential function is the range of the log, which is all real numbers; the range of the exponential function is the domain of the log, which is the interval $(x > 0)$.	Problem 6, Activity 4.10.
Log of a Product: If $A > 0, B > 0, b > 0,$ **and** $b \neq 1,$ **then** $\log_b (A \cdot B) = \log_b A + \log_b B.$ [4.11]	The logarithm of a product is the sum of the logarithms.	$\log_2 4 \cdot 8 = \log_2 4 + \log_2 8$ $= 2 + 3 = 5$
Log of a Quotient: If $A > 0, B > 0, b > 0,$ **and** $b \neq 1,$ **then** $\log_b \left(\dfrac{A}{B}\right) = \log_b A - \log_b B.$ [4.11]	The logarithm of a quotient is the difference of the logarithms.	$\log_3 \left(\dfrac{81}{27}\right) = \log_3 81 - \log_3 27$ $= 4 - 3 = 1$

CONCEPT/SKILL	DESCRIPTION	EXAMPLE
Log of a Sum: If $A > 0, B > 0, b > 0,$ and $b \neq 1$, then $\log_b (A + B) \neq \log_b (A) + \log_b (B).$ [4.11]	The logarithm of a sum is not the sum of the logarithms.	$\log 2 + \log 3 \approx$ $0.3010 + 0.4771 = 0.7781$ $\log(2 + 3) = \log 5 \approx 0.6990$
Log of a Power: If $A > 0$, p is a real number, $b > 0$, and $b \neq 1$, then $\log_b A^p = p \log_b A.$ [4.11]	The logarithm of a power of A is the exponent times the logarithm of A.	$\log_5 x^4 = 4 \log_5 x$ $\log_3 \sqrt{x} = \dfrac{1}{2} \log_3 x$
Change of base formula [4.11]	The logarithm of any positive number x to any base can be found using the formula $$\log_b x = \frac{\log x}{\log b} \text{ or } \log_b x = \frac{\ln x}{\ln b}.$$	$\log_2 2.5 = \dfrac{\log 2.5}{\log 2} \approx 1.3219$
Solving exponential equations algebraically [4.12]	**Step 1.** Isolate the exponential factor on one side of the equation **Step 2.** Take the log (or ln) of each side of the equation **Step 3.** Apply the property $\log b^x = x \log b$ to remove the variable x as an exponent **Step 4.** Solve the resulting equation for the variable	Problem 7, page 568
Solving logarithmic equations algebraically [4.13]	**Step 1.** Rewrite the equation in the form $\log_b (f(x)) = c$, where $b > 0, b \neq 1, c > 0,$ and $f(x) > 0$. **Step 2.** Rewrite the resulting equation from step 1 in exponential form, $f(x) = b^c$. **Step 3.** Solve the resulting equation from step 2 algebraically. **Step 4.** Check the solutions in the original equation.	Problem 4, page 575

1. College tuition is expected to increase 6% each year for the next several years.

   a. If tuition is $300 per credit now, determine how much it will be in five years, in ten years.

   b. Calculate the average rate of change in tuition over the next five years.

   c. Calculate the average rate of change in tuition over the next ten years.

   d. If the growth rate stays at 6%, approximately when will the tuition double?

2. **a.** Determine some of the output values for the function $f(x) = 8^x$ by completing the following table.

x	-1	$-\dfrac{1}{3}$	0	1	$\dfrac{4}{3}$	2	3
$f(x) = 8^x$			1	8	16	64	512

   b. Sketch the graph of the function f.

c. Is this function increasing or decreasing? Explain how you know this by looking at the equation of the function.

d. What is the domain?

e. What is the range?

f. What are the x- and y-intercepts?

g. Are there any asymptotes? If so, write the equations of the asymptotes.

h. Compare the graph of f with the graph of $g(x) = \left(\dfrac{1}{8}\right)^x$. What are the similarities and the differences?

i. In what way does the graph of $h(x) = 8^x + 5$ differ from that of $f(x) = 8^x$?

j. Write the equation of the function that is the inverse of the function $f(x)$.

3. Complete the table for each exponential function. Use your graphing calculator to check your work.

FUNCTION	BASE, b	GROWTH OR DECAY FACTOR	x-INTERCEPT	y-INTERCEPT	HORIZONTAL ASYMPTOTE	INCREASING OR DECREASING
$h(x) = 6^x$						
$g(x) = \left(\dfrac{1}{3}\right)^x$						
$p(x) = 5(2.34)^x$						
$q(x) = 3(0.78)^x$						
$r(x) = 2^x - 4$						

4. Use your graphing calculator to help you determine the domain and range for each function.

Function	$f(x) = 0.8^x$	$h(x) = 6^x + 2$	$t(x) = 3^x - 5$	$q(x) = \log_4 x$	$r(x) = \ln(x - 3)$
Domain					
Range					

5. a. Given the following table, determine whether the given data can be approximately modeled by an exponential function. If it can, what is the growth or decay factor?

x	0	1	2	3	4
y	10	15.5	24	36	55.5

b. Determine an exponential equation that models this data.

6. a. Your salary has increased at the rate of 1.5% annually for the past five years, and your boss projects this will remain unchanged for the next five years. You were making $25,000 annually in 2014. Complete the following table:

2014	2015	2016	2017	2018	2019

b. Write the exponential growth function that models your annual salary during this period of time. Let x represent the number of years since 2014.

c. If your increase in salary continues at this rate, how much will you make in 2022? Is this realistic?

d. You would like to double your salary. How many years must you work before your salary will be twice the salary you made in 2014?

7. Complete the following tables representing exponential functions. Round calculations to two decimal places whenever necessary.

a.

x	0	1	2	3	4
y	3.00	6.12			

b.

x	0	1	2	3	4
y	4.50	3.15			

c.

x	0	1	2	3	4
y	$\frac{1}{4}$	4			

8. Write the equation of the exponential function that represents the data in each table in Problem 7.

9. Determine the value of each of the following without using your calculator.

    a. $\log 1$

    b. $\log_4 4$

    c. $\log 10^3$

    d. $\log_3 9$

    e. $\log_3 \dfrac{1}{9}$

    f. $\log_5 625$

    g. $\log 0.001$

    h. $\ln e^2$

10. Write each equation in logarithmic form.

    a. $6^2 = 36$

    b. $0.000001 = 10^{-6}$

    c. $2^{-5} = \dfrac{1}{32}$

11. Write each equation in exponential form.

    a. $\log_3 81 = 4$

    b. $\log_7 1 = 0$

    c. $\log_{10} 0.0001 = -4$

    d. $\ln e = 1$

    e. $\log_q y = b$

12. Solve each equation for the unknown variable.

    a. $\log_5 x = -3$

    b. $\log_b 256 = 4$

    c. $\log_2 64 = y$

    d. $\log_4 x = \dfrac{3}{2}$

13. a. Complete the table of values for the function $f(x) = \log_5 x$.

x	0.008	0.04	0.2	1	5	25
$f(x)$						

b. Sketch a graph of the function.

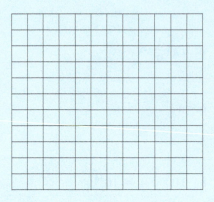

c. Use your graphing calculator to check your result in parts a and b.

d. Determine the x-intercept.

e. What is the domain of the function?

f. What is the range?

g. Does the graph have a vertical or horizontal asymptote?

h. Use your graphing calculator to determine $f(23)$.

i. Use your graphing calculator to determine x when $f(x) = 2.46$.

14. Use the change of base formula and your calculator to approximate the following.

a. $\log_7 21$

b. $\log_{15} \dfrac{8}{9}$

15. Write each of the following as a sum, difference, or multiple of logarithms. Assume that x, y, and z are all greater than 0.

a. $\log_2 \dfrac{x^3 y}{z^{1/2}}$

b. $\log \sqrt[3]{\dfrac{x^4 y^3}{z}}$

16. Rewrite the following as the logarithm of a single quantity.

a. $\log x + \dfrac{1}{4}\log y - 3\log z$

b. $\dfrac{1}{3}(\log x - 2\log y - \log z)$

17. Solve the following algebraically.

a. $3^{3+x} = 7$

b. $\log_2(4x + 9) = 4$

c. $50 + 6\ln x = 85$

18. a. Sketch the graph of the function using the data from the given table.

x	0.1	0.5	1	2	4	16
f(x)	−1.66	−0.5	0	0.5	1	2

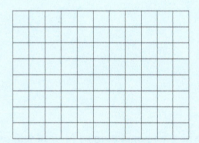

b. Use the table and the graphing feature of your calculator to verify that the equation that defines function f is $f(x) = 0.5\log_2 x$.

c. Use the function to determine the value of $f(54)$.

d. If $f(x) = 2.319$, determine the value of x.

e. Use your graphing calculator to verify that the function $g(x) = 4^x$ is the inverse of f.

19. The populations of New York State and Florida (in millions) can be modeled by the following.

New York State	$P_N = 19.39e^{0.0044t}$
Florida	$P_F = 18.80e^{0.0130t}$

where t represents the number of years since 2010.

a. Determine the population of New York and Florida in 2010 ($t = 0$).

b. Sketch a graph of each function on the same set of coordinate axes.

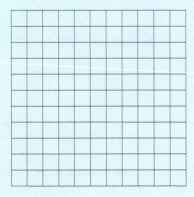

c. Determine graphically the year when the population of Florida was equal to the population of New York State.

d. Determine algebraically the year when the population of Florida will exceed 25 million.

e. Will the population of Florida ever exceed the population of New York? Explain. Assume that population growth given by both populations will continue. Solve algebraically.

20. Atmospheric pressure decreases with increasing altitude. The following data is collected during a science experiment to investigate the relationship between the height of an object above ground level and the pressure exerted on the object.

Atmospheric Pressure, x (in millimeters of mercury)	760	740	725	700	650	630	600	580	550
Height, y (in kilometers)	0	0.184	0.328	0.565	1.079	1.291	1.634	1.862	2.235

a. Use a graphing calculator to determine a logarithmic regression model that best fits the data.

b. Use the log equation in part a to predict the height of the object if the atmospheric pressure is 500 millimeters of mercury.

21. The number of computers infected by a virus t days after it first appears often increases exponentially. Recently, a computer "worm" spread from about 2.4 million computers on June 8 to approximately 3.2 million computers on June 9.

a. Determine the growth factor.

b. Write an exponential equation that can be used to predict the number N of computers infected t days after June 8.

c. Predict the number of computers infected by the virus after three days.

d. Assuming exponential growth, using a graphing approach, estimate how long it will take the computer virus to infect 10 million computers.

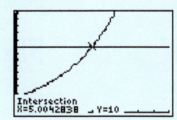

22. Using transformations, sketch a graph of each of the following. Determine any intercepts and asymptotes. Verify the graph using a graphing calculator.

a. $y = 2^x + 3$

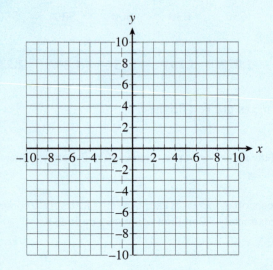

b. $y = 2^{x+3}$

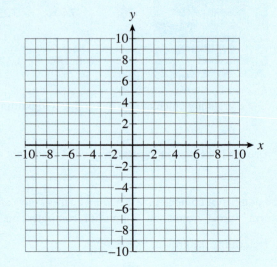

c. $y = 2^{-x}$

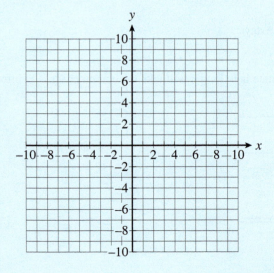

d. $y = -3 \cdot 2^x$

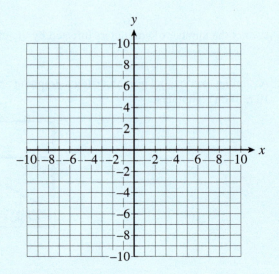

e. $y = e^x + 3$

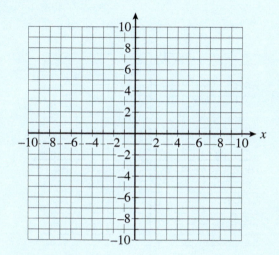

f. $y = -3e^x$

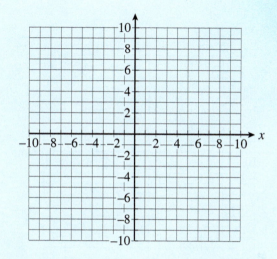

g. $y = 5 + 3\ln x$

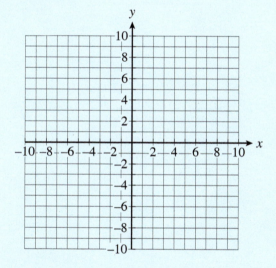

h. $y = \ln(x - 1)$

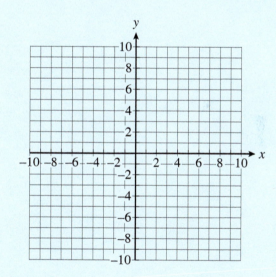

i. $y = -4 \log x$

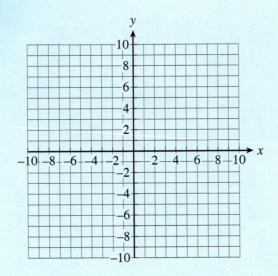

j. $y = \ln(-x)$

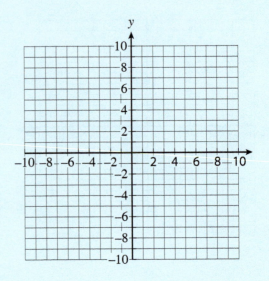

Using Geometric Models to Solve Problems

Activity 5.1

Walking around Bases, Gardens, Trusses, and Other Figures

Objectives

1. Recognize perimeter as a geometric property of plane figures.

2. Write formulas for and calculate perimeters of squares, rectangles, triangles, parallelograms, trapezoids, kites, and polygons.

3. Use unit analysis to solve problems involving perimeter.

4. Write and use formulas for the circumference of a circle.

In this first group of activities, you will explore the properties of geometric figures or shapes that are two dimensional. This means they exist in a **plane**—a surface like the floor beneath your feet or the walls in your classroom. You will begin by reviewing some preliminary definitions that you have studied in previous chapters.

Definition

A **ray** is a portion of a line that starts from a point and continues indefinitely in one direction, much like a ray of light coming from the Sun.

Example:

An **angle** is formed by two rays that have a common starting point. The common starting point is called the **vertex** of the angle.

Example:

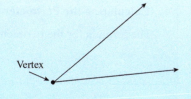

Vertex

Squares

Definition

A **square** is a closed plane figure whose four sides have equal length and are at right angles to each other. A right angle measures 90 degrees.

Examples:

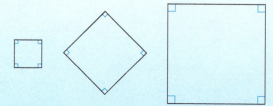

599

1. You are at bat in the middle of an exciting baseball game. You are a pretty good hitter and can run the bases at a speed of 15 feet per second. Further, you know that the baseball diamond has the shape of a square, measuring 90 feet on each side.

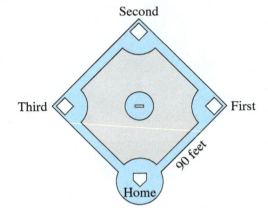

    a. What is the total distance in feet you must run starting from home plate and running the bases back to home? Recall that the total distance around the square is called the **perimeter** of the square.

    b. How many seconds will it take you to run this total distance?

2. Your brother plays Little League baseball. The baseball diamond is still in the shape of a square, but measures 60 feet on each side. What is the total distance if he ran all of the bases?

Definition

The **perimeter of a square** is the total distance around all its edges or sides.

Procedure

Calculating the Perimeter of a Square

The formula for the perimeter, P, of a square whose sides have length s is

$$P = s + s + s + s = 4s.$$

Rectangles

Definition

A **rectangle** is a closed plane figure whose four sides are at right angles to each other.

Examples:

3. You are interested in planting a rectangular garden, 10 feet long by 15 feet wide. To protect your plants, you decide to purchase a fence to enclose your entire garden.

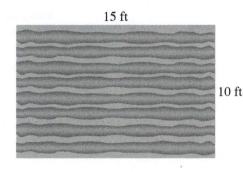

15 ft

10 ft

a. How many feet of fencing must you buy to enclose your garden?

b. A friend suggests first surrounding the garden with a 2-foot-wide path and then enclosing the garden and path with fencing. If you do this, how many feet of fencing must you buy? Explain by including a labeled sketch of the garden and path.

Definition

The **perimeter of a rectangle** is the total distance around all its edges or sides.

Procedure

Calculating the Perimeter of a Rectangle

The formula for the perimeter, P, of a rectangle with length l and width w is

$$P = l + w + l + w = 2l + 2w.$$

Triangles

Definition

A **triangle** is a closed plane figure with three sides.

Examples:

Many houses and garages have roofs supported by trusses. Trusses provide the greatest strength in building design. See accompanying figure.

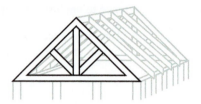

4. Why would a builder or architect need to know about the perimeter of a triangular truss?

5. a. The dimensions of the three sides of a triangular truss are 13 feet, 13 feet, and 20 feet. What is the perimeter of the truss?

   b. If the dimensions of the truss are increased by a scale factor of 2, what is the perimeter of the new truss?

6. If the sides of a triangle measure a, b, and c, write a formula for the perimeter, P, of the following triangle.

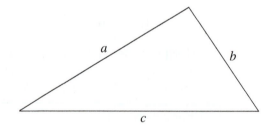

Parallelograms

Definition --

A **parallelogram** is a closed four-sided plane figure whose opposite sides are equal and parallel.

Examples:

7. Where do you see parallelograms in the real world?

8. You decide to install a walkway diagonally from the street to your front steps. The width of your steps is 3 feet, and you determine that the length of the walkway is 23 feet. What is the perimeter of your walkway?

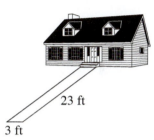

23 ft

3 ft

9. If the sides of a parallelogram measure a and b, write the formula for the perimeter, P, of a parallelogram.

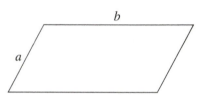

b

a

Trapezoids

Definition

A **trapezoid** is a closed four-sided plane figure that has two sides parallel and two other sides that are not parallel.

Examples:

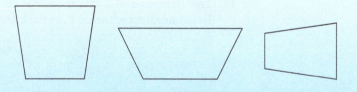

---

10. The sides of a trapezoid measure a, b, c, and d. Write a formula for the perimeter, P, of a trapezoid.

11. You buy a piece of land in the shape of a trapezoid with parallel sides measuring 100 feet, 130 feet, and other sides measuring 40 feet and 50 feet. Draw the trapezoid and calculate its perimeter.

Kites

Definition

A kite is a four-sided plane figure that has two distinct pairs of equal adjacent sides.

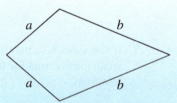

Procedure

Calculating the Perimeter of a Kite

The formula for the perimeter, P, of a kite with short side a and long side b is

$$P = a + a + b + b = 2a + 2b.$$

12. The short side of a kite measures 20 inches and the long side measures 30 inches. Determine the perimeter, P, of the kite.

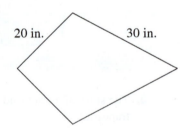

13. You wish to add a decorative thin red cord to the edge of a homemade kite with a short side of 18 inches and a long side of 27 inches. Draw the kite and calculate the length of cord that you need to buy.

Polygons

> **Definition**
>
> A **polygon** is a closed many-sided plane figure.
>
> **Examples:**
>
>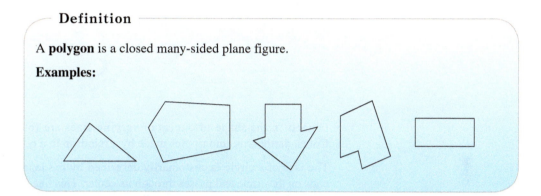

14. Calculate the perimeter of the following polygon.

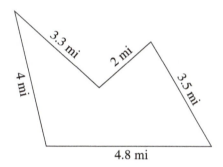

> **Definition**
>
> The **perimeter of a polygon** is the total distance around all its edges or sides.

> **Procedure**
>
> **Calculating the Perimeter of a Polygon**
>
> The perimeter of a polygon is calculated by adding all of the lengths of the sides that make up the figure.

Circles

Definition

A **circle** is a collection of points that are equidistant from a fixed point called the center of the circle.

Examples:

Objects in the shape of circles of varying sizes are found in abundance in everyday life. Coins, dartboards, and ripples made by a raindrop in a pond are just a few examples.

The size of a circle is customarily described by the length of a line segment that starts and ends on the circle and passes through its center. This line segment is called the **diameter** of the circle. The **radius** of a circle is a line segment that connects the center of the circle with any point on the circle. Therefore, the length of the radius is one-half the length of the diameter. The distance around the circle is the perimeter, more commonly called the **circumference**.

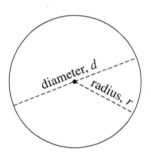

In every circle, the ratio of the circumference, C, to the diameter, d, is always the same $\left(\text{approximately 3.14 or } \dfrac{22}{7}\right)$ and is represented by the Greek letter pi, or π. Most calculators have π keys.

Procedure

Calculating the Circumference of a Circle

The formula for the circumference, C, of a circle with diameter d is

$$\frac{C}{d} = \pi \quad \text{or} \quad C = \pi d.$$

Since $d = 2r$, where r is the radius of the circle, the circumference formula may also be written as

$$C = 2\pi r.$$

---

15. Use the circumference formulas to calculate the circumference of the following circles (not drawn to scale): Use π on your calculator and round your answers to the nearest hundredths.

    a. If the radius is 8 miles, then the circumference = ____.

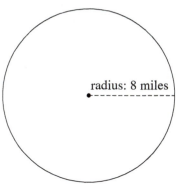

radius: 8 miles

    b. If the diameter is 10 meters, then the circumference = ____.

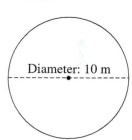

Diameter: 10 m

    c. If the radius of the circle in part a is multiplied by a scale factor 3, what is the circumference of the new circle? How does the circumference compare to the circumference of the original?

SUMMARY: ACTIVITY 5.1

TWO-DIMENSIONAL FIGURE	LABELED SKETCH	PERIMETER FORMULA
Square		$P = 4s$
Rectangle		$P = 2l + 2w$
Triangle		$P = a + b + c$

TWO-DIMENSIONAL FIGURE	LABELED SKETCH	PERIMETER FORMULA
Parallelogram		$P = 2a + 2b$
Trapezoid		$P = a + b + c + d$
Kite		$P = 2a + 2b$
Polygon	A many-sided figure; for example:	P = the sum of the lengths of all the sides.
Circle		$C = \pi d$ or $C = 2\pi r$

EXERCISES: ACTIVITY 5.1

I. During your summer internship at an architecture firm, you learn that many older homes have had additions put on over the years.

    a. In the case of one house that the architects are redesigning, it is discovered that when the house was constructed 100 years ago, it had a very simple rectangular floor plan.

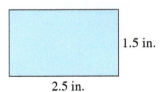

1.5 in.

2.5 in.

If the scale of the blueprint is 1: 120, what were the floor dimensions of the original house?

b. What was the perimeter of the original house in feet?

2. a. Sixty years ago, a rectangular 10-foot by 25-foot garage was added to the original structure.

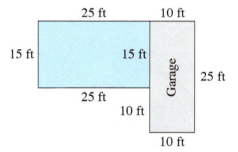

Calculate the perimeter of the floor plan.

b. Twenty-five years ago, a new master bedroom, with the same size and shape of the garage, was added onto the other side of the original floor plan.

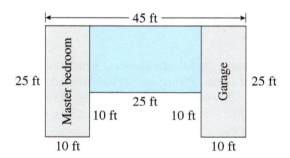

Calculate the perimeter of the remodeled floor plan.

3. Recently, the architects were asked to design a triangular family room as shown in the following scaled floor plan. The scale for the blueprint is 1:120.

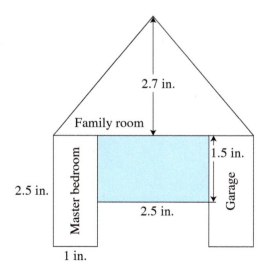

    a. Determine the actual lengths and widths of the rectangular additions and the width and height of the triangular addition.

    b. Determine the total perimeter of the home after the family room addition is completed.

4. A standard basketball court has the following dimensions:

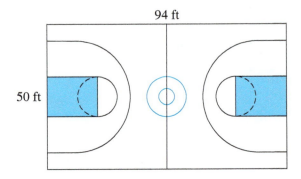

94 ft

50 ft

    a. Calculate the perimeter of the court.

    b. If you play a half-court game, calculate the perimeter of the half-court.

5. The Bermuda Triangle is an imaginary triangular area in the Atlantic Ocean in which there have been many unexplained disappearances of boats and planes. Public interest was aroused by the publication of a popular and controversial book, *The Bermuda Triangle*, by Charles Berlitz in 1974. The triangle starts at Miami, Florida, goes to San Juan, Puerto Rico (1038 miles), then to Bermuda (965 miles), and back to Miami (1042 miles).

    a. What is the perimeter of this triangle?

    b. If you were on a plane that was averaging 600 miles per hour, how long would it take you to fly the perimeter of the Bermuda Triangle?

6. Leonardo da Vinci's painting of *The Last Supper* is a 460 cm by 880 cm rectangle.

    a. Calculate the painting's perimeter.

    b. Would the painting fit in your living room? Explain.

7. Calculate the perimeters of the following figures:

a.

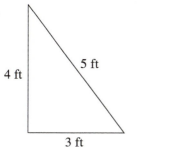

4 ft   5 ft

3 ft

b.   2.5 in.

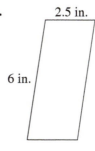

6 in.

c.

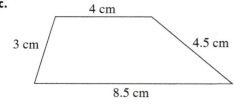

4 cm

3 cm   4.5 cm

8.5 cm

d.   1 mi

1 mi

3 mi

8. If a square has perimeter 64 feet, calculate the length of each side of the square.

9. A rectangle has a perimeter of 75 meters and a length of 10 meters. Calculate its width.

10. The short side of a kite measures 15 inches and the long side is twice as long. Determine the perimeter, P, of the kite.

15 in.

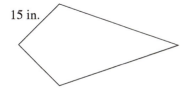

11. You have 124 inches of edging for your kite. The short side of your kite measures 24 inches. What is the length of the long side of your kite?

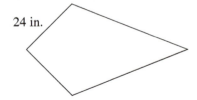

24 in.

12. You would like to build a silver frame for a square-shaped stained-glass ornament to hang in your room. A side of the ornament measures 6 inches. Determine the perimeter of the 6-inch stained-glass ornament.

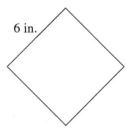

6 in.

For calculations in Exercises 13–17, use π on your calculator and round your answers to the nearest hundredth.

13. You order a pizza in the shape of a circle with diameter 14 inches. Calculate the "length" of the crust (that is, find the circumference of the pizza).

14. United States coins are circular. Choose a quarter, dime, nickel, and penny.

a. Use a ruler to estimate the diameter of each coin in terms of centimeters. Record your results to the nearest tenth of a centimeter.

b. Use a ruler to estimate the circumference of each coin in terms of centimeters. Record your results to the nearest tenth of a centimeter.

c. Check your estimates by using the formulas derived in this lab.

15. Use the appropriate geometric formulas to calculate the circumference for each of the following figures:

a.

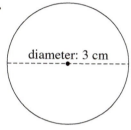

diameter: 3 cm

b.

radius: 3 mi

c.

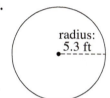

radius: 5.3 ft

d.

radius: 2 in.

16. If a circle has circumference 63 inches, approximate its radius.

17. The number π has an extraordinary place in the history of mathematics. Many books and articles have been written about this curious number. Using the Internet or your local library, research π and report on your findings.

Activity 5.2

Long-Distance Biking

Objectives

1. Calculate the perimeter of many-sided plane figures using formulas and combinations of formulas.

2. Use unit analysis to solve problems involving perimeter.

3. Determine how changes in dimensions affect the perimeter of plane figures.

1. As part of your training for the biking portion of an upcoming triathlon, you choose the following route shown by the solid path in the following figure, made up of partial rectangles and a quarter circle:

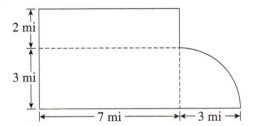

a. Calculate the total length of your bike trip in miles (that is, determine the perimeter of the figure). Use π on your calculator and round to the nearest hundredth.

b. If you can average 9 miles per hour on your bike, how long will it take you to complete the trip?

c. If your bike tires have a diameter of 2 feet, calculate the circumference of the tires to the nearest hundredth.

d. To analyze the wear on your tires, calculate how many rotations of the tires are needed to complete your trip.

e. Participants in the Tour de France bike 3454 kilometers (km). How many days would it take you to complete the race if you average 9 miles per hour?

Predicting the Change in Perimeter

The following problems demonstrate how certain changes in the dimensions of a plane figure affect its perimeter in a predictable way.

2. A local artist has donated a three-tiered sculpture for a July 4th fund-raiser. The top surface of each layer is a square. The square surface of the top layer measures 12 inches on each side. The artist plans to decorate the edges of the top surface of each layer with a ribbon of red, white, and blue around the edges.

a. How long of a ribbon does the artist need for the top layer?

b. Each edge of the top surface of the middle layer will be 18 inches long. How long of a ribbon is needed for the middle layer?

c. In part b, notice that each edge of the top surface of the middle layer is 6 inches longer than the edge of the top surface of the top layer. Is the perimeter of the top surface of the middle layer also 6 inches longer than the perimeter of the top surface of the top layer?

d. Each edge of the top surface of the bottom layer will be 24 inches. How long of a ribbon is needed for the bottom layer?

e. The length of each edge of the top surface of the bottom layer is twice the length of edge of the top surface of the top layer. Is the perimeter of the top surface of the bottom layer twice the perimeter of the top surface of the top layer?

3. In Problem 2e, it appears that if the side of a square is multiplied by a certain scale factor, its perimeter is affected in a predictable way. You can use algebra to prove this observation. Let the side of a square with length s be increased by a scale factor of $k > 0$.

a. Write an expression that represents the perimeter of the new square.

b. Compare the result in part a with the expression for the perimeter of the original square.

c. Complete the following statement: If the side of a square is multiplied by a scale factor k, then its perimeter is _____.

4. Recall that the length of the edge of the top surface of the middle square layer (18 inches) was 6 inches longer than the edge of the top surface of the top layer (12 inches). Although the perimeter of the top surface of the middle layer (72 inches) is not 6 inches more than the perimeter of the top surface of the top layer (48 inches), there is a relationship between the perimeters of these two squares. Determine this relationship. *Hint:* Substitute $s + 6$ for s in the formula for the perimeter of a square.

5. A landscape architect has designed a rectangular garden plot that measures l feet by w feet.

a. Write an expression for the perimeter of the garden.

b. She would like to enlarge the garden by increasing the length by 5 feet and the width by 3 feet. Write an expression that represents the amount of fencing she would need for her new garden.

c. If both the length and width of the rectangular garden are increased by 5 feet, how is the perimeter of the new garden related to the perimeter of the original garden?

d. From your observations in parts a and b, complete the following: If $a > 0$ is added to the length of a rectangle and $b > 0$ is added to its width, then the perimeter increases by _____.

e. Suppose the landscape architect decides to enlarge the rectangular garden by doubling (a scale factor of 2) the length and width of the garden instead. Write an expression that represents the amount of fencing she would need to fence in her new garden.

f. Compare the amount of fencing used in the original garden (see part a) with the amount needed for the garden with dimensions doubled (see part e).

g. If both the length and width of a rectangle are multiplied by a scale factor k, how is the perimeter of the new rectangle related to the perimeter of the original?

SUMMARY: ACTIVITY 5.2

I. A many-sided closed plane figure is a composite of basic plane figures: squares, rectangles, parallelograms, triangles, trapezoids, and circles.

2. To calculate the perimeter of a many-sided plane figure,

  i. Determine the length of each composite part of the perimeter.

  ii. Add the composite lengths to obtain the figure's total perimeter.

3. a. If the side of a square is increased by a factor of $k > 0$, then its perimeter is multiplied by k.

  b. If a number $k > 0$ is added to the length of a square, its perimeter is increased by $4k$ units.

  c. If $a > 0$ is added to the length of a rectangle and $b > 0$ is added to its width, then the perimeter increases by $2a + 2b$.

EXERCISES: ACTIVITY 5.2

1. You plan to fly from New York City to Los Angeles via Atlanta and return from Los Angeles to New York City via Chicago.

Determine the perimeter of the polygon that describes your trip.

2. A Norman window is rectangular on three sides with a semicircular top. You decide to install a Norman window in your family room with the dimensions indicated in the diagram. What is its perimeter?

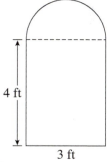

4 ft

3 ft

3. Stonehenge is an ancient site on the plains of southern England consisting of a collection of concentric circles (circles with the same center) outlined with large sandstone blocks. Carbon dating has determined the age of the stones to be approximately 5000 years. Much curiosity and mystery has surrounded this site over the years. One theory about Stonehenge is that it was a ritualistic prayer site. However, even today, there is still controversy over what went on there. The one thing everyone interested in Stonehenge can agree on is the mathematical description of the circles:

The diameters of the four circles are 288 feet for the largest circle, 177 feet for the next, 132 feet for the third, and 110 feet for the innermost circle.

a. For each time you walked around the outermost circle, how many times could you walk around the innermost circle?

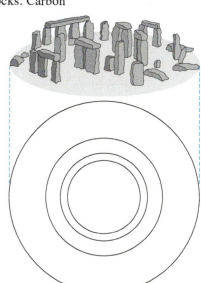

b. What is the ratio of the diameter of the outermost circle to that of the innermost circle?

c. How does this ratio compare to the ratio of their circumferences from part a?

d. What is the ratio of the diameter of the larger intermediate circle to that of the smaller intermediate circle?

e. Predict how much longer the trip around the larger intermediate circle would be than around the smaller intermediate circle.

4. Calculate the perimeter for each of the following figures:

a.

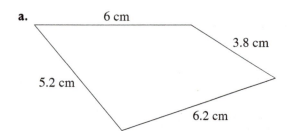

b.   *Note*: All sides of the star are equal in length.

3 ft

c.

10 m $\frac{1}{2}$ m

3 m

2 m

d.   300 yards   *Note*: The sides of the track are semicircles.

150 yards

5. a. Use the perimeter formula to calculate the perimeter of a square with side 5 cm.

b. From your answer to part a, what do you expect to be the perimeter of a square with side 30 cm?

c. Use the perimeter formula to calculate the perimeter of a square with side 30 cm.

d. Was your prediction correct? That is, does your answer to part b match your answer to part a?

6. a. Determine the ratio of the length of the larger rectangle to the length of the smaller rectangle. Then, determine the ratio (larger to smaller) of their widths.

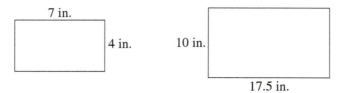

7 in.

4 in.

10 in.

17.5 in.

b. What relationship do you expect the perimeters of the two rectangles to have?

c. Calculate the two perimeters and check your prediction.

7. You enjoy playing darts. You decide to make your own dartboard consisting of four concentric circles (that is, four circles with the same center). The smallest circle, C_1 (the "bull's-eye"), has radius 1 cm, the next largest circle, C_2, has radius 3 cm, the third circle, C_3, has radius 6 cm, and the largest circle, C_4, has radius 10 cm. You decide to compare the circumferences of the circles.

a. What are these circumferences?

b. Compare the radii of circles C_1 and C_2. Compare their circumferences. Is there a connection between the relationship of the radii and the relationship of the circumferences?

c. Compare the radii of circles C_3 and C_2. Compare their circumferences. Is there a connection between the relationship of the radii and the relationship of the circumferences?

d. Compare the radii of circles C_1 and C_4. What would you predict the circumference of C_4 to be? Compare their circumferences. Was your prediction correct?

e. Complete the following: If the radius of a circle is multiplied by a scale factor $k > 0$, then the circumference is _____.

8. a. Calculate the circumference of a circle of radius 4.

b. Use part a to predict the circumference of a circle of radius 6, without using the circumference formula.

c. Use the circumference formula to calculate the circumference of a circle of radius 6.

d. Does your answer to part c match your answer to part b?

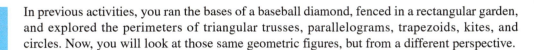

In previous activities, you ran the bases of a baseball diamond, fenced in a rectangular garden, and explored the perimeters of triangular trusses, parallelograms, trapezoids, kites, and circles. Now, you will look at those same geometric figures, but from a different perspective.

Activity 5.3

Walking around, Revisited

Objectives

1. Write area formulas for squares, rectangles, parallelograms, triangles, trapezoids, kites, and polygons.

2. Calculate the area of polygons using appropriate formulas.

3. Determine how changes in dimensions affect the area of plane figures.

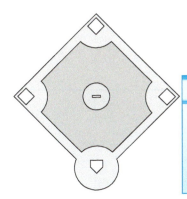

Squares

As groundskeeper for the local baseball team you need to guarantee good-quality turf for the infield, that is, the space enclosed by the base lines. In order to plant and maintain this turf, you need to measure this space. It makes sense to count the number of square units needed to cover the space and label this total number as the area of the figure.

1. Recall that a regulation baseball diamond is a square with 90-foot sides. A square unit (or unit square) in this case is a square with each side measuring 1 foot. How many square units are needed to cover this baseball diamond? Your answer in square feet is called the area of the square.

2. The Little League diamond has sides measuring 60 feet. Calculate the area of the Little League baseball diamond.

Definition

The **area of a square**, or any polygon, is the measure, in square units, of the region enclosed by the sides of the polygon.

Procedure

Calculating the Area of a Square

The formula for the area, A, of a square with sides of length s is

$$A = s \cdot s = s^2.$$

3. An interior designer is helping you select floor treatments for your house. In the living room, there is a square carpet measuring 3 feet on each side under a table.

a. What is the area of the carpet?

b. There is a square carpet with sides 4 times as long under the dining room table. What is the area of the second carpet?

c. Compare the areas of the two carpets. Is there a connection between the relationship of the sides and the relationship of the areas?

d. In general, how does multiplying the length of a square by a scale factor k affect its area? Explain by replacing s with ks in the area formula for a square and comparing the results.

Rectangles

Planting your rectangular 15-foot by 10-foot garden requires that you know how much space it contains.

4. How many square feet are required to cover your garden? Include a sketch to explain your answer.

5. Will doubling the length and width of your garden double its area? Explain.

Procedure

Calculating the Area of a Rectangle

The formula for the area of a rectangle with length l and width w is

$$A = lw.$$

Parallelograms

Once you know the formula for the area of a rectangle, then you have the key for determining the formula for the area of a parallelogram. A parallelogram is formed by two intersecting pairs of parallel sides.

6. If you "straightened up" the parallelogram into the shaded rectangle, does the area change? Explain.

7. Use your observations from Problem 6 to calculate the area of the parallelogram with height 4 inches and base 7 inches. Note that the height of a parallelogram is defined as the perpendicular (right angle) distance from the base to its parallel side.

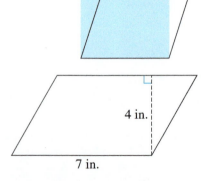

4 in.

7 in.

> **Procedure**
>
> **Calculating the Area of a Parallelogram**
>
> The formula for the area of a parallelogram with base b (the length of one side) and height h (the perpendicular distance from the base b to its parallel side) is
>
> $$A = bh.$$

Triangles

Triangles can always be thought of as one-half a rectangle or parallelogram.

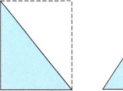

8. For each of the following triangles, draw the rectangle or parallelogram that encloses the triangle.

9. Use the idea that a triangle is $\frac{1}{2}$ of a rectangle to write a formula for the area of a triangle in terms of its base b and height h. Note that the height is the perpendicular distance from the base to the vertex, v, opposite it and that a vertex is a point where two sides of a triangle intersect. Explain.

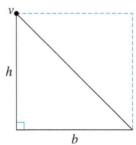
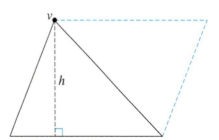

> **Procedure**
>
> **Calculating the Area of a Triangle**
>
> The formula for the area of a triangle with base b (the length of one side) and height h (the perpendicular distance from the base b to the vertex opposite it) is
>
> $$A = \frac{1}{2}bh.$$

10. Use the formula above to calculate the areas of the following triangles:

a.

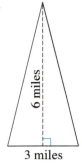

6 miles

3 miles

b.

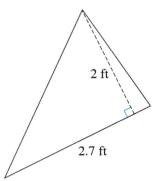

2 ft

2.7 ft

c.

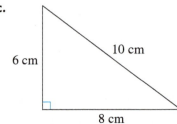

6 cm

10 cm

8 cm

d.

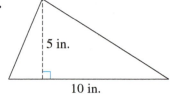

5 in.

10 in.

Trapezoids

Trapezoids may be viewed as one-half of the parallelogram that is formed by adjoining the trapezoid to itself turned upside down, as shown.

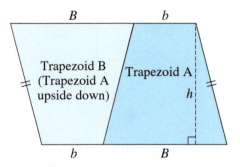

11. a. The parallelogram shown above has base $b + B$ and height h. Determine the area of this parallelogram.

b. Use the idea that a trapezoid is $\frac{1}{2}$ of a parallelogram to write a formula for the area A of a trapezoid in terms of its bases b and B and its height h.

Procedure

Calculating the Area of a Trapezoid

The formula for the area of a trapezoid with bases b and B and height h is

$$A = \frac{1}{2}h(b + B).$$

---

12. Calculate the area of the trapezoid with height 5 feet and bases 6 feet and 11 feet.

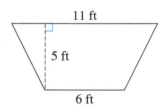

11 ft

5 ft

6 ft

13. If the lower base of the trapezoid in Problem 12 is increased by 2 feet and the upper base is decreased by 2 feet, draw the new trapezoid and compute its new area.

Kites

Procedure

Calculating the Area of a Kite

The formula for the area A, of a kite with short diagonal, d, and long diagonal, D, is

$$A = \frac{1}{2}d \cdot D.$$

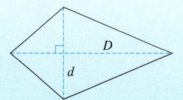

D

d

14. Calculate the area for each of the following kites.

a.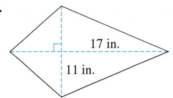

17 in.

11 in.

b.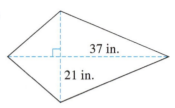

37 in.

21 in.

15. You need to make a kite as part of a school project. You are given two straight sticks measuring 24 inches and 12 inches to use for diagonals. Determine the area of your kite.

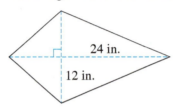

24 in.

12 in.

Polygons

You can determine the area of a polygon by seeing that polygons can be broken up into other more familiar figures, such as rectangles and triangles. For example, the front of a garage shown to the right is a polygon that can also be viewed as a triangle sitting on top of a rectangle:

To determine the area of this polygon, use the appropriate formulas to determine each area, and then sum your answers to obtain the area of the polygon.

16. Calculate the area of the front of the garage:

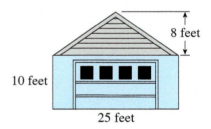

8 feet

10 feet

25 feet

17. You are planning to build a new home with the following floor plan:

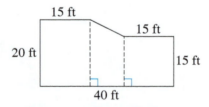

15 ft

20 ft

15 ft

15 ft

40 ft

Calculate the total floor area.

18. Calculate the areas of the following polygons:

a.

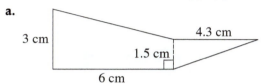

b.

In the following problem, you will explore the effect on the surface area of an object when the dimensions of the object are multiplied by a scale factor k. You will use a rectangle in the exploration, but the result is true for any object.

19. a. The dimensions of a 2-foot by 3-foot poster are tripled. What is the scale factor? What are the dimensions of the enlargement?

b. Determine the surface area of the original poster and the enlargement.

c. Is the surface area of the enlargement three times the surface area of the original poster? Determine a relationship between the surface areas of the poster, its enlargement, and the scale factor.

d. Complete the following table.

Dimension	2 by 3	4 by 6	6 by 9	8 by 12	10 by 15
Dimension Scale Factor	1				
Surface Area	6				

e. If the size of the 2 by 3 poster is increased by a scale factor k, the surface area of the larger poster is _____ times the surface area of the original poster.

In 1638, in his *Two New Sciences*, Galileo Galilei first described the effect on an object's surface area if its size is increased or decreased by a scale factor k. The so-called "Square–Law" states that

> When an object undergoes an increase (or decrease) in size by a scale factor, its new surface area is proportional to the square of the scale factor. If A_1 represents the area of the original object, and A_2 represents the area of the new object, then

$$A_2 = k^2 A_1, \text{ where } k \text{ is the scale factor.}$$

SUMMARY: ACTIVITY 5.3

TWO-DIMENSIONAL FIGURE	LABELED SKETCH	AREA FORMULA
Square		$A = s^2$
Rectangle		$A = lw$
Triangle		$A = \frac{1}{2}bh$
Parallelogram		$A = bh$
Trapezoid		$A = \frac{1}{2}h(b + B)$
Kite		$A = \frac{1}{2}d \cdot D$
Polygon	A many-sided figure that is subdivided into familiar figures whose areas are given by formulas in this summary.	$A =$ sum of the areas of each figure

EXERCISES: ACTIVITY 5.3

1. A high school theater arts teacher has designed a background set for a production of *Hairspray*. She hopes to enter *Hairspray* in the Boston Summer High School Musical Theatre competition. The set is to be 10 feet wide and 8 feet high. It will be made of plywood with a wooden frame.

   a. What area of plywood is needed? How many feet of framing are needed for the perimeter?

   b. After some problems with concealing equipment near the wings, she decides to double the width of the background set, so it is now 20 feet wide. What is the new area of the set, and what is its new perimeter?

   c. Is the change in the area proportional to the change in width?

   d. Is the change in the length of the framing proportional to the change in width?

2. You are carpeting your living room and sketch the following floor plan:

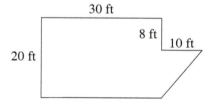

   a. Calculate the area that you need to carpet.

   b. Carpeting is sold in 10-foot-wide rolls. Calculate how much you need to buy. Explain.

3. You need to buy a solar cover for your 36-foot by 18-foot rectangular pool. A pool company advertises that solar covers are on sale for $1.77 per square foot. Determine the cost of the pool cover before sales tax.

4. Calculate the areas of each of the following figures:

a.
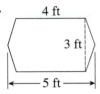

b.
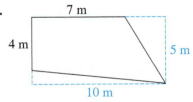

5. How would you break up the following star to determine what dimensions you need to know in order to calculate its area? Explain.

6. A standard basketball court is a rectangle with length 94 feet and width 50 feet. How many square feet of flooring would you need to purchase in order to replace the court?

7. You are planning to build a new garage on your home and need to measure the length and width of your cars to help you estimate the size of the double garage. Your car measurements are:

Car 1: 14 ft. 2 in. by 5 ft. 7 in.
Car 2: 14 ft. 6 in. by 5 ft. 9 in.

a. Based on these measurements, what would be a reasonable floor plan for your garage? Explain.

b. What is the area of this floor plan?

c. You want to produce a model of this project. You want the model to fit comfortably on an 8.5″ by 11″ sheet of paper. What scale would you use?

8. You need to make a kite as part of a community contest. You are given two straight sticks measuring 35 in. and 18 in. to use for diagonals. Determine the surface area of your kite.

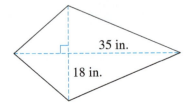

9. You wish to make a kite out of a piece of material that is 2 feet by 4 feet.

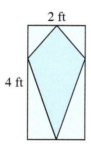

2 ft

4 ft

a. Calculate how much material you have by finding the area of the rectangle.

b. Calculate the area of the kite.

c. How many square feet of material will you have left over?

10. Two neighbors, Ahmad and Marisa, are having their rectangular driveways paved.

    a. Ahmad's driveway measures 10 feet by 75 feet. What is the area of his driveway?

    b. Marisa has decided to pave a large rectangular parking area to accommodate the many cars in the family. This space is twice as long and three times as wide as Ahmad's driveway. Calculate the length, width, and the area of Marisa's new parking area.

    c. What is the ratio of the area of Ahmad's driveway to that of Marisa's parking area?

    d. Can you predict the area of Marisa's parking area from that of Ahmad's driveway by multiplying an amount related to the change in width and length?

    e. The results in part d suggest that if one side of a rectangle is multiplied by $a > 0$, and the other side is multiplied by $b > 0$, then its area is multiplied by ab. Prove this conjecture by comparing the area of a rectangle having length l and width w to the area of a new rectangle formed by multiplying the width by a and the length by b.

11. Your family has purchased a new A-frame vacation home in the mountains. The house contains several triangular windows that match the shape of the roof. In the living room, there is a large picture window shaped like an isosceles triangle (two sides of equal length). The base of the window is 12 feet long and the equal sides measure 10 feet.

    a. What is the perimeter of the living room window?

b. What is the area of the living room window?

c. In the back of the house, there is a bedroom with a smaller isosceles triangle–shaped window that measures 6 feet across the base and 4 feet in height. What is the perimeter of the bedroom window?

d. What is the area of the bedroom window?

e. What is the relationship between the length of the sides of the bedroom window and the sides of the living room window?

f. What is the relationship between the perimeter of the bedroom window and the perimeter of the living room window?

g. What is the relationship between the area of the bedroom window and the area of the living room window?

h. Complete the following: If all sides of a triangle are multiplied by the same scale factor $k > 0$, then the perimeter is _____ and the area is _____.

12. Art students make a 3-foot by 5-foot fully painted model of a mural that they hope will be approved for a wall near the auditorium. Their proposal includes several different size murals for a wall having the following dimensions: 9 feet high and 20 feet long.

a. Complete the following table with lengths that will maintain the width to height ratio of the original mural.

Height	3	4.5	6	7.5	9
Width	5				

b. Write an equation that shows the relationship between the height and the width. Let x represent the height and y the width. Is the height of the mural proportional to its width? Explain.

c. Determine the scale factor for the dimensions of each proposed mural based on the original model. Remember that the dimensions of the smaller rectangle multiplied by the scalar factor equals the dimensions of the larger rectangle.

Dimensions (feet)	3′ by 5′	4.5′ by 7.5′	6′ by 10′	7.5′ by 12.5′	9′ by 15′
Scale Factor of Dimensions	1				

There is some concern about the cost of paint and supplies. The cost of paint and supplies for the 3-foot by 5-foot model mural was $70. Students are asked to determine the cost of the paint and supplies for each size mural based on its surface area.

d. Calculate the surface area of each mural, and record your results in the table below.

e. Determine the scale factor (ratio) of the surface area of each larger mural to the surface area of the original model. Record your results in the appropriate place in the table.

f. Determine the cost of the paint and supplies for each size mural based on its surface area. Assume the ratio of cost to surface area remains proportional to the ratio of the cost to surface area of the original mural. Record your results in the table.

Dimensions (feet)	3′ by 5′	4.5′ by 7.5′	6′ by 10′	7.5′ by 12.5′	9′ by 15′
Surface Area	15				
Scale Factor of Surface Area	1				
Cost	$70				

g. Compare the scale factors of the dimensions of each of the murals in the table in part c with the corresponding scale factors of the surface area of each of the murals in part f. What is the relationship between the dimension scale factors and the corresponding surface area scale factors?

h. What is the largest size proposed mural that can be approved if the school board has approved a budget of $500 for this project?

Activity 5.4

How Big Is That Circle?

Objectives

1. Develop a formula for the area of a circle.

2. Use the formula to determine areas of circles.

Determining the area of a circle becomes a challenge because there are no straight sides. No matter how hard you try, you cannot neatly pack unit squares inside a circle to completely cover the area of a circle. The best you can do in this way is to get an approximation of the area. In this activity you will explore methods for estimating the area of a circle and develop a formula to calculate the exact area.

Definition

The **area of a circle** is the measure of the region enclosed by the circumference of the circle.

1. a. To help understand the formula for a circle's area, start by folding a paper circle in half, then in quarters, and finally halving it one more time into eighths.

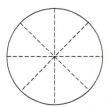

b. Cut the circle along the fold lines into eight equal pieces (called *sectors*) and rearrange these sectors into an approximate parallelogram (see accompanying figure).

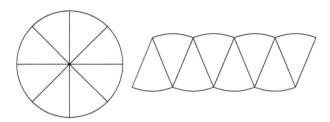

c. What measurement on the circle approximates the height of the parallelogram? Explain.

d. What measurement on the circle approximates the base of the parallelogram?

e. Recall that the area of a parallelogram is given by the product of its base and its height. Use your answers to parts c and d to determine the area of the parallelogram and the approximate area of the circle.

Imagine cutting a circle into more than eight equal sectors. Each sector would be thinner. When reassembled, as in Problem 1, the resulting figure will more closely approximate a parallelogram. Hence, the formula for the area of a circle, $A = \pi r^2$, is even more reasonable and accurate.

> **Procedure**
>
> **Calculating the Area of a Circle**
>
> The formula for the area A of a circle with radius r is
>
> $$A = \pi r^2.$$

2. You own a circular dartboard of radius 1.5 feet. To figure how much space you have as a target, calculate the area of the dartboard. Be sure to include the units in your answer.

3. The diameter of a circle is twice its radius. Use this fact, with the formula $A = \pi r^2$, to derive a formula for the area of a circle in terms of its diameter. Show the steps you took.

4. Calculate the area for each of the following circles:

a.
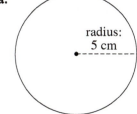
radius: 5 cm

b.
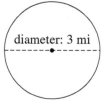
diameter: 3 mi

5. a. True or False: If you double the size of the diameter of a circle, then you double the area of the circle.

b. If the radius of a circle is multiplied by scale factor $k > 0$, then its area is multiplied by _____.

SUMMARY: ACTIVITY 5.4

Two-Dimensional Figure	Labeled Sketch	Area Formula
Circle	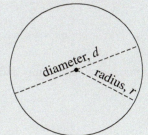	$A = \pi r^2$ (involving the radius) $A = \dfrac{\pi d^2}{4}$ (involving the diameter)

EXERCISES: ACTIVITY 5.4

1. You order a pizza with diameter 14 inches. Your friend orders a pizza with diameter 10 inches. Compare the areas of the two pizzas to estimate approximately how many of the smaller pizzas can fit into the larger pizza.

2. You enjoy playing darts and decide to make your own simplified dartboard consisting of four concentric circles (that is, four circles with the same center, as shown in the diagram below). The smallest circle (the bull's-eye) has radius 1 cm, the next largest circle has radius 3 cm, the third has radius 6 cm, and the largest circle has radius 10 cm. In order to assign point values to the dartboard, you decide to compare the areas of the annular regions (the space between circles) on the board. There are two rules for your game:

    i. The three annular regions and the innermost circle on the board are assigned point values, based on the size of their areas, with the most points given to the smallest area, and

    ii. The winner is the person to score the highest number of points.

a. What are the areas of the four circles?

b. Using the areas from part a, compute the areas of each of the annular areas on the dartboard.

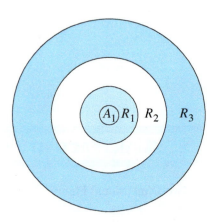

c. Based on the areas of the annular regions, assign reasonable point values to each of the regions. Explain how you used the areas to assign the point values to each of the regions.

3. U.S. coins are circles of varying sizes. Choose a quarter, dime, nickel, and penny and

a. Use a ruler to estimate the diameter of each coin.

b. Calculate the area of each coin.

4. Use an appropriate formula to calculate the area for each of the following figures:

a.

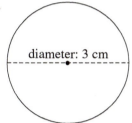

diameter: 3 cm

b.

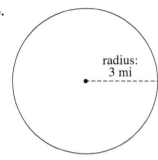

radius:
3 mi

c.

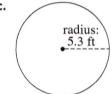

radius:
5.3 ft

d.

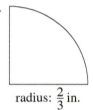

radius: $\frac{2}{3}$ in.

5. You need to polish your circular dining room table whose diameter is 7 feet. The label on the polish can claim coverage for 100 square feet, but you notice that the can is only about one-half full. Will you have enough polish to finish your table? Explain.

6. a. Calculate the area of a circle of radius 4 inches.

b. Use part a to predict the area of a circle of radius 6 inches, without using the area formula.

c. Use the area formula to calculate the area of a circle of radius 6 inches.

d. Does your answer to part c match your answer to part b?

Project 5.5

A New Pool and Other Home Improvements

Objectives

1. Solve problems in context using geometric models.

2. Distinguish between problems requiring area formulas and perimeter formulas.

Your family is the proud owner of a new circular swimming pool with a diameter of 25 feet and you are eager to dive in. However, you quickly discover that having a new pool requires making many decisions about other purchases.

1. The first concern is a solar pool cover. Online research indicates that circular pool covers come in the following sizes: 400, 500, and 600 square feet. Friends recommend that you buy a pool cover with very little overhang. Which size is best for your needs? Explain.

2. You decide to build a circular concrete patio 6 feet wide all around the pool. Draw a sketch of your pool with the patio. What is the area covered by your patio? Explain.

3. State law requires that all pools be enclosed by a fence to prevent accidents. You decide to completely enclose the pool and patio with a stockade fence. How many feet of fencing do you need? Explain.

4. Next, you decide to stain the new fence. The paint store recommends a stain that covers 500 square feet per gallon. If the stockade fence has a height of 5 feet, how many gallons of stain should you buy? Explain.

5. Last, you decide to plant a circular flower garden near the pool and patio but outside the fence. You measure that a circle of circumference 30 feet would fit. What is the length of the corresponding diameter of the flower garden?

Other Home Improvements

Now that the new pool, patio, and flower garden are finished, your parents want to do some other home improvements in anticipation of enjoying the new pool with family and friends.

They have budgeted $1000 for this purpose and the to-do list looks like this:

- replace the kitchen floor

- add a wallpaper border to the third bedroom

- paint the walls in the family room

To stay within the budget, your parents consult with the architect who designed your new family room. Your parents, along with you, expect to do the work yourselves, so the only monetary cost will be for materials.

6. The kitchen floor is divided into two parts. The first section is rectangular and measures 12 by 14 feet. The second section is a semicircular breakfast area that extends off the 14-foot side. The cost of vinyl flooring is $21 per square yard plus 6 percent sales tax. The vinyl is sold in 12-foot widths.

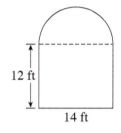

12 ft

14 ft

    a. How long a piece of vinyl flooring will need to be purchased if you want only one seam where the breakfast area meets the main kitchen, as shown? Remember that the vinyl is 12 feet wide.

    b. How many square feet of flooring must be purchased? How many square yards is that? (9 square feet = 1 square yard)

    c. How much will the vinyl flooring cost including tax?

    d. How many square feet of flooring will be left over after it's installed? Explain.

7. The third bedroom is rectangular in shape and has dimensions of $8\frac{1}{2}$ by 13 feet. On each 13-foot side, there is a window that measures 3 feet 8 inches wide and 4 feet high. The door is located on an $8\frac{1}{2}$-foot side and measures 3 feet wide from edge to edge. Your parents plan to put up a decorative horizontal wallpaper stripe around the room about halfway up the wall.

    a. How many feet of wallpaper stripe will need to be purchased?

b. The border comes in rolls 5 yards in length. How many rolls will need to be purchased?

c. The wallpaper border costs $10.56 per roll plus 6% sales tax. Determine the cost of the border.

d. How many feet of wallpaper will be left over after it's installed?

8. The newly constructed family room needs to be painted. It has a cathedral ceiling with front and back walls that measure the same, as shown in the diagram on right. The two side walls measure 14 feet long and 12 feet high. Each of the walls will need two coats of paint. The ceiling will not be painted.

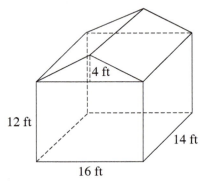

a. How many square feet of wall surface will be painted? (Remember, all walls will need two coats.)

b. Each gallon of paint covers approximately 400 square feet. How many gallons of paint will need to be purchased?

c. The paint costs $19.81 per gallon plus 6 percent sales tax. What is the total cost of the paint you need for the family room?

9. The architect provides you with the cost for all of these home-improvement projects (not including the new pool, patio, and flower garden). What is this amount?

10. Additional costs for items such as paint rollers and wallpaper paste amount to approximately $30. Can you afford to do all the projects? Explain.

1. Area formulas are used when you are measuring the amount of region *inside* a plane figure.

2. Perimeter or circumference formulas are used when you are measuring the length *around* a plane figure.

Additional Applications

11. The driveway is rectangular in shape and measures 15 feet wide and 25 feet long. Calculate the area of the driveway.

12. A flower bed in the corner of the yard is in the shape of a right triangle. The perpendicular sides of the bed measure 6 feet 8 inches and 8 feet 4 inches. Calculate the area of the flower bed. What are the units of this area?

13. The front wall of the storage shed is in the shape of a trapezoid. The bottom measures 12 feet, the top measures 10 feet, and the height is 6 feet. Calculate the area of the front wall of the shed.

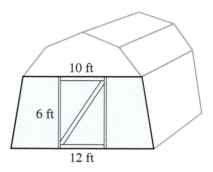

14. Your grandmother lives in a small, one-bedroom condo. Your parents want to surprise her with new wall-to-wall carpeting, but first they need to know the area of the floor space. The bedroom is 10 by 12 feet, the living room is 12 by 14 feet, the kitchen is 8 by 6 feet, and the bathroom is 5 by 9 feet. Calculate the total floor space (area) of her condo.

15. There is a large rectangular window in the living room with dimensions of 10 feet wide by 6 feet tall. In order to fit a new digital TV on the wall, your mother wants to redesign the window so that it admits the same amount of light but is only 8 feet wide. How tall should the redesigned window be?

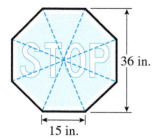

16. A stop sign is in the shape of a regular octagon (an eight-sided polygon with equal sides and angles). A regular octagon can be created using eight triangles of equal area. One triangle that makes up a stop sign has a base of 15 inches and a height of 18 inches. Calculate the area of the stop sign.

17. How does the area of the stop sign in Exercise 16 compare with the area of a circle of radius 18 inches? Explain.

18. Your uncle and aunt are buying plywood to board up their windows in preparation for Hurricane Euclid. In the master bedroom, they have a Norman window (in the shape of a rectangle with a semicircular top). So, they need your help to calculate the area of the window.

    a. If the rectangular part of the window is 4 feet wide and 5 feet tall, what is the area of the entire window?

    b. The master bathroom also has a Norman window whose dimensions are half the size of the one in the master bedroom. Without using the area formula again, determine how much plywood will be needed to board up both windows.

19. The diameter of the Earth is 12,742 km; the diameter of the Moon is 3476 km.

    a. If you flew around Earth by following the equator at a height of 10 km, how many trips around the Moon could you take in the same amount of time, at the same height from the Moon, and at the same speed? Explain.

    b. The circle whose circumference is the equator is sometimes called a "great circle" of the Earth or of the Moon. Compare the areas of the great circles of Earth and the Moon.

Activity 5.6

How Big Is That Angle?

Objectives

1. Measure the size of angles with a protractor.

2. Classify triangles as equiangular, equilateral, right, isosceles, or scalene.

Equipment

In this laboratory activity, you will need the following equipment:

1. A 12-inch ruler

2. A protractor

You may have wondered why a right angle measures 90 degrees. Why not 100 degrees? As with much of the mathematics that we use today, the measurement of angles has a rich history, going back to ancient times when navigation of the oceans and surveying the land were priorities. The ancient Babylonian culture used a base 60 number system, which at least in part led to the circle being divided into 360 equal sectors. The angle of each sector was simply defined to measure 1 degree and so today we say there are 360 degrees (abbreviated 360°) in a circle. In this activity you will be using degrees to measure angles. The illustration shows a sector having an angle of 20 degrees $\left(\dfrac{1}{18} \text{ of a circle} \right)$.

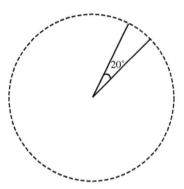

In the following diagram, the circle is divided into four equal sectors by two perpendicular lines. The angles of the four sectors must add up to 360 degrees. Since the angles are equal, dividing 360° by 4 results in each angle measuring 90°. Recall that 90° angles are called **right angles**.

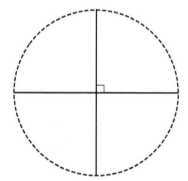

1. How many degrees are in the angle of a sector that is exactly half of a circle? Such an angle is called a **straight angle**.

Measuring Angles

> **Procedure**
>
> **Measuring Angles with a Protractor**
>
> To measure angles, you can select certain tools, such as a protractor, to help you determine the size of angles in degrees. Place the vertex of the angle at the center of the protractor (often a hole in the center of the baseline) and place one side of the angle along the baseline of the protractor. Where the other side of the angle meets the appropriate scale on the semicircle is the measure of the angle in degrees.
>
>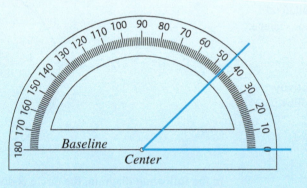

> **Example 1**   *The following angle measures 75°.*
>
>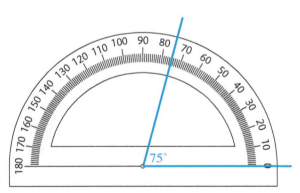

2. Use a protractor to measure the size of each of the following angles.

a.   **b.**

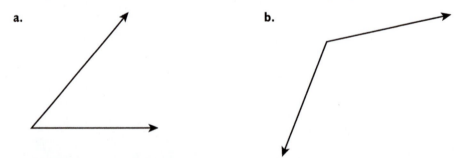

> **Definitions**
>
> An **acute angle** is any angle that is smaller than a right angle. Its degree measure is less than 90°.
>
> An **obtuse angle** is any angle that is larger than a right angle. Its degree measure is greater than 90°.

It is a well-known theorem in geometry that the sum of the measures of the angles of a triangle *must* equal 180°.

3. Verify this theorem by carefully measuring the angles of each of the following triangles, recording the results in the table below. You will have to extend the sides of each triangle to use your protractor effectively.

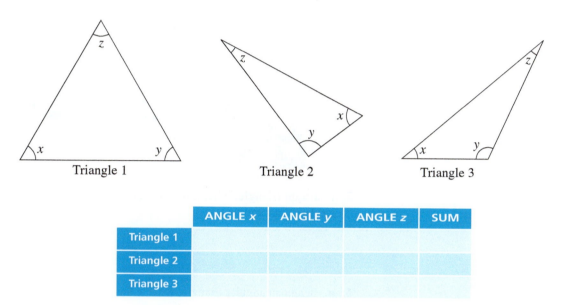

Triangle 1   Triangle 2   Triangle 3

	ANGLE *x*	ANGLE *y*	ANGLE *z*	SUM
Triangle 1				
Triangle 2				
Triangle 3				

Classifying Triangles

Sometimes, it is useful to classify triangles in terms of special properties of their sides or angles. These classifications are summarized in the following table.

TRIANGLE CLASSIFICATIONS	DEFINITION
Equilateral	All three sides have the same length.
Isosceles	Two sides have the same length. The two angles opposite the equal sides will also have the same measure.
Scalene	None of the sides have the same length.
Equiangular	All three angles have the same measure. An equiangular triangle is also an equilateral triangle.
Right	One angle measures 90°.
Acute	All angles measure less than 90°.
Obtuse	One angle measures greater than 90°.

4. Use your protractor to measure the acute angles in the following right triangles, recording your results in the table. What is the sum of the two acute angles in each triangle?

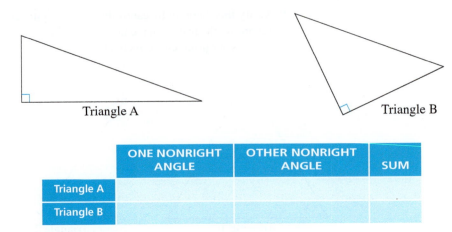

Triangle A   Triangle B

	ONE NONRIGHT ANGLE	OTHER NONRIGHT ANGLE	SUM
Triangle A			
Triangle B			

5. In a right triangle, what must be true about the two angles that are not right angles? Explain your answer.

6. a. In an equiangular triangle, what is the measure of each angle?

b. In an isosceles triangle, if one angle measures 110°, what is the measure of the other two equal angles?

c. Consider the following three triangles:

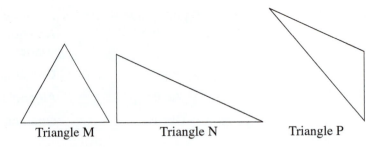

Triangle M   Triangle N   Triangle P

i. Choose the triangles that are scalene, and explain why. Then use a protractor to measure the angles.

ii. Choose the triangle(s) that is (are) obtuse and explain why.

iii. Choose the triangle(s) that is (are) acute and explain why.

SUMMARY: ACTIVITY 5.6

CONCEPT/SKILL	DESCRIPTION	EXAMPLE
1. Using a protractor	Place the vertex of the angle to be measured at the center of the protractor and line up one side with the base line. Read the protractor scale to measure the angle.	See Example 1, page 644
2. Equilateral triangle	A triangle in which all three sides have the same length (also equiangular).	
3. Isosceles triangle	A triangle in which two sides have the same length. The two angles opposite the equal sides will also have the same measure.	
4. Scalene triangle	A triangle in which all the sides have different lengths.	
5. Equiangular triangle	A triangle in which all three angles are the same measure (also equilateral).	
6. Right triangle	A triangle in which one angle measures 90°.	
7. Acute triangle	A triangle in which all angles measure less than 90°.	
8. Obtuse triangle	A triangle in which one angle measures greater than 90°.	
9. Angle sum of a triangle	The sum of the measures of the angles of a triangle equals 180°.	

EXERCISES: ACTIVITY 5.6

1. a. What tool would you use to measure an angle in a triangle?

    b. Use the tool selected in part a to measure the angles in this triangle. Verify that the sum of the angles is 180°.

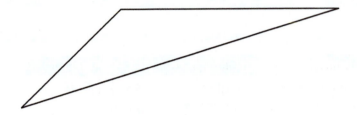

2. In a right triangle, one of the acute angles measures 47°. What is the size of the other acute angle?

3. In an isosceles triangle one of the angles measures 102°. What is the size of the other two angles?

4. What is true about the sizes of the three angles in a scalene triangle?

5. Give an example of the possible sizes of the three angles in a scalene triangle that is also an acute triangle.

Activity 5.7

Not Exactly the Same

Objectives

1. Identify similar geometric figures.

2. Solve problems involving similar figures.

3. Identify congruent geometric figures.

You are planning a surprise anniversary party for your parents and want to prepare a large poster of their wedding photograph. The original picture has width 4 inches and height 6 inches. You want to enlarge the photograph without distorting it. Mathematically, this means you want to create a rectangular enlargement that is **similar** to the original rectangular photograph.

> #### Definition
>
> **Similar** geometric figures have identical *shape* but different *size*. It is often necessary to rotate or flip one similar figure so that it is in the same position as the other. In this case, "corresponding" means in the same relative position.
>
> Similar geometric figures have corresponding angles of equal measure and corresponding sides that have identical ratios (proportional). The constant ratio is a scale factor.

Equivalently, the ratio of any two sides of a given figure is equal to the ratio of the two corresponding sides of a similar figure.

For example, consider the two triangles *ABC* and *XYZ* below. The two triangles are similar because $\angle A = \angle X$, $\angle B = \angle Y$, and $\angle C = \angle Z$. Note that sides *a* and *x* are corresponding sides, sides *b* and *y* are corresponding sides, and sides *c* and *z* are corresponding sides.

Triangle *ABC*

Triangle *XYZ*

The proportionality of the corresponding sides can be described algebraically as

$$\frac{a}{x} = \frac{b}{y}, \quad \frac{a}{x} = \frac{c}{z}, \quad \frac{b}{y} = \frac{c}{z},$$

1. The lengths of the sides of two triangles appear in the following diagram:

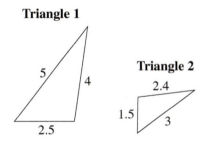

Triangle 1

Triangle 2

a. Rotate triangle 2 so that it is in the same position as triangle 1. What is the ratio of the corresponding sides (of triangle 1 to triangle 2)? Express this ratio in reduced fraction form.

b. What is the perimeter of each triangle?

   c. What is the ratio of the perimeters? Express this ratio in reduced fraction form.

   d. Compare the ratios in parts a and c.

> The scale factor $\frac{5}{3}$ in Problem 1 is also called the **proportionality constant** and applies to *all* corresponding linear measurements associated with the same similar figures—sides, diagonals, perimeters, heights, and so forth. Therefore, the ratio of the altitudes of similar triangles 1 and 2 is also $\frac{5}{3}$.

2. a. The following figure contains two similar triangles. Identify these triangles.

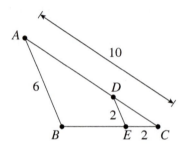

   b. Determine the proportionality constant of the larger triangle to the smaller triangle.

In parts c–f, determine the length of the given side.

   c. *BC*   **d.** *DC*   **e.** *AD*   **f.** *BE*

Let us return to the poster problem.

3. a. Suppose the standard poster size is 36 inches by 48 inches. Are the original 4-inch by 6-inch photograph and the full poster similar rectangles?

   b. If the width and height of the original 4-inch by 6-inch photograph is increased by a factor of 5, what are the dimensions of the enlargement?

c. Are the original photograph and enlargement in part b similar?

d. By what maximum factor can you enlarge the original photograph so that it fits in the standard 36 inch by 48 inch poster?

e. What are the dimensions of the enlarged picture?

f. Is the enlarged picture similar to the original photograph?

Indirect Measurement

4. A 5-foot-tall woman casts a shadow of 4 feet while a nearby tree casts a shadow that measures 30 feet. This situation can be represented by two similar right triangles as shown in the following diagram (not drawn to scale). Let *h* represent the height of the tree.

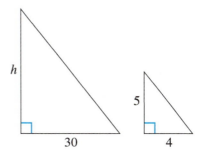

Use the proportionality of corresponding sides to determine the height of the tree.

Purchasing a New TV

5. You are getting ready to buy a new television. You discover that an HD television has a screen width to height ratio of 16:9. The older conventional television set has a screen width to height ratio of 4:3. Do these two screens represent similar rectangles?

6. The displayed size of a TV actually represents the length of the screen diagonal. Suppose you decide to purchase a 32-inch TV. You are curious if the HD TV gives you a larger screen area than the conventional TV screen.

   a. Use the diagram below and your understanding of similarity to determine the width, height, and area of the conventional TV screen.

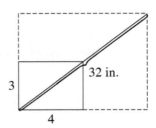

Note that the width-to-height ratio $\frac{4}{3}$ for the conventional television represents a reduced fraction of the actual width and height measurements of the TV. As you encountered in Problem 3, you need to determine the scale factor to multiply 4 and 3 by to acquire the actual width and height.

   b. Determine the width, height, and area of the HD TV.

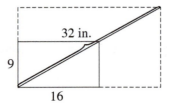

   7. In art, proportion refers to the relative size of the parts within a painting or one element compared to another. Many artists maintain that correct proportion is important to make the painting look realistic. They acknowledge the connection between proportion and the illusion of three dimensions. Search on the Internet for pictures of two famous paintings: *Peasant Wedding* by Pieter Bruegel and *Le Moulin de la Galette* by Auguste Renoir. Identify examples of proportionality in the paintings.

Congruent Figures

> **Definition**
>
> **Congruent** geometric figures have *identical shape* and *size*. As with similar figures, it is sometimes necessary to rotate or flip one congruent figure so that it is in the same position as the other. Once again, *corresponding* means in the same relative position. If you overlap congruent figures, they will coincide.
>
> Congruent geometric figures have corresponding angles of equal measure and corresponding sides of equal length.

8. The following two right triangles are congruent.

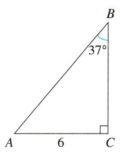
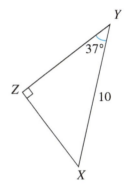

Determine each of the following:

a. The length of the side *AB*.

b. The measure of the angle *YXZ*.

c. The length of side *YZ*.

SUMMARY: ACTIVITY 5.7

1. Similar geometric figures have *identical shape* but *different size*. It is often necessary to rotate or flip one similar figure so that it is in the same position as the other. In this case, *corresponding* means in the same relative position.

2. Similar geometric figures have corresponding angles of equal measure and corresponding sides that have identical ratios (proportional). Equivalently, the ratio of any two sides of a given figure is equal to the ratio of the two corresponding sides of a similar figure.

3. Congruent geometric figures have *identical shape* and *size*. As with similar figures, it is sometimes necessary to rotate or flip one congruent figure so that it is in the same position as the other. Once again, *corresponding* means in the same relative position. If you overlap congruent figures, they will coincide.

4. Congruent geometric figures have corresponding angles of equal measure and corresponding sides of equal length.

EXERCISES: ACTIVITY 5.7

1. Triangle *ABC* is similar to triangle *QPR*.

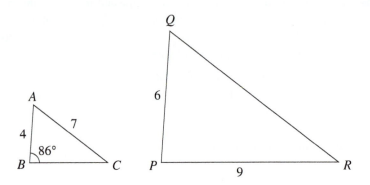

a. Determine the length of the sides *BC* and *QR*.

b. Determine the measure of angle *P*.

2. At 10 A.M. a 6-foot-tall man casts a 4.5-foot shadow. How long a shadow is cast by a 50-foot tree?

3. An architectural blueprint is drawn to a scale of 1:20.

    a. A window on the blueprint measures $2\frac{1}{4}$ inches by $1\frac{3}{8}$ inches. What are the dimensions of the actual window?

    b. What are the dimensions of a square room whose diagonal on the sketch measures $10\frac{3}{4}$ inches?

4. An artist wants to place a 20-inch by 15-inch watercolor painting on a mat so that a 2-inch-wide strip of matting shows on all four sides of the painting. Do the mat and painting form similar rectangles?

5. Determine the length x in each figure:

a.

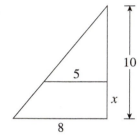

b.

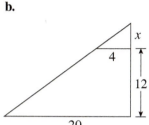

c.

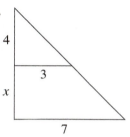

6. Standing 20 feet from a streetlight, a 5-foot-tall woman casts an 8-foot shadow. How tall is the streetlight?

7. You and your friends decide to set up an experiment to estimate the height of the high school gym. You wait until dark and then use a flashlight to project shadows onto the building. One of your friends sets the flashlight on the ground 50 feet from the building and shines it at you as you (6 feet tall) walk away from the flashlight and toward the building. Another friend tells you to stop when the height of your shadow reaches the top of the building; you have walked 12 feet.

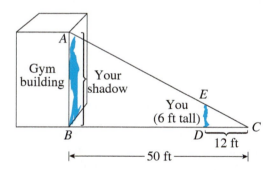

a. Where do you see two similar right triangles in the diagram? Explain.

b. Use the properties of similar right triangles to estimate the height of the gym.

8. When enlarging photographs, the larger image is usually similar to the original photograph. Your original photo is 6 inches wide and 10 inches high. Determine the width of the enlargement if you want the enlargement to be 24 inches high.

9. You wish to enlarge a photograph. The original photograph is 4 inches by 6 inches, but the enlarged photograph you want will be 3 feet by 5 feet.

   a. If you enlarge the original photograph to obtain a 3 feet by 5 feet photograph, will the enlarged photograph be similar to the original photograph? Explain.

   b. In order for the dimensions of the enlarged photograph and the original photograph to be proportional, by how much must you crop the original photo?

10. The following two trapezoids are congruent:

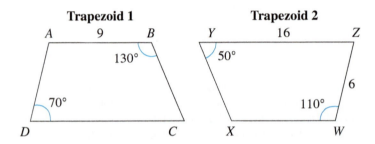

Trapezoid 1   **Trapezoid 2**

   a. Rotate trapezoid 2 so that it is in the same position as trapezoid 1.

In parts b–d, determine the length of the given side.

   b. side *XW*   **c.** side *AD*   **d.** side *DC*

In parts e–g, determine the measure of the given angle.

   e. angle *C*   **f.** angle *Z*   **g.** angle *A*

11. The Wild Goose Pagoda, located in the southern Xi'an, Shaanxi province in China, has been re-built several times since it was originally constructed in the year 652. It currently consists of seven square stories each similar to the previous one. Search the Internet for an image of the pagoda.

   a. The face of the lowest level is 100 feet by 20 feet. If the dimensions of each successive story are 90% of the preceding story, does this represent an enlargement or a reduction? What is the scale factor?

   b. What are the dimensions of the face of the second story? Is the face of the second story similar to the face of the first story?

   c. What are the dimensions of the seventh face?

Activity 5.8

How about Pythagoras?

Objectives

1. Verify and use the Pythagorean theorem for right triangles.

2. Use the Pythagorean theorem to solve problems.

In this activity you will experimentally verify the Pythagorean theorem for right triangles. This formula is used by surveyors, architects, and builders to check whether or not two lines are perpendicular or if a corner truly forms a right angle.

Recall that a right triangle is simply a triangle that has a right angle. In other words, one of the angles formed by the triangle measures 90°. The two sides that are perpendicular and form the right angle are called the **legs** of the right triangle. The third side, opposite the right angle, is called the **hypotenuse**.

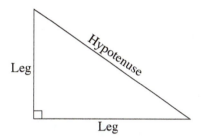

1. Use a protractor to construct three right triangles, one with legs of length 1 inch and 5 inches, a second with legs of length 3 inches and 4 inches, and a third with legs of length 2 inches each.

Triangle 1

Triangle 2

Triangle 3

Pythagorean Theorem

2. For each triangle in Problem 1, complete the following table. The first triangle was done for you as an example. Note that the lengths of the legs of the triangles are represented by a and b. The letter c represents the length of the hypotenuse.

   a. Use a ruler to measure the length of each hypotenuse, in inches. Record the lengths in column c.

   b. Square each length a, b, and c and record in the table.

	a	b	c	a^2	b^2	c^2
Triangle 1	1	5	5.1	1	25	26.0
Triangle 2	3	4				
Triangle 3	2	2				

3. There does not appear to be an equation relating between a, b, and c, but investigate further. What is the relationship between a^2, b^2, and c^2?

The relationship demonstrated in Problem 2 has been known since antiquity. It was known by many cultures but has been attributed to the Greek mathematician Pythagoras, who lived in the sixth century B.C.

> The **Pythagorean theorem** states that, in a right triangle, the sum of the squares of the leg lengths is equal to the square of the hypotenuse length.
>
> Symbolically, the Pythagorean theorem is written as
>
> $$c^2 = a^2 + b^2 \text{ or } c = \sqrt{a^2 + b^2},$$
>
> where a and b are the leg lengths and c is the hypotenuse length.

Note that this relationship is true for any right triangle. Also, if this relationship is true for a triangle, then the triangle is a right triangle.

Example 1   *If a right triangle has legs a = 4 centimeters and b = 7 centimeters, you can calculate the length of the hypotenuse as follows:*

$$c = \sqrt{a^2 + b^2} = \sqrt{(4)^2 + (7)^2}$$
$$= \sqrt{16 + 49} = \sqrt{65} \approx 8.06 \text{ cm}$$

4. A popular triangle with builders and carpenters has dimensions 3 units by 4 units by 5 units.

   a. Use the Pythagorean theorem to show that this triangle is a right triangle.

   b. Builders use this triangle by taking a 12-unit-long rope and marking it in lengths of 3 units, 4 units, and 5 units (see figure below.)

   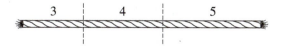

   Then, by fitting this rope to a corner, they can quickly tell if the corner is a true right angle. Another special right triangle has legs of length 5 units and 12 units. Determine the perimeter of this right triangle. Explain how a carpenter can use this triangle to check for a right angle.

5. Suppose you wish to measure the distance across a pond but don't wish to get your feet wet! By being clever and knowing the Pythagorean theorem, you can estimate the distance by taking two measurements on dry land, as long as your two distances lie along perpendicular lines. (See figure below.)

   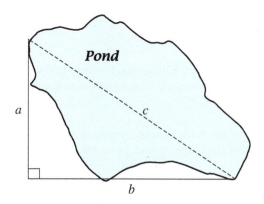

   If your measurements for legs *a* and *b* are 260 feet and 310 feet, respectively, what is the distance, *c*, across the pond?

SUMMARY: ACTIVITY 5.8

Concept/Skill	Description	Example
Pythagorean theorem	In a right triangle, $c^2 = a^2 + b^2$.	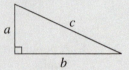

EXERCISES: ACTIVITY 5.8

1. New cell phone towers are being constructed on a daily basis throughout the country. Architects and engineers provide cell tower sites as well as tower and power distribution design. Typically, they consist of a tall, thin tower supported by several guy wires. Assume the ground is level in the following. Round answers to the nearest foot.

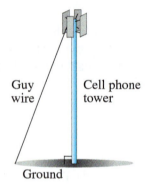

a. The guy wire is attached on the ground at a distance of 100 feet from the base of the tower. The guy wire is also attached to the phone tower 300 feet above the ground. What is the length of the guy wire?

b. You move the base of the guy wire so that it is attached to the ground at 120 feet from the base of the tower. Now how much wire do you need for the one guy wire?

c. Another option is to attach the base of the guy wire 100 feet from the base of the tower and to the tower 350 feet above the ground. How much wire do you need for this option?

2. A building inspector needs to determine if two walls in a new house are built at right angles, as the building code requires. He measures and finds the following information. Wall 1 measures 12 feet, wall 2 measures 14 feet, and the distance from the end of wall 1 to the end of wall 2 measures 18 feet. Do the walls meet at right angles? Explain.

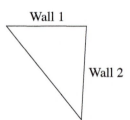

3. Trusses used to support the roofs of many structures can be thought of as two right triangles placed side by side.

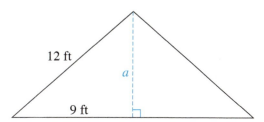

12 ft

9 ft

a

a. If the hypotenuse in one of the right triangles of a truss measures 12 feet and the horizontal leg in the same right triangle measures 9 feet, how high is the vertical leg of the truss?

b. To make a steeper roof, you may increase the vertical leg to 10 feet. Keeping the 9-foot horizontal leg, how long will the hypotenuse of the truss be now?

4. You can consider the truss in Exercise 3 as a single triangle. In this case, it is a good example of an isosceles triangle.

a. If the top angle of the truss is 120°, then what are the measures of the other two angles?

b. If you wish to have a steeper roof, with the base angles of the isosceles truss each measuring 42°, what is the measure of the top angle?

5. For the following right triangles, use the Pythagorean theorem to compute the length of the third side of the triangle:

a.

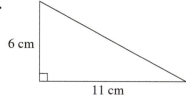

6 cm

11 cm

b.

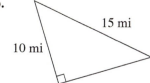

15 mi

10 mi

6. You are buying a ladder for your 30-foot-tall house. For safety, you would always like to ensure that the base of the ladder be placed at least 8 feet from the base of the house. What is the shortest ladder you can buy and still be able to reach the top of your house?

7. Pythagorean triples are three positive integers that could be the lengths of three sides of a right triangle. For example, 3, 4, 5 is a Pythagorean triple because $5^2 = 3^2 + 4^2$.

  a. Is 5, 12, 13 a Pythagorean triple? Why, or why not?

  b. Is 5, 10, 15 a Pythagorean triple? Why, or why not?

  c. Name another Pythagorean triple. Explain.

8. You own a summer home on the east side of Lake George in New York's Adirondack Mountains. To drive to your favorite restaurant on the west side of the lake, you must go directly south for 7 miles and then directly west for 3 miles. If you could go directly to the restaurant in your boat, how far is the boat trip?

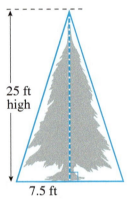

25 ft high

7.5 ft

9. You are decorating a large evergreen tree in your yard for the holidays. The tree stands 25 feet tall and 15 feet wide. You want to hang strings of lights from top to bottom draped on the outside of the tree. How long should the strings of lights be? Explain.

10. As lead engineer, you need to determine the length of the cable support of a bridge deck. The cable is attached at the top of the support tower, 150 meters from ground level, and extends to the bridge deck 100 meters from the base of the tower. What is the length of the support cable?

You are a structural engineer. You are studying the Leaning Tower of Pisa, which is located in Pisa, Italy. Originally the tower was 179 feet high. Although the tower has lost none of its height, it now makes an 85-degree angle with the ground. You want to know how high the top of the tower is above the ground today.

To answer this question, you need a branch of mathematics called **trigonometry**. Although the development of trigonometry is generally credited to the ancient Greeks, there is evidence that the ancient Egyptian cultures used trigonometry in constructing the pyramids. Today, trigonometry is used extensively in architecture and the sciences, especially in physics, engineering, and chemistry.

Right Triangles Revisited

As you will see, the answer to the Leaning Tower question requires the use of **right triangles**. Recall that the hypotenuse is the side opposite the right angle. The legs of a right triangle are identified by their relation (position) to the other two angles of the triangle. Consider the following right triangle with angles A, B, and C and sides a, b, and c.

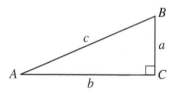

You identify the sides of a right triangle using the following terminology.

> **Definition**
>
> Angle C is the right angle, the angle measuring $90°$. The side opposite the right angle, c, is called the **hypotenuse**. Side a is said to be **opposite** angle A because it is not part of angle A. Side b is said to be **adjacent** to angle A because it and the hypotenuse form angle A. Similarly, side b is the side **opposite** angle B, and side a is the side **adjacent** to angle B.

Example 1   *Consider the following right triangle:*

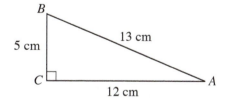

The side opposite angle B is 12 centimeters long. The side opposite angle A is 5 centimeters long. The side adjacent to angle B is 5 centimeters long. The side adjacent to angle A is 12 centimeters long. The hypotenuse is 13 centimeters long.

Activity 5.9

The Leaning Tower of Pisa

Objectives

1. Identify the sides and corresponding angles of a right triangle.

2. Determine the length of the sides of similar right triangles using proportions.

3. Determine the sine, cosine, and tangent of an angle using a right triangle.

4. Determine the sine, cosine, and tangent of an acute angle by using the graphing calculator.

1. Consider the following right triangle:

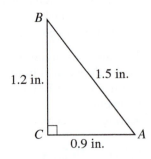

Determine the length of each of the following:

a. The side opposite angle B

b. The side adjacent to angle B

c. The side opposite angle A

d. The side adjacent to angle A

e. The hypotenuse

f. Demonstrate that the lengths of the sides satisfies the Pythagorean theorem.

Similar Triangles

Consider the following right triangles whose angles are the same but whose sides are different lengths. These three triangles are **similar**.

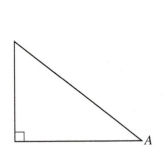

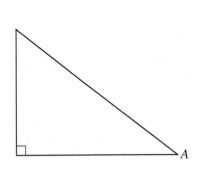

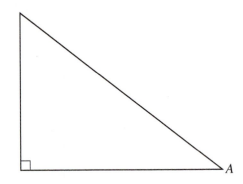

2. a. Using a protractor, estimate the measure of angle A to the nearest degree.

b. Use a metric or English ruler to complete the following table:

	LENGTH OF THE HYPOTENUSE	LENGTH OF THE SIDE OPPOSITE A	LENGTH OF THE SIDE ADJACENT TO A
Small Triangle			
Midsize Triangle			
Large Triangle			

3. a. Use the information in the preceding table in Problem 2b to complete the ratios in the following table with respect to angle *A*. Write each ratio as a decimal rounded to the nearest tenth.

	LENGTH OF THE SIDE OPPOSITE ANGLE *A* / LENGTH OF THE HYPOTENUSE	LENGTH OF THE SIDE ADJACENT TO ANGLE *A* / LENGTH OF THE HYPOTENUSE	LENGTH OF THE SIDE OPPOSITE ANGLE *A* / LENGTH OF THE SIDE ADJACENT TO ANGLE *A*
Small Triangle	0.6	0.8	0.8
Midsize Triangle			
Large Triangle			

b. What do you observe about the ratio $\dfrac{\textit{length of the side opposite angle A}}{\textit{length of the hypotenuse}}$ for each of the three right triangles?

> The table in Problem 3a illustrates the geometric principle that **corresponding sides of similar triangles are proportional**. Recall that the ratios $\dfrac{a}{b}$ and $\dfrac{c}{d}$ are proportional if
>
> $$\frac{a}{b} = \frac{c}{d}.$$

4. Consider another right triangle in which the measure of angle *A* is not the same as the measure of angle *A* in the three similar triangles from Problems 2 and 3.

a. Using a protractor, estimate the measure of angle *A* to the nearest degree.

b. For this new triangle, use your ruler to complete the following table with respect to angle *A*:

	LENGTH OF THE SIDE OPPOSITE ANGLE *A* / LENGTH OF THE HYPOTENUSE	LENGTH OF THE SIDE ADJACENT TO ANGLE *A* / LENGTH OF THE HYPOTENUSE	LENGTH OF THE SIDE OPPOSITE ANGLE *A* / LENGTH OF THE SIDE ADJACENT TO ANGLE *A*
New Triangle			

c. Are the ratios for the new triangle the same as the ratios for the three similar triangles?

d. What changed from the similar triangles to the new triangle to make the ratios change?

Sine, Cosine, and Tangent Functions

As demonstrated in Problem 4, the ratios of the sides of a right triangle are dependent on the size of the angle A. If the angle changes, the ratios change. This fact is fundamental to trigonometry.

> The ratios of the sides of a right triangle with acute angle A are a function of the size of A.

The ratios are given special names and are defined as follows.

Definition

Let A be an acute angle (less than 90°) of a right triangle. The **sine**, **cosine**, and **tangent** of angle A are defined by

$$\text{sine of } A = \sin A = \frac{\textit{length of the side opposite } A}{\textit{length of the hypotenuse}},$$

$$\text{cosine of } A = \cos A = \frac{\textit{length of the side adjacent to } A}{\textit{length of the hypotenuse}},$$

$$\text{tangent of } A = \tan A = \frac{\textit{length of the side opposite } A}{\textit{length of the side adjacent to } A},$$

where sin, cos, and tan are the standard abbreviations for sine, cosine, and tangent, respectively.

- Sine, cosine, and tangent are called **trigonometric functions**.
- Note that the input values for the trigonometric functions are angles and the output values are ratios.
- A mnemonic device often used to help remember the trigonometric relationships is SOH CAH TOA, where SOH indicates that the sine function is the length of the side opposite divided by the length of the hypotenuse, CAH indicates the cosine definition, and TOA indicates the definition of the tangent function.

Example 2 *Consider the following large triangle with the given dimensions:*

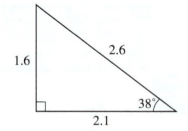

$$\sin 38° = \frac{1.6}{2.6} \approx 0.615 \text{ or } 0.6 \text{ (nearest tenth)}$$

$$\cos 38° = \frac{2.1}{2.6} \approx 0.807 \text{ or } 0.8 \text{ (nearest tenth)}$$

$$\tan 38° = \frac{1.6}{2.1} \approx 0.761 \text{ or } 0.8 \text{ (nearest tenth)}$$

Therefore, for any size right triangle with a 38° angle as one of its acute angles, the sine of 38° is always approximately 0.615, the cosine of 38° is always approximately 0.807 and the tangent of 38° is always approximately 0.761.

You can use your graphing calculator to evaluate sin 38°. Make sure the calculator is in degree mode. Press the (SIN) key, followed by (3), (8), (⟩) and (ENTER).

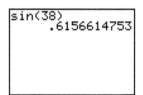

5. a. The sum of the three angles in a triangle is 180°. If one angle measures 90°, the sum of the measures of the other two angles must be 90°. If one of the acute angles measures 38°, the other acute angle is 52°. Use the accompanying figure to determine each of the following.

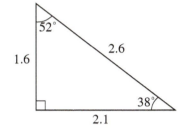

    i. sin 52°

    ii. cos 52°

    iii. tan 52°

    b. Verify your answers in part a using your graphing calculator.

Example 3   *Consider the following right triangle where A and B are acute angles:*

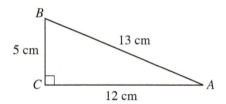

$$\sin A = \frac{5}{13}, \cos A = \frac{12}{13}, \tan A = \frac{5}{12}, \sin B = \frac{12}{13}, \cos B = \frac{5}{13}, \tan B = \frac{12}{5}$$

6. Consider the following right triangle:

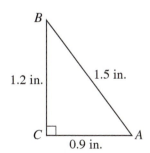

Calculate the following:

a. sin A **b.** cos A **c.** tan A

d. sin B **e.** cos B **f.** tan B

Example 4   *Solve the following equations. Round your answers to the nearest tenth.*

a. $\sin 56° = \dfrac{x}{15}$ **b.** $\cos 13° = \dfrac{24}{x}$ **c.** $\tan 72° = \dfrac{x}{24.7}$

SOLUTION

Use the calculator in degree mode to evaluate the function values,

a. $0.829 \approx \dfrac{x}{15}$ **b.** $0.974 \approx \dfrac{24}{x}$ **c.** $3.078 \approx \dfrac{x}{24.7}$

$15(0.829) = x$ $(0.974)x = 24$ $(3.078)(24.7) = x$

$12.4 \approx x$ $76.0 \approx x$

$$x = \dfrac{24}{0.974}$$

$$x \approx 24.6$$

7. Solve the following equations. Round your answers to the nearest tenth.

a. $\sin 24° = \dfrac{x}{10}$ **b.** $\cos 63° = \dfrac{x}{23.5}$ **c.** $\tan 48° = \dfrac{16}{x}$

Trigonometric Values of Special Angles

To determine the trigonometric function values for 30° and 60°, start with an equilateral triangle, a triangle with three equal sides. Assume for now that all three sides are 2 units in length. Note that all three angles must measure 60°.

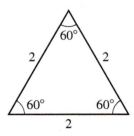

Now bisect one of the angles to form two congruent right triangles with angles measuring 30°, 60°, and 90°.

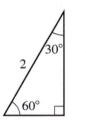 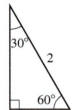

8. a. Considering one of these two new triangles, what are the lengths of the two legs of the right triangle?

b. With respect to the 30° angle, what are the lengths of the opposite side, the adjacent side, and the hypotenuse, respectively?

c. With respect to the 60° angle, what are the lengths of the opposite side, the adjacent side, and the hypotenuse, respectively?

d. Use the results of parts b and c to complete the following table.

θ	sin θ	cos θ	tan θ
30°			
60°			

To determine the trigonometric function values of 45°, start with a 45-45-90 isosceles right triangle.

9. For convenience, now assume that the equal legs have length 1.

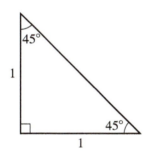

a. Determine the length of the hypotenuse.

b. With respect to the 45° angle, what are the lengths of the opposite side, the adjacent side, and the hypotenuse, respectively?

c. Use the results of part b to complete the following table.

θ	sin θ	cos θ	tan θ
45°			

Knowing the trigonometric function values for 30°, 45°, and 60° can be very helpful in understanding the behavior of these functions. Keeping these values handy or memorizing them is a good idea.

Example 5 *Consider the following two right triangles.*

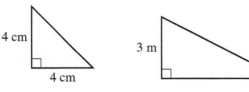

Figure A   **Figure B**

a. Given that the length of one leg of a 45-45-90 triangle is 4 centimeters, determine the exact length of the other two sides. (See Figure A.)

SOLUTION

In Figure A, the two acute angles both measure 45° and the two legs are the same length. So the other leg is also 4 centimeters long. To determine the length of the hypotenuse, use the sine function.

$$\sin 45° = \frac{4}{h}; \quad \frac{1}{\sqrt{2}} = \frac{4}{h}; \quad h = 4\sqrt{2}\text{ cm}$$

b. Given that the length of the shortest side of a 30-60-90 triangle is 3 meters, determine the lengths of the other two sides. (See Figure B.)

SOLUTION

In Figure B, the smallest angle measures 30°. With respect to the 30° angle, the opposite side is given. To determine the length of the adjacent side, the tangent function could be used.

$$\tan 30° = \frac{3}{a}; \quad \frac{1}{\sqrt{3}} = \frac{3}{a}; \quad a = 3\sqrt{3}\text{ m}$$

To determine the length of the hypotenuse, the sine function could be used.

$$\sin 30° = \frac{3}{h}; \quad \frac{1}{2} = \frac{3}{h}; \quad h = 6\text{ m}$$

Tower of Pisa Problem Revisited

You are now ready to answer the original Tower of Pisa problem.

10. a. Recall that the tower was originally 179 feet in length, and it now makes an angle of 85° with the ground. Construct a right triangle that satisfies these conditions.

b. With respect to the 85° angle, is the height of the tower represented by the length of an opposite side, the length of an adjacent side, or the length of the hypotenuse of your triangle?

c. You want to determine how high the top of the tower is above the ground today. Therefore, you want to determine the length of which side of the triangle with respect to the 85° angle?

d. Which trigonometric function relates the side with the length you know and the side with the length you want to know?

e. Write an equation using the information in parts a–d.

f. Using your calculator to evaluate sin 85°, solve the equation in part e.

SUMMARY: ACTIVITY 5.9

1. The **trigonometric functions** are functions whose inputs are measures of the acute angles of a right triangle and whose outputs are ratios of the lengths of the sides of the right triangle.

2. The three sides of a right triangle are the **adjacent** side, the **opposite** side, and the **hypotenuse**. The hypotenuse is always the side opposite the right (90°) angle. The other two sides vary, depending on which angle is used as the input.

3. Corresponding sides of **similar triangles** are proportional.

4. The **sine**, **cosine**, and **tangent** of the acute angle A of a right triangle are defined by

$$\sin A = \frac{\text{length of the side opposite } A}{\text{length of the hypotenuse}}$$

$$\cos A = \frac{\text{length of the side adjacent to } A}{\text{length of the hypotenuse}}$$

$$\tan A = \frac{\text{length of the side opposite } A}{\text{length of the side adjacent to } A}.$$

EXERCISES: ACTIVITY 5.9

1. Triangle *ABC* is a right triangle.

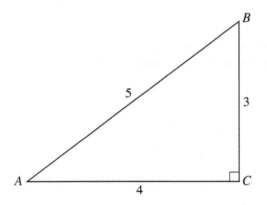

Determine each of the following.

a. sin *A* **b.** sin *B* **c.** cos *A*

d. cos *B* **e.** tan *A* **f.** tan *B*

2. In a certain right triangle, $\sin A = \dfrac{24}{25}$.

a. Determine possible lengths of the three sides of a right triangle. *Hint:* Use the Pythagorean theorem, $c^2 = a^2 + b^2$, to determine the length of any unknown side.

b. Determine cos *A*.

c. Determine tan *A*.

3. In a certain right triangle, $\tan B = \dfrac{7}{4}$.

a. Determine possible lengths of the three sides of the right triangle.

b. Determine sin *B*.

c. Determine cos *B*.

4. Consider the accompanying right triangle.

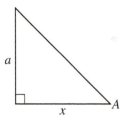

   a. Which of the trigonometric functions relates angle *A* and sides *a* and *x*?

   b. What equation involving angle *A* and side *a* would you solve to determine the value of *x*?

5. Given angle *B* and side *c* in the accompanying diagram, answer the questions in parts a and b.

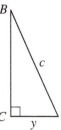

   a. Which of the trigonometric functions relates angle *B* and sides *c* and *y*?

   b. What equation involving angle *B* and side *c* would you solve to determine the value of *y*?

6. Consider the accompanying right triangle.

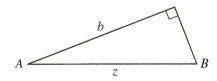

   a. Which of the trigonometric functions relates angle *A* and sides *b* and *z*?

   b. What equation involving angle *A* and side *b* would you solve to determine the value of *z*?

7. Solve the following equations. Round your answer to the nearest tenth.

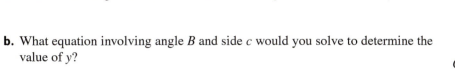

   a. $\sin 49° = \dfrac{x}{12}$ **b.** $\tan 84° = \dfrac{x}{9}$ **c.** $\sin 22° = \dfrac{23}{x}$

8. **a.** Given that the length of the hypotenuse of a 45-45-90 triangle is 7 centimeters, determine the exact length of the two legs. Sketch and label a diagram before calculating.

b. Given that the length of the longer leg of 30-60-90 triangle is 20 feet, determine the exact length of the short leg and the hypotenuse. Sketch and label a diagram before calculating.

9. The chemical formula for water is H_2O. One molecule of water has two hydrogen atoms covalently bonded to one oxygen atom. The structure of the molecule is shown below.

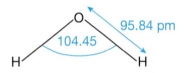

The measure used, pm, is picometer, a very small distance. What is the distance, in picometers, between the hydrogen atoms?

10. An engineer has designed a section of a bridge truss that consists of two triangles (see diagram). You need to determine the total length (nearest tenth) of the beams required to build this section of the truss.

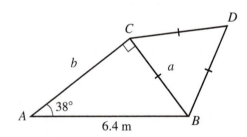

a. Determine the length of side a.

b. Determine the length of side b.

c. Determine the lengths of each side of △BCD.

d. Determine the total length of the beams needed to construct this section of the bridge truss.

11. Frequently, engineers and architects cannot directly measure an object's length (distance), because it is impractical or physically impossible. Trigonometry is a valuable tool in such situations. For example, use the appropriate trigonometric ratio to determine the height of the cliff in the following diagram.

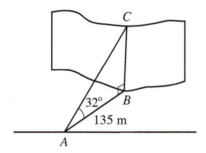

Activity 5.10

Tessellations

Objectives

1. Identify a tessellation.

2. Perform a transformation on a given figure.

3. Reflect a given figure.

4. Translate a given figure.

5. Rotate a given figure.

6. Perform a glide reflection on a given figure.

7. Identify reflective and translational symmetry.

Alhambra Palace

You are on a class trip to Spain and visit the Alhambra Palace in Granada. The Palace is an example of Islamic architecture. Its floors, walls, and ceilings are covered with intricate repeating patterns and geometric shapes. Such designs are called **tessellations**.

> ### Definition
>
> A **tessellation** (or tiling) is a repeating pattern of shapes that covers an entire plane and fits together without gaps or overlaps.

Your tour guide explains that although the kinds of shapes and coloring vary, tessellations can be found in many different cultures. For thousands of years, tessellations have been used in the design of buildings, woodwork, quilts, floors, walls, and gardens. There are many websites that contain spectacular images of tessellating designs that are found in both ancient and modern cultures.

The simplest tessellations use a single **regular polygon** as the repeating shape.

> ### Definition
>
> A polygon is a closed figure having three or more line segments as sides. A **regular polygon** is a polygon having all sides the same length and all interior angles of equal measure.

Only three regular polygons can be used alone to form tessellations: equilateral triangles, squares, or regular hexagons.

REGULAR POLYGON	NUMBER OF SIDES	MEASURE OF AN INTERIOR ANGLE	EXAMPLE
Equilateral Triangle	3	60°	
Square	4	90°	
Regular Hexagon	6	120°	

The following figures are sections of the plane that have been tiled using the given tessellating shape:

A tessellation of equilateral triangles

A tessellation of squares

A tessellation of regular hexagons

Figure 1

1. The point where the sides of the tessellating shape form a corner is called the **vertex**. For example, using equilateral triangles,

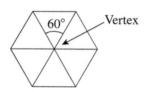

a. Determine the sum of the angles in the indicated vertex. What is the sum of the angles for each of the other vertices in the tessellations of equilateral triangles in Figure 1?

b. Determine the sum of the angles for each vertex using squares as the tessellating shape.

90°

c. What conjecture can you make about the sum of the angles in a vertex in which regular polygons are used as the repeating shape? Test your conjecture in the tessellations in which regular hexagons are used as the tessellating shape.

d. Why can't a regular pentagon be used as a tessellating shape? *Hint:* What is the interior measure of the angles for a regular pentagon?

Tessellating shapes are not limited to geometric shapes. M. C. Escher (1898–1972) was an artist who became fascinated with the Islamic mosaics after visiting the Alhambra Palace. Perhaps his most famous work uses a lizard as a tessellating shape.

Transformational Geometry

Tessellations are an application of a type of geometry called **transformational geometry**. In this type of geometry, figures are moved in various ways without changing the size or shape of the figure. Such a movement is called a **transformation** or **rigid motion**.

> **Definition**
>
> The process of moving a figure from one starting position to some ending position without changing its size or shape is called a **transformation** or **rigid motion**.

Reflections

Perhaps the most familiar type of transformation is a reflection. Reflections occur across a line called the **reflection line** or **axis of reflection**. You reflect a shape across an axis by plotting a special corresponding (reflected) point for every point in the original shape. This corresponding point is the same distance from the axis as is the original point.

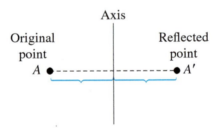

Note that the dashed line containing the original point and the reflected point is perpendicular to the axis. Example 1 contains several examples of reflected shapes across an axis.

Example 1   *Reflect each of the following across the given axis:*

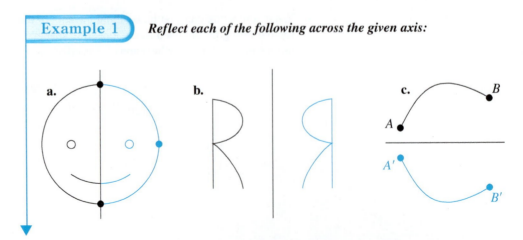

d.

e.

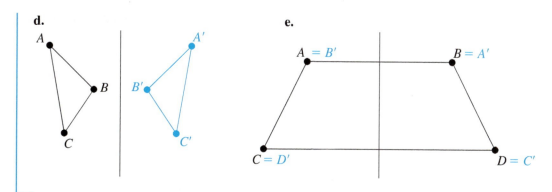

The reflected figure is called the **mirror image** (or simply image) of the original.

2. Reflect each of the given figures across the given axis:

a.   **b.**   **c.**

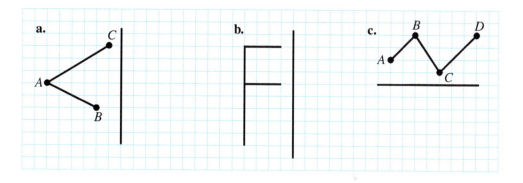

3. Consider the following section of a tessellation of equilateral triangles that was taken from a photograph.

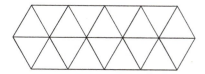

Explain how the tessellation can be created using a reflection.

Translation

A second type of transformation is a **translation**. You move (or glide) a figure along a straight line in a given direction and distance. Example 2 gives several examples of translations of entire shapes in a given direction indicated by an arrow.

Example 2   *Translate each of the following shapes in the direction indicated by the arrow.*

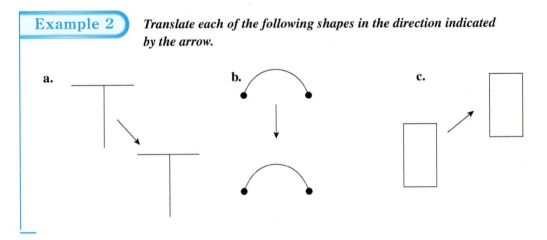

a.   b.   c.

An alternative way to indicate the direction and magnitude of the translation is to specify how many units (*x*) to move to the left or right, followed by how many units (*y*) to move up or down.

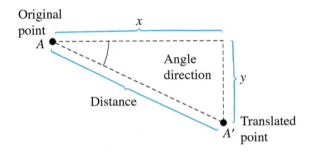

Example 3   *Translate the triangle ABC 4 units to the right and 2 units up.*

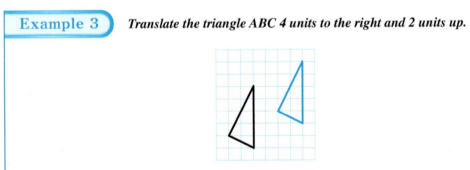

4. Translate each of the following shapes.

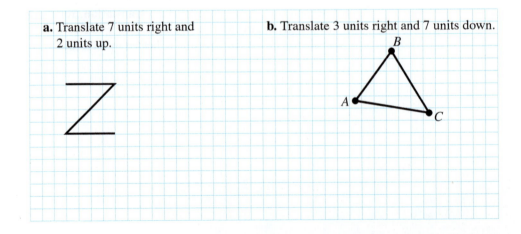

a. Translate 7 units right and 2 units up.

b. Translate 3 units right and 7 units down.

Rotation

A third type of transformation is a **rotation**. When a figure is rotated, it is turned about a specific point called the **center of rotation**. The amount of turning is called the **angle of rotation** and is measured in degrees. Remember, positive degree angles are measured counterclockwise; negative degree angles are measured clockwise.

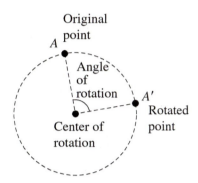

Example 4

a. Rotate the given figure about the given center of rotation through six rotations of 60°.

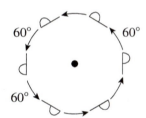

b. Rotate the given figure about the given center of rotation through one rotation of −180°.

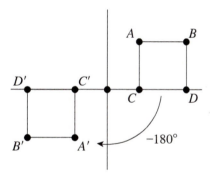

---

5. Determine if the design in the hubcaps of your car were created by the rotation of a given shape. If yes, what is the center of rotation and an estimation of the angle of rotation?

Glide Reflection

The final type of transformation is a **glide reflection**. A glide reflection is essentially a combination of a reflection and a translation (glide). In the following diagram, the translation of a single point is done first, followed by the reflection of the point.

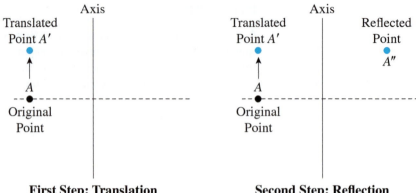

First Step: Translation   **Second Step: Reflection**

> **Example 5** *Perform a glide reflection on triangle ABC by translating the triangle*
> *units to the right and 2 units up, and then reflecting about the y-axis.*
>
> **SOLUTION**

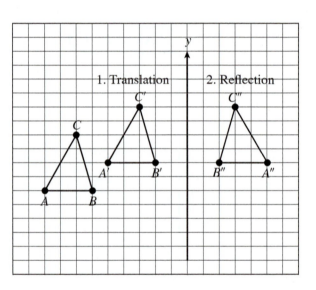

6. Perform a glide reflection on the trapezoid *ABDC* by translating the figure 3 units to
left and 4 units up, followed by a reflection about the *x*-axis.

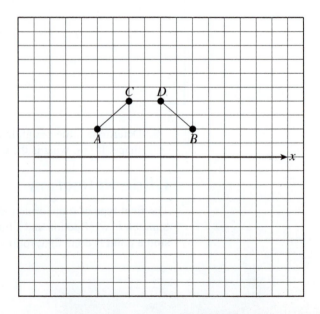

Symmetry

Each of the four transformations of a given figure often results in a transformed figure that is **symmetric** to the original. If the ending position of the transformed figure is exactly the same as the starting position, the figure has symmetry with respect to the transformation done on the figure. For example, consider trapezoid *ABDC*.

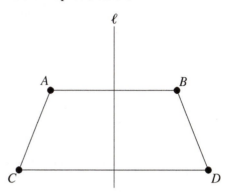

If the trapezoid is reflected across the given axis, the beginning position and the ending position are exactly the same. Therefore, trapezoid *ABDC* is said to have **reflective symmetry** with respect to the line axis *l*. The line *l* is called the **line of symmetry**.

a.

7. a. Does the tessellation to the left have translational symmetry?

    b. Do all tessellations have translational symmetry? To help answer this question, search the internet for examples of Penrose tilings.

8. Consider the following star-shaped figure:

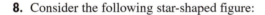

    a. Does the figure have **rotational symmetry** about the point *P* for a 90° rotation?

    b. Does the figure have rotational symmetry about the point *P* for a 180° rotation?

9. Does every tessellation have rotational symmetry?

10. Many textiles display tessellations and symmetry. The image on the left is part of a block print design from a West African adinkra cloth.

    a. Does this design display reflective symmetry? Explain.

b. Does this design display rotational symmetry? Explain.

Symmetry is prized greatly as an ideal in the visual arts. While the symmetry may not be perfect mathematically, apparent symmetry is pleasing to the eye. Photographers take advantage of this fact when deciding upon content and composition.

11. Consider the architecture of the railroad bridge in the photo to the right.

Courtesy of photoneye/Shutterstock.

   a. If you draw a line through the middle of the picture, what type of symmetry is displayed? Explain.

   b. If you consider the same line, what type of transformation can be observed?

   c. What basic polygons make up the tessellation in the photograph?

   d. Does the scene have a vanishing point? If yes, describe its location in the photograph.

Musical Symmetry

12. Notes on a musical scale can be viewed as a geometric design. The whole notes on this treble clef scale form a symmetric pattern. Describe this symmetry.

13. One bar of music is contained between vertical lines on the scale. If one bar, consisting of four notes (below), is translated horizontally to the next two bars, what would you hear when the music is played?

14. Describe the symmetry displayed by these notes.

15. Is it possible for a score (musical selection) written as one sheet of music to sound exactly the same when played normally **and** with the score turned upside down?

SUMMARY: ACTIVITY 5.10

1. A **tessellation** (or tiling) is a repeating pattern of shapes, covering the entire plane that fit together without gaps or overlaps.

2. A **polygon** is a closed figure having three or more line segments as sides.

3. A **regular polygon** is a polygon having all sides the same length and all interior angles equal measure.

4. The process of moving a figure from one starting position to some ending position without changing its size or shape is called a **transformation** or **rigid motion**.

5. There are four types of transformations

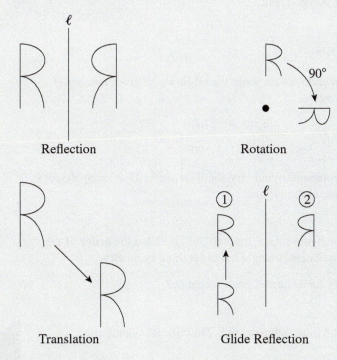

6. If the ending position of a transformed figure is exactly the same as the starting position, the figure has symmetry with respect to the transformation done on the figure.

EXERCISES: ACTIVITY 5.10

1. Graph triangle ABC having coordinates $A(1, 4), B(5, 5), C(3, 1)$.

   a. Graph the reflection of the image, $A'B'C'$, over the y-axis.

   b. What are the coordinates of the reflection triangle $A'B'C'$?

   c. Graph the reflection of the triangle $A'B'C'$ over the x-axis.

   d. Write the coordinates of its reflection $A''B''C''$.

   e. Describe the differences about triangle ABC, triangle $A'B'C'$, and triangle $A''B''C''$.

2. Translate each point left 3 units and up 2 units.

   a. What are the coordinates of each image point?

     $A(3, 2)$

     $B(3, -2)$

     $C(-2, -1)$

     $D(-4, 1)$

   b. Graph each point and its image.

   c. Connect points $ABCD$. What type of polygon is this?

   d. What is the area of polygon $ABCD$?

   e. What is the area of polygon $A'B'C'D'$?

3. Graph each point. Rotate the point counterclockwise about the origin in the given number of degrees. Write the coordinates of the image.

   a. $(-5, -2), 90°$ **b.** $(3, 4), 270°$

   c. $(2, -1), 180°$ **d.** $(-1, 3), 90°$

4. Consider an equilateral triangle. It has rotational symmetry about its center. How many degrees of rotation will produce the same position as the original?

The number of matches that occur as the figure rotates through $360°$ is called the **order** of the rotational symmetry. For example, an equilateral triangle has order three symmetry.

5. What is the order of rotational symmetry for a square? For a rectangle?

6. Here is another block print from a West African adinkra cloth. Describe its symmetry.

7. How many different geometric figures can you think of that display both reflective and rotational symmetry?

8. The following strip designs are similar to those found on Native American pottery. The underlying structure can be defined by the type of transformations used to produce the pattern. The most elemental unit of each design (to the left) is repeatedly transformed. Identify the transformations (there may be more than one, among translation, rotation, vertical reflection, horizontal reflection) that are the basis for each design.

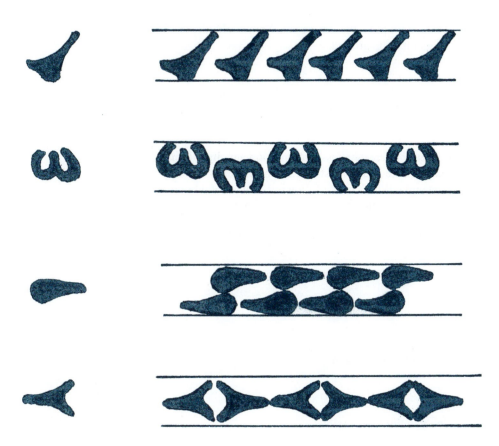

9. In graphic design, the font chosen to display text can make a huge difference visually. In most fonts, individual characters do not display perfect symmetry. But the classic Arial font in the letters below do display several types of symmetry. Determine what types of symmetry each of these letters possess. If there is rotational symmetry, state the order. For reflective symmetry, state the location of the line of symmetry.

A   C   H

N   O   S

W   X   Z

10. With photo editing software, it is very easy to manipulate your own digital photos to create artistic patterns and designs. Many such designs may utilize simple transformations. What transformation would you use to accomplish the following effects?

   a. Your original image is of a person facing to the left. You would like them to be facing to the right.

   b. You have a beautiful image of a mountain lit by the setting sun. You would like to simulate the same mountain with a still lake in the foreground.

   c. You have a great image of your friend airborne on a snowboard. You would like to alter the image so he appears in the middle of a flip, heading in the opposite direction.

11. With photo editing software it is very easy to create a tessellation using any digital image you choose. Of course, the polygonal shape of the image must be chosen carefully, the simplest and easiest polygon being a square or rectangle. Which of the following choices of a boundary for your image would allow you to create a tessellation over your entire new image?

   a. Rhombus   b. Parallelogram   c. Trapezoid

   d. Circle   e. Equilateral triangle   f. Isosceles triangle

   g. Regular pentagon   h. L-shaped polygon

   Make a sketch of each tessellation you think is possible.

12. You can find examples of tessellations in art and architecture.

   a. Visit a Web site that features the work of artist, M.C. Escher. Write a description of some of the tessellations used in his art.

   b. Norman Foster is a British architect who incorporates tessellations in his building designs. Search the Internet for photographs of some of his buildings, including the Hearst Tower in New York City. Write a description of some of the tessellations used in his architectural designs.

   c. Compile a list of examples of tessellations that you observe in your local surroundings, for example, in your home, your school, a local park, or a shopping mall. Be prepared to share your list with the class.

   d. Identify the types of geometric transformations or symmetry that you observe in the examples in parts a–c.

13. You have an online business selling stained glass crafts. Using a compass and a ruler, construct a tessellation shape that you can use in a stained glass window treatment you are doing for a client. Describe the different types of symmetry in the pattern.

14. The mathematical patterns found in types of symmetry and geometric transformations occur in the natural world as well as in the work of artists.

    a. Cut out at least one snowflake from each of the following: a folded circle, square, and equilateral triangle. Describe the type of symmetry present in each snowflake.

    b. Tessellations and symmetric patterns occur in nature. Give examples that occur in each category.

    c. Search the Internet for photographs of artwork of famous artists such as Manet's *A Bar at the Folies-Bergère*. Describe the various types of symmetry and geometric transformations you observe in the artist's work.

15. Create a design that makes a pattern with translational and/or rotational symmetry. Describe the symmetry present in the design.

16. Across all cultures and in all time periods, artists and architects have designed their paintings and buildings with a strong emphasis on symmetry.

    a. Search the Internet for images of the Pantheon in Rome, Italy. Identify at least four examples of symmetry that you observe in the structure.

    b. What type of symmetry is demonstrated in the design of the Alamo?

17. The simplest kind of camera is a pinhole camera. As the illustration here shows, this is a light tight box with a tiny pin prick made on one side and a sheet of film held at the other side. If the box is now aimed at an object, rays of light coming from the scene crisscross each other as they pass through the pinhole, giving rise to a faint upside down image on the film.

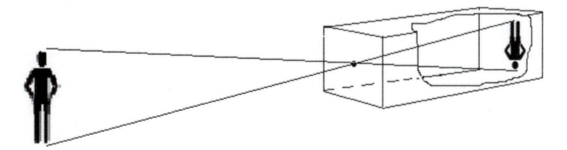

A pinhole camera forms an inverted picture of a scene and forms the basis of the *camera— obscura,* the forerunner of the modern camera.

a. Is the image in the box similar to the original image?

b. Is the image in the box a translation of the original image?

18. There are many Web sites on the Internet where you can find images and history of some of the world's greatest architecture. Select several examples of architecture from various cultures and time periods. Compare the types of symmetry present in the structures.

19. Find examples of photographs that display examples of reflective symmetry, rotational symmetry, and both at the same time.

20. Use a reflective transformation to produce two more bars of music from these two bars.

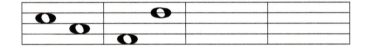

21. Use a 180° rotational transformation, centered on the middle vertical bar, to produce two more bars of music from these two bars.

Activities 5.1–5.10 What Have I Learned?

1. What are the differences and similarities between the area and perimeter of a figure? Explain.

2. If the perimeter of a figure is measured in feet, then what are the units of the area of that figure? Explain.

3. If the area of a circular figure is measured in square centimeters, then what are the units of the diameter of that circle? Explain.

4. A racetrack can be described as a long rectangle with semicircles on the ends.

    a. At a racetrack, who would be interested in its perimeter? Why?

    b. At a racetrack, who would be interested in knowing its area? Why?

5. If different figures have the same perimeter, must their areas be the same? Explain.

6. If different figures have the same area, must their perimeters be the same? Explain.

7. Construct a nonright triangle, measure the sides, and show that the square of the length of the longest side is not equal to the sum of the squares of the lengths of the shorter sides.

8. Explain why the equation $x^2 = -81$ has no solution.

9. You have put on weight, and the radius of your waist has increased by an inch (assume that your waist is approximately circular). How much has your waist measurement increased? Explain.

10. On the high school's campus, there is a very tall tree. Your math teacher challenges the class to devise a way to use similar triangles to indirectly measure the height of the tree. Be specific in explaining your best problem-solving strategy.

11. Classical right-triangle trigonometry was developed by the ancient Greeks to solve problems in surveying, astronomy, and navigation. For purposes of computation, the side opposite the angle θ, side O, is called the opposite, the side opposite the right angle, side H, is called the hypotenuse, and the third side, side A, is called the adjacent.

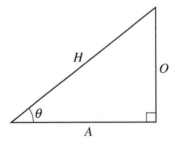

Define the three major trigonometric functions—$\sin \theta$, $\cos \theta$, and $\tan \theta$—in terms of H, A, and O.

12. a. On any right triangle, which trigonometric function would you use to determine the opposite side if you knew the angle measure and the length of the hypotenuse?

b. Which trigonometric function would you use to determine the adjacent side if you knew the angle measure and the length of the hypotenuse?

c. Which trigonometric function would you use to determine the adjacent side if you knew the angle measure and the length of the opposite side?

d. Which trigonometric function would you use to determine the opposite side if you knew the angle measure and the length of the adjacent side?

13. Consider the following right triangle:

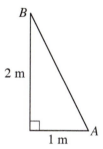

Calculate each of the following:

a. sin *A*   **b.** cos *A*

c. tan *A*   **d.** sin *B*

e. cos *B*   **f.** tan *B*

Activities 5.1–5.10   How Can I Practice?

1. Calculate the area and the perimeter for each of the following figures.

a.

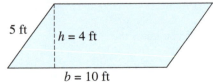

5 ft

h = 4 ft

b = 10 ft

b.

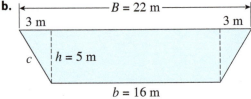

B = 22 m

3 m   3 m

c   h = 5 m

b = 16 m

2. Calculate the area and the perimeter for each of the following figures:

a. Rectangle topped by a semicircle:

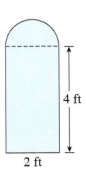

4 ft

2 ft

b. Rectangle topped by a right triangle:

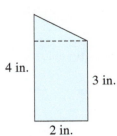

4 in.

3 in.

2 in.

c. Three-quarters of a circle with a square "corner":

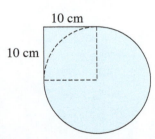

10 cm

10 cm

3. Determine the area of the shaded region in each of the following:

a.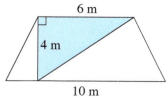

b.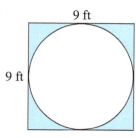

4. What is the length of one side of a square-shaped sign that has a perimeter of 96 inches?

5. a. Calculate the circumference of a circle of radius 2.3 feet.

   b. Use part a to predict the circumference of a circle of radius 9.2 feet, without using the circumference formula.

   c. Use the circumference formula to calculate the circumference of a circle of radius 9.2 feet.

   d. Does your answer to part c match your answer to part b?

   e. Determine the area of the circle having radius 2.3 feet.

   f. Without using the area formula, predict the area of a circle of radius 9.2 feet.

   g. Use the area formula to verify your result in part f.

6. a. To compute the area of a square with side 5 cm, which technique would you use, mental math or estimation? Explain.

  b. To compute the area of a square with side 5.33 cm, which technique would you use, mental math or estimation? Explain.

  c. What do you expect the area of a square with side 30 cm to be?

  d. Use the area formula to calculate the area of a square with side 30 cm.

  e. Was your prediction correct? Does your answer to part d match your answer to part c?

7. A triangle has two angles measuring 42° and 73°.

  a. Make a sketch of the triangle.

  b. Calculate the third angle of the triangle.

  c. Give the angle measurements of a triangle similar to this triangle.

8. You intend to put in a 4-foot-wide concrete walkway along two sides of your house, as shown.

  a. Determine the area covered by the walkway only.

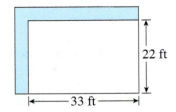

  b. If you decide to place a narrow flower bed along the outside of the walkway, how many feet of flowers should you plan for?

9. Triangle *A* has sides measuring 3 feet, 5 feet, and 7 feet; triangle *B* has sides measuring 4 feet, 4 feet, and 5 feet; triangle *C* has sides measuring 5 inches, 12 inches, and 13 inches.

  a. Which of the three triangles is a right triangle? Explain.

  b. A fourth triangle, *D*, is similar to triangle *B* but not identical to it. What are some possibilities for the lengths of the sides of triangle *D*? Explain.

  c. If 2 feet are added to the lengths of each side of triangle *A*, will the resulting triangle be similar to triangle *A*? Explain.

  d. If the length of each side of triangle *C* is tripled, will the resulting triangle be similar to triangle *C*?

10. a. Calculate the perimeter and area of isosceles triangle 1.

Triangle 2

Triangle 1

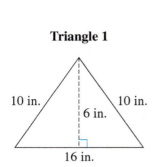
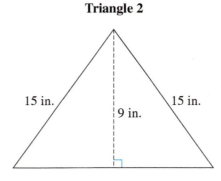

  b. Triangle 1 and triangle 2 are similar. What scale factor is needed to change triangle 1 to the size of triangle 2?

c. Without using the perimeter and area formulas, predict the perimeter and area of triangle 2.

d. How long is the base of triangle 2?

e. Calculate the perimeter and area of triangle 2 and check the prediction made in part c.

11. A softball diamond is in the shape of a square. The bases are 60 feet apart. What is the distance a catcher would have to throw a ball from home plate to second base?

12. You want to know how much space is available between a basketball and the rim of the basket. One way to find out is to measure the circumference of each and then use the circumference formula to determine the corresponding diameters. The distance you want to determine is the difference between the diameter of the rim and the diameter of the ball. Try it!

13. a. Determine the ratio of the sides of the *similar* rectangles shown below.

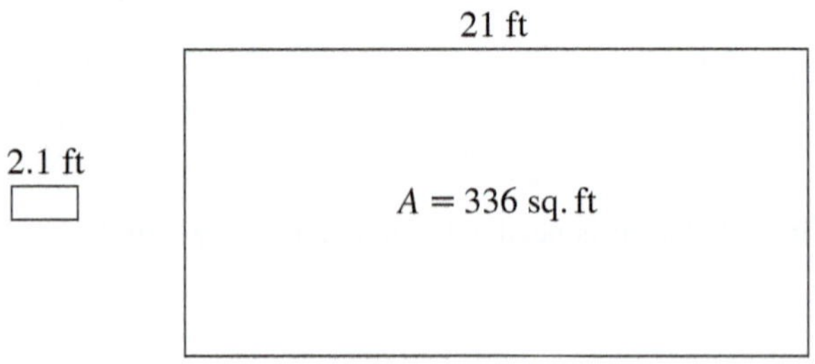

21 ft

2.1 ft

$A = 336$ sq. ft

b. Determine the width of the smaller rectangle.

14. Standing 15 feet from a streetlight, a 5-foot-tall woman casts a 3-foot-shadow. How tall is the streetlight?

15. Triangle *ABC* is a right triangle.

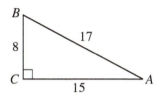

Calculate each of the following. Write your answer as a ratio.

a. tan *A*   **b.** tan *B*   **c.** cos *A*

d. cos *B*   **e.** sin *A*   **f.** sin *B*

16. Use your graphing calculator to determine the values of each of the following. Round your answers to the nearest thousandths.

a. $\sin 47° =$ **b.** $\cos 55° =$

c. $\tan 31° =$ **d.** $\tan 80° =$

17. Given $\sin A = \dfrac{5}{13}$, determine cos *A* and tan *A* exactly.

18. Given $\tan B = \dfrac{7}{4}$, determine sin *B* and cos *B* exactly.

19. Solve the following right triangle. That is, determine all the missing sides and angles.

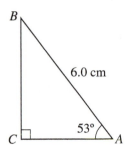

20. Textile strips often display various types of symmetry. For each of the following textile patterns, identify which transformations were used to create the design from the basic element.

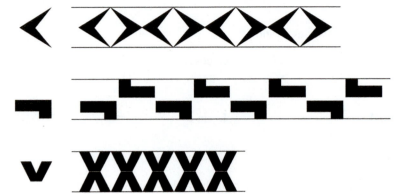

Activity 5.11

Painting Your Way through the Summer

Objectives

1. Recognize geometric properties of three-dimensional figures.

2. Write formulas for, and calculate surface areas of, boxes (rectangular prisms), cans (right circular cylinders), and spheres.

You decide to paint houses for summer employment. To determine how much to charge, you do some experimenting to discover that you can paint approximately 100 square feet per hour. To pay upcoming college expenses, you need to make at least $900 per week during the summer to cover your profit and the cost of paint, and brushes. You figure that it is reasonable to paint for approximately 40 hours per week. Armed with these facts, you are ready to start your painting business.

1. Use the appropriate information from above to determine your hourly fee.

Your first job is to paint the exteriors of a three-building farm complex (a small barn, a storage shed, and a silo) with the following dimensions:

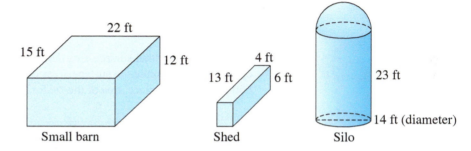

In order to determine your fee to the farmer, you need to estimate how many square feet of surface must be painted. Since there are few windows in the buildings, they may be ignored.

Rectangular Prism

2. To paint the small barn and the shed, use the formula for the surface area S of a box. Such a figure is called a **rectangular prism**, with length l, width w, and height h.

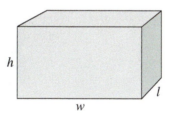

a. To determine the surface area of the rectangular prism, you first write the surface areas of each of the six rectangular surfaces:

Area of front = _____.   Area of back = _____.

Area of one side = _____.   Area of other side = _____.

Area of top = _____.   Area of bottom = _____.

b. Sum these six areas from part a to obtain the formula for the surface area of a rectangular prism.

c. How is the surface area of a rectangular prism affected if you double each of the prism's three dimensions?

d. True or False: If the three dimensions of a rectangular prism are increased (or decreased) by a scale factor, then the surface of the new prism is proportional to the square of the scale factor. Explain.

3. Assume that you will not paint the floors of the small barn and storage shed (but you *will* paint the special flat roofs), and use the appropriate formulas to compute the total surface area for the exteriors of these two buildings.

Right Circular Cylinder

4. To paint the cylinder part of the silo, you need to use the formula for the surface area S of a can, with height h and radius r. Such a figure is called a **right circular cylinder**.

a. To determine the total surface area S of the right circular cylinder, you compute the surface areas of the circular top and bottom, and the surface area of the "sides."

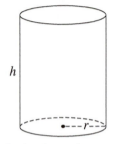

Area of circular top = _____

Area of circular bottom = _____

Area of side = _____

Hint: To determine the area of the side, think about cutting off both circular ends and cutting the side of the can perpendicular to the bottom. Then, uncoil the side of the can into a big rectangle with height h and width $2\pi r$, the circumference of the circle.

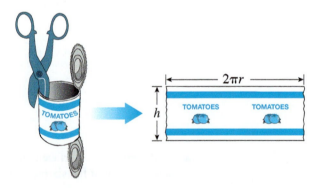

b. Sum the three areas from part a to obtain the formula for the total surface area of a right circular cylinder.

c. In this situation, do you need to determine the area of all three "parts" of the right circular cylinder? Assume you will not paint the floor of the silo.

Sphere

5. a. The roof of the silo is half of a sphere with radius r. The surface area, S, for a ball, or sphere, with radius r, is a little more difficult to derive. You may study the formula in future courses. It is provided here:

$$S = 4\pi r^2$$

The formula seems to be the sum of the areas of four circles of radius *r*. Explain why this formula is reasonable.

b. Determine the surface area of the roof of the silo.

6. You decide it is too dangerous to paint the roof of the silo. Carefully use the surface area formulas for a cylinder to compute the total exterior surface area for the silo. Again, assume you will not paint the floor or the semispherical roof.

7. a. What is the total surface area of the three-building farm complex that you must paint?

b. How long will it take you to paint all three buildings?

8. What will you charge for the job?

SUMMARY: ACTIVITY 5.11

Figure	Labeled Diagram	Surface Area Formula
Box (rectangular prism)	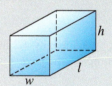	$S = 2wl + 2hl + 2wh$
Can (right circular cylinder)		$S = 2\pi \cdot r^2 + 2\pi \cdot r \cdot h$
Sphere (ball)		$S = 4\pi \cdot r^2$

EXERCISES: ACTIVITY 5.11

1. A basketball has a radius of approximately 4.75 inches.

    a. Compute the basketball's surface area.

    b. Why would someone want to know this surface area?

2. Hot air balloons require large amounts of both hot air and fabric material. A hot air balloon is spherical with a diameter of 25 feet.

    a. Does finding the surface area help you determine the amount of the hot air or the fabric material? Explain.

    b. Compute that surface area. What are the units?

3. Compute the surface area for each of the following figures:

    a. diameter = 7 in.   **b.**   *l* = 9 in.   **c.**

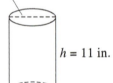 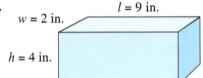

w = 2 in.

h = 4 in.

h = 11 in.

r = 2 m

4. You need to wrap a rectangular box with dimensions 2 feet by 3.5 feet by 4.2 feet. What is the least amount of wrapping paper you must buy in order to complete the job?

5. A can of soup is 3 inches in diameter and 5 inches in height. How much paper is needed to make a label for the soup can?

6. Calculate the height of a right circular cylinder with a surface area of 300 square inches and a radius of 5 inches.

7. As part of your senior project, your class has volunteered to paint all of the fire hydrants in your community. You need to estimate how much paint you will need. As you look at one hydrant, you can see a cylinder with two small cylinders on each side. You think the top of the hydrant resembles a half of a sphere. The town engineer tells you that the diameter of each hydrant is 7 inches and the height to the top of the cylinder is 30 inches. The two side cylinders have a diameter of 3 inches and a length of 4 inches.

   You know that one gallon of paint covers approximately 400 square feet of surface and will cost $23 per gallon. The firefighters have GPS technology in their fire trucks to determine where each hydrant is located. You are told that there are 514 fire hydrants in your community. What will be the cost of the project?

8. Complete the following statement: If the radius of a sphere is multiplied by $a > 0$, then its surface area is _____.

Activity 5.12

Truth in Labeling

Objectives

1. Write formulas for and calculate volumes of boxes and cans.

2. Recognize geometric properties of three-dimensional figures.

When buying a half-gallon of ice cream or a 12-ounce can of Coke, have you ever wondered if the containers actually hold the amounts advertised? In this activity, you will learn how to answer this question. The space inside a three-dimensional figure such as a box (rectangular prism) or can (right circular cylinder) is called its *volume* and is measured in cubic units (or unit cubes).

1. A half-gallon of ice cream has estimated dimensions as shown in the figure. In order to determine the number of unit cubes (1 inch by 1 inch by 1 inch) in this box, explain why it is reasonable to multiply the area of the top or bottom by the height.

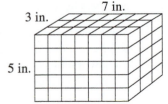

Definition

The **volume** of a box (rectangular prism) is the measure, in cubic units, of the space enclosed by the sides, top, and bottom of a three-dimensional figure.

Procedure

Calculating the Volume of a Box (Rectangular Prism)

The formula for the volume V of a rectangular prism with length l, width w, and height h is

$$V = lwh.$$

2. Use the formula to calculate the volume of the half-gallon of ice cream.

3. One cubic inch contains 0.554 fluid ounces of ice cream. Estimate the amount of fluid ounces of ice cream in the carton.

4. There are 64 fluid ounces in a half gallon. How close is your estimate to the half gallon?

5. In a way similar to the volume of a box, it is reasonable to define the volume of a right circular cylinder (can) as the product of the area of its circular bottom and its height. If a can has height h and radius r, write a formula for its volume. Explain.

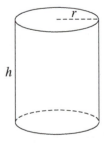

> **Procedure**
>
> **Calculating the Volume of a Can (Right Circular Cylinder)**
>
> The formula for the volume V of a can (right circular cylinder) with height h and radius r is
>
> $$V = \pi r^2 h.$$

6. A 12-ounce can of Coke is estimated to be $2\frac{1}{2}$ inches in diameter and $4\frac{3}{4}$ inches high. Use the volume formula to calculate the volume of a can of Coke.

7. One cubic inch contains 0.554 fluid ounces of Coke. Estimate the amount of Coke in a can. How close to 12 ounces is your estimate?

8. You contact a package designer to develop a container that can store 100 cubic inches of a product.

    a. She designs a long, slim container (like a spaghetti box) to store your product. What are the dimensions? *Hint*: Use a length of 2 inches and a width of 4 inches and solve for the height, h.

    b. The next design for the container is one inch longer, with the same width. What are the dimensions?

    c. Design a cube (length = width = height) to store your 100-cubic-inch volume.

        i. Let x represent the length = width = height. Write an equation for a volume of 100 cubic inches.

        ii. To solve the equation in part i, you need to "uncube," that is, take the cube root of both sides. You are looking for a number whose cube is 100. You note that 4 is too small, because $4^3 = 64$. Is 5 too large or too small?

        iii. Estimate the cube root of 100.

        iv. Most calculators have a cube root key or menu to calculate cube roots. The cube root of 100 is denoted as $\sqrt[3]{100}$. Use your calculator to compute the cube root of 100.

    d. Design a "tall" can (like a juice can) to store your volume. What are the dimensions?

    e. Design a "short" can (like a can of tuna fish) to store your volume. What are the dimensions?

    f. Why do you suppose that there are so many different shapes and sizes of containers?

9. A local sculptor has been commissioned to create a sculpture for the town hall. The original design uses a block of marble in the shape of a rectangular prism having the following dimensions: 4 feet in width, 5 feet in length, and 6 feet in height.

    a. What is the volume and surface area of the block of marble?

    b. The sculptor decides to increase the dimensions by a scale factor of 2. Determine the new volume and surface area.

c. Is the change in volume proportional to the change in width?

d. Is the change in surface area proportional to the change in width?

Therefore, if you double the dimensions of an object, the volume or surface area of the new object is NOT doubled. However, there is a relationship between the scale factor and the volume and surface area of the new object.

10. In this problem, you will explore what happens to the volume of a cube if you increase the length of each edge of the cube by different scale factors.

a. Complete the following table for a cube having 3 units as the measure of each edge.

SCALE FACTOR	LENGTH OF A SIDE, s	VOLUME, v	RATIO OF NEW VOLUME TO THE ORIGINAL VOLUME
2	6	216	$\dfrac{216}{27} = 8$
3	9	729	
4	12	1728	
10	30	27,000	

b. Note that if the length of the side of the cube is doubled, a scale factor of 2, the new volume is $8 = 2^3$ times the volume of the original. Write each ratio in the column 4 in exponential form with exponent 3.

c. What is the relationship between the volume of the new cube and the volume of the original?

d. If k represents the scale factor and s represents the length of a side of a cube, write an algebraic expression that represents the volume of the new cube.

e. Does the resulting formula confirm your results in part c? Explain.

The conclusion reached in Problem 10 can be summarized as follows:

> If V_1 represents the volume of the original cube, and V_2 represents the volume of the new cube, then
>
> $$V_2 = k^3 V_1, \text{ where } k \text{ is the scale factor.}$$

11. a. In Activity 5.3, page 627, what is the important relationship between a scale factor and the surface area of two-dimensional objects?

b. Does Problem 9d demonstrate this property for a three-dimensional object?

c. Verify this relationship for a rectangular prism having width w, length l, height h, and a scale factor k using the formula for the surface area of a rectangular prism (box).

SUMMARY: ACTIVITY 5.12

1. Three-Dimensional Figure | **Labeled Sketch** | **Volume Formula**

Rectangular prism

$$V = lwh$$

Right circular cylinder

$$V = \pi r^2 h$$

2. If the dimensions of a three-dimensional object are multiplied by a scale factor k, then

a. the volume of the new object is k^3 times the volume of the original.

b. the surface area of the new object is k^2 times the surface area of the original.

EXERCISES: ACTIVITY 5.12

1. Compute the volume for each of the following figures:

a. diameter = 7 in.

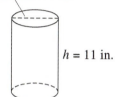

h = 11 in.

b.

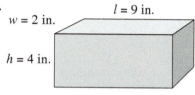

w = 2 in. *l* = 9 in.

h = 4 in.

2. A landscape architect wants to buy a truck and is interested in one with a large carrying capacity. One model features a rectangular prism–shaped cargo space, measuring 6 feet by 10 feet by 2 feet; another has a space with dimensions 5 feet by 11 feet by 3 feet. Which truck provides the landscaper with the most space (that is, the most volume)? Explain.

3. A can of soup is 3 inches in diameter and 5 inches high. How much soup can fit into the can?

4. You construct a box from a rectangular piece of cardboard measuring 2 feet by 3 feet. You do this by cutting out identical squares from each of the corners and folding up the sides. Experiment with several possibilities to determine which dimensions will make a box with the largest volume.

5. Automobile engines come in many different shapes and sizes but the amount of volume occupied by all the engine's cylinders is usually measured in either cubic inches, cubic centimeters, or liters. Research the shapes of engines and describe how their volume is measured. Prepare a report for the class.

6. If the volume of a cube is given as 42 cubic feet, estimate its dimensions.

7. If the volume of a right circular cylinder is given as 50 cubic cm and if its radius measures 2 cm, calculate its height.

8. Which melts faster: a block of ice with dimensions 6 × 3 × 2 inches or 36 one-inch cubes of ice?

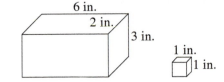

a. Determine the volume of the large ice block.

b. Determine the total volume of the individual cubes.

c. Compare the volume of the cubes with that of the volume of the large block of ice.

It is the exposure of surface area that melts the ice; the more surface area exposed the faster the ice will melt.

d. What is the total surface area of the block of ice?

e. What is the total surface area of the 36 cubes?

f. Compare the surface area of the large block of ice with that of the 36 cubes.

g. Do you think the block of ice or the cubes will melt faster. Explain.

9. There is a special on a cylinder of M&Ms® at the movie theater. You can bring in your own homemade cardboard cylinder. The dimensions of the piece of cardboard that you use to make the cylinder, excluding the top and the bottom, must be the size of a piece of paper $\left(8\frac{1}{2} \times 11\right)$.

You need to experiment. Roll the cardboard into a cylinder making its height 8.5 inches and determine the volume of the cylinder. Now roll the cardboard into a cylinder with the height 11 inches and determine the volume.

Do you think it matters which cylinder you have filled with M&Ms® at the theater?

10. The art of glass blowing is believed to have been invented in first century BC. During the Renaissance, glass blowing techniques were greatly refined by the artisans of Murano, Italy. Today, there are many places in the United States, including Corning, New York, producing artistically exquisite glass-blown items. Suppose a glass-blown vase is in the shape of a right circular cylinder and has a radius of 4 inches and a height of 10 inches.

a. Determine the surface area of the vase. The vase is open at the top.

b. Determine the volume of the vase.

c. If the dimensions of the vase are increased by a factor of 1.5, what are the dimensions of the new vase?

d. Determine the surface area and the volume of the larger vase.

e. Is either the surface area or the volume of the larger vase proportional to the scale factor? Explain.

f. Use the results from part e to determine the surface area and volume of the larger vase using the surface area and volume of the original vase.

g. How do the results from part d compare to the results of part f?

Activity 5.13

Analyzing an Ice Cream Cone

Objective

Write formulas for and calculate volumes of spheres and cones.

A popular summertime treat is the ice cream cone. The geometry of this treat is interesting, since it involves a three-dimensional cone topped with spheres (of ice cream). Let's begin our analysis of the ice cream cone with the geometry of spheres.

Spheres

1. Visually, it seems clear that a golf ball is smaller than a tennis ball, which is smaller than a baseball, which is smaller than a basketball. One way to compare these balls is by their radii, r. Another way is to measure the volume, V, of each ball. Determining a formula for V in terms of r is an involved process, but it can be estimated visually.

a. Think about one-half of a sphere that just fits inside a right circular cylinder.

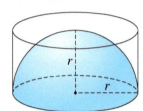

The area of the "great circle" running through the center of the sphere at the base of the cylinder is given by $A = \pi r^2$. The height of the cylinder is r. What is a formula for the volume of the cylinder?

b. Doubling the formula from part a produces a formula for the cylinder that just encloses the entire sphere. What is that formula?

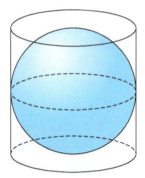

c. It is visually clear that the formula from part b is an overestimate of the volume of the sphere. However, what is not so clear is that the exact volume for the sphere is given by the formula $V = \dfrac{4}{3}\pi r^3$. Explain why this formula for the volume of a sphere is reasonable.

Procedure

Calculating the Volume of a Sphere

The formula for the volume, V, of a sphere with radius r is

$$V = \frac{4}{3}\pi r^3.$$

2. The accompanying table contains the radii of golf balls, tennis balls, baseballs, and basketballs.

	Round and Round	
TYPE OF BALL	RADIUS, *r*	VOLUME, *V*
Golf	0.84 inch	
Tennis	1.28 inches	
Baseball	1.45 inches	
Basketball	4.75 inches	

   a. Use the formula for volume of a sphere to compute the volume for each ball and write your answers in the table.

   b. How many baseballs would "fit" inside a basketball?

3. You notice that the spherical scoop of ice cream on your cone has diameter 2 inches. How much ice cream does the scoop contain?

4. You read an ad for a beach ball that claims it has a volume of 50 cubic inches. Estimate its radius. Explain the procedure you used to determine your estimate.

5. A sphere has a circumference of 20 cm (as measured by the circumference of the "great circle" around the center of the sphere). Estimate its volume.

Cones

6. Determining the formula for the volume of a cone of radius r and height h is also involved, but the formula can be estimated visually.

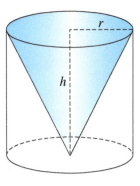

a. Think about a cone of height h and radius r that *just* fits inside a right circular cylinder. The area of the "great circle" at the top of the cone is given by $A = \pi r^2$. The height of the cylinder is h. What is a formula for the volume of the cylinder?

b. It is clear that $V = \pi r^2 h$ is an overestimate for the volume of the cone in the cylinder. Guess what fractional part of $\pi r^2 h$ is the volume of the cone and explain why your guess is reasonable.

c. The correct formula for the volume of a cone is $V = \dfrac{1}{3}\pi r^2 h$. Compare this with your answer to part b.

Procedure

Calculating the Volume of a Cone

The formula for the volume, V, of a cone with radius r and height h is

$$V = \frac{1}{3}\pi r^2 h.$$

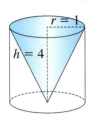

7. Your ice cream cone has a radius of 1 inch and is 4 inches high.

a. Determine its volume.

b. If you filled the cone with soft ice cream so the ice cream is level with the top of the cone, would you have more ice cream than the spherical scoop in Problem 3?

8. Cones are also used to mark highway construction. To stabilize these cones, which measure diameter 1 foot and height 2.5 feet, one option is to fill the cones with various materials. Compute the volume of the cone.

9. If you want to double the volume of the cone from Problem 7 without changing the radius, how high must the new cone be?

SUMMARY: ACTIVITY 5.13

Three-Dimensional Figure	Labeled Sketch	Volume Formula
Sphere		$V = \dfrac{4}{3}\pi r^3$
Cone		$V = \dfrac{1}{3}\pi r^2 h$

EXERCISES: ACTIVITY 5.13

1. Hot air balloons require lots of hot air and lots of fabric material. Suppose a hot air balloon is spherical with a diameter of 25 feet.

   a. Does finding the volume help you determine the amount of hot air or fabric material?

   b. Compute the volume.

2. Earth has a radius of approximately 6378 km; the radius of Mars is approximately 3397 km.

   a. Compute the volumes of Earth and Mars.

   b. The volume of Earth is how many times larger than the volume of Mars?

3. Write the formula for the volume of a sphere in terms of its diameter.

4. Compute the volumes for four spheres of radii 1 foot, 2 feet, 3 feet, and 4 feet and compare the answers. How many of the smallest sphere can be fit into the largest sphere?

5. A basketball has a radius of approximately 4.75 inches. How many cubic inches of air would you need to inflate the ball?

6. Write the formula for the volume of a cone in terms of its diameter and height.

7. You decide to start a soft ice cream cone business. You pack the cones with soft ice cream and then top off each one with a "cone shaped" amount of ice cream swirled on top, measuring about one-third the height of the cone.

  a. If a regular size cone has dimensions diameter 1.5 in. and height 3 in., determine the volume of ice cream inside and on top of the cone.

  b. What size would you make a cone that contains twice as much ice cream as the regular sized cone in part a?

8. Determine the volume for each of the following cones and spheres:

a.

diameter = 2 ft

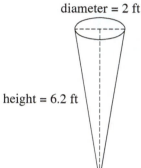

height = 6.2 ft

b.

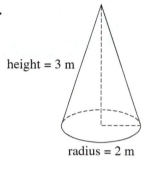

height = 3 m

radius = 2 m

c.

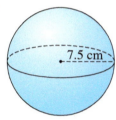

7.5 cm

d.

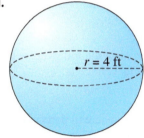

$r = 4$ ft

9. What other real-world examples of cones can you think of? Estimate their volumes.

10. A blown glass sculpture in the shape of a sphere has a radius of 18 inches.

    a. What is the exact volume (in terms of π) of the glass sphere?

    b. What is the exact volume of the glass sphere when the radius is doubled?

    c. What is the exact volume of the glass sphere when the radius is tripled?

d. Complete the following table. Select several of your own values for the radius.

1.	2.	3.	4.	5.	6.
RADIUS	VOLUME	RADIUS DOUBLED	VOLUME	RADIUS TRIPLED	VOLUME
18	7776π	36	62208π	54	209952π

e. Compare columns 2 and 4. When the radius is double (scale factor 2), by how many times did the volume increase?

f. Compare columns 2 and 6. When the radius is tripled (scale factor 3), by how many times is the volume increased?

g. Use your results in parts e–f to complete the following sentence: If the radius of a sphere is multiplied by a scale factor k, then its volume is _____.

h. Does this pattern work with the glass sphere in parts a–c?

Activity 5.14

Summertime

Objectives

1. Use geometry formulas to solve problems.

2. Use scale drawings in the problem-solving process.

Equipment

In this activity, you will need the following equipment:

1. A 12-inch ruler

Swimming Pool

With summer approaching, your neighbor decides to invest in a new in-ground swimming pool. A landscape architect has designed a custom pool and has provided the following scale drawing:

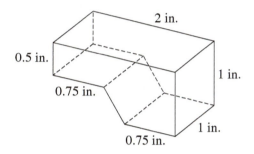

Scale: $\frac{1}{2}$ in. = 6 ft.

1. Use the scale drawing measurements to determine the dimensions (in feet) of your neighbor's new pool.

2. Calculate the perimeter of the top view of the pool.

3. Calculate the surface area of the top of the pool.

4. Calculate the area of the side view of the pool.

5. Calculate the total volume of the pool. (*Hint*: Remember that to obtain the volume of a prism or cylinder, you multiply the area of top or bottom by the height—the distance between the parallel top and bottom.) What is the distance between the two side views?

6. If the pool is to be filled with water to within 6 inches of the pool's top edge, calculate the amount of water needed to fill the pool. What are your units?

7. Determine the number of gallons of water needed to fill the pool. (There are 7.48 gallons in 1 cubic foot of water.)

8. A garden hose can fill the pool at the rate of 4.5 gallons per minute. How many minutes will it take to fill the pool? How many hours? How many days?

9. A pool-filling company charges $0.03 per gallon for water delivered in a big tanker truck. How much will this company charge to fill the pool?

10. Because of the soil conditions in your area, a landscape architect needs to know the weight of the water in the pool. Water weighs 62.4 pounds per cubic foot. Calculate the weight of the water in the pool to the nearest pound. Compare it with the weight of an average car.

11. To prevent accidents, state law requires that all pools be enclosed by a fence. How many feet of fencing are needed, if a fence will be placed around the pool 4 feet from each side?

12. You decide to buy some beach balls of 100 cubic feet volume for playing in the pool. What is the diameter of these balls?

13. To decorate the area around the pool with flowers, your neighbor purchases conical urns of radius 1.5 feet and height 3 feet. How much soil must he buy to fill each urn?

Memorial Park

14. You are asked by a local organization to serve on a committee that is planning a memorial park. An architectural firm has volunteered its services for the design phase. The centerpiece of the park, designed by a local artist, will be a glass monument in the shape of a square pyramid. The architect has a scale model of the pyramid. The scale model measures 1 foot long at the base and stands 8 inches high in the middle.

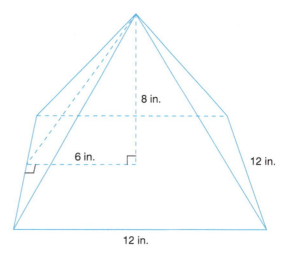

8 in.

6 in.

12 in.

12 in.

a. Describe the symmetry in each face of the monument.

b. The dimension scale factor is $\frac{1}{12}$. What will be the dimensions of the actual pyramid? Give your dimensions in feet.

c. Is the model similar to what the actual monument will be when it is completed?

d. The sides of the pyramid will be glass. How many square feet of glass will be required for the sides of the actual pyramid?

e. How many square feet are there in the 4 faces of the model pyramid?

f. Determine the ratio of the surface area of the actual pyramid to the surface area of the model pyramid. Is the ratio what you expected? Explain.

g. One part of the proposal is to make the pyramid into a fountain with water spilling out of the top of the monument. How many cubic feet of water would be required to fill the actual pyramid? (The volume of a pyramid is given by $V = \dfrac{1}{3} Bh$, where B represents the area of the base of the pyramid and h represents the height of the pyramid.)

h. What is the volume of the model pyramid in cubic feet?

i. Determine the ratio of the volume of the actual pyramid to the volume of the model. Is the ratio the scale factor? Is the ratio what you expected? Explain.

15. The pyramid will sit on a rectangular slab. There are two designs for the base. The first model has a base which is 12 inches square and 1 inch high. In the second design, the model pyramid sits on a base that is $\dfrac{1}{2}$ inch wider than the base of the pyramid on all sides and is 1 inch high.

a. What are the proposed dimensions of the actual base in the first design?

b. What are the proposed dimensions of the actual base in the second design?

c. Is the change from the first base to the second base proportional or non-proportional? Explain.

16. a. The total exposed base including that under the pyramid will have to be sealed with paint or some other substance. What is the surface area, in square feet, of the actual base of the first design?

b. What is the surface area of the actual base of the second design?

c. How many cubic feet of concrete will be necessary to construct the actual base of the first design?

d. How many cubic feet of concrete will be necessary to construct the actual base of the second design?

17. You hire a graphic designer to create a brochure to promote the memorial and solicit donations. The brochure will include a picture of the model of the monument. In the picture, the base of the pyramid is 2 inches long. What is the dimension scale factor from the model to the picture?

18. The huge metal and glass pyramid known as the Louvre Pyramid is the main entrance to the Louvre Museum in Paris, France. This structure is an architectural and engineering marvel. The pyramid's height is about 71 feet. Each side of the square base measures 115 feet. The Louvre Pyramid is made up of 70 triangular and 603 rhombus-shaped glass segments. Search the Internet to obtain a picture of the Louvre Pyramid.

a. Approximate the volume of the pyramid. Recall that the volume V of a pyramid having a square base is $V = \dfrac{1}{3} Bh$, where B is the area of the base and h is the height.

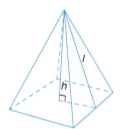

b. The surface area S of the 4 faces of a pyramid having a square base is $S = \dfrac{1}{2} Pl$, where P is the perimeter of the base and l represents the slant height (see diagram). Use the Pythagorean theorem to determine the slant height of the Louvre Pyramid.

c. Determine the surface area of the 4 sides of the pyramid.

d. If the dimensions of the pyramid were increased by a scale factor 1.2, would the surface area or volume be increased by a factor of 1.2? Explain.

e. Determine the surface area of the 4 faces and the volume of the larger Louvre Pyramid if the dimensions were increased by a scale factor 1.2.

Activity 5.15

Math in Art

Objectives

1. Recognize a golden rectangle by using the golden ratio.

2. Recognize Fibonacci numbers and the ratios of successive terms.

3. Understand one- and two-point perspective.

The visual arts have always been influenced by nature and mathematical ideas. In this activity, you will explore one of the classic relationships that has been applied to architecture and design since antiquity. The golden rectangle has been considered to have the most perfect proportion, to be the most pleasing to the eye. An example from ancient Greece, the Parthenon (picture below), was constructed with the golden rectangle as a design guideline.

1. Consider the following rectangles.

   a. Just for fun, which one do you think is the most pleasing to the eye?

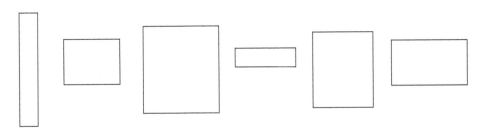

   b. Measure the lengths of the sides of each rectangle, in millimeters. Calculate the ratio of the long to short side for each rectangle, rounding to two decimal places.

The golden rectangle has the interesting property that when a square is cut off, the rectangle that remains is still a golden rectangle. We can use algebra to determine precisely what the ratio of sides must be.

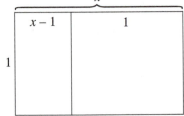

2. Assuming the longer side is x, and the shorter side is one unit, solve the proportion $\dfrac{x}{1} = \dfrac{1}{x-1}$.

(*Hint:* You will need to apply the quadratic formula.)

3. Because the shorter side of the rectangle is one, the value you found for x in Problem 2 is the exact value for the **golden ratio**, namely $\dfrac{1 + \sqrt{5}}{2}$. This is the ratio for every golden rectangle. Approximate the golden ratio to three decimal places. Which rectangle in Problem 1 is closest to a golden rectangle? Was it the one you chose as most pleasing?

4. Calculate the ratio of width to height of Salvadore Dali's *The Sacrament of the Last Supper* (1955). How close is it to the golden ratio? Do you think that it was an accident?

The golden ratio has been utilized for centuries in artistic design, perhaps on purpose or sometimes by accident. But the golden ratio does show up in nature, sometimes quite surprisingly. If the process of cutting off a square from a golden rectangle is repeated indefinitely, a quarter circle drawn in each square will trace out a spiral that appears frequently in nature.

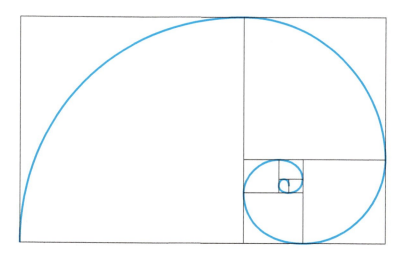

Chambered Nautilus

The Perfect Spiral Galaxy

Such spirals often appear in plant forms, in tightly bound rows. Look carefully at a pine cone or a flower head, like a sunflower. Notice that in the photo in Problem 5 of the sunflower you can pick out rows of spirals going in two directions (or more!).

5. Carefully count the rows of spirals in at least *two* directions. Record what you find, and check that your count agrees with that of your classmates.

You may recall the Fibonacci sequence of numbers: 1, 1, 2, 3, 5, 8, 13, 21, 34, 55, ... (see Exercise 6, Activity 1.2). As a sequence, the numbers in this pattern keep going forever.

6. Do you notice a relationship between the sunflower's spirals and the Fibonacci numbers? Again could it be simply an accident? Check it out with other seed forms, especially pine cones—the rows are easy to count.

Consider again the golden ratio. Sometimes called the *divine proportion*, it appears in nature and throughout mathematics in some unpredictable places, much like the number π. In fact, the Greek letter ϕ (phi) is often used to represent the golden ratio. Here it is again, accurate to thirty digits:

$$\phi = \frac{1 + \sqrt{5}}{2} \approx 1.618033988749894848420458683436$$

7. An interesting pattern develops when you calculate the ratios of successive numbers in the Fibonacci sequence. That is, $\frac{1}{1}, \frac{2}{1}, \frac{3}{2}, \frac{5}{3}, \frac{8}{5}$, and so on. Complete the table, and express the ratio as a decimal accurate to eight decimal places. (Your calculator is essential here!)

FIBONACCI NUMBER	RATIO OF SUCCESSIVE NUMBERS
1	—
1	1
2	2
3	1.5
5	1.6666667
8	1.6
13	1.625

Can you guess what these ratios are getting closer and closer to?

It can be proven in higher mathematics that, in fact, the ratios are indeed getting as close as you like to ϕ. It is the limit of the infinite string of ratios.

One- and Two-Point Perspectives

Let's look at a different aspect of mathematics in art. During the Renaissance in Europe (16th century), artists were very interested in portraying the physical world realistically, to create an illusion of three-dimensional space in a two-dimensional painting. To accomplish this, artists have developed many tricks over the past several centuries. The laws of perspective, an application of geometry, are probably the most significant development. To help understand how perspective works, Renaissance artists developed mechanical devices, such as that illustrated in the wood engraving below by Albrecht Durer (1471–1528), the German Renaissance artist.

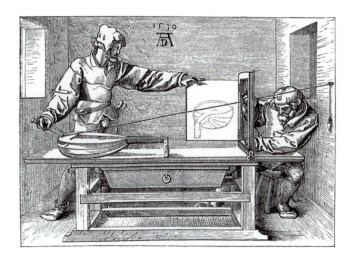

Perspective is most easily illustrated with the aid of vanishing points. Consider the horizontal line at the horizon, where earth meets sky in the distance. **One-point perspective** refers to a single point on the horizon line where all parallel lines moving away from the viewer will appear to converge to that vanishing point.

8. With a straightedge, draw a line segment from point A to point P. Repeat for points B and P. By drawing short horizontal line segments across your two lines, can you visualize a straight railroad track going through the desert? Embellish your drawing with a few cacti to make it appear more "real." Which point is the vanishing point?

9. Now go back and look more closely at Dali's *Last Supper*. The shadow lines and ends of the table are parallel lines moving away from the viewer. Where is the vanishing point located? Do you see how Dali used one-point perspective to emphasize the central point?

Two-point perspective utilizes two vanishing points to create the illusion of depth.

10. With a straightedge, lightly draw line segments connecting point A to both point P and Q. Repeat the process for point B. Also connect points A and B with a line segment. Now draw two line segments, parallel to segment AB, on either side of AB, stopping at the original line segments that you drew.

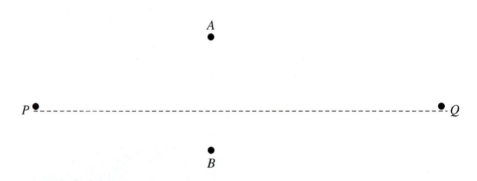

Can you visualize a three-dimensional block, maybe a building on a deserted street? Embellish the illusion by erasing the guideline leading to the vanishing points, darken the horizon line, and use the vanishing points to draw a few lines to represent tops and bottoms of rows of windows.

Definition

If corresponding vertices of two triangles form three lines that intersect in a single point, the two triangles are in **perspective**.

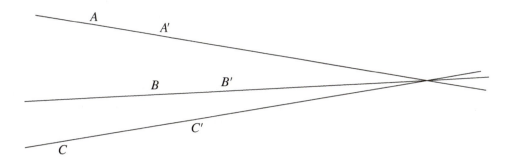

The geometry that underlies these visual tricks was developed shortly after the Renaissance. Here is an example of an important theorem from projective geometry.

Desargues' Theorem (1636)

The following theorem is attributed to the French mathematician, engineer, and architect Girard Desargues.

> If two triangles are in perspective, then the intersection points of corresponding sides of the two triangles will be collinear (be on a single line).

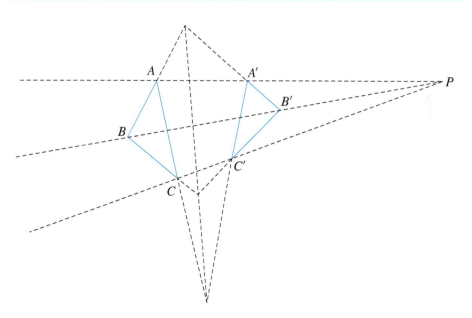

In the illustration you can see the triangles are in perspective, and the intersection points of AC—$A'C'$, AB—$A'B'$, and BC—$B'C'$ all line up.

11. In the illustration below, where the triangles ABC and $A'B'C'$ are in perspective, carefully extend the sides of each triangle. Mark the three intersection points of the corresponding sides to verify that Desargues' theorem works in this case.

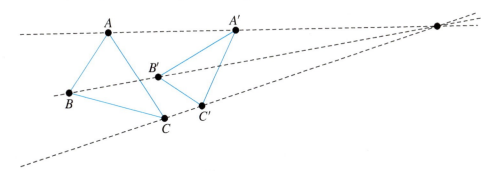

Desargues' theorem was developed, at least in part, to allow architectural drawings to be completed accurately when a vanishing point is well outside the boundary of the paper.

12. In the partially completed two-point perspective drawing of a building shown below, the right hand vanishing point, established by the top of the building and horizon line, is way beyond the edge of the paper. To accurately draw the bottom of the building, the vanishing point is not available. But using two triangles in perspective will allow you to locate the bottom of the building. Triangles ABC and $A'B'C'$ can be drawn so that line CC' will contain the desired line. The construction is partially completed. Determine the location of C' so that the intersection AB—$A'B'$, BC—$B'C'$ and AC—$A'C'$ (the given points X, Y, and Z) are collinear. Then construct the bottom of the building.

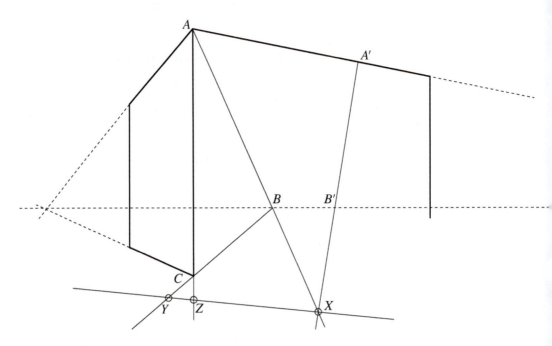

13. The second floor windows of a building are all the same size and shape. In the perspective drawing below, the polygons representing the windows are not the same size.

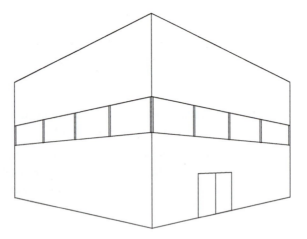

a. Describe the pattern that causes this apparent difference.

b. Do you think the windows in the drawing are all similar polygons? Describe the pattern you see to justify your answer.

c. By creating the illusion of three dimensions, a perspective drawing in architecture allows the viewer to better see and understand the underlying structure of a building. Using a two-point perspective drawing of a typical rectangular building (see drawing), describe how parallel elements will appear.

The vanishing point is a tool used by scenic and architectural photographers to effectively show perspective or scale in large landscapes or cityscapes. The location of vanishing points in a composition can add to the sense of depth needed to depict a three-dimensional reality in a two-dimensional photograph.

14. a. To emphasize a single object in the foreground of your photo, would you utilize one-point or two-point perspective? Why?

b. To emphasize an object or scene receding into the distance, would you utilize one-point or two-point perspective? Why?

15. Give several examples of situations where you would use a single vanishing point in a photographic composition.

16. Describe how you would compose a photograph of a building that would create two vanishing points. Draw a sketch showing those vanishing points.

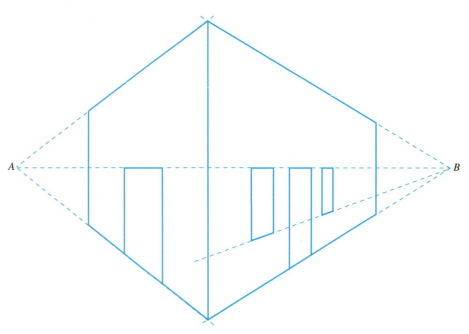

17. You are taking a photograph of a building and wish to emphasize its height. Explain how you could compose the scene to use a three-point perspective. Identify the vanishing points.

SUMMARY: ACTIVITY 5.15

1. A rectangle is golden when removing a square from one end results in a rectangle that has the same proportion as the original rectangle.

2. A golden rectangle's sides are in the ratio, $\dfrac{1 + \sqrt{5}}{2} \approx 1.618$, called the **golden ratio**.

3. The ratios of successive terms of the Fibonacci sequence $\{1, 1, 2, 3, 5, 8, 13, \ldots\}$ approach the golden ratio as a limit.

4. **One-** and **two-point perspective** are geometry techniques that can give the illusion of depth to a two-dimensional figure.

EXERCISES: ACTIVITY 5.15

1. The golden ratio is also sometimes expressed as the golden mean, the ratio of two line segments cut from a single line segment. For example, suppose line segment AB is one unit long. If point C cuts AB so that $AC = 0.618$, what is the ratio of AB to AC? What is the ratio of AC to CB?

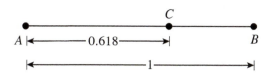

2. The common five-pointed star, as seen on an American flag, gives us another example of the golden mean. Consider one of the five line segments that compose such a star. Segment AB is cut at two points, J and K. Carefully measure, in millimeters, the length of AB, AJ, and AK. Are the ratios $\dfrac{AB}{AJ}$, $\dfrac{AJ}{AK}$, and $\dfrac{AK}{JK}$ the same?

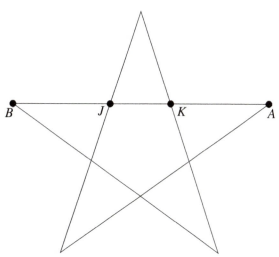

3. You are designing a large mural for the local community center. The wall you will work on is 15 feet wide and $10\dfrac{1}{2}$ feet high. You want the finished work to be a golden rectangle. Using the entire 15 feet width, and assuming your design reaches the top of the wall, how high above the floor should your mural start, to the nearest inch?

4. The Fibonacci sequence is defined by what is called a recursive process. Each number in the sequence is determined by adding the two previous terms. But the sequence is also determined by the first two numbers. In the Fibonacci sequence the first two numbers are both one. Another well-known sequence, the Lucas sequence, starts with 1 and 3.

a. Determine the next five numbers in the Lucas sequence:

1, 3, 4, 7, 11, ＿＿＿ ＿＿＿ ＿＿＿ ＿＿＿ ＿＿＿＿

b. Determine your own sequence by picking different numbers for the first two, and listing the first ten numbers in the sequence.

5. Perspective is used by artists, architects, designers, and photographers to maintain proportions. For example, the ratio of height to width of a building, tree or any object, as it appears in the artist's field of vision, is duplicated in the plane of the drawing. The artist's eye is the point of perspective.

A tree is drawn in perspective, as illustrated here. In the drawing, the tree is 5 inches wide and 9 in. tall. The actual tree is measured to be 60 feet wide. Use proportion to determine the height of the tree. What is the scale factor in this proportion?

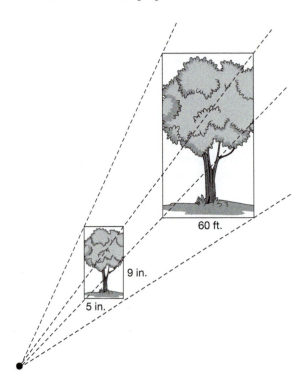

6. Which of the following points are the vanishing points for this two-point perspective drawing?

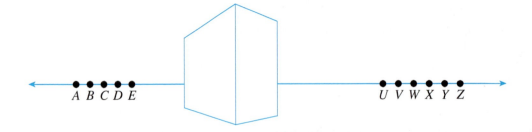

7. By creating the illusion of three dimensions, a perspective drawing in architecture allows the viewer to better see and understand the underlying structure of a building. Using a two-point perspective drawing of a typical rectangular building (see drawing), describe how parallel elements will appear.

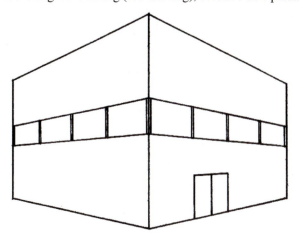

8. Many architectural elements can be viewed as a tessellation, a pattern of shapes that are repeated over a planar area, as in the shingles on a roof, bricks in a wall, or tiles on a floor. Consider the repeating pattern of rectangular bricks in the outer walls of a building. When the building and all its details, including the bricks, are drawn in a two-point perspective, will the bricks, as they appear in the drawing, actually be rectangular? If not, describe the pattern you would observe in the perspective drawing?

9. Many photographers and artists use the rule of thirds as a guideline when composing visual images such as photographs and paintings. The guideline suggests that the image should be imagined as divided into nine equal parts by two equally spaced horizontal lines and two equally spaced vertical lines. The ideal location of the elements of the image being photographed or painted should be placed along these lines or their intersection, with the four intersection points the best focal points for the composition.

    a. If your photo is to be 6 inches by 10 inches, how far will each ideal focal point be from the edge of the photo?

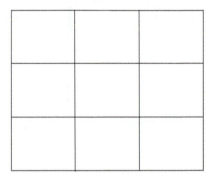

    b. The origin of the rule of thirds is related to the golden rectangle in which the ratio of the long to short side is the golden ratio 1.618 to 1. In the rule of thirds, the distances from each line, whether horizontal or vertical, to both edges is in the ratio of 2 to 1, a bit more than the golden ratio. If the lines did make those distances follow the golden ratio, the resulting composition is thought by many to be even more pleasing to the eye. The rule of thirds is but a rough estimate to the more perfect golden ratio. What would be the approximate positions of the horizontal and vertical lines in your 6 inches by 10 inches photo if they followed the golden ratio?

    c. Rather than placing the vanishing point in the center of the frame, a photographer may compose the scene so that the vanishing point occurs at the intersection of one of the intersection points of the rule of thirds. Search the Internet for an image demonstrating this technique. Describe the composition of the photograph.

10. To emphasize an object or scene receding into the distance, would you utilize one-point or two-point perspective? Why?

Activities 5.11–5.15   **What Have I Learned?**

1. What are the differences and similarities between the *surface area* and *volume* of a figure? Explain.

2. If the surface area of a figure is measured in square feet, then what are the units of the volume of that figure? Explain.

3. If the volume of a spherical figure is measured in cubic cm, then what are the units of the surface area of that sphere? Explain.

4. If different figures have the same surface areas, must their volumes be the same? Explain using an example.

5. If all the dimensions of a statue are multiplied by a scale factor k, then is the surface area of the new object proportional to the scale factor? Explain.

6. If all the dimensions of a sculpture are multiplied by a scale factor k, is the volume of the new object proportional to the original object by the scale factor k? Explain.

7. If different figures have the same volume, must their surface areas be the same? Explain, using an example.

8. If you double the height of a can and keep all other dimensions the same, does the volume double? Explain, using an example.

9. If you double the diameter of a can and keep all other dimensions the same, does the volume double? Explain, using an example.

10. If you triple the radius of a sphere, what effect does that have on the volume? Explain, using an example.

Activities 5.11–5.15   How Can I Practice?

1. Compute the volume and surface area for each of the following figures:

a.

6 in.

b.

diameter = 4.5 ft

height = 6.5 ft

c.

$w = 4\frac{1}{2}$ ft $l = 14.3$ ft

$h = 5\frac{1}{3}$ ft

2. Compute the volume of the cone with dimensions given.

diameter = 2.3 inches

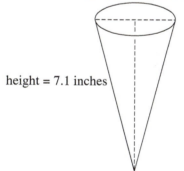

height = 7.1 inches

3. Draw each of the following figures with the desired property. In each case, provide the dimensions for the radius length or height of the figure as appropriate.

a. A sphere with a volume of 20 cubic feet:

b. A sphere with a surface area of 53 square cm:

c. A can with a surface area of 20 square feet and radius 1 foot:

d. A can with a volume of 20 cubic feet:

e. A cone with a volume of 20 cubic feet:

f. A box with a surface area of 20 square feet:

g. A box with a volume of 20 cubic feet:

4. a. You have created a $\frac{1}{8}$ scale model of a monumental sculpture to be cast in bronze. By displacing water in a tub, you determine the volume of the model to be 125 cubic inches. How much bronze, by volume, will be required for the final sculpture?

b. The size of the actual sculpture is increased by a scale factor 3. Determine the volume of the new sculpture.

5. You wish your painting canvas to be a golden rectangle, with the shorter dimension being 15 in. What do you need for the longer dimension?

6. You have set up a still-life to practice your drawing. To appear realistic, you need to accurately produce the same proportions for each object. The vase is 5 in. wide and 9 in. tall. In your drawing, you want the vase to be 4 in. tall. How wide should you draw the vase?

The bracketed numbers following each concept indicate the activity in which the concept is discussed.

CONCEPT/SKILL	DESCRIPTION	EXAMPLE
Perimeter formulas [5.1]	Perimeter measures the length around the edge of the figure.	
Square [5.1]	$P = 4s$	$P = 4 \cdot 1 = 4$ ft
Rectangle [5.1]	$P = 2l + 2w$	$P = 2 \cdot 2 + 2 \cdot 3 = 10$ in. 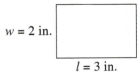
Triangle [5.1]	$P = a + b + c$	$P = 3 + 4 + 6 = 13$ meters
Parallelogram [5.1]	$P = 2a + 2b$ 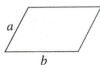	$P = 2 \cdot 7 + 2 \cdot 9 = 32$ inches 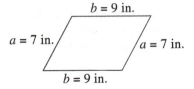
Trapezoid [5.1]	$P = a + b + c + d$ 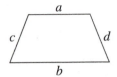	$P = 2 + 8 + 3 + 5 = 18$ feet 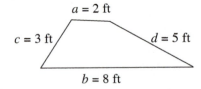
Polygon [5.1]	$P =$ sum of the lengths of all the sides $= a + b + c + d + e$ 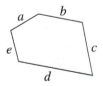	$P = 1 + 4 + 5 + 7 + 2 = 19$ cm 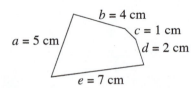

CONCEPT/SKILL	DESCRIPTION	EXAMPLE
Circle [5.2]	$C = 2\pi r$ For circles, perimeter is usually called *circumference*. 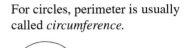 $C = \pi d$ 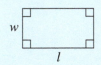	$C = 2\pi \cdot 3 = 6\pi \approx 18.85$ ft $C = \pi \cdot 5 \approx 15.71$ m 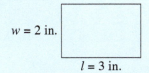
Area formulas [5.3]	Area is the measure of the region enclosed by a polygon or circle.	
Square [5.3]	$A = ss = s^2$	$A = 1 \cdot 1 = 1$ sq. ft
Rectangle [5.3]	$A = lw$	$A = 2 \cdot 3 = 6$ sq. in. 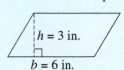
Triangle [5.3]	$A = \frac{1}{2}bh$	$A = \frac{1}{2} \cdot 4 \cdot 6 = 12$ sq. m 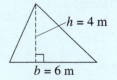
Parallelogram [5.3]	$A = bh$ 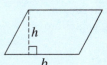	$A = 6 \cdot 3 = 18$ sq. in.

CONCEPT/SKILL	DESCRIPTION	EXAMPLE

Trapezoid [5.3]

$$A = \frac{1}{2}h(b + B)$$

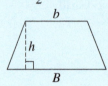

$$A = \frac{1}{2} \cdot 6 \cdot (4 + 9) = 39 \text{ sq. ft}$$

Polygon [5.3]

$A = $ sum of the areas of all the parts of the figure.

$$A = 2 \cdot \frac{1}{2} \cdot 4 \cdot (6 + 12) = 72 \text{ sq. m}$$

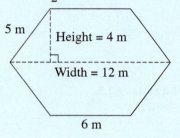

Circle [5.4]

$$A = \pi r^2$$

$$A = \pi \cdot 3^2 = 9\pi \approx 28.27 \text{ sq. ft}$$

Circle [5.4]

$$A = \frac{\pi d^2}{4}$$

$$A = \pi\left(\frac{5}{2}\right)^2 = \frac{\pi \cdot 5^2}{4} \approx 19.63 \text{ sq. m}$$

Classification of triangles [5.6]

Equilateral [5.6]

All three sides have the same length.

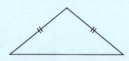

Isosceles [5.6]

Two sides have the same length and the corresponding angles are identical.

Scalene [5.6]

None of the sides have the same length.

CONCEPT/SKILL	DESCRIPTION	EXAMPLE
Equiangular [5.6]	All three angles have the same size.	
Right [5.6]	One angle measures 90°.	
Acute [5.6]	All angles measures less than 90°.	
Obtuse [5.6]	One angle measure greater than 90°.	
The sum of the angles of a triangle [5.6]	The sum of the angles of a triangle is 180°.	 $a + b + c = 180°$
Similar triangles [5.7]	Two triangles are similar provided their corresponding angles are equal; the lengths of their corresponding sides must be proportional (that is, their ratios must be equal).	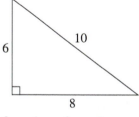 $\dfrac{3}{6} = \dfrac{4}{8}$ and $\dfrac{3}{6} = \dfrac{5}{10}$
Pythagorean Theorem [5.8]	$c^2 = a^2 + b^2$ This important formula for right triangles is useful for indirect measurement. 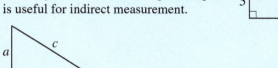	$13^2 = 5^2 + 12^2$, since $169 = 169$

CONCEPT/SKILL	DESCRIPTION	EXAMPLE
The sine function of the acute angle *A* of a right triangle [5.9]	$\sin A = \dfrac{\text{length of the side opposite } A}{\text{length of the hypotenuse}}$	Example 2, Activity 5.9.
The cosine function of the acute angle *A* of a right triangle [5.9]	$\cos A = \dfrac{\text{length of the side adjacent to } A}{\text{length of the hypotenuse}}$	Example 2, Activity 5.9.
The tangent function of the acute angle *A* of a right triangle [5.9]	$\tan A = \dfrac{\text{length of the side opposite } A}{\text{length of the side adjacent to } A}$	Example 2, Activity 5.9.
Tessellation [5.10]	A tessellation (or tiling) is a repeating pattern of shapes that covers an entire plane and fits together without gaps or overlaps.	M. C. Escher is an artist who used tessellating shapes in his art.
Transformational geometry [5.10]	Figures are moved in various ways without changing the size or shape of the figure.	Reflections, translations, rotations, and glide reflections are types of transformations.
Reflections [5.10]	A type of transformation in which a figure is reflected across a line called the axis of reflection.	See Example 1 in Activity 5.10.
Translation [5.10]	A figure is moved along a straight line in a given direction and for a given distance.	See Examples 2 and 3 in Activity 5.10.
Rotation [5.10]	A figure is turned about a specific point called the center of rotation.	See Example 4 in Activity 5.10.
Glide reflection [5.10]	Transformation of a figure by first translating the figure, followed by a reflection.	See Example 5 in Activity 5.10.
Symmetry [5.10]	The ending position of the transformed figure is exactly the same as the starting position.	Reflective symmetry and rotational symmetry.
Reflective symmetry [5.10]	If a figure is reflected across a given axis, the beginning position and the ending position are exactly the same.	See Example 1, part e, in Activity 5.10.
Rotational symmetry [5.10]	The ending position of a rotated figure is exactly the same as the beginning position.	Rotating a figure 360° about a given point.

CONCEPT/SKILL	DESCRIPTION	EXAMPLE

Surface area formulas [5.11]

Surface area is the measure in square units of the area on the outside of the three-dimensional figure.

Box (rectangular prism) [5.11]

$S = 2lw + 2lh + 2wh$

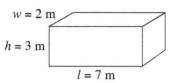

$S = 2 \cdot 7 \cdot 2 + 2 \cdot 7 \cdot 3 + 2 \cdot 2 \cdot 3$
$S = 28 + 42 + 12 = 82$ sq. m

Can (right circular cylinder) [5.11]

$S = 2\pi r^2 + 2\pi rh$

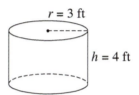

$S = 2\pi \cdot 3^2 + 2\pi \cdot 3 \cdot 4 = 42\pi$

$S \approx 131.95$ sq. ft

Sphere [5.11]

$S = 4\pi r^2$

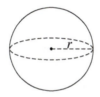

$S = 4\pi \cdot 6^2$
$= 144\pi \approx 452.39$ sq. in.

Volume formulas [5.12, 5.13]

Volume is the measure in cubic units of the space inside a three-dimensional figure.

Box (rectangular prism) [5.12]

$V = lwh$

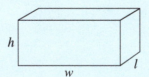

$V = 2 \cdot 4 \cdot 9 = 72$ cu. m

Can (right circular cylinder) [5.12]

$V = \pi r^2 h$

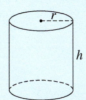

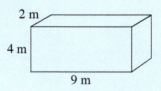

$V = \pi \cdot 3^2 \cdot 2 = 18\pi \approx 56.55$ cu. ft

CONCEPT/SKILL	DESCRIPTION	EXAMPLE

Sphere [5.13]

$$V = \frac{4}{3}\pi r^3$$

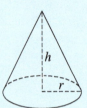

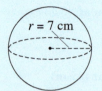

$r = 7$ cm

$$V = \frac{4}{3} \cdot \pi \cdot 7^3 \approx 1436.76 \text{ cu. cm}$$

Cone [5.13]

$$V = \frac{1}{3}\pi r^2 h$$

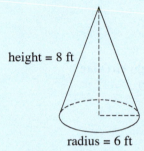

height = 8 ft

radius = 6 ft

$$V = \frac{1}{3}\pi \cdot 6^2 \cdot 8 \approx 301.59 \text{ cu. ft}$$

Golden rectangle [5.15]

A golden rectangle's sides are in the ratio $\frac{1 + \sqrt{5}}{2} \approx 1.618$, called the golden ratio.

The ratio of width to height of Dali's painting (see Problem 4 in Activity 5.15).

One- and two-point perspective [5.15]

Geometry techniques that can give the illusion of depth to a two-dimensional figure.

See Problems 8–13 in Activity 5.15.

1. Consider the following two-dimensional figures:

Figure A   **Figure B**   **Figure C**

a. The area of figure A = _____.

b. The perimeter of figure B = _____.

c. The area of figure B =
   _____.

d. The area of figure C =
   _____.

2. Draw a circle of area 23 square inches and label the length of its radius. Explain.

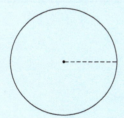

3. You buy a candy dish with a 2-inch by 2-inch square center surrounded on each side by attached semicircles.

   a. Draw the candy dish described above and label its dimensions.

   b. Determine its perimeter and area.

c. If you place the dish on a 1 foot by 1 foot square table, how much space is left on the table for other items? Explain.

4. You are in the process of building a new home and the architect sends you the following floor plan for your approval:

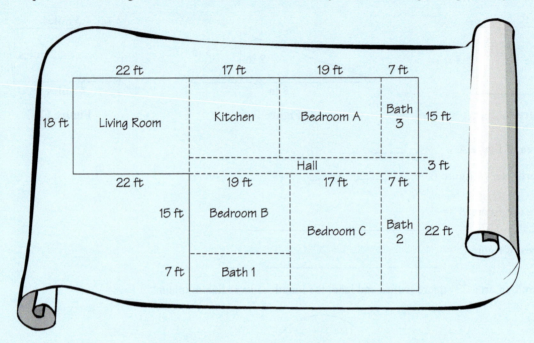

a. What is the perimeter of this floor plan?

b. What is the floor space, in square feet, of the floor plan?

c. Which bedroom has the largest area?

d. If you decide to double the area of the living room, what change in dimensions should you mark on the floor plan that you send back to the architect?

5. Your home is located in the center of a 300-foot by 200-foot rectangular plot of land. You are interested in measuring the diagonal of that plot. Use the Pythagorean theorem to make that indirect measurement.

6. Construct a triangle with its shortest side measuring 3 inches and similar to the following triangle:

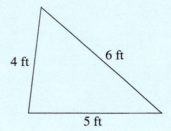

Label the lengths of each of the sides.

7. You plan to decorate a banner for a school play. The banner has a height of 3.8 feet and each side measures 4 feet long.

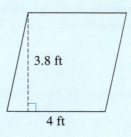

a. Determine the perimeter and area of the banner.

b. Suppose that you double the length of each side of the banner so that it now has a height of 7.6 feet and a base of 8 feet. Determine the new perimeter and area.

c. Compare the perimeter and area of the larger banner to the perimeter and area of the smaller banner.

8. a. Calculate the perimeter and area of triangle T1.

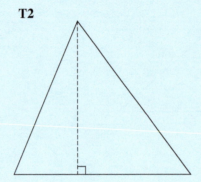

T1

13   15

5   9

T2

  b. The sides of the second triangle, T2, are formed by multiplying the sides of T1 by a scale factor of 2.5. Without using the perimeter and area formulas, predict the perimeter and area of T2.

  c. Determine the lengths of the sides and the height of T2 and write them in on the drawing above.

  d. Calculate the perimeter and area of T2 and check the prediction made in part b.

9. At 5 P.M., a 6-ft-tall man casts a 7.5-ft shadow. How long a shadow is cast by a 40-ft tree?

10. An architectural sketch is drawn to a scale of $\frac{1}{2}$ inch : 1 foot. A room on the drawing measures $8\frac{1}{4}$ inches by $10\frac{3}{4}$ inches. What are the dimensions of the actual room?

11. You walk seven miles in a straight line 63° north of east.

  a. Determine how far north you have traveled.   **b.** Determine how far east you have traveled.

12. Solve the following triangles:

a.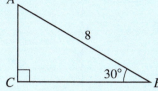

b.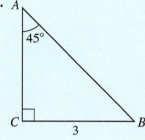

13. a. Given $\sin \theta = \dfrac{6}{10}$, determine $\cos \theta$ and $\tan \theta$ without using your calculator.

b. Given $\tan \theta = \dfrac{8}{5}$, determine $\sin \theta$ and $\cos \theta$ without using your calculator.

14. Calculate the volumes of the following three-dimensional figures:

a.

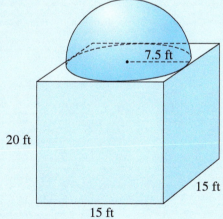

b.

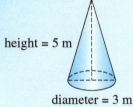

height = 5 m

diameter = 3 m

c.

diameter = 5 inches

height = 8 inches

15. Biologists studying a lake determine that the lake's shape is approximately circular with a diameter of 1 mile and an average depth of 187 feet. Estimate how much water is in the lake.

Note that for consistency in units, 1 mile = 5280 feet.

16. The Alaska pipeline is a cylindrical pipeline with diameter 48 inches that carries oil for 800 miles through Alaska. It is an engineering marvel, with many safety and business concerns.

    a. What is the total capacity of the pipeline for oil at any given time, assuming that it could be filled to capacity? Explain.

    b. Estimate the amount of material needed to construct the pipeline. Explain.

Problem Solving with Graphical and Statistical Models

Activity 6.1

Visualizing Trends

Objectives

1. Recognize how scaling of the axes of a graph can misrepresent the data.

2. Analyze data displayed in print and electronic media.

The overwhelming amount of data available to the American public has become an integral part of our information society. Newspapers, magazines, television news, and especially the Internet provide us with data collected from a wide variety of sources. The United States Census Bureau collects a vast amount of data. For example, each month the United States Census Bureau's American Community Survey collects data from 250,000 households. The survey asks a wide range of questions about the household and each person in the household. (See www.census.gov/acs/www.)

Statistics

Statistics is the science of gathering, analyzing, and making predictions from data (numerical information). Statistics has become an indispensable tool in the study of such diverse areas as medicine and health issues, the economy, marketing, manufacturing, population trends, and the environment.

An important part of a statistical study is organizing and displaying the data collected. Throughout this textbook, you will observe how useful graphs can be for identifying patterns and trends. However, graphs can also be misleading. For example, consider the following three linear graphs:

Graph I

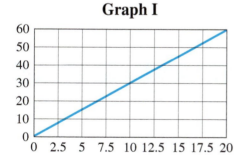

Graph II

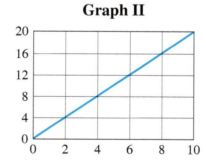

Graph III

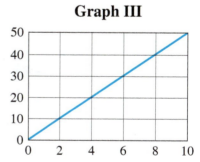

At first glance, the three graphs appear to represent the same line. However, with closer examination, the scales along the *x*- and *y*-axes are not the same. This results in the deception that the lines are identical. The lines rise at different rates as shown below:

Line I: Rate of increase $= \dfrac{up\ 60}{right\ 20} = 3$ units up per one unit to the right.

Line II: Rate of increase $= \dfrac{up\ 20}{right\ 10} = 2$ units up per one unit to the right.

Line III: Rate of increase $= \dfrac{up\ 50}{right\ 10} = 5$ units up per one unit to the right.

1. You see two online ads for brokerage companies that claim they will help build your stock portfolio. One ad shows a graph for stock ABC. The other company shows a graph for stock XYZ. In statistical terminology, these graphs are called **line graphs** because straight line segments connect adjoining data points. They are also known as **broken-line graphs**.

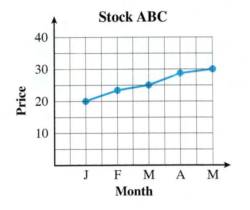
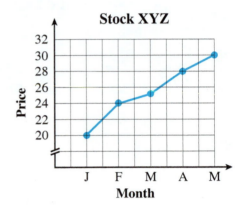

   a. Which ad promotes the company's ability to increase your stock portfolio?

   b. The graphs have different scales along the *y*-axis. Why does the scale for stock ABC make the graph appear to be rising more slowly?

As a matter of fact, the graphs in Problem 1 display the same information.

2. The following table shows the life expectancy for women in the United States for selected years in the period 1975–2010.

Living Longer

YEAR	1975	1980	1985	1990	1995	2000	2005	2010
Women	76.6	77.4	78.2	78.8	78.9	79.3	79.9	81.0

Data Source: U.S. Bureau of the Census

   a. Construct a line graph that makes the increase in the life expectancy from 1975 to 2010 appear to be relatively small.

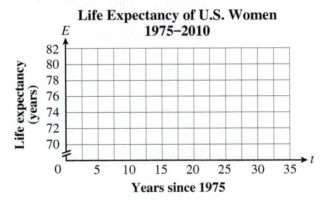

   b. How would you change the vertical scale to make the increase seem larger?

3. The following graphs are examples of **bar graphs**. Bar graphs are a very popular way of displaying data in print media such as newspapers and magazines. Each graph shows the median annual incomes of print publication readers.

Median Annual Income for Print Publication Readers

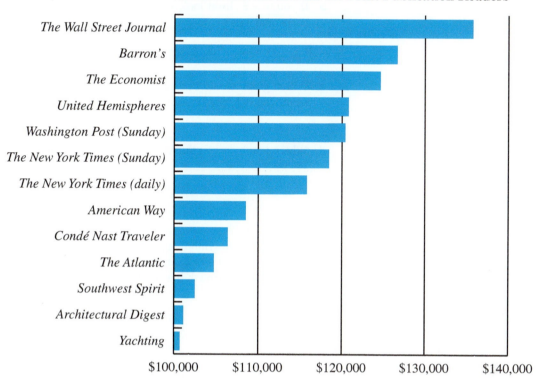

Data Source: BtoBonline, Media Business section, Crane Communications

Median Annual Income for Print Publication Readers

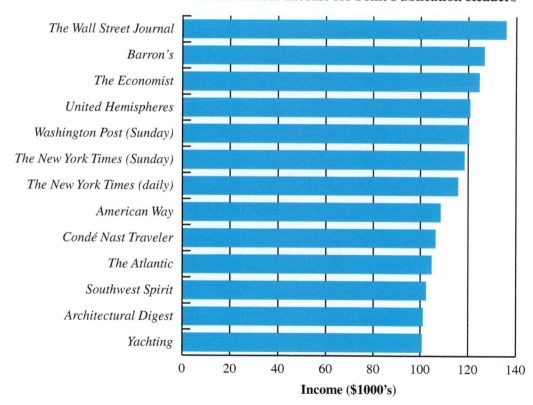

Data Source: BtoBonline, Media Business section, Crane Communications

    a. At first glance, what does each graph tell you about the incomes of print publication readers?

    b. If you take a closer look at each graph, what do you notice about the horizontal axis?

    c. Which graph would best support the claim that *The Wall Street Journal* has significantly higher-income readers than the other print publications?

    d. Which graph do you think best represents the data on reader incomes?

4. Sometimes a single symbol or picture is used instead of a bar, its size indicating the quantity being measured. For example, the graph below (sometimes called a **pictograph**) is meant to show the growth in sales for a carpet retailer.

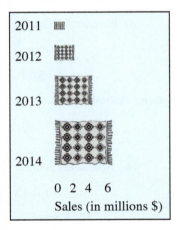

The length of the symbol represents the amount of sales, just like the length of a bar in a bar graph. Approximately how much did sales grow from 2011 to 2014?

5. There is a possible deception, however, in the graph of Problem 4. The size or area of the picture can be misleading. For example, if the width and length of a rectangle both increase by a factor of three, by how much does the area of the rectangle increase?

Applying this idea to the smallest and largest symbol in the pictograph above, it appears visually that sales have increased nine times as much from 2011 to 2014.

Another illusion may appear when bars are represented in 3-D instead of as simple rectangles. Since representing three dimensions requires perspective, a graph may be manipulated to appear different than it really is. This situation is demonstrated in Problem 6.

6. The following three variations of bar graphs display municipal waste collection data for three Florida counties. These graphs are all based on exactly the same data. The first graph is looked at directly and appears two-dimensional, or flat. The second and third graphs are viewed from angles different from straight ahead and therefore appear three-dimensional. Which graph could be misinterpreted in comparing the waste collections of the three counties? Explain how someone could make false claims because of the visual effect.

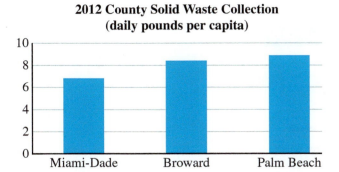

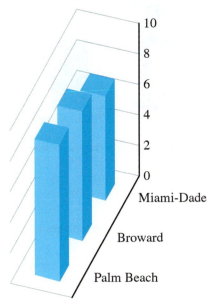

Data source: Florida Dept. of Environmental Protection, *Solid Waste Management in Florida 2012 Annual Report*, Table 1A, *County Municipal Solid Waste Collected Per Capita (2012)*

SUMMARY: ACTIVITY 6.1

1. Line graphs, also known as broken-line graphs, use straight line segments to connect adjoining data points to display data.

2. A **pictograph**, uses a single symbol or picture to display data. The size of the symbol or picture indicates the quantity being measured.

EXERCISES: ACTIVITY 6.1

1. A 2014 survey examined the attitudes of students about the use of mobile devices in learning. The results of the survey are summarized in the following table and the corresponding bar graph. The percentages given are for the number of high school students who reported using the technology at least a few times a week in doing schoolwork.

TECHNOLOGY	USE IN SCHOOLWORK
Tablet	27%
Laptop*	66%
Smartphone	43%

*"Laptop" also includes notebooks and Chromebooks.

Data Source: Harris Interactive, *Pearson Student Mobile Device Survey: Grades 4 through 12, May 9, 2014.*

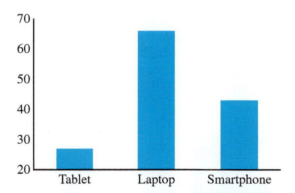

a. Use the table to determine the actual difference in percentages between laptop use and smartphone use for schoolwork.

b. Use *only* the graph shown to estimate the difference in percentages between laptop and smartphone use for schoolwork.

c. From your work in parts a and b and for other reasons you may have, how well do you think the graph presents the data in the table?

2. The following pictograph, showing the sales for a local furniture store, appeared in an ad in the local newspaper.

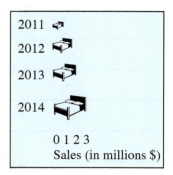

2011

2012

2013

2014

0 1 2 3

Sales (in millions $)

a. How much did sales actually grow from 2011 to 2014?

b. The furniture store owner claimed that the sales of the store quadrupled in the past three years. Did the graph support the claim?

3. Look through a newspaper or magazine and search the Internet for advertisements containing a line graph or bar graph. Determine if the ad's claim or implied conclusion is supported by the graph. Does the scaling of the axes misrepresent the data? Explain.

When the American bald eagle was adopted as the national symbol of the United States in 1782, there were an estimated 25,000 to 75,000 nesting bald eagle pairs in what are now the contiguous 48 states. Bald eagles were in danger of becoming extinct about 50 years ago, but efforts to protect them have worked. On June 28, 2007, the Interior Department took the American bald eagle off the endangered species list.

Even though the bald eagle was removed from the endangered species list in 2007, many states have continued to monitor their eagle population and report large increases of their eagle population. For instance, Wisconsin recorded 1337 nesting pairs in 2012.

Bar Graphs

The following bar graph displays the numbers of nesting bald eagle pairs in the lower 48 states for the years from 1963 to 2006. The horizontal direction represents the years from 1963 to 2006. The vertical direction represents the number of nesting pairs.

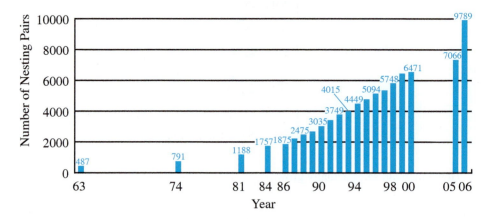

Data Source: U.S. Fish and Wildlife Service

1. **a.** Estimate the number of nesting pairs in 1963? In 1986? In 2000?

   b. Explain how you located these numbers on the bar graph.

2. **a.** Estimate the number of nesting pairs in 1989. In 1997.

   b. Explain how you estimated the numbers from the bar graph.

   c. To what place value did you estimate the number of nesting pairs in each case?

   d. Compare your estimates with the estimates of some of your classmates. Briefly describe the comparisons.

3. **a.** Estimate the number of nesting pairs in 1983 and 1985.

   b. Explain how you determined your estimate from the graph.

c. Estimate to the nearest thousand the number of nesting pairs in 1977.

d. From 1986 to 1998, what do the bars indicate about the growth trend in the number of eagle pairs?

Bar graphs can be oriented either vertically or horizontally.

4. Approximately how many more bald eagle pairs were observed in Louisiana compared to Texas in 2000?

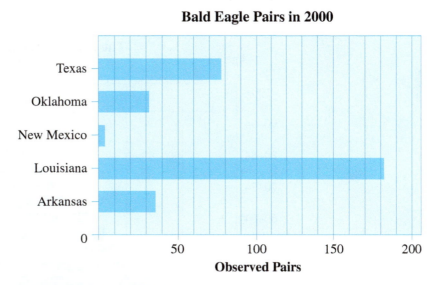

Bald Eagle Pairs in 2000

Data Source: U.S. Fish and Wildlife Service

Grouped Bar Graph

Bar graphs can also show paired data for each category. The following a **grouped bar graph** shows the number of endangered and threatened species for four animal categories (groups) for each of three years: 1995, 2005, and 2013.

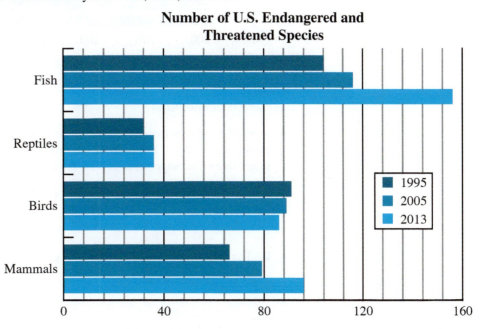

Number of U.S. Endangered and Threatened Species

Data Source: U.S. Fish and Wildlife Service

5. a. Which category of animal actually saw a drop in the number of endangered and threatened species between 1995 and 2013? Explain your reasoning.

b. Which category of animal showed the largest increase between 1995 and 2005, and by approximately how many species?

The following is an example of a **paired bar graph** based on U.S. Census Bureau data. It shows the age-gender structure of the global population, a chart commonly called a population pyramid from its shape.

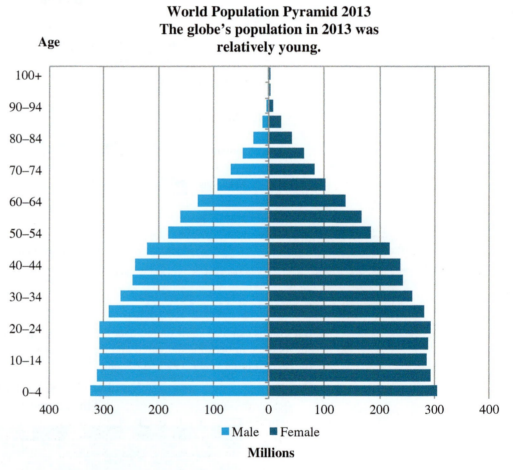

World Population Pyramid 2013
The globe's population in 2013 was relatively young.

Data Source: U.S. Census Bureau

6. A number of conclusions can be drawn about the age-gender structure of the global population in 2013. For each of the following claims, give specific data from the bar graph that either supports or refutes the claim.

a. No matter what the age group, there are more females than males.

b. The largest age group, regardless of gender, was 0–4-year-olds.

c. Males and females are very close in number for all age groups between 35 and 54.

d. The worldwide birth rate appears to have increased dramatically during the 1989–2003 period.

Stacked or subdivided bar graphs are another way to show paired data. The U.S. Census Bureau uses mathematical models to project what the world's population will look like in the coming decades. The projections are illustrated in the stacked bar graph that follows. Note that a percentage given within a bar represents the proportion of the population that is between 0 and 14 years of age in that year.

**Global Population 0–14 Years of Age Compared to
Total Global Population: 2002–2050**
The percent of children across the globe is projected to decline
by one-third over 5 decades.

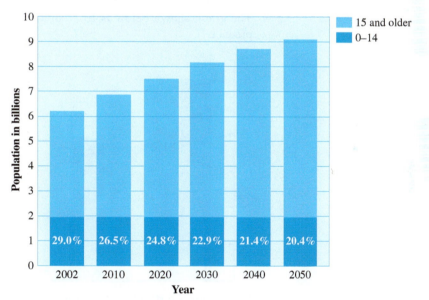

Data Source: U.S. Census Bureau, International Programs Center, International Database

7. The preceding stacked bar graph provides information about the global populations both directly and indirectly. For example, you can read *directly* the total population counts for the listed years between 2010 and 2050. You can obtain information *indirectly* by reasoning from specific information that you read directly. Keep this in mind as you answer the following questions.

a. What was the total population for 2002 and the total population predicted for the year 2050?

b. The graph shows the percentage of the population between the ages of 0 and 14. Use this information and your answers from part a to determine the number of children in the years 2002 and 2050.

c. What do you observe about the percents of children, ages 0–14, in the years from 2002 to 2050?

d. The caption above the graph states that the percent of children 0–14 years of age will decrease by one-third over 5 decades. Do you agree or disagree with this statement? Support your answer by reasoning from the information provided in the graph.

8. The graph shown below is from an advertisement for ClaraCutis, a new skin care product.

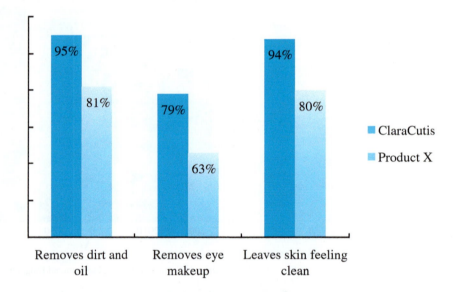

The ad states that this product removes 99% of dirt, oil, and makeup and leaves your skin feeling clean. The product was tested by a sample of 400 women chosen by mail. Respondents volunteered to try either the advertised product or a competitor's product in a blinded, two-week study. The graph shows the percentage of women who agreed with the statement below the bars.

a. What type of graph is this?

b. The bars present percentages. What do these mean?

c. Are the bars misleading in any way? Explain.

d. Do the bar values agree or disagree with the 99% claim in the text below the bars?

SUMMARY: ACTIVITY 6.2

1. A **bar graph** presents data in the form of bars or columns, drawn either horizontally or vertically on a scale of values. Each bar or column represents a category and the height or length of the column or bar represents the data value for the category. See the graph in Problem 1.

2. Sometimes, bar graphs compare categories in more than one aspect such as populations in two different years or smoking habits of men and of women in different age categories. These bar graphs have a number of different styles.

   a. One style is called a **grouped bar graph**. This style allows the comparison of categories in two (or three or more) aspects at the same time. This is done by drawing the category bars for each aspect attached next to each other. See the graph in Problem 5.

   b. Another style of grouped bar graphs is called a **paired bar graph**. These graphs compare data categories in two respects at the same time by placing the bars for each aspect in each category opposite from each other, base to base. See the graph for Problem 6.

   c. A third type of grouped bar graph is called a **stacked** or **subdivided bar graph** where each category bar is divided into two or more component parts, each of which represents a different aspect. Most often the largest or most important component is put next to the zero line. See the graph for Problem 7.

EXERCISES: ACTIVITY 6.2

1. The following chart shows the number of Superfund hazardous waste sites designated in various regions of the United States by October 2013.

Superfund Sites by Region

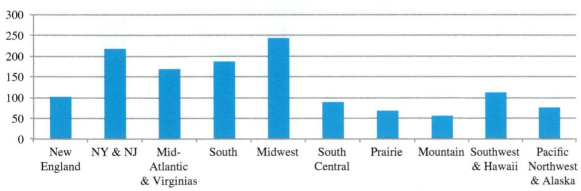

Data Source: Environmental Protection Agency

a. How many regions are represented in the chart?

b. Which region has the greatest number of Superfund sites? Estimate the number.

c. Which region has the least number of Superfund sites? Estimate the number.

d. What feature of this chart aids in estimating the number of Superfund sites in each region?

e. Which three regions have the fewest Superfund sites?

f. From the chart, estimate the total number of Superfund sites.

2. The following bar graph provides the annual rainfall in Baltimore, Maryland, from 1950 to 2013.

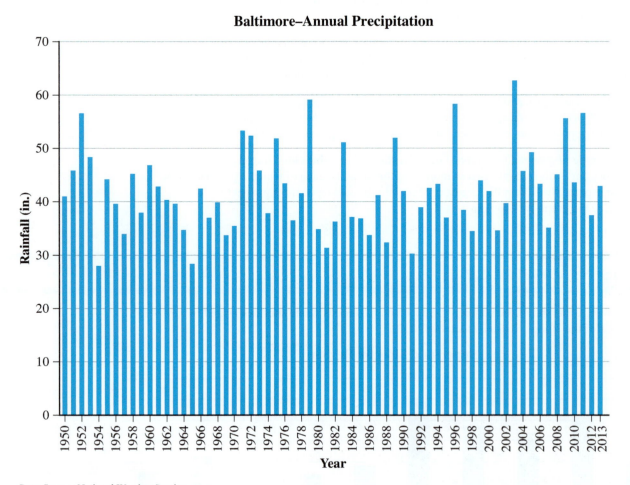

Baltimore–Annual Precipitation

Data Source: National Weather Service

a. In which year(s) was there the greatest rainfall? Estimate the number of inches.

b. In which year(s) was there the least rainfall? Estimate the number of inches.

c. In how many years did Baltimore have more than 50 inches of rain? Less than 30 inches?

3. The bar graph below shows some interesting projections for the world's population.

Pyramids of Global Population: 2002 and 2050
The globe's population is expected to grow at progressively higher rates at higher ages.

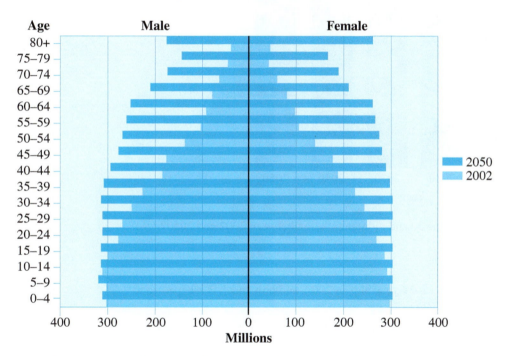

Data Source: U.S. Census Bureau, International Programs Center, International Database

a. Which age group of females shows the greatest increase from 2002 to 2050, and by approximately how much?

b. In 2002, the age groups for males decreased in size consistently from age 10 and older. The projections for 2050 show decreases in size from age 35 to 79, but the 80+ age group shows an increase. What does this mean, compared to 2002?

c. In 2050, over what age groups would there be very close to the same number of women?

d. Which age group shows the smallest change in size between 2002 and 2050?

4. Use this stacked bar graph of population projections to answer the following questions.

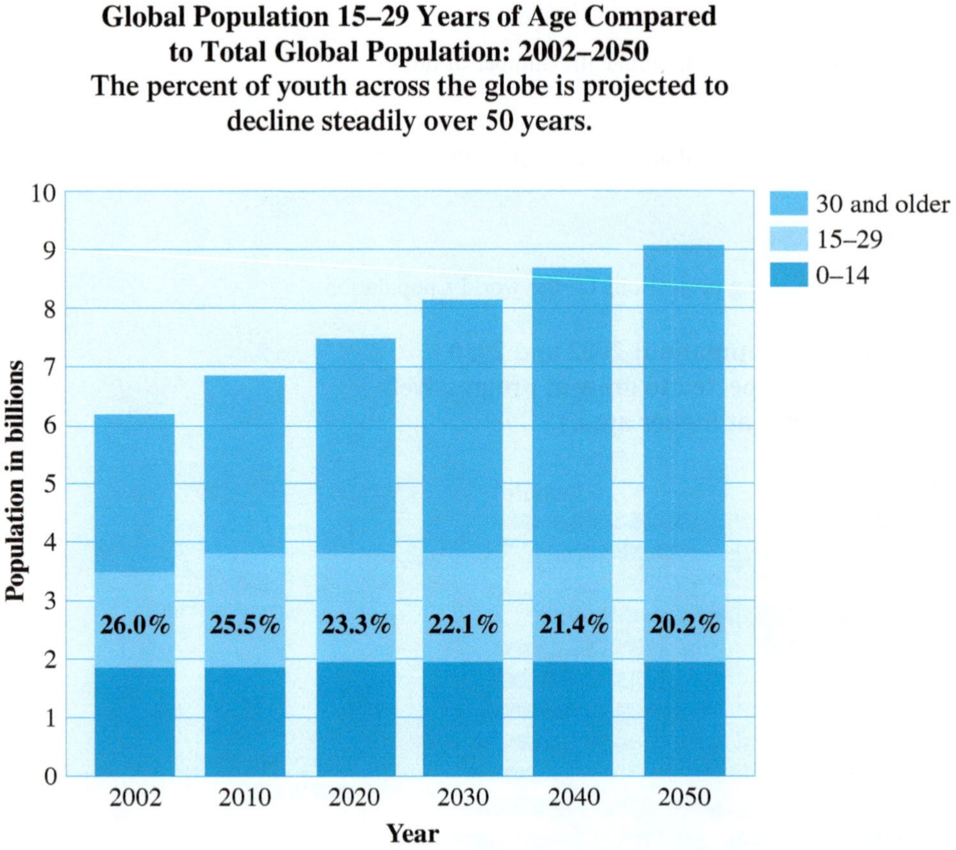
Global Population 15–29 Years of Age Compared
to Total Global Population: 2002–2050
The percent of youth across the globe is projected to
decline steadily over 50 years.
Population in billions
0
1
2
3
4
5
6
7
8
9
10
26.0%
25.5%
23.3%
22.1%
21.4%
20.2%
2002
2010
2020
2030
2040
2050
Year
30 and older
15–29
0–14

a. What is the projected increase in the 30 and older population between 2002 and 2050?

b. What was the actual population of 15–29-year-olds in 2002, and how much is it expected to change by 2050?

c. In light of your answer to part b, why are the percents decreasing for the 15–29-year-olds?

d. The 0–14-year-old population is projected to change very little through 2050. Which of the following statements provides the best explanation?

i. The birthrate (births per year) will level off and remain constant.

ii. The birthrate will actually be decreasing.

iii. Due to overpopulation, famine, and disease, the death rate in this age group will increase.

iv. It is not possible to know the individual birth rate and death rate trends, but the combined rates will result in no significant change.

Activity 6.3

People and Places

Objectives

1. Construct line, bar, picture, and circle graphs.

2. Determine an appropriate graph to display data.

Many government agencies and research institutions depend upon statistical reports to understand the demographics of a population. Demographics refers to the characteristics of a human population. Good demographic information allows for an informed decision making process.

The Texas Department of State Health Services conducts the Texas Youth Tobacco Survey every even year and collects law enforcement data annually. In January 2011, the organization submitted their report to the Texas State Legislature regarding the sale of tobacco to Texas minors.

Table 6.1   Texas Tobacco Sales to Minors

YEAR	1998	1999	2000	2001	2002	2003	2004	2005	2006	2007	2008	2009	2010
Texas Tobacco Sales to Minors	24.0	13.0	14.6	13.4	12.9	15.7	23.7	15.5	12.4	7.2	13.4	11.3	11.3

Data Source: Texas Dept. of State Health Services, *Texans and Tobacco: A Report to the 82nd Texas Legislature,* Jan. 2011

1. **a.** Using the data in Table 6.1, construct a line graph and a bar graph that displays the Texas tobacco sales to minors during the years from 1998 to 2010.

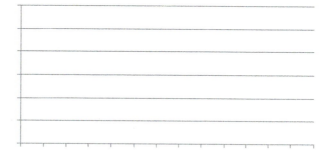

b. What trend do you observe from your graphs?

    c. Which graph do you think better displays the data?

    2. The Texas report also included data on the use of tobacco by students in grades 7–12, as shown in Table 6.2. The percentages given are for those who used tobacco within the past month and those who have ever used it.

Table 6.2   Texas Youth Tobacco Use

YEAR	1990	1992	1994	1996	1998	2000	2002	2004	2006	2008	2010
Ever Used (%)	56	54	54	55	55	52	45	38	35	32	31
Within Month (%)	24	21	25	27	27	24	18	17	16	14	13

Data Source: Texas Dept. of State Health Services, *Texans and Tobacco: A Report to the 82nd Texas Legislature,* Jan. 2011

This table could be plotted as two separate graphs, for lifetime use and for recent use. But, because the lifetime and recent use of the respondents may be related, a single graph to compare is preferred for this information.

    a. Plot the percentage of students who have ever used tobacco products with the percentage who used them within the past month together on the same line graph.

    b. Each line displays a trend over the same time period. How does the lifetime use (ever used) trend compare with the recent use (within a month) trend?

    3. a. Plot the data from Table 6.2 as a paired bar graph, by using side-by-side bars for each biennial survey. Clearly indicate which is which.

b. Compare this paired bar chart to your paired line graph in Problem 2. Do you find one display easier to read and/or interpret? Is there any significant difference?

In Problems 1–3, the **actual number** of students who used tobacco and the sales of tobacco to minors were not represented in your graphs, but rather the percents or portions of the students who had used tobacco were represented.

Sometimes actual population counts are included, as in the data for Table 6.3 for the United States. The U.S. Census asked questions about language use at home to locate people who speak a language other than English. This table is based on data collected in 1990 and 2011.

Table 6.3   Language Spoken at Home in the United States, 1990–2011

LANGUAGE	1990		2011	
	NUMBER	PERCENT	NUMBER	PERCENT
English	198,600,798	86%	230,947,071	79%
Spanish	17,239,172	7%	37,579,787	13%
Asian	3,653,668	2%	9,485,464	3%
Other	10,852,139	5%	13,511,769	5%

Data Source: U.S. Census Bureau

The number column in each category refers to the actual number of people in each language category. The percent column refers to the percent of the total number of people in the United States that make up that language group.

4. a. Using the data in Table 6.3, make a side-by-side bar graph for the actual number of people in each language category for the given years.

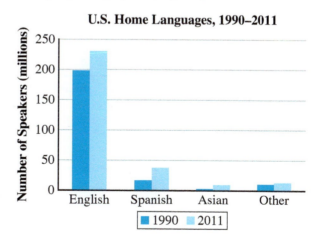

b. Draw another side-by-side bar graph for the two given years, this time using the percentages.

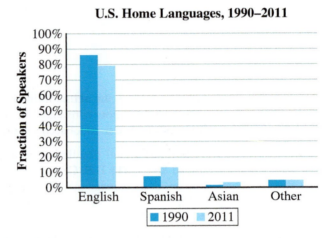

c. How do your graphs compare?

A line graph is usually used when a trend over time is being observed (this graph is sometimes called a **time plot**). When the counts do not take place over time, connecting points to create a line graph may not have as much meaning or significance. For example, the following two line graphs display the same information for the 2011 home language data in Table 6.3.

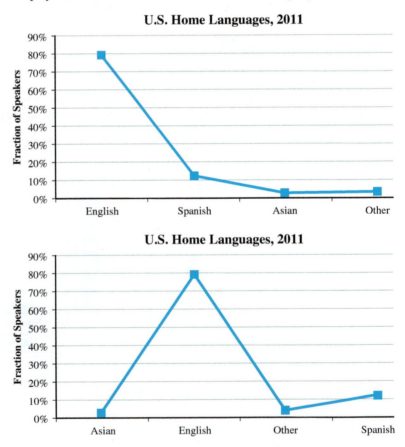

5. For the two graphs above, does it matter which order you use to place the languages on your axis? Is there any significance to the shape of the line segments making up the graphs? Explain.

Circle Graphs

Circle graphs, also called **pie charts**, are a convenient way to display data when you are interested in showing the relative sizes of different categories. The size of each piece ("slice") of the pie is proportional to the percentage of the whole for that category. For example, the 1990 home language data in Table 6.3 can be displayed by the following circle graph:

U.S. Home Languages, 1990

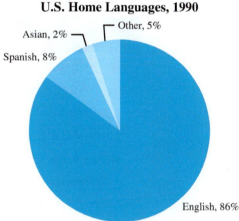

The percentage gives the relative size of each piece (or sector) in the circle and the fractional part of each group compared to the total number of people in the population. For example, the fraction of people from Spanish-speaking homes in 1990 was $17{,}339{,}172 \div 230{,}445{,}777 \approx 0.07524 = 7.524\%$ or approximately 8%. Therefore, the sector representing Spanish-speaking homes is about 8% of the total area of the circle. The central angle of the sector representing the fraction of people from Spanish-speaking homes is 7.524% of 360 degrees or $0.07524(360°) \approx 27.1°$, or approximately 27 degrees.

6. Using the 2011 home language data from Table 6.3, label each sector of the following pie chart accordingly.

U.S. Home Languages, 2011

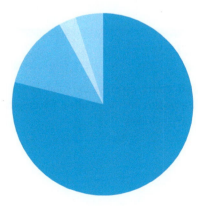

7. Using the preceding side-by-side bar graph (Problem 4b) and the circle graphs, describe how home language use in the United States has changed from 1990 to 2011. In your opinion, which graphs were most useful in answering this question, and why?

Circle graphs must always represent the whole collection, or population, of whatever is being displayed. When expressed as percents, those percents must add up to 100%, the whole amount.

8. Determine the appropriate percents for each region of the world, and label each sector of the pie chart accordingly. The sectors for each category are already drawn.

WORLD REGION	APPROXIMATE 2015 WORLD POPULATION (in thousands)
Africa	1,151,242
Asia	4,347,642
Europe	742,234
South America	409,770
North America	562,074
Oceania	37,143

Data Source: U.S. Census Bureau

Approximate 2015 World Population

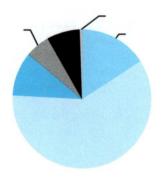

SUMMARY: ACTIVITY 6.3

1. A **circle graph**, or **pie chart**, displays the relative number of data values in each category by the size of the corresponding sector, or slice. Circle graphs require that all the categories that make up the whole are included.

EXERCISES: ACTIVITY 6.3

1. a. Use the data in the following table to construct a bar graph and a line graph.

MEAN 2009 SALARIES BY EDUCATIONAL LEVEL: WORKERS 18 YEARS OF AGE AND OLDER								
Educational Level	Some high school	High school graduate	Some college	Associate's degree	Bachelor's degree	Master's degree	Doctoral degree	All workers
Mean Salary	$20,241	$30,627	$32,295	$39,771	$56,665	$73,738	$103,054	$42,469

Data Source: U.S. Census Bureau

b. Which graph is the most appropriate display of the data? Explain.

2. In 2012, there were approximately 9,300,113 residents of Ohio who were at least 15 years of age. Use the circle graph below to estimate the number of residents in each category to the nearest ten thousand.

Marital Status: Ohio Residents 15 and Over, 2012

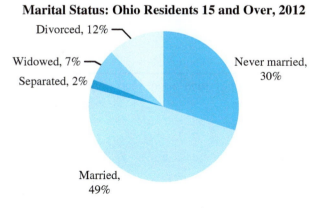

Divorced, 12%

Widowed, 7%

Separated, 2%

Never married, 30%

Married, 49%

Data Source: U.S. Census Bureau

3. A pictograph can use symbols or pictures of the same size, instead of lines or bars, to give the visual image more meaning. A key is given to indicate how many items are represented by each symbol. In the pictograph below, each symbol or picture represents one million vehicles.

Registered Vehicles in Texas

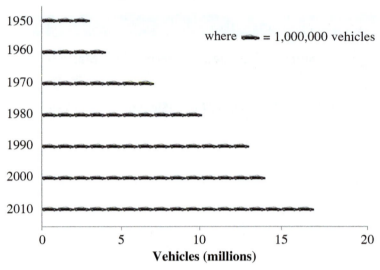

where  = 1,000,000 vehicles

Data Source: Federal Highway Administration, U.S. Department of Transportation, *Highway Statistics Series*

a. Approximately how many more vehicles were registered in Texas in 2010 compared with 1950?

b. Do you think a pictograph is more like a bar graph or line graph?

4. The line graph below shows population growth in the city of Seattle, Washington.

Population of Seattle

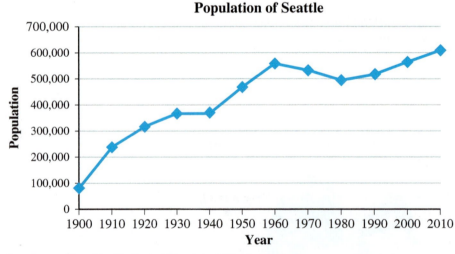

Data Source: City of Seattle, Dept. of Planning and Development

a. Could a bar graph be used just as effectively? Explain.

b. Could a circle graph be used just as effectively? Explain.

5. A 2014 survey of student attitudes about the use of mobile devices in learning provided the results shown in the table below.

TECHNOLOGY	USE IN SCHOOLWORK
Tablet	27%
Laptop*	66%
Smartphone	43%

*"Laptop" also includes notebooks and Chromebooks.

Data Source: Harris Interactive, *Pearson Student Mobile Device Survey: Grades 4 through 12*, May 9, 2014.

The following circle graph displays the percentages of students who use the technologies at least a few times per week in doing schoolwork.

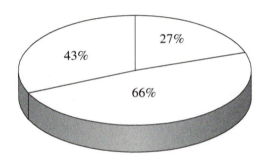

List the errors in the circle graph.

6. California is a major orange-producing state. The table gives orange production for several growing seasons. Which type of graph (line, bar, or circle) do you think would best display the data and why? Produce a graph and interpret what you see.

SEASON	2001–2	2002–3	2003–4	2004–5	2005–6	2006–7	2007–8	2008–9	2009–10	2010–11	2011–12	2012–13	2013–
Production	103	124	101	129	122	92	122	91	112	122	115	112	10

Production is in millions of cartons.

Data Source: National Agricultural Statistics Service, U.S. Department of Agriculture

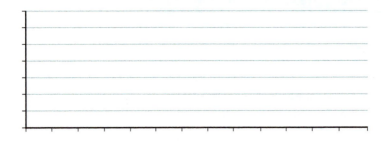

Activity 6.4

The Class Survey

Objectives

1. Organize data with frequency tables, dotplots, and histograms.

2. Organize data using stem-and-leaf plots.

Decisions that are made in business, government, education, engineering, medicine, and many other professions depend on analyzing collections of data. As a result, data analysis has become an important topic in many mathematics classes.

In this activity, you will collect and organize data from your class.

1. Record the requested data for your entire class.

GENDER	FAMILY SIZE	MILES FROM SCHOOL	TIME DOING HOMEWORK YESTERDAY (TO THE NEAREST HALF HOUR)

2. Using the data collected in Problem 1, determine the following characteristics of your entire class.

   a. The most common family size

   b. The average number of miles from school

   c. More females or males

   d. The most hours studied last night

The data can be organized in several different ways to help you arrive at your answers. One visual approach is to produce a **dotplot** for each category. To illustrate, consider the following data representing the size of 20 families.

Family sizes: 4, 6, 2, 8, 3, 5, 6, 4, 7, 2, 5, 6, 4, 6, 9, 4, 7, 5, 6, 3

The data values appear on the horizontal scale of the dotplot. The number of occurrences of each data value (called the **frequency**) is recorded on the vertical scale. Each dot represents one data value.

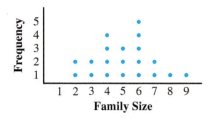

The following frequency table for these data shows another way to organize the data.

Family Size	2	3	4	5	6	7	8	9
Frequency	2	2	4	3	5	2	1	1

3. Draw dotplots for each of the four data categories from Problem 1.

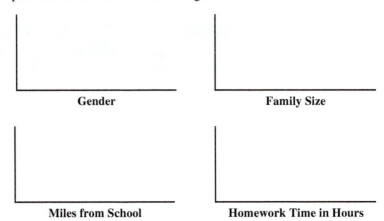

Gender   Family Size

Miles from School   Homework Time in Hours

4. Explain how you can use the dotplots to answer each of the following questions.

   a. What is the most common family size?

   b. What is the average number of miles from school?

   c. Are there more girls or boys in your class?

   d. What was the most number of hours studied?

In the study of statistics, the dotplot and frequency table are usually called **frequency distributions**. They describe how the data are distributed over all possible values.

5. Looking at your dotplots in Problem 3, describe in a few sentences how the frequency distributions compare with each other visually. Note any similarities and differences.

6. Of the four frequency distributions, which one is really quite different because of the nature of the data values?

Histograms

Histograms are statistical graphs that can be used to display frequency distributions. The data values are located on the horizontal axis of the histogram. The frequencies are located on its vertical axis. A rectangle is constructed above each data value and its height indicates the frequency of the specific data value. Therefore, in a histogram, the column of data in a dotplot is replaced by a rectangle.

Example 1   *Construct a histogram of the following frequency distribution.*

Family Size	2	3	4	5	6	7	8	9
Frequency	2	2	4	3	5	2	1	1

SOLUTION

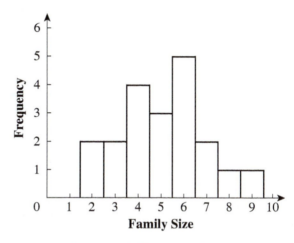

Note that the vertical scale should start at 0. The rectangles are centered at the data value and have the same width. The height of a rectangle represents the frequency of the data value.

The rectangles in a histogram should always touch.

7. Construct a histogram to show the frequency distribution from Problem 3 of the number of miles from school.

Your graphing calculator can also be used to generate a graph of a frequency distribution. Using the sample family size data in Example 1, the following **histogram** can be generated.

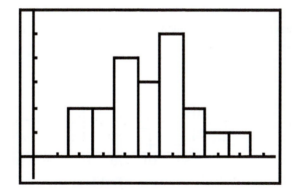

8. Use a graphing calculator to produce the histogram for the class miles from the school distribution. Refer to Appendix A for detailed instructions for the TI-83/84 plus calculator.

Grouped Data

In the Class Survey problems, you learned that a histogram is a very useful method to display numerical data in order to visualize how individual values of data are distributed. Often, data are grouped together in order to provide information about the distribution that would be difficult to observe if the data were treated individually.

For example, suppose your mathematics teacher wants to analyze the grades on the last test for the students in his two math classes. In this situation, rather than determining the frequency for a specific test grade, such as 82, your teacher is interested in the number of 80's, as well as the number of 60's, 70's, and so forth. Therefore, the test scores are **grouped** into intervals. The interval for the number of 80's is 80–89. The frequency table for the test scores is given as follows:

FREQUENCY INTERVAL	FREQUENCY
50–59	2
60–69	6
70–79	16
80–89	18
90–100	8

Note that the **frequency intervals**, also called **classes**, for the grouped data should be of the same width. The numbers 50, 60, 70, 80, and 90 are called the **lower class limits**. The numbers 59, 69, 79, 89, and 100 are called the **upper class limits**. The difference between any two consecutive lower class (or upper class) limits is the class width. In this case, the **class width** is 10. The frequency intervals should not overlap, and each piece of data should belong to only one interval.

9. a. How many students are in the two courses?

b. How many students had a test grade in the 80's?

The grouped test scores for the students in the two mathematics sections is displayed by the following histogram:

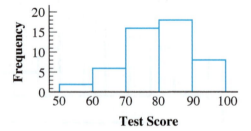

10. What specific information can you gather by examining the preceding histogram?

Although you are able to easily identify the frequency of test scores within each frequency intervals, the original test scores cannot be identified from the frequency table of the histogram. For example, you know that there are eight scores in the 90's, but there is no way to determine what they are. As a result, you are not able to calculate the exact values for statistical measures.

Stem-and-Leaf Plots

A stem-and-leaf plot is a way to organize and display groups of data as well as displaying the actual data values. In a stem-and-leaf plot, each data value is split into two parts called the **stem** and the **leaf**. The *stem* is the digit or group of digits *with the greatest place value*. The remaining digits on the right of the stem form the leaf for the data value. For example, the number 517 can be split into a stem and leaf as follows:

Stem: 5   Leaf: 17

The number 517 could also be split with the stem 51 and the leaf 7. The single-digit number, 5, which can be written as 05, would have stem 0 and leaf 5.

The following example demonstrates the procedure for constructing a stem-and-leaf plot.

Example 2

Construct a stem-and-leaf plot for the following test scores on a 50-point quiz:

26 29 32 34 35 37 39 40 42 43 43 46 47 48 50 50

SOLUTION

Step 1. Write the data in numerical order. Although this step is not required, it makes working with the data more convenient.

The data is already in numerical order.

Step 2. Identify which part of the number is the *stem* and which part is the *leaf*.

Since the scores consist of two digits, use the first digit, the tens digit, as the stem and the units digit as the leaf. For example, the test score 26 has stem 2 and leaf 6.

Step 3. List the stems in numerical order as follows. Note that a vertical line separates the stem from the leaf.

Stem	Leaf
2	
3	
4	
5	

Step 4. Proceeding in numerical order, place the leaf of each data value on the right of the vertical line next to the appropriate stem, lining up each leaf entry vertically. Proceed until all the leaves of the data values are displayed.

Stem	Leaf
2	6 9
3	2 4 5 7 9
4	0 2 3 3 6 7 8
5	0 0

Step 5. Determine a legend (key) to indicate how the numbers are represented by the stem and leaves.

Legend: 2│6 represents the number 26

Note that if the data in Example 2 is listed line by line, in groups, you have

26 29
32 34 35 37 39
40 42 43 43 46 47 48
50 50

The stem-and-leaf plot of this data reduces the repetition of the tens digits and displays the data in a more compact fashion. The stem-and-leaf plot gives the same visual display as a sideways histogram.

11. The test scores from the two math classes are listed below:

57 58 62 64 64 68 69 69 70 70 71 71 71 72 73 73 75 76 76 78 78
78 79 79 81 81 82 82 83 83 84 84 85 85 85 85 85 86 86 87 88 89
91 92 92 94 95 95 98 99

Construct a stem-and-leaf plot for this data. Use the tens digit for the stem and the ones digit for the leaf.

Suppose your teacher wants to analyze the test scores from each of the math classes separately. When comparing two similar collections of data, a **back-to-back stem-and-leaf plot** can be constructed.

12. The test scores for each math class are as follows:

Class 1: 57 58 64 68 69 70 70 71 72 73 73 76 78 79 82 85 85 85 88
91 92 95

Class 2: 62 64 69 71 71 75 76 78 78 79 81 81 82 83 83 84 84 85
85 86 86 87 89 92 94 95 98 99

Complete the following back-to-back stem-and-leaf plot for the two classes.

CLASS 1		CLASS 2
LEAF	STEM	LEAF
8 7	5	
9 8 4	6	2 4 9

13. Construct a stem-and-leaf plot for the following data collected in a science experiment:
7.6 8.3 9.4 6.7 5.4 6.3 6.1 5.7 7.5 8.4 8.6 7.3 7.7 8.8 9.1

a. Write the data values in ascending order.

While these values have a decimal point, a stem-and-leaf plot would display these numbers without a decimal point. The legend (key) of the plot is used to show the position of decimal points in the original data.

b. Ignore the decimal point in the data values, and construct a stem-and-leaf plot for these numbers. Use the ones digit for the stem and the tenths digit for the leaf.

c. Write a legend for this plot.

SUMMARY: ACTIVITY 6.4

1. A **frequency distribution**, usually displayed in a frequency table, or visually in a dotplot or histogram, describes how frequently each of the data values occurs.

2. A **grouped histogram** is useful when there is a wide range of data. In a grouped histogram, data are grouped into intervals of the same class width.

3. A **stem-and-leaf plot** organizes data by splitting each data value into two parts called the stem and leaf.

EXERCISES: ACTIVITY 6.4

1. A large publishing company wants to review the ages of its sales representatives. The ages of a sample of 25 sales reps are as follows:

50   42   32   35   41   44   24   46   31   47   36   32   30

44   22   47   31   56   28   37   49   28   42   38   45

Construct a histogram for the ages, using eight classes with a class width of five.

2. The scores on the last math quiz are summarized in the following frequency table.

Score	10	9	8	7	6	5	4	3	2	1	0
Frequency	5	7	6	4	2	0	1	0	0	0	0

Construct a dotplot for the scores.

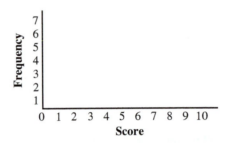

3. The frequency distribution for a collection of data is given in the following table:

CLASS	FREQUENCY
9.5–11.4	3
11.5–13.4	7
13.5–15.4	10
15.5–17.4	15
17.5–19.4	9
19.5–21.4	2

a. Determine the total number of data.

b. What is the width of each class?

c. If an additional class were to be added, what are its class limits?

d. Construct a histogram of the frequency distribution.

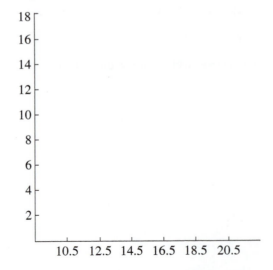

4. The following collection of data gives the weights (in pounds) of 45 male high school gym students:

105 112 115 118 125 129 130 132 135 135 137 138 138 139 141 144 145
148 148 151 154 154 156 158 158 160 161 161 162 165 166 166 169 172
174 174 176 181 184 185 192 195 206 209 218

In parts a–c, create representations for the collection of data, such as a frequency table, histogram, and stem-and-leaf plot, to communicate your mathematical ideas.

a. Use this data to construct a frequency table having the first interval (class) 100–109.

FREQUENCY INTERVAL	FREQUENCY

b. Construct a histogram of the frequency distribution from the table above.

c. Construct a stem-and-leaf plot of the original data. Use the hundreds and the tens digit for the stem and the ones digit for the leaf.

5. The scores on a 60-point exam are given in the following stem-and-leaf plot. List the original data.

Stem	Leaf
0	9
1	3 7
2	1 6 8 9
3	3 7 7 8
4	2 2 3 5 5 6 6 6 9
5	1 3 4 5 5 8 9
6	0 0

Legend: 1│3 represents 13

6. Subjects in a psychology study were timed while completing a certain task. Complete a stem-and-leaf plot for the following list of times (to the nearest tenth of a minute):

7.5 8.2 9.3 6.8 5.9 6.4 6.1 7.9 5.8 7.3 8.2 8.7 7.4 7.8 8.2 9.2 7.7

Objectives

1. Determine measures of central tendency, including the mean, median, mode, and midrange.

2. Recognize symmetric and skewed frequency distributions.

3. Distinguish between percentiles and quartiles.

In Activity 6.4, The Class Survey, you worked with the following data representing the size of 20 families.

Family sizes: 4, 6, 2, 8, 3, 5, 6, 4, 7, 2, 5, 6, 4, 6, 9, 4, 7, 5, 6, 3

The frequency distribution of the data was displayed graphically using a dotplot and a histogram.

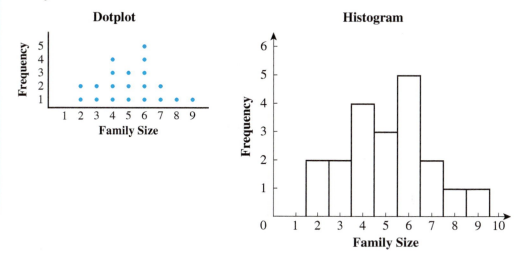

1. By observing the graphs, estimate where you think the center of the frequency distribution lies.

2. Look at the three frequency distributions with numerical data values in Problem 3 in Activity 6.4. Estimate on each dotplot where you think the center of the frequency distribution lies. Record your estimates and compare with your classmates.

FREQUENCY DISTRIBUTION	ESTIMATE OF THE CENTER
Family size	
Miles from school	
Time doing homework	

Measuring the center of a frequency distribution is a basic starting point in many statistics applications. There are several different ways to measure the center, including the mean, the median, the mode, and the midrange. Since each will result in a number near the center of the data, these statistics are said to measure the **central tendency** of a frequency distribution. The term "average" is often used to describe these measures of central tendency.

Measures of Central Tendency

The **arithmetic mean**, or simply **mean**, is the most commonly used "average." The mean of a sample of a population is represented by the symbol $\bar{x}$, read "x bar." The mean of the entire population is represented by the Greek letter mu, μ. Unless specified, you can assume that the data used in this activity represent samples.

The mean is calculated by adding the data values and dividing the sum by the number of values. The notation $\sum x$, read the "sum of x," is used to represent the sum of all the data. The symbol $\sum$ is the Greek letter sigma, used to indicate summation. The formula for calculating the mean $\bar{x}$ is

$$\bar{x} = \frac{\sum x}{n},$$

where n represents the number of data values in the sample.

3. Determine the mean of the 20 family sizes in the introduction of this activity.

A second average is the **median**. The median is in the exact middle of the numerically ordered (ranked) data. Half of the data lies above (is greater than) the median. Half of the data lies below (is less than) the median.

To calculate the median:

Step 1. Rank the data from the smallest to the largest (or largest to the smallest).

Step 2. i. If there is an odd number of data values, the median is the number in the middle of the ranked data.

ii. If there is an even number of data values, the median is halfway between the two middle data values. Expressed as a formula,

$$\text{median} = \frac{\text{sum of middle data values}}{2}.$$

4. Determine the median of the 20 family sizes in the introduction of this activity.

The other commonly used measures of central tendencies are the **mode** and **midrange**.

- The mode is the data value that occurs most frequently.
- The midrange is the exact midpoint between the lowest and highest data values. Expressed as a formula,

$$\text{midrange} = \frac{\text{lowest value} + \text{highest value}}{2}.$$

5. Determine the mode and midrange of the 20 family sizes.

6. a. Now determine the following statistics for the class survey distributions, where possible.

DISTRIBUTION	MEAN	MEDIAN	MODE	MIDRANGE
Family size				
Miles from school				
Time studying				

b. On your original dotplots in Problem 3, page 784, clearly mark with a vertical line each of the above statistics from the preceding table.

c. For each distribution, circle the statistic (mean, median, mode, midrange) that is closest to the estimate of the center that you made in Problem 2. Which type of statistic did you have in mind when you made your estimates?

7. Which statistic would you choose to best describe the "average" family size among your classmates, and why?

8. Do any of these measurements of central tendency make sense for the gender distribution? Explain.

The commonly used "averages" can give very different results for the same set of data. Problem 9 demonstrates how the averages can be used to misrepresent the given situation.

9. A small company pays its president $200,000. The vice president receives $150,000. There are five employees in the company: the foreman who earns $30,000; three workers, earning $25,000, $22,000, $22,000, respectively; and a secretary who earns $19,000.

a. Determine the mean, median, and mode income in this company.

b. The union representative wants to negotiate new wages for the workers. In the negotiation, should the union representative emphasize the mean, median, or mode? Explain.

c. Which of the three measures would the president of the company stress in the negotiations? Explain.

d. Which do you think best represents this sample? Explain why in a paragraph.

e. Do you think that the measure you chose in part d is always the best to use? Explain.

10. Which measure of central tendency is most affected by extreme data values? Explain.

11. Consider the following set of scores from a science test:

10, 16, 21, 62, 62, 62, 68, 70, 71, 72, 74, 77, 78, 79, 80, 85, 88, 93

a. Determine the mean, median, and mode of the science test scores.

b. Which of the "averages" in part a best represents the grades on the science exam? Explain.

Using Technology to Determine Measures of Central Tendency

12.

Appendix

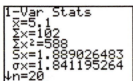

Once the data are stored in your calculator, you can generate a variety of statistics for the data, including the measures of central tendency. For example, the windows of the TI-83/84 Plus calculator that appear on the left result from the 20-sample family size data. Refer to Appendix A for detailed instructions for the TI-83/84 Plus calculator.

The mean ($\bar{x}$) and median (Med) are given, n is the number of data values, and minX and maxX are the smallest and largest data values respectively. From these you can calculate the midrange.

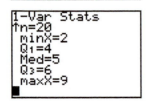

12. Use a graphing calculator to produce the measures of central tendency for the family size distribution from the Class Survey.

Types of Frequency Distributions

13. a. The three frequency distributions I, II, and III pictured on the left have the same spread of data values. What is it?

b. Describe in your own words the visual differences among the distributions in part a.

I.

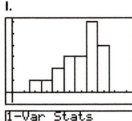

c. The means for the distributions in part a are as follows:

$$\bar{x}_I = 6.0 \qquad \bar{x}_{II} = 5.0 \qquad \bar{x}_{III} = 4.2$$

Compare these to the median for each distribution.

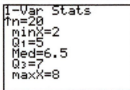

II.

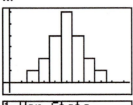

Some distributions are symmetrical (like distribution II in Problem 13). Others are slightly "off," as if they have been stretched to one side or the other. Distribution I is **skewed to the left**, while distribution III is **skewed to the right**.

14. a. If a distribution is perfectly symmetrical, what do you think will be true about the relationship between the mean and median?

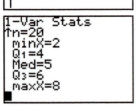

b. Based on your results in Problem 13, what is the relation between the mean and median of a skewed distribution?

III.

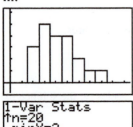

c. Enter in your calculator a skewed distribution with ten data values. Record your data, sketch the histogram (or dotplot), and calculate the mean and median. Does this result help confirm your answer in part b?

Measure of Position

Two measures of position within a distribution are **quartiles** and **percentiles**. Quartiles divide a set of data into four equal parts. Percentiles divide a set of data into 100 equal parts.

For a given set of data, your calculator generates the data values corresponding to the quartiles. For example, for the 20-sample family size data, the 1-Var Stats screen of the TI 83/84 Plus calculator gives the following information:

$$Q_1 = 4$$
$$\text{Med} = 5$$
$$Q_3 = 6$$

Q_1 represents the first quartile and has a value of 4. This means that 25% of the data has a value less than 4.

15. a. Explain the meaning of $Q_3 = 6$.

b. What is the second quartile?

Such measures are often used to make comparisons of scores of individuals when the amount of data is large. For example, suppose you scored 600 on the math portion of the Scholastic Aptitude Test (SAT). If this score is at the 80th percentile, denoted P_{80}, it does **not** mean that you answered 80% correct on the test. It means that 80% of the high school students who took the test scored less than you. That is, 80% of the students scored less than 600.

SUMMARY: ACTIVITY 6.5

A **frequency distribution**, usually displayed in a frequency table, or visually in a dotplot or histogram, describes how frequently each of the data values occurs.

There are four typical measures of the **central tendency** of a frequency distribution:

1. The **mean** is the usual average found by adding the data values and dividing by the number of values.

2. The **median** is in the exact middle of the numerically ordered data—half the data values lie above and half lie below the median. If there is an even number of data values, the median is the mean of the two middle values.

3. The **mode** is the data value that occurs most frequently.

4. The **midrange** is the exact midpoint between the lowest and highest data values.

The following are measures of position within a distribution:

5. Quartiles divide a set of data into four equal parts.

6. Percentiles divide a set of data into 100 equal parts.

EXERCISES: ACTIVITY 6.5

1. A large publishing company wants to review the ages of its sales representatives. The ages of a sample of 25 sales reps are as follows:

41 33 25 27 38 26 43 35 52 48 30 22 31

48 42 37 26 54 39 44 28 41 36 29 56

a. Construct a histogram with eight classes and a class width of five.

b. Calculate the mean and median ages.

c. Determine the midrange.

d. Is there a mode for this distribution?

2. Suppose the scores of seven players in a men's golf tournament are 68, 62, 60, 65, 76, 66, and 72. Determine the mean, median, and midrange.

3. The scores on the last math quiz are summarized in the following frequency table:

Score	10	9	8	7	6	5	4	3	2	1	0
Frequency	5	7	6	4	2	0	1	0	0	0	0

Calculate the mean, median, mode, and midrange of this quiz distribution.

4. a. Create a set of ten data values whose distribution is perfectly symmetrical.

b. Verify that the mean, median, and midrange are identical.

c. Change two data values so that the resulting distribution is skewed to the left.

d. Predict how this will change the mean and median. Then verify by doing the calculations.

5. A set of data has a mean of 32.4 and a median of 38.9. Would you expect the distribution for this data to be symmetrical, skewed to the left, or skewed to the right?

6. a. What tool would you select to create a histogram of a collection of data?

b. Find a collection of at least 25 data values from a real-world source. Use the tool you selected in part a to create a histogram.

c. What tool would you select to determine the four measures of central tendency of the data in part b?

d. Use the tool you selected in part c to determine the four measures of central tendency of the data you collected.

7. The mean is usually most affected by any change in the data.

a. Determine the mean, median, mode, and midrange of the following data:

$$10, 11, 12, 14, 14, 15, 16, 20$$

b. Change the 16 to 19 in part a. Determine the mean, median, mode, and midrange of the resulting set of data.

c. Which measure of central tendency was affected by changing the 16 to 19?

d. In addition to the mean, which measure would be affected if the 20 was changed to 19 in part a?

8. The salaries of ten employees of a small company are listed below:

$$33,000, 28,000, 36,000, 31,000, 31,000,$$
$$68,000, 29,000, 32,000, 86,000, 34,000.$$

a. What technology tool would you select to determine the mean, median, mode, and midrange of the ten salaries?

b. Determine the mean, median, mode, and midrange of the ten salaries. Verify your results using the statistics feature of the graphing calculator.

c. Which average would the employees use to demonstrate that they were underpaid? Explain.

d. Which average would management use if they did not want to give a raise to the employees? Explain.

9. Reports about incomes and other highly skewed distributions usually give the median rather than the mean as the measure of the "average." Explain.

10. Locate an article in a newspaper or magazine that discusses an "average."

   a. Is the average used the mean, median, or mode? Explain.

   b. Is the choice of the average used in the article appropriate for the situation? Explain.

Statistical analysis is useful when important decisions need to be made. The following sets of data are the result of testing two different switches that can be used in the life-support system on a submarine. Two hundred of each type of switch were placed under continuous stress until they failed, the time recorded in hours. Switch A and switch B have approximately the same means and medians, as displayed by the following histograms. (The displays are from a TI-83/84 Plus.)

Activity 6.6

A Switch Decision

Objectives

1. Determine measures of variability including range, interquartile range, and standard deviation.

2. Construct and compare boxplots.

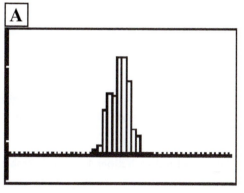
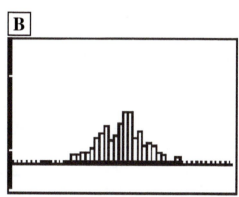

x̄=82.04405225	minX=76.464516
Σx=16408.81045	Q₁=80.77875085
Σx²=1347130.55	Med=82.2356903
Sx=2.109137206	Q₃=83.45049998
σx=2.103857755	maxX=86.993903
n=200	

$\bar{x}=82.04405225$ $minX=76.464516$
$\Sigma x=16408.81045$ $Q_1=80.77875085$
$\Sigma x^2=1347130.55$ $Med=82.2356903$
$Sx=2.109137206$ $Q_3=83.45049998$
$\sigma x=2.103857755$ $maxX=86.993903$
$n=200$

$\bar{x}=81.95269223$ $minX=68.606254$
$\Sigma x=16390.53845$ $Q_1=78.78661495$
$\Sigma x^2=1348001.07$ $Med=81.9755658$
$Sx=4.886817056$ $Q_3=85.03659857$
$\sigma x=4.874584704$ $maxX=96.897925$
$n=200$

1. Verify that the means and medians for both switches are approximately the same. What does this tell you about the general shapes of the distributions?

2. Which distribution is the most spread out?

3. Which distribution is packed more closely around its center?

4. Based on these tests, which switch would you choose to use and why?

Like the central tendency of a distribution, the spread, or **variability**, of a distribution is also measured several different ways. The **range** is simply the difference between the minimum and maximum data values.

5. Determine the range for each set of switch data to the nearest hour. Do these ranges confirm your answers to Problems 2 and 3?

6. The range is of limited value in describing the variability (or spread) of a distribution. Consider these two data sets.

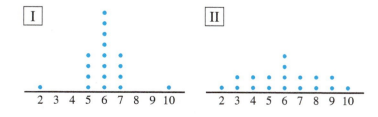

   a. Which distribution has greater variability?

   b. Calculate the range for each distribution.

   c. What is your conclusion about the range as a measure of variability?

Standard Deviation

The most common measure of spread is the **standard deviation** (symbol σ, the Greek lowercase letter sigma). It measures how much the data deviates from the mean.

7. Calculate the mean for distributions I and II of Problem 6.

Deviation refers to how far a data value is from the mean. For example, if the mean is 7, then a data value of 10 is a distance of 3 from the mean, and you would say that the data value 10 has a deviation of 3 from the mean. If the data value is 4, it is still a distance of 3 from the mean, but its deviation is -3, indicating that it is below the mean.

8. a. Calculate the deviations for every data value in both distributions I and II of Problem 6. Record your results in the column having heading $x - \bar{x}$. Then add all the deviations for each distribution.

DISTRIBUTION I				DISTRIBUTION II		
x	$x - \bar{x}$	$(x - \bar{x})^2$		x	$x - \bar{x}$	$(x - \bar{x})^2$
Total				**Total**		

b. Is the sum of the deviations, denoted by $\Sigma(x - \bar{x})$, a good measure of the spread of the data? Explain.

To overcome the deficiency of the total deviation, you could simply take the absolute value of each deviation. The more conventional method is to calculate the **standard deviation**. The standard deviation of a distribution involves summing the squares of the deviations, denoted by $\Sigma(x - \bar{x})^2$.

9. For each distribution, calculate the standard deviation using the following procedure.

a. Square each deviation in both of the tables of Problem 8 and then record your results in the column $(x - \bar{x})^2$.

b. Add the squares of all the deviations: _____   _____

c. Divide by the number of data values: _____   _____

d. Take the square root: _____   _____

In a single formula, the standard deviation is $\sigma = \sqrt{\dfrac{\Sigma(x - \bar{x})^2}{n}}$, where σ is the standard deviation, Σ is short for "the sum of," x is a single data value, $\bar{x}$ is the mean, and n is the number of data values.

e. Does distribution I or distribution II have greater variability? Explain using the standard deviation of each distribution.

10. Verify that your standard deviations in Problem 9d agree with those in the following display, from the TI-83/84 Plus. (The TI-83/84 Plus uses σx for σ.)

Distribution I   Distribution II

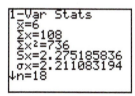

```
1-Var Stats
x̄=6
Σx=108
Σx²=688
Sx=1.533929978
σx=1.490711985
↓n=18
■
```

```
1-Var Stats
x̄=6
Σx=108
Σx²=736
Sx=2.275185836
σx=2.211083194
↓n=18
```

11. What do you think a standard deviation of zero would mean?

There is another way to calculate a standard deviation. When the data is only part of a larger collection or population, it is called a **sample**. The standard deviation for a sample is denoted by the letter s (note in the TI calculator display, sx, which appears above σx, is used for s).

The formula for s only differs in the division by $n - 1$ instead of n; $s = \sqrt{\dfrac{\Sigma(x - \bar{x})^2}{n - 1}}$.

12. a. Return to the testing of switches A and B. Since the calculations were based on 200 switches, a sample of the larger population of all switches, s is the correct standard deviation, as given in the calculator output screen in the beginning of this activity. Record the standard deviation here.

   b. Which distribution has the greater variability? Explain using the standard deviation of each distribution.

Boxplots

Another type of statistical graph, the **boxplot**, helps you to visualize the variability of a distribution. Five statistics form a boxplot: the minimum data value, the **first quartile (Q_1)**, the median, the **third quartile (Q_3)**, and the maximum data value, in order from smallest to largest. These are often referred to as the **five-number summary** for the distribution.

Quartiles refer to values that separate the data into quarters. The lowest 25% of all the data falls between the minimum and the first quartile. The next lowest 25% of the data falls between the first quartile and the median. Hence, 75% of the data is above the first quartile.

13. a. How much of the data is below the third quartile?

   b. How much of the data is between the first and third quartiles?

The five-number summaries are displayed here for the switch A and B data.

Switch A
```
minX=76.464516
Q1=80.77875085
Med=82.2356903
Q3=83.45049998
maxX=86.993903
```

Switch B
```
minX=68.606254
Q1=78.78661495
Med=81.9755658
Q3=85.03659857
maxX=96.897925
```

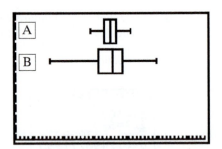

Sometimes called box-and-whisker plots, the preceding boxplots show visually how the quartiles for the switch data compare. The box that begins at Q_1 and ends at Q_3 represents the middle 50% of the data. The spread of the middle 50% of the data is called the **interquartile range**, or **IQR**. To calculate the interquartile range, you subtract Q_1 from Q_3, so IQR $= Q_3 - Q_1$. The whiskers represent the extent of the lower and upper quarters of the data.

The interquartile range (IQR) for switch A is $Q_3 - Q_1 = 83.45049998 - 80.77875085 \approx$ 2.67 rounded to the hundredths place.

See Appendix A for details to obtain a boxplot using the TI-83/84 Plus calculator.

14. a. Calculate the interquartile range for switch B.

b. Which switch has the larger interquartile range?

15. Consider the following boxplots for three distributions.

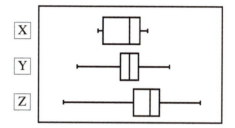

a. Compare the minimum, first quartile, median, third quartile, and maximum for the three distributions.

b. Compare the ranges for each of the three distributions.

c. Which distribution do you think has the least variability? Explain.

SUMMARY: ACTIVITY 6.6

1. The **variability** of a frequency distribution refers to how spread out the data is, away from the center.

2. The **range** of a frequency distribution is the difference between the lowest and highest data values.

3. The **deviation** of a data value is how far it is from the mean of the data set.

4. The **standard deviation** is a measure of how far all the data collectively is from the mean.

5. A **boxplot** is a graph that displays the **five-number summary** (the minimum, first quartile, median, third quartile, and maximum data value) for a distribution.

6. The **interquartile range** (IQR) is the spread or variability of the middle 50% of the data and equals $Q_3 - Q_1$.

EXERCISES: ACTIVITY 6.6

Do the calculations in each of the following exercises. Fill in the table as part of your calculation. Then check your work by using the statistics features of your calculator.

1. Consider the following data: 5   8   2   1   7   1

Determine the sample standard deviation by filling in the table as part of your calculation.

x	$\bar{x}$	$x - \bar{x}$	$(x - \bar{x})^2$

2. Eight adults are surveyed and asked how many credit cards they possess. Their responses are 4, 0, 3, 1, 5, 2, 2, and 3. Determine the sample standard deviation for these data. Fill in the table as part of your calculation.

x	$\bar{x}$	$x - \bar{x}$	$(x - \bar{x})^2$

3. A student takes ten exams during a semester and receives the following grades: 90, 85, 97, 76, 89, 58, 82, 102, 70, and 67. Find the five-number summary.

4. In a year-long music course, you are required to attend 12 concerts. Last season, the lengths of the concerts were 92, 101, 98, 112, 80, 119, 92, 90, 116, 106, 78, and 65 minutes.

   a. Find the five-number summary for these data. Decide on a scale and draw a boxplot.

b. Calculate the range for these data.

c. Calculate the interquartile range for these data.

5. The weights, in pounds, of a group of workers are as follows:

173   123   171   175   188   120   177   160   151   169   162   128   145

140   158   132   202   162   154   180   164   166   157   171   175

Find the five-number summary for these data. Decide on a scale and draw a boxplot.

6. Joe DiMaggio played center field for the New York Yankees for 13 years. Mickey Mantle, who played for 18 years, succeeded him. Here is the number of home runs by DiMaggio and Mantle.

DiMaggio: 29, 46, 32, 30, 31, 30, 21, 25, 20, 39, 14, 32, 12

Mantle: 13, 23, 21, 27, 37, 52, 34, 42, 31, 40, 54, 30, 15, 35, 19, 23, 22, 18

a. Compute the five-number summary for each player.

	Min	Q_1	Median	Q_3	Max
DiMaggio:	_____	_____	_____	_____	_____
Mantle:	_____	_____	_____	_____	_____

b. Using the same scale, draw a box-and-whisker plot for each player. (Draw one box-plot above the other.)

c. Discuss the similarities or differences of the two boxplots.

7. Given the following boxplots of three distributions, which do you think will have the smallest standard deviation?

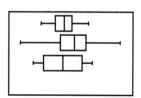

Activities 6.1–6.6   What Have I Learned?

1. Bar graphs and circle graphs can be used to visualize the same collection of data. What advantage does each display have over the other?

2. Determine if the mean and median accurately represent an "average" of the following set of numbers:

   a. Zip codes: 12303, 13601, 32703, 52104

   b. Ranks of different brands of a product in a taste test: 1, 3, 5, 2

3. A small supply company employs a supervisor at $1200 a week, an inventory manager at $700 a week, six stock boys at $400 a week, and four drivers at $500 a week.

   a. Determine the mean and median wage.

   b. Which "average" best describes a typical wage at this company? Explain.

4. If a distribution is skewed to the right, then which average is generally larger, the mean or the median? Explain.

5. Which do you think has more variation: the IQ scores of 20 randomly selected graduating college seniors or the IQ scores of 20 randomly selected adults entering the local mall? Explain.

6. Does adding the same number to each value of a set of data affect the measures of variability for that set of data?

7. Which distribution has greater variability?

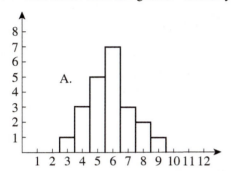

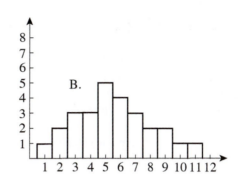

8. a. Use a graphing calculator to determine the mean and standard deviation of the following two data sets:

Data 1	9.14	8.14	8.74	8.77	9.26	8.10	6.13	3.10	9.13	7.26	4.74
Data 2	6.58	5.76	7.71	8.84	8.47	7.04	5.25	5.56	7.91	6.89	12.50

b. Construct a stem-and-leaf plot of each distribution and compare the shapes of the distribution.

c. Compute the five-number summary for each data set.

	MIN	Q_1	MEDIAN	Q_2	MAX
Data 1					
Data 2					

d. Compute and compare the interquartile ranges (IQR).

9. For a given data set, is the second quartile, Q_2, always the same as the median of the data? Explain.

Activities 6.1–6.6   How Can I Practice?

1. The number line in the following graph represents annual net exports in billions of dollars for eight countries. Net exports are obtained by subtracting imports from exports; a negative net export means that the county imported more goods than it exported.

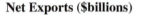

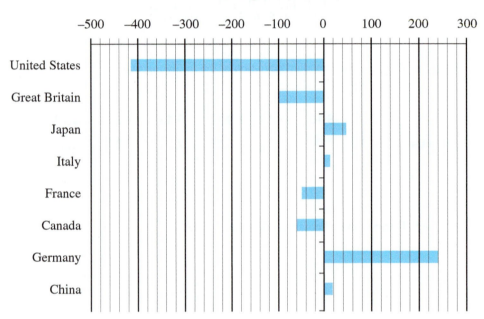

Data Source: The Economist, Nov. 9, 2013, p. 96, "Economic and Financial Indicators." Data are for mid-2013.

a. Estimate the net amount of exports for Japan. Is your answer positive or negative? Explain what this number tells you about the imports and exports of Japan.

b. Estimate the net amount of exports for the United States. Is your answer positive or negative? Explain what this number tells you about the imports and exports of the United States.

c. What is the difference between net exports for Japan and Italy?

d. What is the difference between net exports from Britain and Germany?

e. What is the total sum of the exports of the eight countries listed?

2. In the bar graphs below, a projection of world population by age groups and gender for six different regions is given as a percent of the region's population at the time.

Population Pyramids for Regions and Selected Countries: 2002 and 2050
In 2050, the elderly are expected to be a substantially larger part of national and regional populations.

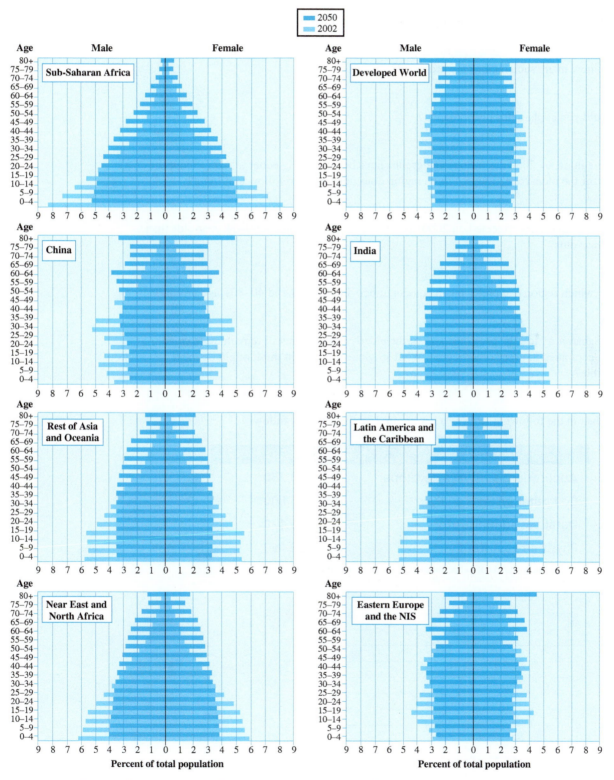

Data Source: U.S. Census Bureau, International Programs Center, International Database

a. In 2002, which region had the largest portion of its population in the 0–4-year-old age group, and what was that percent?

b. What is the 2050 projection for the 10–14 age group of the region you identified in part a?

c. In 2002, which region's 80+ age group is larger than the 75–79 age group?

In 2050, how many such regions are projected?

How could you explain this phenomenon?

d. In most regions, by 2050 the various age groups have become fairly equal in size, except perhaps for the oldest age groups. Which region does not appear to follow that trend?

e. Which two regions are projected, in 2050, to have larger populations of 60–64-year-olds than 45–49-year-olds?

f. In 2002, which region had pretty much equal numbers in the 0–39 and 40–80+ age groups?

g. In Sub-Saharan Africa, the projected size of the elderly population will remain quite small in 2050, compared to every other region. What single factor do you think might be most responsible?

3. Graph the data in the following table as a bar graph. Use the grid that follows.

OCCUPATION	MEAN SALARY (2012)
Applications software developer	$93,280
Computer systems analyst	$83,800
Database administrator	$79,120
Physician's assistant	$92,460

Data Source: Bureau of Labor Statistics, U.S. Dept. of Labor

4. The educational attainment of adults age 25 and older in the United States in 2013 is summarized below.

Education in 2013: U.S. Residents 25 Years or Older

EDUCATIONAL LEVEL	NUMBER (THOUSANDS)
Less than 9th grade	9,922
Some high school	14,595
High school diploma	61,704
Some college	34,805
Associate's degree	20,367
Bachelor's degree	41,575
Graduate degree	23,931

Data Source: U.S. Census Bureau

Identify each category on the following circle graph.

U.S. Educational Attainment, Adults 25 and Over, 2013

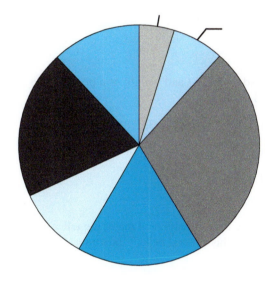

5. Wayne Gretzky scored 50% more points than anyone who ever played professional hockey. He accomplished this while playing in 280 fewer games than Gordie Howe, the second leading record holder. The number of games played by Gretzky during each season of his career are listed:

79, 80, 80, 80, 74, 80, 80, 79, 64, 78, 73, 78, 74, 45, 81, 48, 80, 82, 82, 70

Create a stem-and-leaf display.

6. The traveling times in minutes for 15 workers randomly chosen at High Tech Manufacturing are as follows:

$$30, 20, 15, 15, 25, 20, 35, 60, 20, 40, 10, 30, 15, 35, 20$$

a. Create a dotplot of the frequency distribution of the traveling times.

b. Determine the mean, median, mode, and midrange of the set of traveling times.

c. Which measure best represents the set of data? Explain.

7. The body mass index (BMI) is a measure based on height and weight of a person. The BMI of randomly selected males and females are listed below:

Males: 23.6, 23.4, 24.6, 26.5, 24.2, 31.0, 28.1, 26.3, 22.5, 27.9, 29.2

Females: 19.6, 22.8, 20.6, 28.1, 25.5, 21.1, 27.5, 20.5, 27.9, 18.7, 29.8

a. Determine the mean of each group of BMI measures.

b. In general, it is believed that males weigh more than females, and males are taller than females. Do the sample means in part a support this general belief? Explain.

8. The following data is the number of cancellations on a certain commuter flight on 12 randomly selected days in the past 6 months.

$$5, 5, 6, 9, 9, 11, 12, 12, 13, 13, 13, 14$$

a. Construct a dotplot of the data.

b. Does the distribution seem skewed? Explain.

c. Determine the mean and median of the number of cancellations.

d. When the data is skewed, the mean is "pulled" from the center to the direction of the skew. The median tends to remain near the center of the distribution. The following diagram illustrates the relationship between the mean and median in a distribution that is skewed to the right.

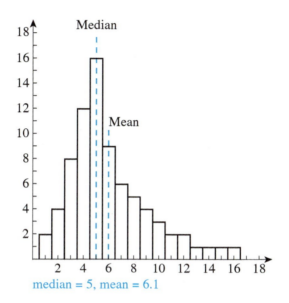

median = 5, mean = 6.1

Approximate the location on the *x*-axis of the mean and median in the following distribution that is skewed to the left.

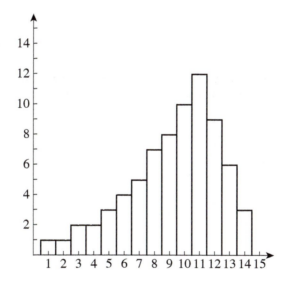

e. Do the results in parts b and c confirm the observation in part d?

9. Consider the following data: 12, 8, 9, 10, 12, 9

Determine the sample standard deviation by filling in the table as part of your calculation.

x	$\bar{x}$	$x - \bar{x}$	$(x - \bar{x})^2$

10. The following speeds of 21 cars are recorded as they pass a concealed marker on an interstate highway that has a legal speed limit of 65 mph: 73, 67, 69, 65, 64, 77, 59, 74, 71, 67, 70, 58, 67, 69, 71, 70, 69, 64, 85, 67, and 64.

   a. Determine the five-number summary and produce the boxplot.

   b. Compute the interquartile range (IQR).

11. The management of Disney World would like to determine the amount of waiting time for the rides at Space Mountain. The following data represents a random sample of waiting times in minutes (rounded to the nearest half-minute):

$$4.0 \quad 5.5 \quad 9.5 \quad 5.0 \quad 13.5 \quad 6.5 \quad 8.5 \quad 9.0 \quad 9.5 \quad 12.0 \quad 8.0 \quad 7.5$$

   a. What is the mean waiting time for the ride?

   b. What is the sample standard deviation?

Activity 6.7

What Is Normal?

Objectives

1. Identify a normal distribution.

2. List the properties of a normal curve.

3. Determine the z-score of a given numerical data value in a normal distribution.

4. Identify the properties of a standard normal curve.

5. Solve problems using the z-scores of a standardized normal curve.

The following collection of data gives the heights, in inches, of 35 randomly selected eleventh grade male students. The measurement was made to the nearest inch.

63 64 65 65 66 66 66 67 67 67 67 67 68 68 68 68 68 68

69 69 69 69 69 70 70 70 70 71 71 72 72 73 74 75 76

1. a. Complete the following frequency distribution.

HEIGHT (in inches)	NUMBER OF ELEVENTH GRADE MALES

b. Construct a histogram from the frequency distribution in part a.

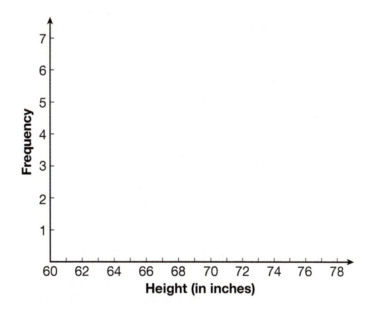

c. Does the distribution appear to be skewed or symmetrical? Explain.

d. Determine the mean, median, and mode of the given set of data. Do your results verify your conclusion in part c?

Normal Distribution

One of the most important distributions is called the **normal distribution**. The histogram for a normal distribution has the following general shape:

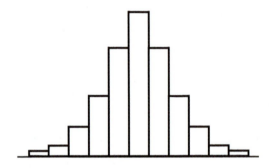

The normal distribution is important because many sets of data are normally distributed, or nearly so. Such distributions include heights and weights of males and females, measures of reading ability, intelligence quotients, scores on standardized tests, and wear-out mileage of car tires. Much of the theory in the study of statistics is based on normal distributions.

2. Is the distribution in Problem 1 normal? Explain.

When the histogram of a normal distribution is smoothed to form a curve, the curve is **bell-shaped**. This curve is called a **normal curve** and is used to model the normal distribution.

The bell can vary in size. It can be high and narrow, or short and wide. For example, each of the following graphs represents a normal curve.

Although normal curves have a variety of sizes, all normal curves have the same basic properties. Use the normal curve given in Problem 3 to help identify some of these properties.

3. Assume the heights of all 16-year-old males are normally distributed with mean $\mu = 68$ and standard deviation $\sigma = 2$. The following normal curve represents this distribution, where x represents the height in inches.

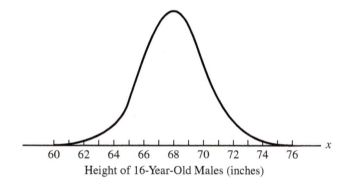

Height of 16-Year-Old Males (inches)

a. For what value of x does the maximum or highest point of the curve occur? What does this value of x correspond to in this distribution?

b. Each tick mark along the x-axis presents one inch. Locate the point on the x-axis that is one standard deviation greater than the mean.

c. If $\mu + \sigma$ represents the x-value that is one standard deviation greater than the mean, what does $\mu - \sigma$ represent? Locate the point on the graph.

d. The curve is symmetric about a vertical line through the mean μ. Describe this symmetry property of the given normal curve. What is the equation of the vertical line in this situation?

e. Since the curve is symmetrical about the mean, what is the median and mode of the height distribution? What percentage of the data points is greater than the mean? Less than the mean?

f. Approximately 68% of the data values in any normal distribution fall between $\mu - \sigma$ and $\mu + \sigma$. What percent of the data values are between μ and $\mu + \sigma$?

g. The horizontal axis is an asymptote for every normal curve. Describe what this means.

Summary of the properties of a normal curve:

1. The curve is bell-shaped with the highest point at the mean μ.

2. The curve is symmetrical about a vertical line $x = \mu$.

3. The mean, median, and mode are all equal.

4. 50% of the data values of the distribution are to the right of the mean μ; 50% of the data values are to the left of the mean μ.

5. Approximately 68% of the data values fall between $\mu - \sigma$ and $\mu + \sigma$; that is, 68% of the data values in the normal distribution are between $x = \mu - \sigma$ (one standard deviation less than the mean) and $x = \mu + \sigma$ (one standard deviation greater than the mean).

6. The normal curve model approaches the horizontal axis, but never touches or crosses the axis.

4. Use the normal curves A and B below to answer parts a–f.

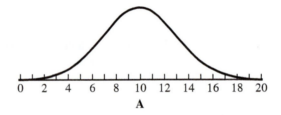

A

a. Determine the mean, median, and mode of each normal distribution.

b. One of the curves corresponds to a normal distribution with $\sigma = 3$ and the other with $\sigma = 1$. Match each curve with the correct standard deviation.

c. For curve A, what value corresponds to each of the following:

    i. one standard deviation less than the mean, represented by $\mu - \sigma$

    ii. one standard deviation greater than the mean, represented by $\mu + \sigma$

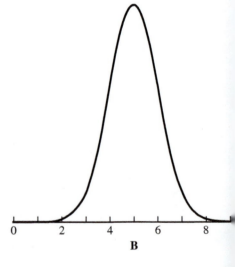

B

d. What percentage of the data in the distribution represented by curve A is between the values determined in part c?

   e. The middle 50% of the data values in a normal distribution falls approximately between $Q_1 = \mu - 0.67\sigma$ and $Q_3 = \mu + 0.67\sigma$. Determine the interquartile range (IQR) for each normal distribution.

   f. Approximately 99.7% of the data values in a normal distribution fall between $\mu - 3\sigma$ and $\mu + 3\sigma$. Calculate $\mu - 3\sigma$ and $\mu + 3\sigma$, and then approximate the practical range for each normal distribution.

   g. What effect does the size of the standard deviation, σ, have on the practical range and the interquartile range of a normal distribution?

5. You and a friend compare your last chemistry test grades. In your friend's class the mean was 73 and the standard deviation was 8. In your class the mean was also 73 but the standard deviation was 6. The grades in each class were normally distributed. Both of you scored a grade of 91. Your grade of 91 was the-highest grade in your class. Your friend's grade of 91 was only the third highest in the class.

   a. Calculate the practical range for each class. Show your work.

   b. Analyze the practical range for each class, and then explain why your friend's grade was not the highest grade in his class.

Standard Scores

Different normal distributions vary from each other based on the values of the mean μ and standard deviation σ. The mean μ may be located anywhere on the x-axis. The bell-shape may be wide or narrow, depending on the value of the standard deviation σ. But no matter how large or small the standard deviation might be, about 68% of the data values in the normal distribution will fall within one standard deviation of the mean. Therefore, the standard deviation σ is often used as a standard unit of measurement that will allow you to compare different normal distributions.

Standard scores, also called *z*-scores, give the number of standard deviations that the original measurement *x* is from the mean, μ. The *z*-score formula is

$$z = \frac{x - \mu}{\sigma},$$

where z = *z*-score,
x = given data value,
μ = mean of the *x*-score distribution,
σ = standard deviation of the *x*-score distribution.

Example 1

The heights of 16-year-old males are normally distributed with mean 68 inches and standard deviation 2 inches. Determine the z-score for:

a. 70 inches.

b. 66 inches.

SOLUTION

a. Let $x = 70$, $\mu = 68$, and $\sigma = 2$. Substituting, you have

$$z = \frac{x - \mu}{\sigma} = \frac{70 - 68}{2} = \frac{2}{2} = 1.$$

b. Let $x = 66$, $\mu = 68$, and $\sigma = 2$.

$$z = \frac{x - \mu}{\sigma} = \frac{66 - 68}{2} = \frac{-2}{2} = -1$$

Therefore, in the 16-year-old male height distribution with mean 68 and standard deviation 2, a height of 70 has a *z*-score of 1. This means that a height of 70 inches is one standard deviation greater than the mean of 68 inches. Similarly, a height of 66 inches has a corresponding *z*-score of −1. This indicates that 66 inches is one standard deviation less than the mean of 68 inches.

6. Determine the *z*-score for each of the following heights of 16-year-old males. Recall that this population is normally distributed with mean 68 inches and standard deviation 2 inches.

a. 71 inches   **b.** 74 inches

c. 68 inches   **d.** 64 inches   **e.** 62 inches

7. The mean of the normal 16-year-old male height distribution is 68. The corresponding z-score (see Problem 5c) for 68 is 0. Will the mean of every normal distribution have a z-score of 0? Explain.

> Data below the mean will always have negative z-scores. Data above the mean will always have positive z-scores. The mean will always have a z-score of 0.

8. Suppose the test scores on the last exam in a Liberal Arts Mathematics course are normally distributed. The z-scores for some of the students in your class were:

$$1.5, 0, -1.2, -2, 1.95, 0.5.$$

a. List the z-scores of students that scored below the mean.

b. List the z-scores of students that scored above the mean.

c. If the mean of the exam was $\mu = 80$, did any of the students selected above have an exam score of 80? Explain.

d. Which z-score represents the student that scored the highest among the students selected above? Explain.

e. If the standard deviation of the exam was $\sigma = 5$, what was the actual test score for the student having z-score 1.95?

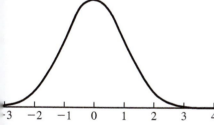

If the original distribution of x-values is normal, then the corresponding z-values will be normally distributed. The z-distribution will have a mean of $\mu = 0$ and a standard deviation of $\sigma = 1$. The resulting curve, shown on the left, is called the **standard normal curve**.

9. You have been assigned the task of selecting a switch to use in a piece of heavy machinery that your company is building for export. A warranty service call will be a significant expense, so reliability is a concern. You need a failure rate of under 80 per million hours of service to keep costs down.

The manufacturers have a lot of marketing claims on their Web sites; you find their reliability data and copy it down in a table. Switch failure rates have a normal distribution.

SWITCH	A	B
Failure rate (per million hours)	64	71
Standard deviation (per 10^6 hours)	16	4

   a. What are your first thoughts on which switch is better?

   b. How can you compare the failure rates more accurately using the table information?

   c. Compute the z-score for switch A.

   d. Compute the z-score for switch B.

   e. Which switch would be the better choice, assuming their cost difference is small?

SUMMARY: ACTIVITY 6.7

1. A **normal curve** is a **bell-shaped** curve that is symmetric about its mean. The mean, mode, and median in a normal distribution are the same. Each normal distribution can be modeled by a corresponding normal curve.

2. 50% of the data values of a normal distribution are greater than the mean μ; 50% of the data values are less than the mean μ.

3. Approximately 68% of the data values fall between $\mu - \sigma$ and $\mu + \sigma$; that is, 68% of the data values in the normal distribution are between $x = \mu - \sigma$ (one standard deviation less than the mean) and $x = \mu + \sigma$ (one standard deviation greater than the mean).

4. The normal curve model approaches the horizontal axis, but never touches or crosses the axis.

5. A normal curve can have any mean and any standard deviation; however, a **standard normal curve** has $\mu = 0$ and $\sigma = 1$.

6. **Standard scores**, also called z-scores, indicate the number of standard deviations that the original measurement x is from the mean μ of the normal distribution.

7. The z-score formula is: $z = \dfrac{x - \mu}{\sigma}$,

where $z = z$-score,

x = given data value,

μ = mean of the x-score distribution,

σ = standard deviation of the x-score distribution.

8. The mean of a normal distribution always has z-score 0. The values in any normal distribution that are below the mean have negative z-scores; those above the mean have positive z-scores.

EXERCISES: ACTIVITY 6.7

1. A normal distribution has a mean $\mu = 100$ and standard deviation $\sigma = 10$. Determine the z-score for each of the following x-values.

a. 110　　　　　　　**b.** 125　　　　　　　**c.** 100

d. 90　　　　　　　**e.** 85

2. The high school physical education department offers an advanced first-aid course. The scores of the first exam were normally distributed. The z-scores for some of the students are shown below:

$$1.10, 1.70, -2.00, 0.00, -0.80, 1.60$$

a. Which of these students scored above the mean?

b. Which of these students had a test score equal to the mean of the exam?

c. Which of these students scored below the mean?

d. If the mean score was $\mu = 150$ with standard deviation $\sigma = 20$, what was the exam score for the student having z-score 1.70?

    e. Given $\mu = 150$ and $\sigma = 20$, find Q_1 and Q_3. Then determine the interquartile range.

    f. Use the fact that approximately 99.7% of the data values in a normal distribution fall between $\mu - 3\sigma$ and $\mu + 3\sigma$ to approximate the practical range of exam scores.

3. You and a friend compare your last algebra quiz grades. The quiz was worth 30 points. In your friend's class the mean was 21 and the standard deviation was 3. In your class the mean was also 21 but the standard deviation was 2. The quiz grades in each class were normally distributed. Both of you scored a grade of 27. Your grade of 27 was the highest grade in your class. Your friend's grade of 27 was only the second highest in the class.

    a. Calculate the practical range for each class.

    b. Analyze the practical range for each class, and then explain why your friend's grade was not the highest grade in his class.

4. Students who drive to school are randomly selected. Each participant is asked the "arrival" time (in minutes) it takes to drive from their home, park the car, and enter the high school building. The data is normally distributed with mean 20 minutes and standard deviation 3 minutes. In this problem, you will analyze this information to make inferences regarding arrival time.

    a. If the arrival time is 26 minutes, determine the corresponding z-score.

    b. If the arrival time is 14 minutes, determine the corresponding z-score.

    c. If the arrival time is 20 minutes, determine the corresponding z-score. Is this answer to be expected? Explain.

    d. If the arrival time is 25 minutes, determine the corresponding z-score.

    e. If the arrival time is less than 20 minutes, is the corresponding z-score positive or negative?

    f. If the z-score for a certain arrival time is positive, then what can you determine about the value of the arrival time?

5. Your z-score on a college entrance exam is 1.3. If the x-scores on the exam have a mean of 480 and a standard deviation of 70 points, what is your x-score?

6. If a z-score is less than -2 ($z < -2$) or more than 2 ($z > 2$), the corresponding x-value is considered "unusual." If $z < -3$ or $z > 3$, then the corresponding x-value is considered "very unusual."

  a. Assume the white blood cell count per cubic millimeter of whole blood has a distribution that is approximately normal with mean $\mu = 7500$ and standard deviation $\sigma = 1750$. If someone had a white blood count of 3500, would that be considered unusually low?

  b. If the red blood cell count for women (in millions per cubic millimeter) has an approximately normal distribution with mean $\mu = 4.8$ and standard deviation $\sigma = 0.3$, would a red blood count of 5.9 or higher be considered unusually high?

7. a. You score 680 on the mathematics portion of the SAT. If the math scores on the SAT are normally distributed with mean $\mu = 518$ and standard deviation $\sigma = 114$, determine the corresponding z-score for $x = 680$.

  b. Your friend scores 27 on the ACT assessment mathematics test. If the ACT math scores are normally distributed with mean $\mu = 20.7$ and standard deviation 5.0, determine the corresponding z-score for $x = 27$.

  c. Assume the tests both measure the same kind of ability. Compare the z-scores in parts a and b to determine who scored higher.

Activity 6.8

Sampling a Population

Objectives

1. Understand the difference between a census and a sample.

2. Identify the characteristics of a simple random sample.

3. Understand what bias in a sample means.

4. Be able to select a simple random sample, when possible.

5. Identify how the size of a sample affects the result.

Suppose you have a need to know the average weight of every adult in the United States, or the mean age of every blue crab in the Chesapeake Bay, or the median income of every household in Denver. To determine such measures of central tendency, it is not usually practical to obtain the data for every individual in the population. If every individual in the population is measured it is called a **census**.

1. Do you think it is possible to conduct a census to determine each of the measures mentioned above? Give specific reasons for each case.

Conducting a census to determine the characteristics of an entire population is usually either too costly or time consuming, if even possible. Other ways have been devised to achieve the same objective with a high degree of certainty. By taking a relatively small sample from the entire population, a great deal can be determined about that population. A **sample** consists of any portion of a population, no matter how large or small. Once a sample is chosen, this smaller collection can be measured and a central tendency can be recorded.

2. Since a sample can be of any size, how would you compare the practicality and usefulness of samples of size 1, 25, and 2 million when attempting to estimate the income of Colorado households?

The process of collecting data from some fraction of a population is called **sampling**. It allows us to learn about the entire population by studying a relatively small portion of the population. The size of the sample along with the method used to select individuals for the sample can have a big effect on the results.

3. Suppose you are interested in knowing the mean average height for everyone in your school. A census would give the exact value while a sample will result in an estimate.

   a. Assume you have a sample the size of the population minus one. How close do you think the mean height for the sample will be to the mean of the population?

   b. If you got another sample, this time with ten fewer students than the population, do you think this sample mean would be as close to the population mean as the sample in part a?

   c. If you took a sample of size 100 and another sample of size 10, which sample's mean do you think will be closer to the population's mean?

In general, a larger sample size will result in a more accurate estimate of the population. But when the population is very large, larger samples will not produce significantly better results and are usually more expensive or difficult to obtain. A decision is normally made that will balance these two tendencies, not too small to sacrifice accuracy, and not too large to sacrifice cost.

As a general rule, for very large populations, a sample in the size of hundreds is normally sufficient.

To help illustrate some of these ideas you will explore some relatively small populations. The table below shows the current grades (0–100) for all the 20 students in Mr. Horton's economics class.

ID NUMBER	LAST NAME	CURRENT GRADE
1	Adams	78
2	Baker	84
3	Cooper	82
4	Davenport	95
5	Elacqua	71
6	Flanagan	83
7	Grant	97
8	Haught	65
9	Jacobs	80
10	Kim	78
11	Lee	67
12	Medina	91
13	Naraparaju	69
14	O'Brochta	74
15	Park	81
16	Reeves	79
17	Sanchez	62
18	Thoreson	88
19	Vilece	93
20	Whiting	63

4. Consider this class as the entire population. In this case, a census is quite feasible. Determine both the median and mean for this population.

5. Now suppose you did not have access to all the grades, but could be given the grades for five students. In other words, you can take a sample of size five. Choose any five students you like and calculate the mean and median grade for your sample. Such measurements taken from a sample are called **statistics**. How far are your statistics from the population mean and median?

6. Now compare your statistics to those of others in your group, or even the entire class. If you found the average of all your sample means, would you expect it to be close to the population mean? Find this average and check.

7. Describe how you chose the five students for your sample, and compare your method with others in your group and/or class.

If you chose your sample by looking at the grades and then tried to pick a balanced set of grades (not too high, not too low), or purposely picked all high scores or all low scores, you would introduce a bias to the selection process. **Bias** simply means that some individuals are somehow favored over others in the population. This is a critically important idea when attempting to select a sample that is representative of the entire population.

There is a classic example that helps illustrate this point. In 1936, *Literary Digest* magazine conducted a survey to predict the outcome of the U.S. presidential election. They had conducted similar surveys the five previous presidential elections, and had correctly predicted the winner each time. In 1936, ten million postcards were distributed to a list of voters, generated primarily from owners of automobiles and telephones. Over two million of the postcards were returned. On that basis, the Republican governor of Kansas, Alf Landon, was predicted to beat the incumbent Democratic president, Franklin Roosevelt, by an overwhelming margin, with 57% of the popular vote. The election was actually won by Roosevelt, with over 61% of the popular vote!

8. There was a bias in this presidential election survey. What do you think could have been the cause?

Simple Random Sample (SRS)

Many methods have been devised to select samples in a way that eliminates, or at least minimizes, any form of bias. To achieve this, the most basic idea is the random selection of the individuals in the sample. To select randomly means every individual in the population has an equal chance of being selected. A **simple random sample** (SRS for short) is the result of a sampling method that assures every possible sample of the same size has an equal chance or probability of being selected.

9. Describe a method that would ensure your sample of five students from Mr. Horton's class will be a simple random sample. Compare and discuss your method with others in your class.

Appendix

10. Use your calculator to generate five random integers between 1 and 20. Use these numbers to identify the five students for your sample, by simply counting each student's position in the alphabetical class list.

a. Suppose one of your random numbers is repeated. Since you need five different numbers, what could you do next?

b. Record your five random numbers and the student associated with each number.

c. Does the same chance or probability apply to every possible sample of size five using this method? Why?

d. Calculate the mean and median grade for your sample.

e. As you did in Problem 5, use the mean and median from part d for all the samples to find the average mean and average median for your class. Compare your averages this time to what you found in Problem 5. Which are closer to the population mean and median?

In general, all the possible sample means will naturally vary, both above and below the population mean. But the average of several sample means will generally be very close to the population mean, an idea that is useful for further statistical analysis.

Random numbers are a primary tool in selecting simple random samples. There are several ways to get random numbers, in addition to using your calculator. Books of random numbers have been published (especially useful before computers became commonplace), and websites exist that provide truly random numbers, based upon physical phenomena, like the weather. (See www.random.org.)

SUMMARY: ACTIVITY 6.8

1. A **census** is the result of measuring or counting every individual in a population.

2. A **sample** is the result of measuring or counting any portion of a population.

3. **Sampling** is the process of collecting a sample from a population.

4. The larger the sample size, the more accurate the estimate for the population.

5. A **bias** occurs when some individuals in a population have a greater chance of being selected for a sample than others.

6. A **statistic** is a measurement derived from a sample, usually intended to be representative of an entire population.

7. A **simple random sample** (**SRS** for short) is a sample that is selected by a method that ensures every possible sample of a given size has an equal chance of being selected.

EXERCISES: ACTIVITY 6.8

I. The following list of digits was taken from a table of random numbers.

39 63 46 23 49 74 08 86 55 64 16 37 91 97 13 39 15 39 45 91 79 86 45 37

59 53 50 50 40 46 92 74 78 44 52 66 73 31 93 36 55 45 26 22 35 69 08 32

30 73 47 15 71 83 72 27 97 12 25 77 56 51 78 07 76 32 92 81 13 13 01 96

62 88 91 26 91 25 42 40 90 25 75 20 30 91 39 41 17 31 46 06 08 91 56 30

83 19 51 13 43 51 14 20 82 15 14 03 47 33 68 07 61 82 92 69 48 68 04 68

70 58 37 03 61 41 04 72 67 92 84 66 90 43 24 33 01 49 39 09 86 54 59 06

54 09 20 83 00 19 11 60 76 75 52 48 79 25 31 23 17 84 12 07 77 72 50 10

95 83 62 25 30 91 78 58 02 10 34 36 15 22 28 33 86 99 43 32 38 68 61 67

a. Since there are 20 students in Mr. Horton's class, think of each student's ID number as a two digit sequence (the ID number for Baker is 02). Start at the beginning of the table, looking at each two-digit number, and record the first eight numbers that are between 01 and 20. These give the student ID numbers. (If there is a repeated number, simply ignore the repetition.) What students would make a sample of size 8, using this table?

b. If you wanted a different sample, you would of course need to use different random numbers. But there is another way to use the table without bypassing so many numbers. Since there are 100 possible two-digit numbers in the table (00 through 99), and 100 is a multiple of 20, each student could be assigned five numbers, their student ID plus a multiple of 20. So, for example, Jacobs would be identified by 09, 29, 49, 69 and 89. Use the first row of the table to choose a different sample of size 8, and identify the students by name.

2. A sample of size 100 is to be selected from each of the following populations. In which cases do you think a simple random sample is possible? Give a reason for your answer.

  a. All the adult blue crabs in the Chesapeake Bay

  b. All the students enrolled in your school

  c. All the passenger cars registered in the state of Colorado

  d. All the passenger cars on the road in Texas on December 25, 2014

  e. All the homeless people in the United States

3. The data below (10 rows by 20 columns) gives the weights, in pounds, of 200 chickens at a poultry farm.

5.9	6.4	4.3	5.9	5.3	6.0	5.5	5.4	5.3	4.8	6.0	4.0	5.1	4.9	5.5	6.0	4.6	4.1	4.8	6.7
5.7	6.1	6.1	6.2	5.9	4.1	4.7	5.2	6.7	6.5	5.1	5.4	5.8	4.1	5.8	5.1	5.0	6.0	5.3	6.2
5.8	6.5	6.9	6.7	5.9	4.6	4.9	6.3	6.6	5.4	5.3	5.2	6.9	6.7	4.7	6.9	5.5	6.6	5.7	5.2
6.8	5.2	4.9	4.9	6.8	6.7	6.7	6.7	5.6	6.9	6.9	6.8	5.1	5.3	7.0	4.9	4.2	4.5	6.7	4.4
5.7	6.2	6.8	5.8	4.5	4.7	5.8	4.4	4.7	5.1	4.5	5.1	6.1	4.6	5.2	4.7	4.2	4.9	6.2	6.8
4.4	4.9	6.1	6.9	6.1	6.9	4.3	4.8	4.8	4.9	6.2	4.3	4.4	4.4	5.2	5.0	6.2	6.3	6.7	6.0
6.1	5.3	6.3	5.8	4.8	7.0	6.7	4.3	6.6	6.7	5.3	6.2	6.6	6.4	6.4	5.8	5.2	5.3	4.0	6.6
4.2	4.8	5.9	4.5	6.7	5.5	5.7	6.2	4.1	6.5	5.8	4.8	5.3	6.2	5.4	5.8	6.9	5.5	6.4	5.0
6.1	4.1	4.3	6.9	6.3	4.2	6.6	5.8	6.6	5.1	4.6	4.8	4.1	6.9	4.4	6.5	6.7	4.8	5.5	6.1
5.5	4.7	5.9	5.3	5.8	5.9	4.8	6.1	6.9	6.4	5.0	4.3	5.7	7.0	5.4	6.2	6.4	5.2	4.9	5.6

  a. Generate random numbers from your calculator to select a SRS of size 10. (Use integers between 1 and 200.) Explain exactly how you chose your sample, showing your random numbers and the associated weights. Calculate the mean weight for your sample.

  b. Now select a SRS of size 25 and calculate the mean weight for this sample.

  c. Which of your samples do you think has a mean closer to the population mean, and why?

  d. Calculate the population mean to check your answer to part c.

  e. If the mean from your larger sample was closer to the population mean, that was to be expected, but was not a certainty. Hand pick ten weights that result in a sample mean closer to the population mean than your size 25 sample mean.

4. The random numbers generated on your calculator are not quite as random as you might think. It is possible that two different calculators could generate the same random list of numbers in the same order. The random number generator (rand) used in the calculators relies on what is known as a seed value. The default seed value is 0. With each rand execution, the calculator will generate the same random number sequence for a given seed value. To obtain a different random number sequence, store any nonzero number to rand.

This sort of random number is called a **pseudorandom number**. For most random sampling purposes, pseudorandom numbers work fine. Truly random numbers can be generated from truly random processes, like the weather or the physical behavior of atomic particles. Such random numbers are available on the web, at sites such as www.random.org. Visit this website, generate a list of ten different random integers between 1 and 200, and report on the process that was used to generate the numbers.

Objectives

1. Identify the characteristics of a well-defined sample survey.

2. Understand how to collect a stratified sample, a systematic sample, a cluster sample, and a convenience sample.

3. Identify ways in which a sample might be biased.

Opinion polls are everywhere these days, on TV and the Internet, in all print media, and at meetings of decision makers in business and government. Such polls generally attempt to measure the percentage of a population that holds a certain opinion or will make a certain decision (like for whom they plan to vote). Such a poll is an example of a **sample survey**, a methodical process of collecting a sample that is representative (hopefully) of the larger population. A simple random sample, as defined in the previous activity, results from one of several different methods that can be employed to attain an unbiased sample. In this activity you will explore different sampling methods and sources of bias. As you complete this activity, consider the various purposes for research methods, such as a survey.

Highway Proposal

In your capacity as an administrative assistant for your county government, you need to assess the attitude of your community for a new highway proposal. Some people will favor the highway in the belief it will provide greater mobility and promote business. Others will argue that it will bring even more congestion and will negatively impact the environment. An opinion poll will be helpful for government leaders in making final decisions.

1. Since you have been asked to submit your preliminary report as soon as possible, there is no time to even consider a census of the 100,000 plus adult residents of your county. So you turn to one of the more convenient tools at your disposal, the phone book. Describe a plan you could use to come up with a sample of 50 adult residents using the phone book. Do you think your method results in a simple random sample?

Even if you did truly randomize the process of selecting names from the phone book, the overall process of using a phone book is flawed. For example, will every adult resident of your community have an equal chance of being included in the sample?

2. Give several reasons why your sampling plan of using a phone book to create the sample and then calling each person in the sample is biased and will not result in a SRS of the entire population of adult residents.

One of the biggest difficulties with whatever plan you might employ in your poll is the same as the bias created with the *Literary Digest* poll in 1936. A significant number of adult residents didn't even have a chance to be included. It is possible that all the individuals in the phone book really are representative of the entire adult population, but you don't know that. Therefore, you have to assume there is a bias in the sampling results.

3. Can you think of other ways to obtain a complete list of adult residents of your county, so a true SRS could be attained?

Whatever list is used for sampling purposes, usually not the entire population, is called the **sampling frame**. The closer the sampling frame comes to the actual population, the better. At best, it should be representative of the entire population.

Nonrespondents

The issue of **nonrespondents** also will create a bias. No matter how you select individuals for the sample, some will not answer their phone, or will refuse to participate, or will simply forget to return the questionnaire. That group of individuals may not be representative of the whole population, so their exclusion will definitely result in inaccurate results.

4. It is important for any sampling plan to include how to deal with nonrespondents. What do you think are some possibilities?

Mean of the Sample Means

Now let's turn to a sample survey that you can do with your classmates. You may wonder how much Facebook time the students in your school have each week. It would probably be feasible to do a census in this case, but for the sake of understanding the sampling process, let's assume it is best to take a sample. In fact, let's first limit the population to your math class.

5. Using a list of all students in your math class, devise and describe a plan to take a SRS of size ten. Take the poll by asking each selected student to estimate how many hours they use Facebook each week, and then calculate the mean for your sample.

6. Now determine the mean of all the sample means from the samples in your class. Compare your sample mean to the class's mean of all the sample means.

7. Since the population (your math class) is relatively small, you can take a census to determine the population mean number of Facebook hours used each week. How close is the class mean of the sample means to this population mean?

You should have discovered that the mean of all the sample means comes very close to the population mean. This basic fact of statistics is very powerful and is used to make inferences and decisions about entire populations based upon relatively small samples. This phenomenon will be studied in great detail if and when you take an elementary statistics course.

Stratified Sampling

Now suppose you suspect there may be a real difference in boys' and girls' Facebook use habits. (If you happen to be in a class with all girls, or all boys, you will need to use a different distinction, maybe based on age or ethnicity.) If there is a real difference, and your sample just happened to be almost all girls, or almost all boys, it will not be very representative of the population.

8. What is the ratio of boys in your class to the total number of students in the class (the population)? Express as a decimal number.

9. **a.** What fraction of your sample of size ten, from Problem 5, consisted of boys?

   b. If you round the population ratio you determined in Problem 8, is it the same as the sample ratio?

You want the ratios in Problems 8 and 9 to be nearly the same. Then your sample would be proportional to the population. But it is entirely possible your SRS could result in almost all boys, or almost all girls. To try to remove any gender bias in your sample, you could devise your sampling method to assure the sample ratio of boys and girls is the same as, or at least as close as possible to, the population ratios.

10. With a sample size of ten, how many boys should be in your sample to closely match (as close as possible) the ratio of boys to students in your class?

Once you know how many boys and how many girls need to be in your sample, you can then separate the class into two groups, called **strata**, the boys and the girls. From each group you then randomly select the required number. Such a method is called **stratified sampling**.

11. Use your result in Problem 10 to produce a stratified sample, and then calculate the mean.

> Additional sampling methods are presented in the Exercises: cluster sampling, systematic sampling, and convenience sampling.

Self-selection

You will end this activity by considering one of the most prevalent types of opinion polls. Whether on the web, TV, the radio, or in a print magazine or newspaper, you often are offered the chance to make your opinion known.

12. You hear on a TV news show that the results of a phone-in poll revealed that 83% of the callers prefer text-messaging over in-person talking. This seems like an overwhelming majority.

   a. How would you describe this method of sampling?

   b. Why do you think this sample is biased?

c. Do you think there is a way to design this method to eliminate a bias?

The previous problem illustrates a classic form of bias, called **self-selection**. The individuals in the sample who chose to take part in the poll were motivated to do so. That motivation will most likely be more prevalent among those that hold one opinion over another. Of course, it also means everyone in the population will not have the same probability of being included, even if everyone hears of the poll. There will be some individuals with no motivation to respond, hence no likelihood of being included in the sample. Self-selection as a method may be cheap and easy to do (hence its prevalence), but as a sampling method, it is deeply flawed.

SUMMARY: ACTIVITY 6.9

1. A **sample survey** is a methodical process of collecting a sample from a larger population.

2. A **sampling frame** is the collection of individuals from which a sample is drawn. It may or may not be the entire population.

3. There are several common sources of **bias** that can occur, even with a well established sampling method.
Selected individuals for a sample may be unavailable or choose to not respond. Such **nonrespondents** do not have the same likelihood of being included, hence creating a bias. A sample composed of individuals that voluntarily respond, a **self-selected** sample, creates a bias since everyone in the population is not equally likely to be included.

4. A **stratified sample** results when a population is separated into two or more subgroups, called **strata**, and simple random samples are selected from each stratum in proportion to the relative size of the stratum.

EXERCISES: ACTIVITY 6.9

1. Use your local phone book to select a sample of ten people in your community.

    a. Precisely describe the method you used.

    b. Is your sample random? Explain.

2. One way to reduce the number of nonrespondents in a survey is to make face-to-face contact. This is more expensive than phoning or mailing, due to additional travel time, especially if the population is spread over a large geographic area. One method that helps reduce this expense is called **cluster sampling**. For example, to personally interview a representative sample of all residents of Florida, a simple random sample of 500 residents would most likely be spread far and wide across the state. Creating a plan to efficiently visit 500 different locations would be a costly challenge. With cluster sampling, the plan would be to randomly select a small number of geographic regions, or clusters, and then find a fixed size SRS within each cluster.

    a. In designing a cluster sampling plan, what do you think would be a reasonable choice for the clusters?

    b. If you chose the counties of Florida for the clusters, how many clusters would there be? (A resource for the geography of Florida: www.floridacountiesmap.com)

    c. If your plan is to include ten clusters, what size SRS should be selected from each cluster?

d. Find a list of all the counties in Florida and randomly select ten counties for your clusters. Describe your process. On a map of Florida like the one below, shade in the selected counties.

e. Assuming you then generate the required SRS for each cluster, in what ways might your resulting overall sample have a bias? Could a bias be fixed?

3. When a sample is to be drawn from a list, such as a phone book, it is sometimes easiest to select one individual at random, and then count from that first individual a fixed number to pick the next individual. The fixed number is also randomly chosen, and is used to continue counting until the sample size is completed. Such a method is called **systematic sampling**. For example, the random number 134 might be chosen, to select the 134th individual on the list for the sample. Then if the random number 20 was chosen, the 154th individual on the list would be the next one chosen.

a. With these numbers, what positions would the third, fourth, and fifth chosen individuals have?

b. Suppose you desire a sample of size 50, chosen from a list of students at your school. If you randomly generate 58 for the first student chosen, and 15 for the count to the next student, what will be the positions of the second, third, and fourth students chosen?

c. By the time the last student is chosen, what would their position be?

d. Suppose there are not that many students in your school? How could you still complete your sample of size 50 using this method?

e. As long as the order of the list will in no way influence the response, as in an alphabetic listing, a representative sample will be acquired. Suppose the listing is by age. Can you think of a question and random numbers that would result in a bad sample (meaning not representative of the population)?

4. The principal of a high school wants to determine the "average" number of children in the families of students who attend the high school. The principal visits each classroom in the building and selects the four students who are seated closest to the corners of the classroom. The principal then asks each student how many children are in his family. Will this sampling technique result in an unbiased sample? Explain.

5. A **convenience sample** uses data that is easily or readily available. However, this type of sampling can be extremely biased. For example, suppose the city council wants to raise taxes in order to build a new science wing on the high school. In order to obtain the opinion of the city residents on the issue, the first 50 adults leaving the local shopping mall are asked if they are in favor of raising taxes to build the new wing.

a. Discuss the possible cause of a bias in the sample.

b. Describe a sampling technique that would be less biased.

6. The student council at your high school wants to survey students about the quality of food in the school cafeteria. It is decided that students leaving the cafeteria will be given a short questionnaire.

a. Would students be more interested in the purpose of this survey or the survey in Problem 4? Explain.

b. Is this method of sampling unbiased? Explain.

7. Your student council desires a quick and representative poll of student attitudes toward a dress code proposed by the administration. In case there are differences in opinion among students in the various grades, you recommend that a stratified sample be used. You know there are 212 seniors, 273 juniors, 255 sophomores, and 320 freshmen in your high school. To get a sample of 50 students, using the four classes as your strata, what number of students from each class should be in your sample?

8. Pearson Learning Solutions wants you to obtain a sample of 100 students who use this textbook. Describe a procedure to obtain the required data for parts a–d.

   a. a simple random sample

   b. a stratified sample

   c. a cluster sample

   d. a systematic sample

9. *Buyer's Market* magazine mailed a questionnaire to its subscribers about cars and other consumer products. Also included were a request for a voluntary contribution and a ballot for the board of directors. Responses were to be mailed back in envelopes that required postage. Discuss potential problems with this sampling technique.

10. Identify each of the following as simple random sampling, stratified sampling, systematic sampling, cluster sampling, or convenience sampling.

   a. Randomly select high schools from different geographic regions and survey all students at each of the chosen schools.

   b. Survey each 10th student who walks in the front entrance of the school.

   c. Obtain a list of students enrolled in a high school. Number these students and then use a random number table to obtain the sample.

   d. Obtain a list of students enrolled at a high school. Draw a random sample from each of the freshman, sophomore, junior, and senior classes, proportional to the size of each class.

   e. At a local tire manufacturing plant, every 20th tire coming off the assembly line is checked for defects.

   f. Survey the first five students who enter the classroom.

   g. Elementary school children in a large city are classified based on the neighborhood school they attend. A random sample of five elementary schools is selected. All the children from each selected school are included in the sample.

   h. A teacher at the high school teaches five classes of Spanish. Two of the classes are chosen by the principal and all the students in those classes fill out a teacher evaluation form about that Spanish teacher.

   i. The first 50 people entering a museum are asked if they support the city council's increase in funding for the arts.

   j. Students in your senior class obtain the voting records for the county. You randomly select 50 Democrats, 40 Republicans, and 8 Independents. (The numbers chosen are based on the percentage of registered voters in each category in the county.) You survey them concerning their favorite candidate for county commissioner in the upcoming fall election.

k. A state is divided into regions using zip codes. A random sample of 15 zip code areas is selected.

l. Every 10th person in line to purchase tickets to a rock concert is asked their age.

m. At a Key Club meeting, the names of all members are written on identical slips of paper and placed in a bowl. The names of five members are drawn. These five people will participate in face-to-face interviews with a pollster.

n. You are approached by a person at a local mall who asks you to answer a few questions about the security at the mall.

o. Numbered ping-pong balls are placed in a bin. The bin is shaken. A ball is then selected from the bin.

p. A marketing expert for MTV is planning a survey in which 500 people will be randomly selected from each of the following age groups: 10–14, 15–19, 20–24, 25–29.

Project 6.10

Statistical Survey

Objective

Design and execute a statistical survey.

Many different sampling methods were described in Activities 6.8 and 6.9. Efficiency in cost and time is usually an important factor in deciding which sampling method is appropriate. Designing a sample survey requires a well-defined, step-by-step process, as outlined below.

1. Formulate the problem and devise a plan.

   • State what you need to know.

   • Identify the population and what is to be measured.

   • Decide on an appropriate sampling method and sample size.

   • Think carefully about the exact form of all questions and responses.

2. Execute the plan.

   • Follow the plan to gather the data.

   • Specify all the details of the methodology.

3. Analyze and organize the results in a report.

   • Discuss all elements of the sampling process.

   • Interpret the results and state final conclusions.

To illustrate this process, let us return to the highway proposal situation introduced in Activity 6.9. Suppose your county executive decides that you should survey only registered voters in your community to determine whether the majority of county residents favor a new highway proposal. You do have a complete list of registered voters, with their party affiliation (34% are Republicans, 45% are Democrats, and 21% are independent). In Problems 1–3 use complete sentences to record your design for this sample survey.

1. You can save money by using a sample of size 100 instead of doing a census. Record a plan for this sample survey.

   a. State what you need to know.

   b. Identify the population that is to be measured and the sampling frame.

   c. Which sampling method, simple random sample or stratified sample, do you think would be best? Give a reason for your choice.

2. Since you are only discussing a hypothetical sample survey, you are not actually going to execute the survey. However, describe precisely all the details of what you would need to do.

3. The report for your survey will include all the difficulties that were encountered, the actual results of the survey, how the results might be biased, and an interpretation of those results. Use your imagination to write a feasible report for this sample survey. (*Hint*: Start by assuming you used a stratified sample, and fill in the table to help create your report.)

RESPONSE	REPUBLICAN	DEMOCRAT	INDEPENDENT	TOTAL
Favors proposal				
Doesn't favor				

In a real sample survey, the results will be subjected to more advanced statistical analysis, which include the notion of margins of error and confidence levels. For example, if you concluded that 58% of residents favor the highway proposal, the sample size would yield an error of $\pm 10\%$, at the .05 confidence level. This means that you expect the real percentage of residents favoring the proposal to be between $58 - 10 = 48\%$ and $58 + 10 = 68\%$, with a .05 probability that you are wrong. These topics are studied fully in a statistics course.

4. Conduct your own sample survey to determine how many hours per week students at your school spend using the Internet during the week. In your sampling plan, be sure to follow the step-by-step process outlined in this activity:

    a. Formulate the problem and devise a plan.

        State what you need to know.
        Identify the population and what is to be measured.
        Decide on an appropriate sampling method and sample size.
        Think carefully about the exact form of all questions and responses.

    b. Execute the plan.

        Follow the plan to gather the data.
        Specify all the details of the methodology.

    c. Analyze and organize the results in a report.

        Discuss all elements of the sampling process.
        Interpret the results and state final conclusions.

5. Depending upon what sampling method you used in Problem 4, discuss how you could have applied either stratified or systematic sampling to conduct your survey.

6. With a group of your classmates, come up with an issue concerning some particular population that you would like to know about.

   a. Clearly state the following, as a preliminary proposal.

      i. What would you like to find out?

      ii. Precisely what population would be involved?

      iii. How would you go about collecting a sample?

   b. Is your proposal feasible? (That is, is it something you and your classmates could actually do?)

      i. If feasible, clear your idea with your teacher, and proceed to conduct your statistical survey. Be sure to follow all steps in the process outlined in this activity.

      ii. If your proposal is not feasible (or your teacher does not approve it), go back and revise the proposal, or start all over.

Objectives

1. Understand the purpose and principles of experimental design.

2. Know when a causal relationship can be established.

3. Understand the importance of randomization, control treatments, and blinding.

4. Understand the purpose of an observational study.

5. Understand the difference between an experiment and an observational study.

Experimental Design

Scientists and statisticians have developed methods to determine causal relationships. **Experimental design** involves a very meticulous procedure for controlling the environment surrounding two variables. The goal is to determine whether changing one variable actually causes a change in the second variable.

1. The mathematics lab at your school has started using a new computerized study program for Algebra I. After the first semester, the average scores on the midterm test were 5% higher than last year. The principal thinks that the computer program caused the increase in test scores.

a. Do you think there is enough evidence to justify this claim? Explain the reason for your answer.

b. What other factors might be involved that could also result in higher scores?

c. What groups of students are being compared, and how might that affect the results?

Treatment and Control Groups

When attempting to establish a causal relationship, a good experimental design will always try to control outside variables. In this example, any factor that might influence test scores, other than using the computer program, should be kept the same for all students, as much as possible. For example, to better establish the principal's claim, all Algebra I students could be divided into two groups. One group would use the computer program (or receive the **treatment**, as it is usually called). The other group would not (usually called the **control group**).

2. If a control group is used for the Algebra I experiment, what outside influences (that you may have identified in Problem 1) could be controlled so they do not vary between the groups?

Another critically important aspect of a good experimental design is randomization. Recall how collecting a sample randomly is essential to assure unbiased results. In a similar way, selecting members for the control group must be done randomly to help even out any unforeseen variables that could affect the results.

3. Suppose the group to receive the computer program treatment in Algebra I is selected on a voluntary basis instead of being selected by a random process. How might this invalidate the results of the experiment?

> To summarize, a **good experimental design** will consist of a **control group** and at least one **treatment group** that will receive the treatment being studied. The selection of subjects for the groups must be **selected through a random process**.

The real point of a control group is to try to minimize the possibility that some unanticipated factor might be at least part of the cause for the results you see. If everyone receives the treatment, maybe the real cause of the improved test scores is a subtly improved test. Students may do better on the test simply because they now understand the questions. The average scores would improve but not necessarily because of the treatment. With a control group, the improvement would be seen in both the treatment group and the control group, so there would be no reason to believe the computer program was the cause of the improved test scores.

This methodology has been used for research in the sciences for many years, credited to the psychologist Gustav Fechner in the mid-1800s. Medical research has benefited greatly, whether in testing the effectiveness and safety of new drugs or searching for new innovative treatments of dangerous diseases. A classic example of the latter occurred in the early 1950s, as medical researchers were seeking a cure for polio, a devastating disease that has since been nearly defeated.

4. Dr. Jonas Salk is credited with discovering a polio vaccine. This successful discovery depended upon many years of experimentation with control groups. To further control for unseen variables, subjects in both the control group and the treatment group were not allowed to know which group they were in. This is called **blinding**.

   a. What kind of influence do you think blinding would eliminate?

   b. During the course of this lengthy experiment, many children participated. Some were treated and some were not. The treatment eventually proved to be effective in preventing the spread of polio. What ethical problem comes to mind as a result of this experiment?

You may have heard of the **placebo effect**. A placebo is a false treatment. In other words, if the treatment is a medicine that is given as a pill, a fake pill (a sugar pill) would be given to the control group so they would not know they were not given the treatment (they would think they were given the treatment). The placebo effect is the real power of the human mind/body to affect positive results without a treatment, just by believing a treatment has been received! Compare this with your answer in Problem 4a.

5. Suppose you are involved in a study to determine whether a new drug is effective in lowering blood pressure. A placebo is administered to the control group.

   a. How important is it that the control group not know they are given a placebo?

   b. Should the treatment group be allowed to know that they are given the real treatment? Why or why not?

   c. What should be done so the administrator/technician does not inadvertently tip off the subject?

To assure that the subject does not know which group they are in, the personnel administering the treatment (or placebo) are also blinded (don't know which they are giving). This makes the experiment **single-blind**. To further assure there is no subjective influence among any involved parties, the evaluators are also blinded to group membership. This will make an experiment **double-blind**.

   6. There have been cases where experimenters have influenced the results by evaluating the subjects without double-blinding. Suppose a researcher is trying to establish a new herb from the Amazon rainforest as a remedy for dry skin. The experiment is single-blind, but not double-blind. The study results in the claim that there was a remarkable improvement in skin condition. What might be wrong with this conclusion?

Replication

It is critical that an experiment include enough participants for the results to be statistically significant. Providing a treatment to one or two individuals will prove nothing. But to truly establish a causal relationship the entire experiment must be repeatable by others. It is very possible that positive results occurred by chance. Repeating the experiment will help verify, or deny, the claim. This feature is usually called **replication**. If researchers in another laboratory can't duplicate the results independently, then the entire study is brought into question. A classic example occurred in 1989 when physicists at the University of Utah produced what they called cold fusion. Fusion is a nuclear process that produces vast amounts of energy but is very difficult to control. After many years of research, it is not currently feasible as an energy source. But the claim of cold fusion (fusion that could take place very inexpensively) appeared to be a breakthrough and caused quite a stir in the mass media. As it turned out, no other physicists could replicate the results, so the excitement was short lived.

Observational Study

Studies often focus on a specific characteristic of a population. For instance, health and fitness clubs collect data on how many times their members visit the club each week. Schools collect data on student academic achievement, such as GPA scores, PSAT scores, and AP exam scores. Lunch programs might collect data to study the impact of a new snack bar on lunch time sales of fruits. Such studies are called **observational studies**, because there is no treatment involved. An observational study does not involve a control group and a treatment group, which differentiates it from an experimental study.

   7. For each of the following studies, distinguish the differences between the experimental and the observational study, and explain why.

     a. The student council surveyed students to see if they approved of the current dress code. Of the students surveyed, 85% approved of the current dress code.

b. A clinical study gave a new pain reliever to 20 arthritis patients and a placebo pill to 20 other arthritis patients. Two hours later the patients were asked to rate their pain relief.

c. A pre-school teacher asked her three-year-olds if they were tired. Then, she logged how long it took them to fall asleep at naptime.

8. What are the similarities and differences between an experiment and an observational study?

9. How is a double-blind experiment different from a single-blind experiment?

SUMMARY: ACTIVITY 6.11

1. To establish a **cause-and-effect relationship** between two variables an **experimental design** must include a **control group**, have **randomization**, and be **replicable**.
A **control group** includes individuals that do not receive the treatment under study. To be valid, assignment of members to the control and treatment groups must be done through a **random** process. An experiment should include as many subjects as possible, and the results should be **replicated** with further experiments with different study groups.

2. Most good experiments are **double-blind**, meaning neither the participants (subjects and administrators) nor the evaluators know which subjects received the treatment and which are in the control group.

3. A **placebo** is a treatment intended to have no effect (a false treatment) given to the control group so all subjects experience the same conditions.

4. Many humans will respond positively to a placebo, even though they are not receiving the treatment. This **placebo effect** could fool the researcher into thinking the treatment caused a positive result unless a placebo is used with the control group.

5. An **observational study** focuses on a characteristic of a population and collects data but does not apply a treatment.

EXERCISES: ACTIVITY 6.11

1. a. Suppose a city passes a gun control law, and two years later the number of violent crimes increases. Can you conclude that the gun control law caused the crime rate to go up? Explain.

b. If the crime rate had decreased, can you conclude that the gun control law caused the crime rate to go down? Explain.

2. A study skills company claims their SAT prep program will increase your overall SAT score by 45 points. Their claim is based upon 2350 students that completed their course and scored on average 45 points higher than their first attempt at the SAT. Do you accept their inferred claim that their program caused the higher scores?

3. In the 1990s, a study was designed to determine if listening to Mozart's music would improve performance on an IQ test. Subjects were randomly assigned to two groups. One group listened to Mozart; the second group was not given any instructions or music. The sample mean scores on the IQ test were 119 for the Mozart group and 110 for the no instructions group. Describe how well this experiment was designed, and whether you think a causal relationship was established. How could the experiment be improved, in your opinion?

4. Over the past 20 years, mathematics educators have been very busy reforming the way algebra is taught. The goal has been to improve the basic conceptual understanding required to successfully apply algebra in other courses and in work. Suppose you wish to determine how well a particular reformed algebra course accomplishes this goal, compared to a traditionally taught algebra course. In other words, does taking a reformed algebra course cause a higher level of learning than the traditional course?

a. Classes using the reformed course materials take a standardized algebra test. Standardized means the test has been given to many other algebra students and the mean and standard deviation are known. If the reformed class scores higher than the mean for this test, would you be able to conclude the reformed course caused the higher scores? Explain.

b. Describe how you could design an experiment to better assess whether the goal is being met.

5. A research doctor has developed a new medication to relieve the joint swelling of arthritis. Twenty patients have volunteered to take part in an experiment to determine if the new medication is more effective than their old medication.

    a. Describe how this experiment should be conducted. Carefully describe how each aspect of a good experiment can be accomplished.

    b. How can the control group and placebo effect be managed so patients still receive some medication?

    c. A family doctor receives samples of the new arthritis medication. The doctor gives 20 of his arthritis patients a 30-day test sample of the new medication and asks each patient to schedule an appointment for one month later. At the follow-up appointment, the doctor asks the patient whether or not they preferred the new arthritis medication over their old medication. How does this study differ from the experiment that you set up in part a?

6. Does filling the gas tank of a car with premium gasoline result in better fuel efficiency than with regular unleaded gasoline? Design an experiment to determine whether premium gasoline causes a car to get better gas mileage.

7. Sometimes a very strong belief can influence how the results of an experiment are interpreted, hence the need for double-blinding. Many popular beliefs and superstitions have not been subject to rigorous testing with good experimental design. Dowsing is a widely held belief that the use of a forked stick (called a dowsing rod) can be used to detect underground water. The dowser holds the two ends of the forked part, leaving the main stem pointing in front. When walking along with the dowsing rod held horizontally in front, if the pointer dips down, that is interpreted to mean underground water has been detected. Many people have employed dowsers over the years to help find the best location to dig a well, for example.

In attempting to prove that hidden water is causing the dowsing rod to dip the following experiment is performed. Twenty buckets are laid out in a field, half are filled with water, the other half are empty. Boards are placed over each bucket, so no one can see inside. The dowser proceeds to walk around, passing the dowsing rod over each bucket. The evaluator, who also was responsible for filling and laying out the buckets, records the reaction of the rod in each case. The dowser goes over all the buckets a second time and reconfirms the choices from the first round.

The dowser has picked out 12 buckets that he claims contain water, and in fact nine of them do. This means the dowser has correctly identified the contents of 16 out of 20, or 80%, of the buckets. A pretty good score, since the expected result by randomly guessing would be 50%. Are you convinced? Explain.

8. Think of something that interests you. Try to find two variables that are related to your interests and to each other. Pose a question about how one variable might cause a change in the other variable. Design an experiment that could test this situation, being sure to address how your experiment would address all aspects of a good design. If possible, conduct the experiment and report on the results.

9. For each of the following studies, determine whether it is an experiment or an observational study, and explain why.

   a. A math teacher was curious about the amount of time students had studied for their last test, so she asked them to write the number of hours that they had studied for their test. She then compared each student's study time to his/her test grade.

   b. The cross-country coach randomly divided his team into two groups. One group wore a new sneaker designed for cross-country running, and the other group wore their old sneakers. The coach then recorded each runner's time in a 5000-meter race and compared it with their previous best time for a 5000-meter race.

   c. A cross-country coach thought that the type of sneaker that students wear might influence their running time. The coach recorded the type of sneaker worn by each student, and then compared each runner's time in a cross-country race with his/her type of sneaker.

   d. Distinguish the differences in the studies in part b and c.

Activities 6.7–6.11   **What Have I Learned?**

I. Does the following histogram represent a distribution that is approximately normal? Explain.

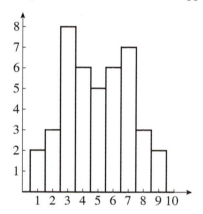

2. Identify the following distributions as skewed to the left, normal, or skewed to the right.

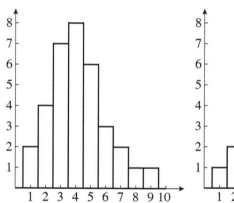 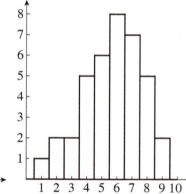 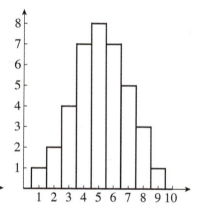

3. A value from a large data set has a corresponding z-score of -2. Is the data value greater than or less than the mean of the data set?

4. For a standard normal distribution:

   a. What are the mean μ and standard deviation σ of the distribution?

   b. Determine the positive z-score that is two standard deviations from the mean on the z-axis.

   c. Determine the percentage of z-scores that are within 1 standard deviation of the mean. That is, the percentage of z-scores between $z = -1$ and $z = 1$?

5. Heights of males are normally distributed. If the heights (x-values) are measured in units of inches, what are the units for the z-scores that correspond to specific heights?

6. A voluntary response sample (or self-selected sample) is one in which individuals decide to be included in the sample. Will such a sampling technique result in a biased example? Explain.

7. You want to determine the average family size of households in Texas. You collect data consisting of family size from students in your high school. Regardless of the sampling technique used, is the sample representative of all households in the state?

8. Determine if the following sampling techniques result in a random sampling. Explain why or why not.

 a. Your teacher obtains a sample by selecting the first six students entering the classroom.

 b. Your math classroom consists of 30 students in six different rows. Your teacher rolls a die to determine a row, then rolls the die again to select a particular student in the row. This process is repeated until a sample of six students is obtained.

9. Your spring project in your biology class is to determine if a new type of fertilizer for a specific type of tomato plant is better than the fertilizer used in past years. Design an experiment that would test the effectiveness of the new fertilizer. Assume that your class is given 40 tomato seedlings grown under identical conditions.

10. For each of the following studies, determine whether it is an experiment or an observational study.

 a. Many taste tests have been done over the years to compare Coke and Pepsi products. The total number of sales of Coke drinks and Pepsi drinks at the last home football game were compared to see which brand was more popular.

 b. A study involved 20 women and 20 men recruited as pairs. In a questionnaire, 90% of the participants stated that they texted at least three times daily. A driving simulator was used for a 32-mile drive of freeway and multilane rural road. Each participant used his or her own cell phone for text messaging.

 Participants were tested on just their driving ability, and then they were also tested on their driving ability while texting. The order of participation varied among the participants. Each driver followed a pace car that randomly braked 42 times during the drive. The times required for the driver to depress the brake pedal were recorded. Data sourced from "Text Messaging During Simulated Driving" by Frank A. Drews et al., University of Utah, Salt Lake City, first published in *Human Factors: The Journal of the Human Factors and Ergonomics Society Online*, December 16, 2009.

11. You have heard about a new, organically formulated energy drink (Organo-Erg). Their webpage says that traditional energy drinks ramp up your body's nervous system with excessive amounts of sugar and artificial caffeine. You get a big boost of energy, but it's short-lived and ends in a crash after your body has used up the ingredients in the drink. Organo-Erg contains only natural caffeine and less harmful sugars, such as fructose. These don't burn up rapidly in the body, so they will increase your energy more gently and for a longer time.

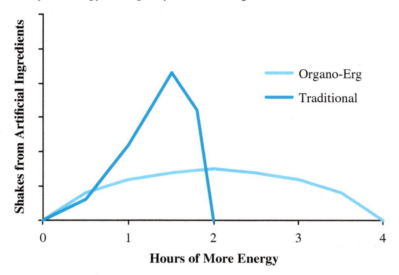

The Web site's graph contains two segmented lines: one for the advertised Organo-Erg and another for traditional energy drinks.

a. Analyze the horizontal axis and vertical axis of this graph.

b. What does the graph suggest about the advertiser's product (Organo-Erg)?

c. How reliable would you consider this graph to be? Explain.

Activities 6.7–6.11   How Can I Practice?

1. The heights of 18-year-old females in the United States are normally distributed with a mean of $\mu = 63.5$ inches and a standard deviation of $\sigma = 2.5$ inches. Determine the z-score for each of the following heights of 18-year-old females.

   a. 61 inches

   b. 66 inches

   c. 63.5 inches

   d. 56 inches

2. The last chapter test in your math class had a mean of $\mu = 72$ and a standard deviation of $\sigma = 5$.

   The last test in your science class had a mean of $\mu = 75$ and a standard deviation of $\sigma = 7$. Suppose you scored an 80 on both tests.

   a. Compute the z-score for each test.

   b. Compared to the rest of your class, which score was the better score?

3. A high school's student council has decided to conduct a survey on extracurricular sports. Assume that the high school's student body consists of 960 students and that the student council has decided to choose 100 as its minimum sample size. Identify the type of sampling (simple random sample, stratified, cluster, systematic, convenience) described by the following proposals.

   a. The three-digit numbers 001 through 960 are assigned to a list of student names. One hundred three-digit numbers are randomly generated.

   b. Twenty-five students are randomly selected from the freshman, sophomore, junior, and senior classes.

    c. Five homerooms are randomly selected, and all students from each homeroom are given the survey.

    d. Student council members survey the first 100 students entering the cafeteria for lunch.

    e. Fifty girls and fifty boys are randomly selected.

4. A marketing analyst conducts a taste study on a brand of protein bars at a local whole foods club. Shoppers are offered a sample and asked to rate the bar's flavor. Is this study biased? Explain.

5. For each of the following studies, determine whether it is an experiment or an observational study.

    a. During an election year, a pollster randomly phoned voters in his district to see if they preferred the Republican candidate or the Democratic candidate.

    b. A student studies the weights of babies born at a local hospital for the past year.

    c. A local veterinarian tried several drugs before controlling a dog's allergies.

    d. A clinic gives a new blood pressure drug to a group of 20 patients and a placebo to another group of 20 patients to find out if the drug lowers a person's blood pressure.

Chapter 6   Summary

The bracketed numbers following each concept indicate the activity in which the concept is discussed.

CONCEPT/SKILL	DESCRIPTION	EXAMPLE
Line graph [6.1]	When data points are plotted and connected with line segments to better show a trend.	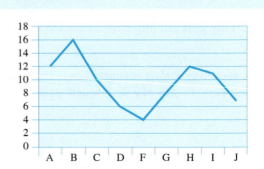
Pictograph [6.1]	Uses a single symbol or picture to display data. The size of the symbol or picture indicates the quantity being measured.	 Figure 2. Purchasing power of the Canadian dollar, 1980 to 2000 1980 = $1.00 1985 = $0.70 1990 = $0.56 1995 = $0.50 2000 = $0.46 Courtesy of Statistics Canada, 2001
Bar graph [6.1], [6.2], [6.3]	A plot of data where the number of data values falling in a category is represented by the height or length of a rectangle, or bar.	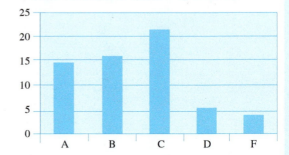
Stacked bar graph [6.2]	When paired data separates a bar into two parts, to show relative sizes for each category.	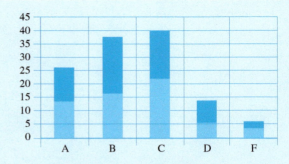

860

CONCEPT/SKILL	DESCRIPTION	EXAMPLE
Circle graph [6.3]	Displays the relative number of data values in each category by the size of the corresponding sector, or slice.	
Frequency table [6.4]	A table that displays the number of times (the frequency) a data value occurs.	
Dotplot [6.4]	The frequency is measured on the vertical scale, one dot for each data value.	
Histogram [6.4]	Displays the frequency of each category of data, with each rectangle's height giving the frequency.	
Stem-and-Leaf plot [6.4]	Organizes data by splitting each data value into two parts called the stem and the leaf. The stem is the digit or group of digits with the greatest place value. The remaining digits on the right of the stem form the leaf for the data value.	Stem-and-Leaf Plot for {10, 15, 18, 21, 23, 27, 29, 32, 33, 35} where the tens digit is the stem and the ones digit is the leaf: 1 | 0 5 8 2 | 1 3 7 9 3 | 2 3 5
Frequency distribution [6.4]	Any collection of data that has been organized, according to how frequently each data value occurs, in a table or graph.	
Measures of the central tendency of a frequency distribution [6.5]	An indication, or measurement, of the approximate location of the center or middle of a collection of data.	Mean, median, mode, and midrange are all different ways of measuring the center.
Mean [6.5]	The usual average found by adding the data values and dividing by the number of values.	The mean of $\{1, 3, 4, 8, 10\}$ is $(1 + 3 + 4 + 8 + 10)/5 = 5.2$.

CONCEPT/SKILL	DESCRIPTION	EXAMPLE
Median [6.5]	The exact middle of the numerically ordered data, half the data values lie above and half lie below the median. (If there is an even number of data values, the median is the mean of the two middle values.)	The median of $\{1, 3, 4, 8, 10\}$ is 4.
Mode [6.5]	The data value that occurs most frequently.	The mode of $\{1, 3, 3, 3, 5, 7, 7\}$ is 3.
Midrange [6.5]	The exact midpoint between the lowest and highest data values.	The midrange of $\{1, 3, 4, 8, 10\}$ is $(1 + 10)/2 = 5.5$.
Variability of a frequency distribution [6.5]	Refers to how spread out the data is, away from the center.	$\{1, 4, 7, 13, 25, 34\}$ has greater variability than $\{3, 3, 3, 4, 4, 5\}$.
Range of a frequency distribution [6.6]	The difference between the minimum and maximum data values.	The range of $\{1, 3, 4, 8, 10\}$ is $10 - 1 = 9$.
Deviation of a data value [6.6]	Refers to how far a data value is from the mean.	In the data set $\{1, 3, 4, 8, 10\}$, the deviation of 10 is $10 - 5.2 = 4.8$.
Sample standard deviation [6.6]	A measure of how far all the data collectively is from the mean. A standard deviation of zero means all the data is the same number (the mean), there is no variation.	$s = \sqrt{\dfrac{\sum (x - \bar{x})^2}{n - 1}}$ For $\{1, 3, 4, 8, 10\}$, $s \approx 3.70$.
| **Interquartile Range (IQR)** [6.6] | The interquartile range (IQR) is the spread or variability of the middle 50% of the data and equals $Q_3 - Q_1$. | ```
1-Var Stats
↑n=20
 minX=18
 Q1=19
 Med=21
 Q3=26
 maxX=35
```<br>$IQR = Q_3 - Q_1 = 26 - 19 = 7$ |
| **Five-number summary for a distribution** [6.6] | The minimum, first quartile, median, third quartile, and maximum data value, for a distribution. | ```
1-Var Stats
↑n=20
 minX=18
 Q1=19
 Med=21
 Q3=26
 maxX=35
``` |
| **Boxplot** [6.6] | A graph that displays the five-number summary for a distribution. | |
| **Standard normal curve (distribution)** [6.7] | A bell-shaped curve representing a normal distribution with $\mu = 0$ and $\sigma = 1$. | |

CONCEPT/SKILL	DESCRIPTION	EXAMPLE
Standard (z) scores [6.7]	The number of standard deviations a data value is from the mean of normal distribution.	$z = \dfrac{x - \mu}{\sigma}$, where x = data value, μ = mean, σ = standard deviation
Census [6.8]	The result of measuring or counting every individual in the population.	
Sample [6.6], [6.8]	The result of measuring or counting any portion of the population.	
Statistic [6.8]	A measurement derived from a sample.	
Simple random sample (SRS) [6.8]	A sample that is selected where every possible sample of a given size has an equal chance of being selected.	
Bias in a sample [6.8], [6.9]	When some individuals in a population have a greater chance of being selected for a sample than others.	
Sampling plan [6.8], [6.9], [6.10]	A detailed step-by-step procedure to collect data through a specific sampling method.	
Stratified sample [6.9]	When a population is separated into two or more strata, with a SRS selected from each, proportional to the relative size of each stratum.	
Self-selected sample [6.9]	A sample composed of individuals who voluntarily respond.	
Experimental design [6.11]	To establish a cause-and-effect relationship between two variables through control, randomization, and replication.	
Control group [6.11]	The individuals in an experiment who do not receive the treatment.	
Double-blind [6.11]	Desirable in a good experiment, when all individuals involved in the study, subjects, administrators and evaluators, have no knowledge of which subjects are receiving the treatment.	
Observational study [6.11]	A study that is focused on a characteristic of the population and collects data but does not apply a treatment.	
Placebo effect [6.11]	When a human responds positively to a false treatment (normally in the control group).	

I. The histogram below shows the sizes of incorporated cities in California, in square miles. Approximately what percentage of these cities is larger than 40 square miles? (Show your calculation.)

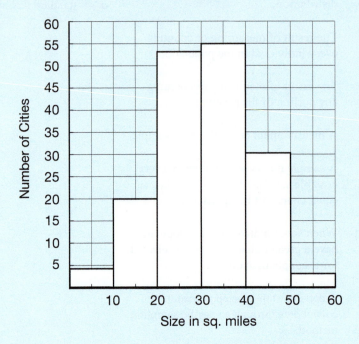

2. The weights, in pounds, of a group of students are as follows:

173 123 171 175 188 120 177 160 151 169
162 128 145 140 158 132 202 162 154 180
164 166 157 171 175

Determine the mean, standard deviation, and five-number summary for these data. Decide on a scale and draw a box and whisker plot. Check your answers with your grapher.

3. a. Determine the mean, median, mode, and midrange for this collection of class test scores:

88 82 97 76 79 92 65 84 79
90 75 82 78 77 93 88 95 73
69 89 93 78 60 95 88 72 80
94 88 74

b. From these measures of central tendency, would you guess that the distribution is symmetrical, skewed to the left, or skewed to the right? Check your answer by displaying the histogram and/or boxplot on your grapher.

4. The ages of 10 randomly selected NASCAR drivers are listed.
33, 48, 41, 29, 40, 48, 44, 42, 49, 28

    a. Determine the mean age of the 10 drivers.

    b. Determine the median age.

    c. Determine the mode, if it exists.

    d. Determine the midrange of the ages of the 10 drivers.

    e. Determine the standard deviation of the ages of the 10 NASCAR drivers.

    f. Verify your results in parts a–d using a graphing calculator.

5. A company is receiving complaints from customers about the amount of time waiting on hold when attempting to contact customer service. A sample of 40 customers are randomly selected and the amount of time on hold when calling customer service is recorded (in minutes).

0.5	4.6	5.5	6.3	6.8	7.8	8.9	10.1
1.6	4.6	5.6	6.3	6.9	8.0	9.2	10.5
3.2	4.9	6.0	6.4	7.0	8.3	9.3	10.6
3.7	5.1	6.0	6.5	7.2	8.6	9.4	11.0
4.3	5.3	6.2	6.5	7.5	8.6	9.7	11.3

    a. Using 0–1.9 as the first class width, construct a histogram of the data.

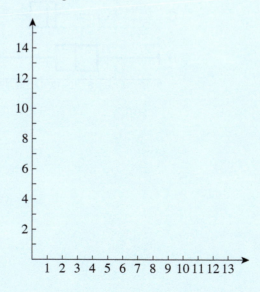

b. Describe the shape of the distribution.

c. Determine the mean of the sample.

d. Determine the standard deviation.

6. If the mean of a distribution is greater than the median, predict the shape of the distribution.

7. Use complete sentences to describe each distribution in terms of shape and any unusual features. Estimate the median and range for each.

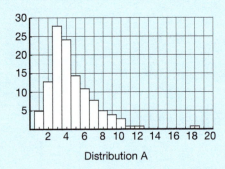

Distribution A

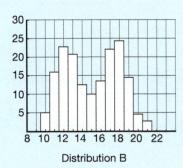

Distribution B

8. Match each histogram with its corresponding boxplot. Label each histogram A, B, C, or D.

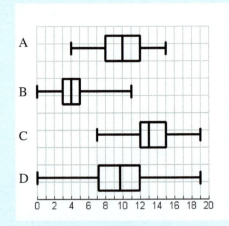

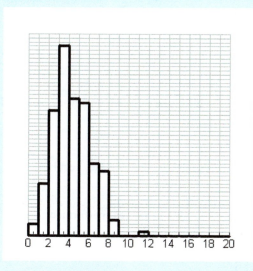

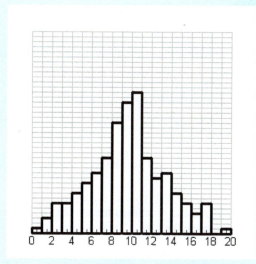

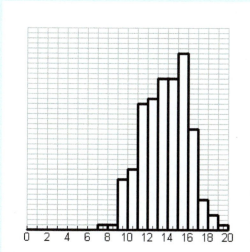

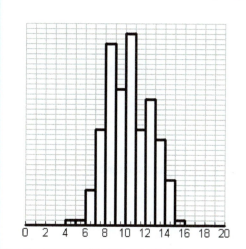

9. In which of the following cases do you think it is possible to select a simple random sample of size 50? Give a reason for your answer.

    a. All fishing boats currently in the Gulf of Mexico.

    b. All registered Republicans in the city of Dallas.

10. A convenience sample uses data that is easily or readily available. However, this type of sampling can be extremely biased. For example, suppose the city council wants to raise taxes in order to build a new wing on the high school. In order to obtain the opinion of the city residents on the issue, the first 50 adults leaving the local shopping mall are asked if they are in favor of raising taxes to build the new wing. Discuss the possible cause of a bias in the sample.

11. Some examples of sampling are described below. State the kind of sampling method used: cluster, stratified, systematic, or simple random sample.

    a. 200 rural and 200 urban persons of age 65 or older were asked about their health and experience with prescription drugs.

    b. After a hurricane, a disaster area is divided into 200 equal grids. Thirty of the grids are selected, and every occupied household in the grid is interviewed to help focus relief efforts.

    c. Chosen at random, 1819 hospital outpatients were contacted and asked their opinion of the care they received.

12. A study in Switzerland examined the number of caesarean sections (surgical deliveries of babies) performed in a year by doctors. Here are the data for 15 male doctors:

27	50	33	25	86
25	85	31	37	44
20	36	59	34	28

The study also looked at 10 female doctors. The numbers of caesareans performed by these doctors were:

5	7	10	14	18
19	25	29	31	33

    a. Calculate the mean and standard deviation for the male doctors.

    b. Find the median and the IQR for the male doctors.

    c. Calculate the mean and standard deviation for the female doctors.

    d. Find the median and the IQR for the female doctors.

    e. Make two boxplots using the same scale.

```
1-Var Stats
↑n=15
 minX=20
 Q₁=27
 Med=34
 Q₃=50
 maxX=86
■
```

```
1-Var Stats
↑n=10
 minX=5
 Q₁=10
 Med=18.5
 Q₃=29
 maxX=33
■
```

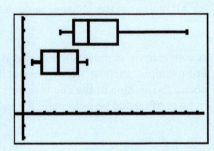

f. Would it be more appropriate to use the mean and standard deviation or the median and IQR when comparing the center and spread for the two groups? Explain your answer.

g. In context, what can you conclude from the boxplots?

13. What is the **standard** normal distribution?

14. Two people are on a reducing diet. The first belongs to an age group for which the mean weight is 146 pounds with a standard deviation of 14 pounds. The second belongs to an age group for which the mean weight is 160 pounds with a standard deviation of 17 pounds. If their respective weights are 178 and 193 pounds, which of the two is more overweight for their age group? Use z-scores, and explain your answer.

15. For each of the studies below, indicate whether it is an experiment, an observational study, or neither.

a. A research team studies a new blood pressure medication using two similar groups of 500 people. One group receives the new drug, and the other group receives an older prescription. The pills are identically marked, shaped, and colored. The progress of the patients is recorded and analyzed.

b. A pharmaceutical company records data on side effects received from patients taking a new drug that they have brought to market. The data are analyzed and possible causes of the side effects investigated.

c. A car dealership hires a new advertising agency for their new sales campaign. They also open at a new location and have a party.

d. Ornithologists record the bald eagle population in each state in a series of annual surveys.

e. Physicists at a university publish a study indicating that hydrogen fusion can occur at low temperatures when catalyzed by palladium. Their study varied (in a controlled manner) the temperatures, hydrogen concentrations, and amount of catalyst used. Researchers at other universities repeat the study unsuccessfully, and the idea is discredited.

16. While waiting for a doctor's appointment, you spot a magazine ad for a new over-the-counter pain reliever. The statistics are mostly in a table that compares the new product with aspirin in the relief of "certain types of pain" to quote the ad. The aspirin results are "from a previous survey." The ad text notes the improved performance in "excellent" pain relief.

	TEST GROUP SIZE	EXCELLENT	GOOD	FAIR	POOR
Test 1	45	47%	46%	6%	1%
Test 2	413	22%	50%	20%	8%
Test 3	346	24%	55%	12%	9%
Test 4	136	27%	49%	10%	14%
Average		30%	50%	12%	8%
Aspirin		26%	55%	11%	8%

a. Find the number of people saying the pain relief was excellent in the four groups.

b. Across all four groups, how many people gave an excellent rating? What percentage is this? How does this compare with the number in the table?

c. Based on the statistics in the ad and your response to part b, analyze the marketing claim. Does the table support the marketing claim? Explain.

17. The following online advertisement presents some statistics resulting from a study of Camdyl, a hair care product.

Herbal Complex
CAMDYL
Clinically Tested*

97%   More than 97% of subjects reported their hair became thicker and fuller when using shampoo and serum combined.

83%   More than 83% of subjects reported their hair became thicker and fuller when using only CAMDYL Herbal Shampoo.

*Results of 6 month, placebo controlled, independent clinical study of

CAMDYL BOTANICALS Shampoo and Serum for Thin Hair

Analyze the claim made in the advertisement. Write a response about whether the claim is justified by the information presented in the ad. What factors make this a good or poor experimental design?

Chapter 7

Problem Solving with Probability Models

Activity 7.1

Chances Are!

Objectives

1. Determine relative frequencies for a collection of data.

2. Determine both theoretical and experimental probabilities.

3. Simulate an experiment and observe the law of large numbers.

4. Identify and understand the properties of probabilities.

The study of probability began with mathematical problems arising from games of chance. In 1560, Italian Gerolamo Cardano wrote a book about games of chance. This book is considered to be the first written on probability. Two French mathematicians, Blaise Pascal and Pierre de Fermat, are credited by many historians with the founding of probability theory. They exchanged ideas on probability theory in games of chance and worked together on the geometry of the die.

Although probability is most often associated with games of chance, probability is used today in a wide range of areas, including insurance, opinion polls, elections, genetics, weather forecasting, medicine, and industrial quality control.

Relative Frequency

The following table gives the years of service of the 789 workers at High Tech Manufacturing.

SENIORITY	LESS THAN ONE YEAR	BETWEEN ONE AND TEN YEARS	MORE THAN TEN YEARS	TOTAL
Male	82	361	47	
Female	49	202	48	
Total				

1. **a.** Complete the table to display the total number of workers in each category.

   b. What is the total number of males?

   c. What is the total number of workers having between one and ten years of service?

   d. Should the sum of the right-most column be the same as the sum of the bottom row?

Each of the numbers in the above table can be considered a frequency for that category. If you consider this frequency a fractional part of the whole workforce, it becomes a **relative frequency**. For example, the relative frequency of male workers with less than one year of experience is $\frac{82}{789} \approx 0.104 = 10.4\%$.

2. Calculate the relative frequency, expressed as a decimal, for each category. Record your answer in the following table.

SENIORITY	LESS THAN ONE YEAR	BETWEEN ONE AND TEN YEARS	MORE THAN TEN YEARS	TOTAL
Male	0.104			
Female				
Total				

Now consider choosing one of the 789 workers by some random process, like drawing names out of a hat. The **probability** of picking a male worker with less than one year of experience is 0.104, the relative frequency for that particular outcome. Choosing a male worker with less than one year experience is called an **event**.

3. a. What is the probability of choosing a female with more than ten years of experience?

b. What is the probability of choosing a male?

c. What is the probability of choosing a worker with less than one year of experience?

Definition

$$\text{Probability of event} = \frac{\text{number of outcomes for which the event is true}}{\text{total number of equally likely possible outcomes}}$$

For example, the probability of choosing a male is $\frac{490}{789} \approx 0.621$. The event is "choosing a male." Of the 789 possible outcomes, 490 of the outcomes make the event true.

If x represents a particular outcome, then $P(x)$ represents the probability of that outcome occurring. Therefore, the probability of choosing a male can be written as

$$P(\text{choosing a male}) \approx 0.621$$

One important assumption is necessary for this definition to hold true. All the possible outcomes must be equally likely to occur. Selecting one worker at **random** implies that each worker has the same likelihood, or probability, of being chosen.

4. a. If your uncle is one of the 789 workers, what is the probability that he will be chosen?

b. What is the probability that the person selected works at High Tech?

c. What is the probability that your aunt (who does not work at High Tech) will be chosen?

5. Now consider a very random experiment, rolling a typical six-sided die. Each side contains a unique number of dots, from 1 to 6.

  a. What makes this experiment random?

  b. What is the probability of rolling a five?

  c. What is the probability of rolling an even number?

  d. For this experiment, think of an event that has a probability of zero.

  e. Complete this table for the die roll **probabilities**.

OUTCOME	PROBABILITY
1	
2	
3	
4	
5	
6	

  f. Letting $P(x)$ be the probability of rolling x on the die, what does $P(3)$ equal?

  g. What is the sum of all the possible probabilities for the die roll? Explain why this makes sense.

The collection of all possible single outcomes for an experiment is called a **sample space**. All the possible probabilities make up a **probability distribution**.

6. Consider another random experiment, flipping a coin.

   a. To be a random process, you must use a fair coin. What do you think is meant by a fair coin?

   b. What is the probability of getting heads on a single flip of the coin?

   c. What is the sample space for this experiment?

The probability definition you used to answer the questions in Problems 5 and 6 is sometimes called the **theoretical probability** of an event. Theoretical probability is determined through a study of all possible outcomes that can occur.

Experimental Probability

Sometimes theoretical probabilities are not possible to calculate. For example, suppose you want to estimate the probability of rain today or to predict the likelihood of living past the age of 80. There is another way to define probability that depends on experimental observation.

7. Suppose the weather forecaster claims that there is a 60% chance of rain tonight. How do you think such a probability could be calculated?

8. a. If you flip a coin ten times, how many heads would you expect to get?

   b. Would you be surprised to get ten heads in your ten flips?

   c. How about nine heads or only one head?

   d. Have each member of the class flip a coin ten times. Summarize the results for the entire class in the following frequency table.

NUMBER OF HEADS	FREQUENCY
0	
1	
2	
3	
4	
5	
6	
7	
8	
9	
10	

e. Does the frequency distribution confirm your answers in parts a–c?

f. You know that the theoretical probability of getting a head on a single coin flip is $\frac{1}{2}$.

For this to be true, what assumption are you making?

9. When you flip a coin, you should expect to get heads approximately one-half the time. Determine the relative frequency of getting a head for the class experiment in Problem 8d. Would you be surprised if the relative frequency was not close to one-half?

Based on the results of the experiment, you could use the corresponding relative frequency to define the probability of getting a head, especially if you did not know the theoretical probability. This probability would be an **experimental probability**.

10. Repeat the coin-tossing experiment as a class, with each member of the class tossing the coin 20 times and recording the number of heads observed. How does the resulting relative frequency compare with the theoretical probability?

Law of Large Numbers

If more is better, you could be flipping coins for a long time to get a truly large number of observations. You can accomplish this by employing technology. Your calculator or a computer can be used to **simulate** the flipping of a coin a very large number of times. This can be done by randomly generating a large number of zeros and ones (ones representing heads and zeros representing tails). The probability of getting a one must be 50%. (See Appendix A for TI-83/84 Plus instructions.)

Appendix

11. a. Use technology to simulate flipping a coin 100 times and then 500 times. Record your results in the following table. Then combine all your classmates' results to get as many observations as you can.

NUMBER OF FLIPS	NUMBER OF HEADS	RELATIVE FREQUENCY OF HEADS
100		
500		

b. What conclusion can you make about the relative frequency as the number of observations increases?

In Problems 9–11, as the number of coin flips increased, the experimental probability of obtaining a head got closer to the theoretical probability of getting a head. This is usually called the **law of large numbers**.

SUMMARY: ACTIVITY 7.1

1. An **experiment** is a controlled procedure that gives a set of recordable results.

2. The possible results of an experiment are called **outcomes**.

3. An **event** is a single outcome or collection of outcomes of the experiment.

4. The **sample space** of an experiment is the collection of all possible single outcomes of the experiment.

5. Randomly selecting an object from a collection means that all objects in the collection have an equal chance of being selected.

6. The probability of a particular event (or combination of events) measures the likelihood of that event occurring.

7. Theoretical Probability

$$\text{Probability of event} = \frac{\text{number of outcomes in event}}{\text{total number of equally likely possible outcomes}}$$

8. Experimental Probability

$$\text{Probability of event} = \frac{\text{number of observed occurrences in event}}{\text{total number of observations}}$$

9. Law of Large Numbers

The larger the number of observations, the closer the experimental probability of an event will get to the theoretical probability of that same event.

10. Properties of Probabilities

If $P(x)$ is the probability that outcome x occurs, then the following special properties of probabilities must be true.

- $0 \leq P(x) \leq 1$ for all x in the sample space.
- If x is impossible, then $P(x) = 0$.
- If x is a certainty, then $P(x) = 1$.
- The sum of the probabilities of all x in the sample space must equal 1.

EXERCISES: ACTIVITY 7.1

1. Some teens belong to the 4-H club. A random survey of 100 teens in your community yields the following results.

	4-H MEMBER	NON 4-H MEMBER	TOTAL
Female	24	32	
Male	24	20	
Total			

a. Complete the totals in the table.

b. Calculate the relative frequency for each category.

	4-H MEMBER	NON 4-H MEMBER	TOTAL
Female			
Male			
Total			

c. Using these relative frequencies as the basis for defining probabilities, what is the probability that a teen in your community is a 4-H member (4-Her)?

d. If you choose one teen at random from your community, what is the probability that he or she is a female non 4-H member?

e. Given the fact that you have already chosen a female at random, what is the probability that she is a 4-H member?

f. Given the fact that you have already chosen a male at random, what is the probability that he is a 4-H member?

g. From your results in parts e and f, is it more likely that a randomly selected male teen is a 4-Her or that a randomly selected female teen is a 4-Her? Explain.

2. Twenty uniformly shaped stones are placed in a vase. Five of the stones are white, eight of the stones are gray, and the remaining stones are black. One stone is drawn at random from the vase. The outcome of this experiment is to note the color of the selected stone.

a. What is the sample space for this experiment?

b. Make a table showing the probability of each individual outcome.

x	white	gray	black
$P(x)$			

c. Determine each of the following: P(white), P(gray), P(black), and P(red).

3. If $P(x)$ is the probability of living to age x, interpret the statement $P(80) = 0.16$.

4. State whether you think each of the following is a theoretical or experimental probability.

   a. The chance of rain tonight is 35%.

   b. The probability of living past 100 years of age is 0.03.

   c. The probability of getting two tails on two flips of a coin is 0.25.

   d. The probability that the Sun will come up tomorrow is 1.

   e. The probability of selecting a green stone from the vase of Exercise 2 is zero.

5. You are blindfolded and then throw a dart at a square dartboard, as illustrated. Assuming that the dart does hit the board, what is the probability that you will hit region II?

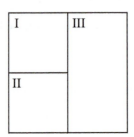

6. If you and your friend separately choose a number between 1 and 10 at random, what is the probability that you will choose the same number?

7. Which of the following tables could be a probability distribution? For those that cannot be a probability distribution, give a reason.

a.

x	1	2	3	4	5
$P(x)$	0.2	0.2	0.2	0.2	0.2

b.

x	3	4	5	6	7	8	9	10	11	12
P(x)	0.1	0.15	0.2	0.25	0.12	0.1	0.05	0.02	0.01	0.01

c.

x	blue	green	red	white	black	gray
P(x)	0.3	0.4	0.2	−0.3	0.2	0.2

8. The spinner pictured at the right is used in a board game.

a. Assuming that it is a fair spinner, construct a table for the probability distribution where x represents the set of individual outcomes.

x	1	2	3	4
P(x)				

b. What is the probability of getting a number greater than 1?

c. What is the probability of getting an odd number?

9. In your class of 28 students, there are 12 girls. Of these 12 girls, only 3 are under 16 years old. There are 13 students in class that are 16 years old or older. This information is summarized in the following table. Complete the table.

	UNDER 16	16 OR OVER	TOTAL
Girls	3		12
Boys			
Total		13	28

A student is selected at random.

a. What is the probability that the student is a boy?

b. What is the probability that the student is a girl who is also at least 16 years old?

c. What is the probability that the student is under 16 years old?

10. You flip a coin 500 times and get 500 heads. What conclusion would you make about this coin?

11. A standard deck of 52 playing cards consists of four suits: hearts, clubs, diamonds, and spades. Hearts and diamonds are red cards; clubs and spades are black cards. Each suit has 13 cards, including numbered cards ace (1) through 10 and three picture (or face) cards: the jack, the queen, and the king. There are 12 picture cards, consisting of 4 jacks, 4 queens, and 4 kings.

One card is randomly selected from the deck of cards. Determine the probability that the card selected is

a. a 10 of hearts.

b. a 10.

c. a club.

d. a red card.

e. a picture card.

f. a card greater than 4 and less than 7.

g. a diamond and a club.

12. In an election for senior class president at your high school, 45 seniors were polled and asked for whom they planned to vote. The results of the poll are given in the following table.

	VOTES
Candidate A	18
Candidate B	12
Candidate C	15

a. If one student who was polled is randomly selected, what is the probability that the student planned to vote for Candidate A?

b. According to the poll, estimate the probability that a senior not polled will vote for Candidate A in the election. Is this an experimental probability or a theoretical probability? Explain.

13. A pharmaceutical company is testing a new drug to help people lose weight. The drug is given to 300 adults. The results of the test are as follows:

WEIGHT	NUMBER OF ADULTS
Reduced	190
Unchanged	60
Increased	50

a. If the drug was given to an adult **not** in the testing program, estimate the probability that the person will lose weight.

b. Is the probability in part a experimental or theoretical?

14. Most individuals have seen the "taste test" between Coke® and Pepsi®.

a. Design an experiment to determine the experimental probability that a person randomly selected will choose Pepsi® over Coke® when given samples of both Pepsi® and Coke®.

b. Perform the experiment in part a. Construct a table showing the results of the experiment.

c. Determine the experimental probability that a randomly selected person will prefer Pepsi® over Coke®.

15. Insurance companies use experimental probabilities in determining insurance premiums. Use a variety of references, such as the Internet or your family insurance agent, to gather information on this topic. Write a report on your findings. Include any graphs, numerical information, or formulas that can help demonstrate the use of experimental probabilities in determining insurance premiums.

16. Certain physical traits are not evenly distributed in the general population. For example, there are more right-handed than left-handed people. Conduct a survey of persons chosen at random at your high school. Ask at least 50 students the following questions:

1. Are you right or left-handed?

2. What color is your hair?

3. What color are your eyes?

Complete the following table by filling in all of the data collected.

HANDEDNESS	NUMBER OF PERSONS
Right-handed	
Left-handed	

EYE COLOR	NUMBER OF PERSONS
Brown	
Blue	
Green	
Other	

HAIR COLOR	NUMBER OF PERSONS
Brown	
Blond	
Black	
Red	

a. What is the total number of students in your survey?

Using the survey results, estimate the probability that a randomly selected student in your high school

b. is left-handed.

c. has blue eyes.

d. has brown hair.

Suppose you are looking for a new car and have narrowed your decision down to a Mustang, but can't decide on the exact color, transmission, engine, or options package. There are three sizes of engine (3.0 liters, 3.8 liters, and 4.6 liters), two transmissions (standard and automatic), five colors you like (black, silver, red, yellow, and green), and three option packages (GL, Sport, and XL). With all these possible choices, you want to know how many different Mustangs there are from which you must choose.

1. From the choices given, how many different Mustangs are possible? (Solve this any way you can. A diagram or list may help.)

In the Mustang problem, the outcomes you wish to count consist of a sequence of choices. In such a case, you could have used three different methods to solve the problem. Example 1 demonstrates these methods.

Activity 7.2

Choices

Objectives

1. Apply the Fundamental Counting Principle.

2. Determine the sample space for a random experiment.

3. Display a sample space with a tree diagram or a table.

4. Determine complementary probabilities.

5. Use Venn diagrams to illustrate relationships between events.

Example 1   *Suppose you have three sweaters: one cotton, one wool, and one alpaca. You also have four hats, colored red, green, black, and purple. If you want to wear one sweater and one hat, how many different combinations are possible?*

SOLUTION

Option 1: You could simply list all the possibilities:

cotton + red	wool + red	alpaca + red
cotton + green	wool + green	alpaca + green
cotton + black	wool + black	alpaca + black
cotton + purple	wool + purple	alpaca + purple

You could shorten the process with a **tree diagram** (and abbreviations), where each level of branching represents the next choice.

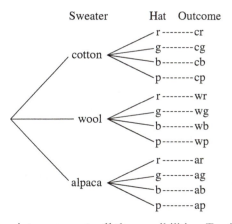

The total number of end points represents all the possibilities. Tracing each branch displays all the possible outcomes.

Option 2: You could use a table approach if there are only two levels of choices.

	RED	GREEN	BLACK	PURPLE
Cotton	cr	cg	cb	cp
Wool	wr	wg	wb	wp
Alpaca	ar	ag	ab	ap

Option 3: You could multiply the 3 choices for a sweater by the 4 choices for a hat to get a total of 12 combinations. This is called the **Fundamental Counting Principle**.

> **The Fundamental Counting Principle** If some choice can be made in M ways and a subsequent choice can be made in N ways, then there are M times N ways these choices can be made in succession.

2. Apply the fundamental counting principle to the Mustang problem to verify your answer in Problem 1.

3. **a.** You want to create an ID code for all your customers based on three characters. The first character must be a letter of the alphabet, and the second and third must each be a digit between 1 and 9, inclusive. How many such codes are there? (Use the fundamental counting principle.)

   b. Would a list or tree diagram be practical in this case?

4. **a.** Suppose you flip a penny and a dime. Determine the number of outcomes.

   b. Use a tree diagram to display all possible outcomes.

   c. Assume that each single outcome is equally likely. Use the tree diagram to help calculate probabilities. What is the probability of getting two heads?

   d. What is the probability of getting one head and one tail, on either coin?

e. If the variable x represents the number of heads, complete the table for this **probability distribution**.

x	P(x)
0	
1	
2	

f. What must be true about the sum of the probabilities for this (and any other) probability distribution?

> The collection of all the possible single outcomes displayed by a tree diagram is the **sample space** of the random experiment.

5. Two fair dice are rolled.

a. Determine the number of possible outcomes.

b. Display the sample space of all possible outcomes. You may use a tree diagram or a table. It may help to think of the dice as being different colors.

c. What is the probability of getting a total of two dots on the dice (snake eyes)? That is, what is the probability of rolling a 1 on each die?

d. What is the probability of rolling a total of 6 on the two dice?

e. If the variable x represents the total number of spots on the two dice, refer to your sample space to complete the table for this probability distribution.

x	2	3	4	5	6	7	8	9	10	11	12
P(x)											

f. What must be the sum of all probabilities?

g. Are you as likely to roll a sum of 2 as you are of rolling a sum of 7? Explain.

6. To win the jackpot in a large state lottery, the winner must pick correctly all six numbers from among 1 through 54. The theoretical probability of doing this is $\dfrac{1}{25,827,165} \approx .0000000387$. Stated another way, a single pick has a 1 in 25,827,165 chance to win the jackpot (and even then it might be shared with other winners). The theory of probability originated out of a desire to understand games of chance. Show how the 25,827,165 is determined.

7. What is the probability of getting ten heads on ten flips of a coin?

8. Two tennis balls are randomly selected from a bag that contains one Penn, one Wilson, and one Dunlop tennis ball.

a. If the first tennis ball picked is replaced before the second tennis ball is selected, determine the number of outcomes in the sample space.

b. Construct a tree diagram and list the outcomes in the sample space.

c. Determine the probability that a Penn followed by a Wilson tennis ball is the outcome.

d. Suppose the first tennis ball picked is not replaced before the second tennis ball is selected. Determine the number of outcomes in the sample space.

e. Construct a tree diagram and list the outcomes in the sample space.

f. Determine the probability that a Penn followed by a Wilson tennis ball is the outcome.

Complementary Events

In some situations, you are interested in determining the probability that an event A does **not** happen. The event "not A," denoted by $\overline{A}$, is called the **complement** of A. For example,

- The complement of success is failure.
- The complement of "a selected voter is a Democrat" is "the selected voter is not a Democrat."
- The complement of "a 6 is rolled" is "a 6 is not rolled."

Since the sum of the probabilities for all outcomes of an experiment is 1, it follows that

$$P(A) + P(\overline{A}) = 1.$$

Rewriting, you have

$$P(A) = 1 - P(\overline{A}) \text{ or } P(\overline{A}) = 1 - P(A).$$

9. a. A die is rolled. What is the probability that the number 6 does not show?

b. A card is randomly selected from a standard deck of 52 cards. What is the probability that the card is not a king?

Complementary probabilities are very useful when trying to determine the probability of "at least one." The statement that "an event happens at least once" is equivalent to "the event happens one or more times." Therefore,

$$P\left(\begin{array}{c}\text{event happens}\\\text{at least once}\end{array}\right) = P\left(\begin{array}{c}\text{event happens}\\\text{one or more times}\end{array}\right). \tag{1}$$

The complement of "an event happens one or more times" is "the event does not happen." Therefore,

$$P\left(\begin{array}{c}\text{event happens}\\\text{one or more times}\end{array}\right) + P(\text{event does not happen}) = 1 \qquad (2)$$

or

$$P\left(\begin{array}{c}\text{event happens}\\\text{one or more times}\end{array}\right) = 1 - P(\text{event does not happen}). \qquad (3)$$

Substituting the results of equation (3) into equation (1), you have the following formula

$$P\left(\begin{array}{c}\text{event happens}\\\text{at least once}\end{array}\right) = 1 - P(\text{event does not happen}).$$

10. a. A fair die is rolled. What is the probability that the number is at least two?

b. Two fair dice are rolled. What is the probability that the sum is at least 3?

Venn Diagrams

Complementary events can be illustrated using Venn diagrams. These diagrams were invented by English mathematician John Venn (1834–1923) and first appeared in a book on symbolic logic in 1881. In a Venn diagram, a rectangle generally represents the sample space. The items (all possible outcomes of a particular experiment) inside the rectangle can be separated into subdivisions, which are generally represented by circles. The items in a circle represent the outcomes in the sample space that make a certain event true.

For example, suppose a die is rolled. The rectangle would contain all the possible outcomes of whole numbers 1 through 6. If A represents the event that the number rolled is greater than 4, then the following Venn diagram represents the relationship between the sample space and the event A.

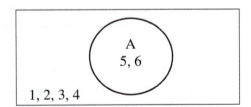

11. a. Determine the complement of event A in the roll-a-die experiment.

b. Describe what portion of the Venn diagram above represents the complement of A.

SUMMARY: ACTIVITY 7.2

1. The **Fundamental Counting Principle** says that if some choice can be made in M ways and a subsequent choice can be made in N ways, then there are M times N ways these choices can be made in succession.

2. A **tree diagram** displays all possible outcomes for a sequence of choices, one outcome for each branch of the tree.

3. The **sample space** of a random experiment is the collection of all possible outcomes.

4. The sum of the probability that an event A will occur and the probability that the event will not occur is 1. The event "not A," denoted by $\overline{A}$, is called the **complement of A**. Stated symbolically,

$$P(A) + P(\overline{A}) = 1 \text{ or } P(A) = 1 - P(\overline{A}).$$

EXERCISES: ACTIVITY 7.2

1. Phone numbers consist of a three-digit area code followed by seven digits. If the area code must have a 0 or 1 for the second digit, and neither the area code nor the seven-digit number can start with 0 or 1, how many different phone numbers are possible?

2. You have four sweaters, five pairs of pants, and three pairs of shoes. How many different combinations can you make, wearing one of each?

3. If you flip a coin ten times, how many different sequences of heads and tails are possible?

4. If you roll a die three times, how many different sequences are possible?

5. You want to order a triple-scoop ice cream cone, with a different flavor for each scoop. If there are 23 flavors available, how many different cones are possible?

6. In a single experiment, a die is tossed and a spinner with the letters A, B, and C is spun. Each letter is equally likely.

   a. Determine the number of possible outcomes.

   b. Determine the resulting sample space.

    c. Determine the probability of getting a 2 on the die *and* a B on the spinner.

    d. Determine the probability of getting a B.

    e. Determine the probability of getting a 2 or a B.

7. A computer is programmed to generate a sequence of three digits, where each digit is either 0 or 1, and each of these is equally likely to occur.

    a. Construct a tree diagram that shows all possible three-digit sequences of 0s and 1s.

    b. What is the probability that a sequence will contain exactly one 0?

    c. What is the probability that a sequence will contain at least one 0?

8. Two teams are playing a best-of-three series. For each game there are only two possible outcomes: One team wins or the other team wins. Construct a tree diagram showing all the possible outcomes for such a series. As soon as one team wins a second game, the series ends (and that branch of the tree ends also).

9. You roll two fair dice. Use the sample space you recorded in Problem 5 of the activity to determine the probabilities of the following events.

a. P(a sum of 7)

b. P(both dice show an even number)

c. P(one die is even, the other odd)

d. P(one die is 1 more than the other)

e. P(the sum is less than 6)

f. P(the product of the two numbers on the dice is even)

10. You have decided to purchase a new computer system. After researching the cost and features of different models of computers, printers, and monitors, you have narrowed your choices to the following:

Computer: Dell, Asus, and Lenovo

Printer: Epson, Hewlett-Packard

Monitor: Samsung, LG

a. Determine the number of possible computer system combinations that you can purchase.

b. Construct a tree diagram and list the possible computer systems that can be purchased.

c. Determine the probability that a Dell computer is part of the system purchased. Assume that all outcomes are equally likely.

d. What is the probability that a Dell computer is not part of the computer system you select?

e. Determine the probability that an Epson printer is part of the computer package purchased.

f. Determine the probability that the system purchased contained a Lenovo computer and Samsung monitor.

11. Two thumbtacks are dropped, one after another, onto a hard surface. Assume that the thumbtack can only land point up or point down. Let

- U represent the event that the thumbtack will land point up.
- D represent the event that the thumbtack will land point down.

a. Are the events U and D equally likely? Explain.

b. Design and conduct an experiment involving dropping thumbtacks. Use the results to determine the experimental probability that a thumbtack will land point up when dropped and the experimental probability that the thumbtack will land point down.

Consider the following two probability experiments.

Experiment 1. You roll two fair dice. What is the probability that a 4 shows on each die?

Experiment 2. You pick two cards from a standard deck of 52 cards without replacing the first card before picking the second. What is the probability that the cards are both red suited?

These experiments are very similar. In each case, you are determining the probability that two events occur **together**. In Experiment 1, both dice must show a 4 for the event to be successful. In Experiment 2, both cards must be red suited for a favorable outcome to occur. The word *and* is used to indicate that both events must occur.

The word *and* is generally interpreted to mean intersection. Venn diagrams can be used to illustrate the overlap or common outcomes between two events. The shaded region in the following Venn diagram represents the outcomes that satisfy both events A and B.

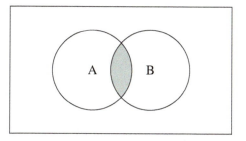

The probability question in Experiments 1 and 2 can be restated as follows:

Experiment 1. Determine P (4 on first die and 4 on second die)

Experiment 2. Determine P (red on first card and red on second card)

Although similar, Experiments 1 and 2 have a very important difference. These differences are investigated in Problems 1 and 2.

1. a. In the dice experiment, what is the probability that a 4 is rolled on the first die?

b. Assuming that a 4 was rolled on the first die, what is the probability that a 4 is rolled on the second die?

In the dice experiment, the outcome of a 4 on the first die has no effect on the probability of a 4 showing on the second die. In such a case, the two events are said to be **independent**.

Definition

Events A and B are independent events if the occurrence of either event does not affect the probability of the occurrence of the other event.

2. a. In the card experiment, what is the probability of a red suited card on the first draw?

b. When calculating the probability of picking a red suited card on the second draw, you must assume that a red card was picked on the first draw. Since the first card is not returned to the deck, there are only 51 cards remaining and only 25 of these are red suited. What is the probability of picking a red card on the second draw?

c. Are the events *a red suited card on the first draw* and *a red suited card on the second draw* independent events? Explain.

Therefore, in the card experiment, the two events are not independent. The outcome of the first draw changes the probability of getting a red card on the second draw. In such a case, the events are said to be **dependent**.

Definition

Events A and B are dependent events if the occurrence of either event affects the probability of the occurrence of the other event.

---

3. You draw two cards from a standard deck of 52 cards. The first card is replaced before the second card is drawn.

a. What is the probability that an ace is drawn on the first draw?

b. What is the probability that an ace is drawn on the second draw?

c. Are the events in parts a and b dependent or independent? Explain.

Calculating Probabilities Involving "*and*" Statements

The probability of an outcome composed of a sequence of independent events can be determined by multiplying the probabilities of the individual events. This is formally called the **multiplication principle of probability**.

Multiplication Principle of Probability

The probability that two independent events both occur is equal to the product of the probabilities of the individual events. Stated symbolically, if A and B are independent events, then

$$P(A \text{ and } B) = P(A) \cdot P(B)$$

When using this formula, you must be sure that A and B are independent events.

You will now use the multiplication principle to answer the probability problem in Experiment 1.

4. Experiment 1. You roll two fair dice.

    a. Use the multiplication formula to determine the probability that a 4 shows on each die.

    b. **i.** How many outcomes are in the sample space for Experiment 1?

        ii. List the outcomes in the sample space for which the condition is satisfied.

        iii. Using the results from parts i and ii, determine the probability that a 4 shows on each die.

    c. Compare the results from parts a and b.

In most real-world applications, it is often too time-consuming or even impossible to solve probability problems by first constructing the sample space. That is one of the reasons why formulas were developed.

As you will discover in Problem 5, the multiplication formula must be adjusted if the consecutive events are dependent.

  5. Experiment 2. You pick two cards from a standard deck of 52 cards without replacing the first card before picking the second. What is the probability that both cards are red?

    Be careful! Recall that the events in this situation are dependent. The probability of picking a red card on the second draw is affected by not replacing the red card picked on the first draw. When the events are dependent, the multiplication formula is written as follows:

$$P(A \text{ and } B) = P(A) \cdot P(B) \text{ becomes}$$
$$P(A \text{ and } B) = P(A) \cdot P(B, \text{ given } A \text{ has occurred})$$

    a. Determine $P(\text{red on first})$.

    b. Determine $P(\text{red on second, given red on first, not replaced})$.

c. Determine P(red on first and red on second) using the modified multiplication rule.

6. Experiment 3. A fair coin is tossed three times. What is the probability that the coin shows heads on all three tosses?

Calculating Probabilities Involving "*or*" Statements

Consider the following experiment:

Experiment 4. Each of the numbers 1, 2, 3, 4, 5, 6, 7, 8, 9 and 10 are written on index cards. The ten cards are then put in a box and one card is randomly selected. Determine the probability that the number is odd or a number less than 5.

7. Referring to Experiment 4, let A represent the number picked is odd and B represent the number picked is less than 5.

a. List the outcomes in the sample space of Experiment 4.

b. List the outcomes that make each event A and B true.

c. Complete the following diagram by writing the remaining outcomes in the appropriate region of the Venn diagram.

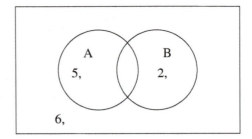

d. What relationship do you observe between the outcomes that make events A and B true?

8. a. In order for an outcome in Experiment 4 to satisfy the condition, must the number showing be both odd **and** less than 5? Explain.

b. What is the probability that the number is odd?

c. What is the probability that the number is less than 5?

In general, $P(A \text{ or } B) = P(A) + P(B)$. However, in Experiment 4, the two events–number is odd, number is less than 5–overlap. You do not want to count the outcomes that satisfy both events twice. Therefore,

$$P(A \text{ or } B) = P(A) + P(B) - P(A \text{ and } B).$$

d. List the outcomes in Experiment 4 for which the number selected is both odd and less than 5. What is the probability of the number selected is both odd and less than 5?

e. Determine the probability that the number selected is odd or less than 5.

9. **Experiment 5.** A single card is drawn from a standard deck of 52 cards.

In parts a–d, determine the indicated probability of selecting the given cards.

a. $P(\text{ace})$ **b.** $P(\text{jack})$

c. $P(\text{ace and jack})$ **d.** $P(\text{ace or jack})$

In Problem 9c, it is not possible for the card drawn to be both an ace and a jack. Such events are said to be **mutually exclusive**.

Definition

Two events A and B are said to be **mutually exclusive** if it is not possible for both events to occur at the same time.

---

10. Construct a Venn diagram that illustrates events A and B are mutually exclusive.

11. With regard to Experiment 5, in which a single card is drawn from a standard deck of 52 cards,

  i. Determine if the following events are mutually exclusive.

  ii. Determine $P(A \text{ or } B)$.

  a. A: a red card, B: a black card

  b. A: a king, B: a red card

12. Experiment 1 revisited: You roll a fair die twice.

  a. What is the probability that you will get an odd number on the first roll and a number greater than 2 on the second roll?

b. What is the probability of obtaining a sum of 10 on the two rolls? *Hint*: The sample space for rolling twice has 36 equally likely outcomes.

c. What is the probability of rolling a sum of less than 5?

SUMMARY: ACTIVITY 7.3

1. Events A and B are **independent events** if the occurrence of either event does not affect the probability of the occurrence of the other event.

2. Events A and B are **dependent events** if the occurrence of either event affects the probability of the occurrence of the other event.

3. The probability that two independent events both occur is equal to the product of the probabilities of the individual events. Stated symbolically, if A and B are independent events, then

$$P(A \text{ and } B) = P(A) \cdot P(B)$$

4. If A and B are dependent events, then $P(A \text{ and } B) = P(A) \cdot P(B, \text{ given } A \text{ has occurred})$.

5. Two events A and B are said to be **mutually exclusive** if it is not possible for both events to occur at the same time.

6. If A and B are mutually exclusive events, then

 i. $P(A \text{ and } B) = 0$

 ii. $P(A \text{ or } B) = P(A) + P(B)$

7. If A and B are not mutually exclusive, then

$$P(A \text{ or } B) = P(A) + P(B) - P(A \text{ and } B)$$

EXERCISES: ACTIVITY 7.3

1. One of your classmates is selected at random. Let *A* represent the event that the person selected owns a computer, and *B* represent the event that the person selected owns an iPod®. Are *A* and *B* mutually exclusive events? Explain.

2. You are one of ten finalists in a radio station contest for tickets to a concert. Names of the finalists are written on index cards, placed in a hat, and a name is randomly selected. Suppose that three sets of tickets are to be given away and that a single finalist can win only one set of tickets. Are the three events of selecting the three winners from the group of finalists independent events? Explain.

3. The spinner pictured at the right is used in a board game. Assume that it is a fair spinner and that the pointer cannot land on a line. The pointer is spun twice. Assuming independence, determine the probability that the pointer lands on

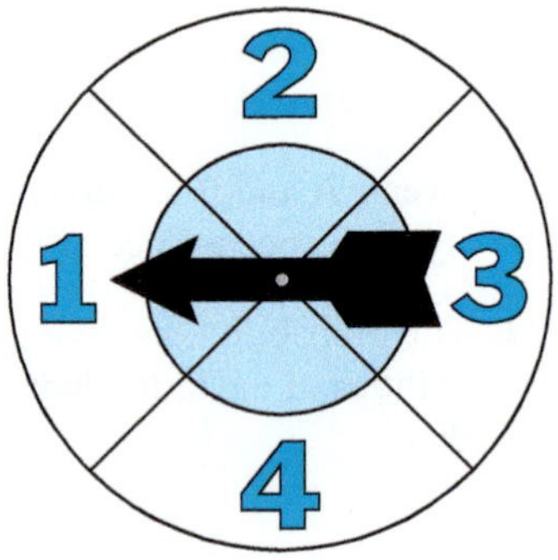

   a. 2 both times.

   b. an even number first and then 4.

   c. an even number and then an odd number.

4. Suppose the pointer in Exercise 3 is spun once. What is the probability that the pointer lands on

   a. 2 or an even number.

b. 2 or 3.

c. Are the events "lands on 2" and "lands on 3" mutually exclusive? Explain.

5. You pick two cards one at a time from a standard deck of 52 cards.

  a. Determine the probability that both cards are aces if the first card is replaced before picking the second card.

  b. What is the probability that both cards are aces if the first card is not replaced before picking the second card? Compare your results to the probability in part a.

  c. Determine the probability that the first card is a heart and the second card is a club. Assume the first card is not replaced before picking the second card.

6. A young couple with three children has purchased the house next to you. How likely is the occurrence that all three children are boys? Assuming independence and the probability of a boy is $\frac{1}{2}$, determine the probability that

  a. all three children are boys.

b. all three children are girls.

c. the youngest child is a boy and the older children are girls.

7. a. Referring to Exercise 6, what is the probability of any specific boy-girl combination of three children?

b. If a family has three boys, what is the probability that a fourth child will be a boy? Explain.

8. One card is drawn from a standard deck of 52 cards. Determine the probability that the card selected was

a. an ace or king.

b. a 10 or a red-suited card.

c. a club or a red card.

d. a face card (jack, queen, king) or a king.

9. A survey was conducted to determine the number of television sets in a household. The results are summarized in the following table.

NUMBER OF TELEVISION SETS				
0	1	2	3	4 or more
Probability 0.05	0.22	0.35	0.20	0.18

Determine each of the following:

a. $P(1 \text{ or } 2 \text{ sets})$

b. $P(1, 2, \text{ or } 3 \text{ sets})$

c. $P(2 \text{ or more sets})$

10. Two golf balls are randomly selected from a bag that contains 3 Titleist, 5 Top Flite, and 2 Pinnacle golf balls.

a. If the first golf ball is replaced before the second golf ball is selected, determine the probability that you select a Titleist and then a Top Flite.

b. If the first golf ball selected is not replaced before the second ball is picked, determine the probability that you select a Titleist and then a Top Flite.

c. If the first golf ball is not replaced, determine the probability that both golf balls selected are Top Flite.

d. Suppose only one golf ball is selected from the bag. What is the probability that the ball is a Pinnacle or Titleist?

11. A five-question multiple-choice quiz is given in science class. Each question has four possible answers. If you randomly pick an answer for each question, determine the probability that you answer

a. any given question correctly.

b. any given question incorrectly.

c. all five questions correctly.

d. only the first two questions correctly.

12. There are 18 sophomores and 15 juniors in your math class. Of the 18 sophomores, 10 are male. Of the 15 juniors, 8 are male.

   a. Complete the following table.

	MALES	FEMALES	TOTAL
Sophomores			
Juniors			
Total			

A student is randomly selected. Determine the probability that the student selected is

b. a male.

c. a sophomore or female.

d. a junior or male.

13. **a.** The style and manufacturer of automobiles in a local used car dealership is summarized in the given table. Complete the table.

MODEL	UNITED STATES	EUROPEAN	JAPANESE	TOTAL
SUV	8	4	6	
Sedan	13	11	15	
Hatchback	12	7	9	
Total				

A vehicle is randomly selected. Determine the probability that the vehicle selected is

b. a Japanese model.

c. a Japanese model and a hatchback.

d. a Japanese model or a hatchback.

e. European model or a sedan.

14. Assume the probability that a person in a certain income bracket will be audited is 0.35. The tax return of a person in this income bracket is randomly selected. What is the probability that the person selected will be audited in three successive years. Assume that each year's selection of who to be audited is independent of last year's selection. Also assume that a person stays in the same income bracket for three consecutive years.

Activity 7.4

Conditional Probabilities

Objectives

1. Identify a conditional probability problem.

2. Determine conditional probabilities using the sample space or the data from a table.

3. Determine conditional probabilities using a formula.

It has rained for the past three days. However, the weather forecast for tomorrow's football game is sunny and no chance of rain. You decide not to take a rain jacket, only to get drenched when there is a sudden downpour during the game.

Although not perfect, the level of accuracy of weather forecasting has increased significantly through the use of computer models. These models analyze current data and predict atmospheric conditions at some short period of time from that moment. Based on these predicted conditions, another set of atmospheric conditions are predicted. This process continues until the forecast for the day and the extended forecast for the next several days are completed.

    1. a. What does the level of accuracy in the weather forecasting model described above depend upon?

      b. The level of accuracy decreases as the forecasts extend several days ahead. Explain.

The accuracy of weather forecasting is based upon **conditional probabilities**; that is, the probability that one event happens, given that another event has occurred. For example, the probability that it will rain this afternoon, given that a low pressure system moved into the area this morning, is a conditional probability.

Weather forecasting is revisited in Situation 2 of Activity 7.5.

Calculating Conditional Probabilities

How do you determine conditional probabilities? First, consider the following problem to see how the condition affects the sample space.

    2. A family with two children is moving into your neighborhood.

      a. Determine the sample space of possibilities for the gender of the first and second child using a tree diagram.

      b. Assume that the probability of a boy and girl are each $\frac{1}{2}$. Are the outcomes in the sample space equally likely? Explain.

c. What is the probability that both children are boys?

d. What is the probability that both children are boys, given that the first child is a boy?

In Problem 2c, you selected outcomes from the entire sample space for which the event (both boys) was true. In Problem 2d, you are given that the first child is a boy. Therefore, the sample space is **reduced** to only include outcomes that make the given event true. The outcomes in which the first child is not a boy are neglected.

The probability that both children are boys, given that the first child is a boy, is a **conditional probability**. There is a special notation for conditional probabilities.

> In general, $P(A \mid B)$ denotes the probability of event A, given that event B has occurred.

In Problem 2d, the probability that both children are boys, given that the first child is a boy, is written as

$$P(\text{both boys} \mid \text{first is a boy})$$

3. Referring to Problem 2, what is the probability that both children are boys, given at least one of the children is a boy?

Calculating Conditional Probabilities Using a Formula

In general, a formula is used to calculate conditional probabilities. The following gives an intuitive justification for this formula.

4. Suppose a person is selected at random in the United States. Now, let

A represent the event that the person selected is a student at the University of Virginia, and

B represent the event that the person selected is a resident of Charlottesville, VA.

a. What can you say about $P(A)$ and $P(B)$? Explain.

b. How does the $P(A)$ in part a compare to the probability of attending the University of Virginia given that the person selected lives in Charlottesville, VA? That is, how does $P(A)$ in part a compare to $P(A|B)$? Explain.

The situation in Problem 4 can be demonstrated by the following Venn diagram. All the people who live in the United States are contained in the rectangle; all the people who attend the University of Virginia are contained in circle A; all people who live in Charlottesville, VA, are in circle B.

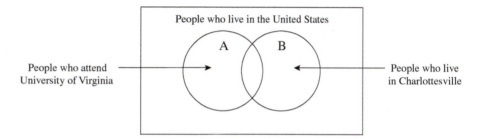

5. a. Describe the region that represents the new sample space, given that the person selected lives in Charlottesville.

b. Describe the region that represents the outcomes that satisfy the event that the person selected is a student at the University of Virginia, given the person lives in Charlottesville.

Therefore, the $P(A|B)$ should be a fraction of the number of people that are both a student at University of Virginia and live in Charlottesville divided by the number of people who live in Charlottesville. Stated symbolically,

If the outcomes for events A and B are equally likely, then

$$P(A|B) = \frac{n(A \text{ and } B)}{n(B)},$$

where $n(A \text{ and } B)$ represents the number of outcomes common to both events A and B, and $n(B)$ is the number of outcomes for which event B is true. Note that $n(B) \neq 0$.

6. In Problem 2d, you used the reduced sample space to determine the probability that the two children are both boys, given the first child is a boy. Redo the problem using the conditional probability formula. Compare the results.

7. A taste test is given to 200 people at a local mall. The results are summarized in the following table.

	PREFERS COKE	PREFERS PEPSI	TOTAL
Male	51	40	91
Female	34	75	109
Total	85	115	200

If one person who participated in the taste test is randomly selected, use the data in the table to determine the probability that the person selected

a. prefers Coke.

b. is a male.

c. is a male and prefers Coke.

d. prefers Coke, given that a male is selected.

e. prefers Pepsi, given that a female is selected.

In many applications, you don't know the actual numbers, but you do know the percentages and probabilities. In such situations, the formula used to calculate conditional probabilities is

$$P(A \mid B) = \frac{P(A \text{ and } B)}{P(B)}, \text{ where } P(B) \neq 0.$$

The probabilities $P(A \text{ and } B)$ and $P(B)$ are determined using the original sample space.

8. Recently in Europe, 88% of all households had a television and 51% of all households had a television and a DVD player.

a. If a household in Europe is randomly selected, what is the probability that the household had a television?

b. Determine the probability that a randomly selected household had a television and a DVD player.

c. What is the probability that a household has a DVD player given that it had a television?

9. Motor vehicles sold in the United States are classified as either cars or light trucks, and as either domestic or imported. Light trucks include SUVs and minivans. The following table gives the probabilities for a randomly selected vehicle recently sold in the United States. For example, the probability of selecting a domestic light truck is 0.47.

	DOMESTIC	IMPORTED	TOTAL
Light truck	0.47	0.07	
Car		0.13	
Total			

a. Complete the table.

In parts b–g, determine the probability that a randomly selected vehicle sold in the United States is

b. a car.

c. imported.

d. imported car.

e. a car or imported.

f. a car, given that the vehicle is imported.

g. light truck, given that the vehicle is domestic.

SUMMARY: ACTIVITY 7.4

1. The **conditional probability** of an event A is when the probability of the event A is affected by the knowledge that other events have already occurred.

2. $P(A|B)$ denotes the probability of event A, given that event B has occurred.

3. If the outcomes for events A and B are equally likely, then

$$P(A|B) = \frac{n(A \text{ and } B)}{n(B)},$$

where $n(A \text{ and } B)$ represents the number of outcomes common to both events A and B, and $n(B)$ is the number of outcomes for which event B is true. Note that $n(B) \neq 0$.

4. The formula used to calculate conditional probabilities is

$$P(A|B) = \frac{P(A \text{ and } B)}{P(B)}, \text{ where } P(B) \neq 0.$$

The probabilities $P(A \text{ and } B)$ and $P(B)$ are determined using the original sample space.

EXERCISES: ACTIVITY 7.4

1. A family has three children.

a. Determine the sample space of possible outcomes for the gender of the first, second, and third child using a tree diagram.

b. What is the probability that all three children are girls? Assume the probability of a girl and a boy are each $\frac{1}{2}$.

c. What is the probability that all three children are girls, given the first two are girls?

d. Does your result in part c support the argument that the events "a boy" and "a girl" are independent?

2. A card is randomly selected from a standard deck of 52 cards. Determine the probability that the card selected is

a. a king.

b. a king, given the card is a heart.

c. a king, given the suit is black.

d. a heart, given the card is a queen.

e. an even number, given the card is a picture card.

3. Two fair dice are rolled one after the other. The sample space for this experiment is given in Problem 5 in Activity 7.2. Determine the probability that the sum of the dots on the dice total

   a. 5.

   b. 5, given the first die is a 2.

   c. 5, given the first die is 5.

   d. 6, given the first die is even.

4. A bag contains 6 blue marbles and 4 red marbles. A marble is randomly selected from the bag. A second marble is then randomly selected without replacing the first marble.

   a. What is the probability that you pick a blue marble and then a red marble?

   b. What is the probability that you pick a red marble, *given* the first marble selected is blue?

5. A total of 200 people were surveyed to determine which evening news they watch most frequently. The results of the survey are given in the table.

	ABC	NBC	CBS	OTHER	TOTAL
Male	38	13	26	12	89
Female	32	37	29	13	111
Total	70	50	55	25	200

One of the participants in the survey is randomly selected. What is the probability that the person selected watches

   a. CBS.

   b. NBC or CBS.

c. ABC, given the viewer is a male.

d. ABC or NBC, given the person is a female.

6. A child psychologist conducted a survey of 500 children in grades 4, 5, and 6 in elementary school. The aim of one of the questions in the survey was to see if boys and girls at this age had similar goals in school. They were asked whether their primary goal was to get good grades, to be popular, or to be good in sports. The results are given in the table.

GOALS				
	GRADES	POPULAR	SPORTS	TOTAL
Boy	128	50	60	238
Girl	141	91	30	262
Total	269	141	90	500

A child that participated in the survey is randomly selected. Determine the probability that the child selected is

a. a girl.

b. a girl and primary goal was to be popular.

c. a girl, given the goal in school was to be popular.

d. a boy, given the goal in school was to get good grades.

7. At your high school 8.7% of students take Technology and Spanish; 68% take Technology. What is the probability that a student takes Spanish given that the student is taking Technology?

8. At a local college, 85% of incoming freshmen nursing students are female and 15% are male. Recent records show that 70% of entering female students will graduate with a BSN degree, 90% of male students will graduate with a BSN degree. An incoming freshman is selected at random.

   a. Draw a tree diagram of the possible outcomes for freshmen nursing students.

Determine the following probabilities:

   b. P(student will graduate | student is female)

   c. P(student will graduate | student is male)

   d. P(female and student will graduate)

   e. P(student is female and not graduate)

Project 7.5

Weather Forecasting

Objective

1. Use conditional probabilities to solve problems.

The following three Problems illustrate how conditional probabilities are used in the problem solving process.

Problem 1. Suppose the local weather station has collected and analyzed data about the probability of precipitation in your region. The results predict that if it rains on one day, then the chance it will rain on the next day is 50%. If it is not raining, it will rain on the next day only 20% of the time.

a. Is the statement "If it rains on one day, then the chance it will rain on the next day is 50%." a conditional probability? Explain.

b. Determine P(does not rain next day | rains today).

c. If it does not rain today, what is the probability that it will rain the next day?

d. Determine P(does not rain the next day | does not rain today).

e. The weather forecast for today predicts a 90% chance of rain. Starting with today as the first day, the following tree diagram gives the sample space for three consecutive days of weather outcomes. Each branch of the tree indicates whether it rains (denoted by R) or does not rain (denoted by N). Complete the tree diagram.

f. Determine the probability that it rains on all three days.

g. Determine the probability that it rains only on the third day.

h. Determine the probability that it rains on the third day.

    i. You are given that it rains on the first day. What is the probability that it rains on the third day?

Problem 2. Suppose your high school basketball team is playing in a best-of-three holiday tournament against another local team.

  a. Construct a tree diagram to determine the sample space for this situation.

  b. The coach feels the team has a 50–50 chance of winning the first game against the very strong opponent. If the team wins the game, the probability of winning the next game increases to $\frac{2}{3}$. If the team loses the game, the probability of winning the next game is $\frac{1}{3}$.

    i. If the team wins the game, what is the probability of losing the next game?

    ii. If the team loses the game, what is the probability of losing the next game?

c. What is the probability that the team wins the first game, but then loses the next two games?

d. What is the probability that the team wins the first two games?

e. What is the probability that the team wins the tournament, given it wins the first game?

f. What is the probability that the team wins the tournament?

Problem 3. On each point in tennis, a player is allowed two tries to successfully serve the ball into the service box of the opponent's side of the court.

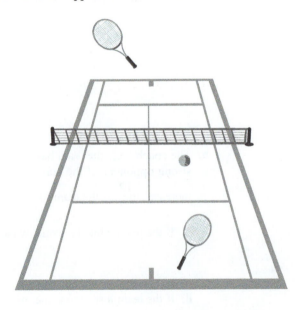

If the player is not successful in these two tries, the player automatically loses the point. This is called a double fault.

Suppose while playing in a tournament, a certain professional player gets his first serve in the service box about 75% of the time. When he gets his first serve in, he wins the point about 80% of the time. If he misses his first serve, his second serve goes in the service box about 90% of the time. When this happens, he wins the point on the second serve about 40% of the time.

    a. Draw a tree diagram of all the possible outcomes.

Determine each of the following probabilities

    b. *P*(misses first serve)

    c. *P*(wins the point | first serve in)

    d. *P*(loses the point | first serve in)

    e. *P*(first serve in and wins the point)

    f. *P*(second serve in | first serve misses)

    g. *P*(misses the second serve)

    h. *P*(wins the point | second serve in)

  i. *P*(loses the point | second serve in)

  j. *P*(first serve out and second serve in and wins the point)

  k. *P*(wins the point)

Activity 7.6

Selecting and Rearranging Things

Objectives

1. Determine the number of permutations.

2. Determine the number of combinations.

3. Recognize patterns modeled by counting techniques.

4. Use permutations and combinations to compute probabilities.

In working through Activity 7.2 (Choices), you may have noticed there are many different varieties of counting problems. Sometimes a tree diagram or list may be feasible to display all the possibilities. But usually the overwhelming number of possibilities makes constructing a list or tree diagram impractical. In this activity you will explore two types of counting problems, **permutations** and **combinations**, both of which use the Fundamental Counting Principle.

Permutations are arrangements of objects, from first to last, where the order in which the objects are selected is most important.

Example 1 *You have five textbooks and wish to arrange them on your bookshelf in the usual way. In how many ways can this be done?*

SOLUTION

Think of the five positions on the shelf, left to right. Consider how many ways the first space can be filled (5), then how many ways to fill the second space (4, since there are four books remaining), and so on. You could illustrate with a tree diagram if you wish, but a simple application of the Fundamental Counting Principle will give you $5 \cdot 4 \cdot 3 \cdot 2 \cdot 1 = 120$. So there are 120 different arrangements or orderings of your five books. Equivalently, you say there are 120 **permutations** of any five objects.

1. Suppose you have seven different sweatshirts, and wish to wear a different one each day of the week. In how many ways can this be done?

2. There are 26 letters in the standard English alphabet. If you consider using all the letters to create a 26-letter word, how many different words are possible? (You will notice this is a very large number, so you will need scientific notation to give an approximate answer.)

Factorial Notation

The product of any positive integer with all smaller positive integers down to one has a special name and notation. For example, five factorial is written as $5! = 5 \cdot 4 \cdot 3 \cdot 2 \cdot 1 = 120$.

3. Use the factorial notation on your calculator (see Appendix A) to determine 12!

Note: 0! is defined to be equal to one. (Check this on your calculator.) You can think of it as the number of ways to arrange zero things.

Sometimes you are interested in arrangements of only some of the objects from a larger collection. Words that are 26 letters long are a bit unrealistic, but using only five different letters might make more sense.

Example 2   *Suppose exactly five different letters are used to make a word (in this case a word does not have to be one found in a dictionary, but simply be a sequence of five letters). How many different words can be made?*

SOLUTION

As with the bookshelf problem, imagine five spaces, to hold each of the five letters. But now, there are 26 choices for the first space, 25 for the second, and so on. $26 \cdot 25 \cdot 24 \cdot 23 \cdot 22 = 7,893,600$ different words are possible. Notice that since there are only five letters, this is not a full factorial. These are still called permutations. There are 7,893,600 permutations of 26 objects, taken five at a time.

Permutations of n objects taken r at a time = the first r factors of n!

Notation: $_nP_r = n(n - 1)(n - 2) \cdots (n - r + 1) = \dfrac{n!}{(n - r)!}$

4. Redo Example 2 using the permutations formula above. Then, verify your answer by using the permutation feature on your calculator (see Appendix A).

5. The governor is visiting your school and you need to assign seating for 14 students. Unfortunately there are only six seats in the front row. In how many different ways can you select and arrange the six lucky students?

Combinations

Sometimes the order in which selections are made is not important. For example, in Problem 5, if the arrangement of the six chosen students is not important, you would simply want to count how many ways a group of six students could be chosen for the first row. For this revised situation you are counting the number of **combinations** of 14 objects taken six at a time.

Combinations are collections of objects selected from a larger collection, where the order of selection is not important. The number of ways to select r objects from a collection of n objects is called the number of combinations of n objects taken r at a time.

6. In Problem 5, for each possible group of six students, how many seating arrangements are possible?

7. If you could list all 2,162,160 permutations in Problem 5, you could also group them according to the actual six students selected. Each group would have the number you found in Problem 6. Divide accordingly to determine the number of ways to select the six students.

Combinations of n objects taken r at a time: $\quad {}_nC_r = \dfrac{{}_nP_r}{{}_rP_r} = \dfrac{n!}{(n-r)!\,r!}$

8. Apply the combinations formula above to verify your answer to Problem 7.

9. From a student body of 230 students, how many different committees of four students are possible? (*Hint:* The order in which the four students are selected is not important.)

An interesting symmetrical property of combinations appears when you consider not only the objects selected from a collection, but also those left behind. Does it make sense that the number of ways to select 3 objects from a collection of 8 objects is the same as counting the number of ways to leave 5 behind?

10. Verify the above statement by computing ${}_8C_3$ and ${}_8C_5$.

Many times the most important part of the problem is determining whether or not the order of the selections is significant.

11. In each situation, determine whether you are asked to determine the number of permutations or combinations. Then do the calculation.

   a. How many ways are there to pick a starting five from a basketball team of twelve members?

   b. How many ways are there to distribute nine different books among 15 children if no child gets more than one book?

   c. How many ways are there to pick a subset of four different letters from the 26-letter alphabet?

   d. How many ways can the three offices of chairman, vice chairman, and secretary be filled from a club with 25 members?

Applications

Many variations on these basic counting techniques arise in the study of probability and computer science. Here are a couple of examples, with problems for you to try.

Example 3   *How many different ten-digit binary sequences (only 0s and 1s) are there with exactly seven zeros and three ones?*

SOLUTION

While the order of the binary digits is important, this counting problem is most easily solved by determining the number of combinations of ten digits taken seven at a time. Think of the ten positions for the digits, and ask: How many ways can you select seven positions for the zeros?

$$_{10}C_7 = \frac{10 \cdot 9 \cdot 8 \cdot 7 \cdot 6 \cdot 5 \cdot 4 \cdot 3 \cdot 2 \cdot 1}{(3 \cdot 2 \cdot 1) \cdot (7 \cdot 6 \cdot 5 \cdot 4 \cdot 3 \cdot 2 \cdot 1)} = \frac{10 \cdot 9 \cdot 8}{3 \cdot 2} = 120.$$

Or equivalently, how many ways can you select three positions for the ones?

$$_{10}C_3 = \frac{10 \cdot 9 \cdot 8 \cdot 7 \cdot 6 \cdot 5 \cdot 4 \cdot 3 \cdot 2 \cdot 1}{(7 \cdot 6 \cdot 5 \cdot 4 \cdot 3 \cdot 2 \cdot 1) \cdot (3 \cdot 2 \cdot 1)} = \frac{10 \cdot 9 \cdot 8}{3 \cdot 2} = 120.$$

12. If a coin is flipped 20 times, how many different ways are there to get exactly five heads?

13. What is the probability that a coin flipped twenty times will come up heads exactly five times?

Example 4   *How many ways are there to arrange the letters of the word SYSTEMS?*

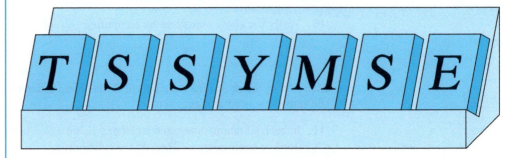

SOLUTION

The repeated *S* makes this more difficult than a simple permutation problem. If you imagined the letters on tiles (like the game Scrabble), with the *S* tiles labeled with numbers: S_1, S_2, S_3, so you can tell them apart, then there would be simply 7! ways to arrange the seven tiles. But without the numbers, or tiles, the *S*'s all look the same, so you would be way over-counting the different words. Each unique word will appear 3! times, the number of ways the three *S*'s can be ordered. So if you divide out this multiplicity, you will have an accurate count:

$$\frac{_7P_7}{_3P_3} = \frac{7!}{3!} = 7 \cdot 6 \cdot 5 \cdot 4 = 840 \text{ truly unique words.}$$

14. How many different words can be formed by rearranging all the letters of the word *MIRROR*?

SUMMARY: ACTIVITY 7.6

1. Permutations are arrangements of objects, from first to last, where the order in which the objects are selected is important.

2. Factorial Notation: $n! = n(n-1)(n-2)\cdots 3\cdot 2\cdot 1$

3. Permutations of n objects taken r at a time = the first r factors of $n!$

Notation: $_nP_r = n(n-1)(n-2)\cdots\cdot (n-r+1) = \dfrac{n!}{(n-r)!}$

4. Combinations are collections of objects selected from a larger collection, where the order of selection is not important.

5. Combinations of n objects taken r at a time = $_nC_r = \dfrac{_nP_r}{_rP_r} = \dfrac{n!}{(n-r)!\,r!}$

EXERCISES: ACTIVITY 7.6

1. Compute each of the following:

    a. $5!$

    b. $\dfrac{100!}{98!}$

    c. $_9P_9$

    d. $_{12}P_9$

    e. $\dfrac{_{10}P_4}{_6P_4}$

    f. $_{12}C_5$

    g. $_8C_8$

    h. $_{285}C_1$

    i. $_8C_3 \cdot _5C_2$

2. How many different ways are there to arrange twelve books on a shelf?

3. How many ways are there to select a committee of five from a club with 30 members?

4. How many different four-digit numbers can be made using the digits 1, 2, 3, 4, 5, 6 if no digit can be used more than once?

5. How many ways are possible to select a president and vice president from an association with 5400 members?

6. How many different words can be made by rearranging the letters of the word *SYMMETRY*?

7. If a coin is flipped 50 times, in how many ways could there be exactly two tails?

8. A lottery ticket has 54 numbers, from which the player chooses six.

   a. What is the probability that all six numbers are correct, and the player wins the big jackpot?

   b. What is the probability that the player gets five of the six numbers correct?

9. A club with 46 members, of which 20 are girls, needs to form a committee of six, to be composed of the same number of boys as girls. In how many ways can this committee be selected? (*Hint:* Calculate two different numbers of combinations, then apply the multiplication principle.)

10. The number of combinations for any size collection makes an interesting pattern, usually called Pascal's triangle, after the French mathematician/philosopher Blaise Pascal (1623–1662). This pattern was introduced back in the exercises for Activity 1.2. Each row of the triangle has one more number than the previous row, each number the result of adding the two numbers immediately above it. The first and last numbers are always one.

   a. Identify the pattern and then complete the next three rows of Pascal's triangle.

$$
\begin{array}{ccccccccccc}
& & & & & 1 & & & & & \\
& & & & 1 & & 1 & & & & \\
& & & 1 & & 2 & & 1 & & & \\
& & 1 & & 3 & & 3 & & 1 & & \\
& 1 & & 4 & & 6 & & 4 & & 1 & \\
1 & & 5 & & 10 & & 10 & & 5 & & 1 \\
\end{array}
$$

   b. Each row actually contains the number of combinations of n objects taken r at a time, where n is the number of the row (starting with 0) and r is the position in the row (again, starting with 0). For example, 1, 3, 3, 1 are $_3C_0$, $_3C_1$, $_3C_2$, $_3C_3$. Which number in Pascal's triangle corresponds to $_7C_4$? Verify with your calculator.

   c. When displayed in this form, the symmetry of the combinations is plain to see. If you know the value of $_{30}C_{18}$, what else is it equal to?

Activity 7.7

How Many Boys (or Girls)?

Objectives

1. Recognize components of a binomial experiment.

2. Calculate binomial probabilities.

3. Recognize the components of a geometric experiment.

4. Calculate geometric probabilities.

In a family with five children, how unusual would it be for all the children to be boys? Or by the same token, all girls? Such questions are relevant for many families in cultures around the world, especially when one gender is favored over another.

Suppose a family does have all boys. What is the probability of such an event?

1. Assume the birth of a boy or a girl is equally likely. What is the probability that a single child is born a boy?

2. Now suppose a family has one boy already. What is the probability that their next child will also be a boy?

The probabilities you just determined are for a single child. You are really interested in the probability that for a family with two children, both are boys.

3. For a family with two children, what are the possible outcomes? Provide a tree diagram or simply list the possibilities.

4. **a.** Use the sample space in Problem 3 to determine the theoretical probability of the event that a family with two children has two boys.

   b. Assume that the individual outcomes of the two births are independent (i.e., the outcome of the second birth in no way depends upon the outcome of the first birth). Use the multiplication principle of probability to determine the probability that a two-child family has two boys.

5. What is the theoretical probability of the event that a family with two children has two girls?

6. What is the theoretical probability of the event that a family with two children has one boy and one girl?

7. At this point, you have determined the probabilities of all the possible outcomes for a two-child family. Letting the random (independent) variable x represent the number of boys in a family with two children, complete the following table for and verify the properties of the probability distribution.

Recall: $0 \leq P(x) \leq 1$ for all x in the sample space and the sum of the probabilities of all x in the sample space must equal 1.

x	P(x)

Note that $P(1 \text{ boy and } 1 \text{ girl}) \neq \frac{1}{2} \cdot \frac{1}{2}$. This is because there are two ways the family could have one boy, either first born or second born. If you wanted the probability of "the first born is a boy and the second born is a girl," the probability is the product of the probabilities of independent events.

Binomial Probability Distribution

There are many situations where each individual outcome can go one of two ways: boy or girl, head or tail, right or wrong, success or failure. Assuming the two outcomes have well defined and fixed probabilities, and each occurrence of an outcome is independent, combinations of such outcomes define a **binomial probability distribution**. To determine an individual binomial probability, it is necessary to count how many ways an event can occur (like the two distinct ways a family with two children could have one boy).

For the moment, let's consider a different binomial (two outcome) situation, flipping a coin. Assuming fairness (head and tail equally likely, as you had assumed for boy and girl), $P(\text{head}) = P(\text{tail}) = 0.5$.

8. Consider flipping a fair coin three times. Let x = the number of heads, record the sample space for this experiment (as a tree, chart or list), and fill in the table for this binomial probability distribution.

x	P(x)

9. Repeat the same process for flipping a coin four times.

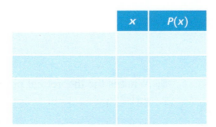

x	P(x)

There are several shortcuts for determining the number of ways a certain number of heads can come up. In flipping a coin four times, it is not too difficult to simply list the different ways exactly two heads could result: tthh, htht, htth, thth, thht, and hhtt. If you can agree that these are all different, and there are no more possible, then the theoretical probability of getting exactly two heads when four coins are flipped is $\dfrac{6}{16} = \dfrac{3}{8}$. (Is this what you got in Problem 9?)

But suppose you flipped a coin 100 times, or even only ten times. To determine the number of ways to get exactly five heads by listing them, or drawing a tree diagram, would be tedious. There is, however, a formula for calculating these numbers, which are called **binomial coefficients**. They are the very same **combinations** you learned about in Activity 7.6.

Example 1 *Consider Problem 9, in which you flipped a coin four times. How many ways can exactly two heads occur?*

SOLUTION

This is sometimes called the "number of combinations of four things taken two at a time." There are two types of notation used, $\dbinom{4}{2}$ and $_4C_2$. The computation is $_4C_2 = \dfrac{4 \cdot 3 \cdot 2 \cdot 1}{(2 \cdot 1)(2 \cdot 1)} = 6$.

The number of ways to get exactly three heads on five flips would be
$$_5C_3 = \frac{5 \cdot 4 \cdot 3 \cdot 2 \cdot 1}{(3 \cdot 2 \cdot 1)(2 \cdot 1)} = 10.$$

In general, $\dbinom{n}{r} = _nC_r = \dfrac{n!}{r! \cdot (n-r)!}$

$$= \frac{n(n-1) \cdots 3 \cdot 2 \cdot 1}{(r(r-1) \cdots 3 \cdot 2 \cdot 1)((n-r)(n-r-1) \cdots 3 \cdot 2 \cdot 1)}$$

This is a rather complicated formula, but binomial coefficients can be easily computed on your calculator. (See Appendix A.)

10. Calculate the number of ways to get exactly four heads when a coin is flipped five times, by applying the formula and by checking on your calculator.

11. Now determine the binomial probabilities for the number of heads when a coin is flipped five times.

x	P(x)

12. Experimentally, you can see if your probability distribution seems reasonable by actually flipping coins. Instead of flipping an individual coin five times, use five coins, shake them up in your cupped hands, and drop them on the table, recording the number of heads. Repeat this ten times, recording your results in a frequency table. How close are your relative frequencies to the theoretical probabilities you calculated in Problem 11?

	10 REPETITIONS		ENTIRE CLASS	
x	FREQ.	REL. FREQ.	FREQ.	REL. FREQ.

13. Record the frequencies and relative frequencies for the entire class. Did you observe what you expected, with regard to the theoretical probabilities? Explain.

14. It is also possible to simulate many repetitions of the experiment on a computer or your calculator. (See Appendix A.) Simulate flipping five coins 500 times on your calculator, by generating a list of 500 random integers between 0 and 5, from a binomial distribution. Record your results in the table, and then record the collective total for your entire class. Comment on the results, compared to your previous work.

	500 REPETITIONS		ENTIRE CLASS	
x	FREQ.	REL. FREQ.	FREQ.	REL. FREQ.

15. It is also possible to determine any binomial probability directly from a table or your calculator. (See Appendix A.) Use your calculator to verify the theoretical probabilities you calculated in Problem 11.

In general, a binomial probability is calculated using the following formula:

$$P(x) = \binom{n}{x} p^x q^{n-x}, \text{ where}$$

> n is the number of independent trials,
>
> x is the random variable, what is being counted,
>
> p is the probability of a success (the outcome x is counting),
>
> q is the probability of a failure (the other outcome)

For example, in the previous Problems 11–15, $n = 5, p = q = 0.5$. The probability of exactly three heads (successes) out of five coin flips is

$$P(3) = \binom{5}{3} 0.5^3 0.5^{5-3} = \frac{5 \cdot 4 \cdot 3 \cdot 2 \cdot 1}{(3 \cdot 2 \cdot 1) \cdot (2 \cdot 1)} 0.5^5 = 0.3125.$$

16. How close did your final relative frequency for three heads come to this theoretical probability?

17. a. Apply the formula to determine the probability that a two-child family has both boys.

b. Apply the formula to determine the probability that among eight children exactly five will be boys. Verify your result by finding directly with your calculator.

Because you assumed a fair coin, and the birth of a boy and girl are equally likely, the probability of a success and the probability of a failure were both 0.5. Usually these probabilities are different, as you might have guessed from the formula.

18. Suppose that 10% of all electrical switches are defective.

a. What is the probability of selecting a defective switch in a single selection?

b. What is the probability of selecting a good switch in a single selection?

c. If you select five switches, what is the probability that exactly one is defective? Apply the formula and then verify directly on your calculator.

Geometric Probability

Geometric probabilities are similar to binomial probabilities. The trials are independent, and the outcome of each trial is either "success" or "failure." A geometric probability is calculated using the following formula. It represents the probability that the first success occurs on the nth trial.

> $$P(x = n) = q^{n-1} \cdot p,$$
>
> where p is the probability of success and q is the probability of failure.

For example, suppose you flip a coin until you get a head. Remember that the probability of getting heads on any one toss is $\dfrac{1}{2}$, so $p = \dfrac{1}{2}$. The probability of not getting a head is also $\dfrac{1}{2}$, so $q = \dfrac{1}{2}$. The probability that you get your first head on say the fourth toss is

$$P(x = 4) = \left(\frac{1}{2}\right)^{4-1} \cdot \frac{1}{2} = \left(\frac{1}{2}\right)^3 \cdot \left(\frac{1}{2}\right) = \frac{1}{16} = 0.0625$$

The probability that you get your first head on the fifth toss is

$$P(x = 5) = \left(\frac{1}{2}\right)^{5-1} \cdot \frac{1}{2} = \left(\frac{1}{2}\right)^4 \cdot \left(\frac{1}{2}\right) = \frac{1}{32} = 0.03125$$

19. Apply the geometric probability formula to determine the probability of getting the first six on the third roll of a die.

You could also use the multiplication principle of probability to determine the probability of getting the first six on the third roll.

ROLL	1st	2nd	3rd
Outcome	Not 6	Not 6	6
Probability	$\dfrac{5}{6}$	$\dfrac{5}{6}$	$\dfrac{1}{6}$

Using the multiplication principle, the probability of getting the first six on the third roll is $\dfrac{5}{6} \cdot \dfrac{5}{6} \cdot \dfrac{1}{6} = \dfrac{25}{216} \approx 0.115741$.

20. Draw a card from a standard deck of 52 cards, look at the card, and return it to the deck. Calculate the probability of getting your first ace during the second time you draw a card.

SUMMARY: ACTIVITY 7.7

1. **Binomial Coefficients:** $\binom{n}{r} = {}_nC_r = \dfrac{n!}{r! \cdot (n-r)!}$

$$= \dfrac{n(n-1)\cdots 3\cdot 2\cdot 1}{(r(r-1)\cdots 3\cdot 2\cdot 1)((n-r)(n-r-1)\cdots 3\cdot 2\cdot 1)}$$

(Also called the number of combinations of n things taken r at a time.)

2. **Binomial Probabilities:** When an experiment is repeated n times (called n trials), each trial having only two possible outcomes (success or failure), and all trials being independent, then the probability of having x successes in n trials is called a binomial probability, given by the formula $P(x) = \binom{n}{x}p^x q^{n-x}$,

where n is the number of trials,

x is the random variable, what is being counted,

p is the probability of a success (the outcome x is counting),

q is the probability of a failure (the other outcome)

3. **Geometric Probabilities:** When an experiment has trials with only two possible outcomes (success or failure), and all trials are independent, then the probability that the first success occurs on the nth trial is called a geometric probability, given by the formula $P(x = n) = q^{n-1} \cdot p$, where

p is the probability of success, and

q is the probability of failure.

EXERCISES: ACTIVITY 7.7

1. In a family with eight children, what is the probability that exactly six are boys?

2. Again with eight children, what is the probability there are at least six are boys?
(*Hint:* Add the three probabilities for either six, seven, or eight boys.)

3. What is the probability that when a fair coin is flipped 20 times, there will be exactly five tails?

4. In a randomly generated sequence of 24 binary digits (0s and 1s), what is the probability that exactly half of the digits are 0?

5. According to the U.S. Census Bureau, 49.6% of Texas residents were male in 2010.

    a. If one Texan was selected at random in 2010, what is the probability he was male?

    b. If one Texan was selected at random in 2010, what is the probability she was female?

    c. Consider the experiment of selecting ten Texans at random. If you are interested in determining the probabilities of selecting x males, is this a binomial probability distribution? And if so, what are the values of n, p, and q?

    d. Determine the probability that exactly two men are chosen, out of the ten people, by applying the formula. Verify by also going directly to the binomial probabilities on your calculator.

6. Suppose the probability of your hitting the bull's-eye on a dartboard is 0.06.

    a. What is the probability of getting exactly two bull's-eyes when you throw five darts?

Appendix

    b. Empirically verify that your probability in part a is reasonable by simulating the throwing of five darts 500 times on your calculator. (See Appendix A.) Display a histogram for your simulation, and record the relative frequencies for all the possible values of x, the number of bull's-eyes.

x	REL. FREQ.

    c. Now calculate the remaining theoretical probabilities, to see that they are close to the relative frequencies of your simulation.

x	P(x)

    d. About how many times might you expect to have to repeat this experiment to get five bull's-eyes on all five throws?

7. In taking a multiple-choice test, there are five possible answers for each problem, only one of which is correct. Assume that someone guesses randomly on all 25 problems.

    a. What is the probability of getting the first problem correct?

    b. What is the probability of getting the first problem wrong?

    c. What number of correct answers is needed to achieve a passing grade of 70%?

    d. What is the probability of getting a passing grade with the minimum number correct?

8. Suppose a player on your basketball team makes 70% of his free throws. Then, the probability of him making any one free throw is 0.7.

    a. Determine the probability that he makes his second free throw but not his first.

    b. Determine the probability that he does not make a free throw until his fourth try.

Activities 7.1–7.7   What Have I Learned?

1. If you flip a coin 50 times and get 22 tails, should you assume that the probability of getting tails with this coin is 0.44? Explain.

2. What does the law of large numbers say about the relationship between theoretical and experimental probabilities?

3. If a random experiment consists of rolling a six-sided die ten times, how many individual outcomes make up the sample space, and what is the probability of getting all sixes?

4. Could this table represent a probability distribution? Explain.

x	1	2	3	4	5
$P(x)$	0.2	0.3	0.4	0.1	0.05

5. You shuffle a standard deck of 52 cards, and then start turning them over one at a time. The first one is a red card. So is the second card turned over. The third card turned over is also a red card. This continues so that the first 10 cards turned over are all red.

   a. What is the probability that you get 10 red cards in a row?

   b. Is the probability that the eleventh card will be black higher than the likelihood of getting another red card? Explain.

6. If A and B are mutually exclusive events, then what is $P(A \text{ and } B)$? Explain.

7. If A and B represent independent events, what is $P(A \mid B)$? Explain.

8. Is it possible for mutually exclusive events to be independent? Explain.

9. In selecting objects from a collection, the order or arrangement of the selected items is not always important. If the order is important, are you counting permutations or combinations?

10. In each of these counting problems, state whether you would find the number of permutations or combinations.

a. In how many ways can 20 delegates be arranged in a row of 20 chairs?

b. In how many ways can a committee of four be selected from a club with 235 members?

c. In how many ways can the president, vice president, and treasurer be selected from an association of 2550 members?

d. In how many ways can you pick three different flavors of ice cream for your sundae when there are 53 flavors from which to choose?

11. What are the defining characteristics of a binomial probability experiment, or distribution?

12. In each of these experiments, determine whether or not the probabilities are binomial probabilities. If they are, identify the values for n, p, and q.

a. A die is rolled 20 times. What is the probability of getting x even numbers?

b. A pair of dice are rolled 50 times. What is the probability of getting doubles x times?

c. Five marbles are selected from an urn with 20 marbles (10 white and 10 black). What is the probability of selecting x white marbles?

13. Your county has been in a dry spell, but the latest weather forecast predicts three fronts will pass through on Monday, Wednesday, and Saturday. Each has an 80% probability of precipitation.

a. What is the chance that the first rain will fall on Saturday?

b. What is the chance that the first rain will fall on Monday?

Activities 7.1–7.7   **How Can I Practice?**

I. The following table displays the annual salaries and years of education for a cross section of the population. Complete the totals for each category. Then calculate the relative frequency for each entry.

EDUCATION	LESS THAN $20,000	$20,000 SALARY $50,000	$50,000 SALARY $100,000	MORE THAN $100,000	TOTAL
12 years	130	146	41	4	
14 years	105	198	52	10	
16 years	31	252	187	24	
More than 16	6	161	241	60	
Total					

    a. If an individual is selected at random, use the corresponding relative frequencies to estimate the probability that the person has:

        i. 12 years of education and makes more than $100,000.

        ii. 14 years of education and makes less than $20,000.

        iii. more than 16 years of education and makes more than $50,000.

    b. Given that the individual chosen has 14 years of education, what is the probability that he or she makes more than $50,000?

    c. Given that the individual chosen has 16 or more years of education, what is the probability that she or he makes more than $50,000?

2. A vase contains 35 marbles: 10 blue, 8 green, 7 yellow, 5 red, and 5 white. One marble is selected at random. Find the following probabilities.

    a. P(red)　　　　　　　　　**b.** P(not blue)

    c. P(green or white)　　　　　**d.** P(black)

    e. P(not black)

3. If a six-sided die is rolled four times, how many different four-digit sequences are possible?

4. If a six-sided die is rolled three times, how many outcomes will be in the tree diagram that displays the sample space?

5. If a coin is flipped four times, determine the probability distribution for the number of tails.

x, The Number of Tails	0	1	2	3	4
P(x)					

6. A coin is flipped and a fair die is rolled.

   a. Construct a tree diagram showing all the possible outcomes.

   b. Determine the probability that a head shows and a 2 is rolled.

   c. What is the probability that a tail shows or an even number is rolled?

7. The school's Guidance Counselor submitted a report to the School Board summarizing the percentage of students in grades 9–12 that were studying one language other than English. The results are given in the table.

SPANISH	FRENCH	GERMAN	OTHER	NONE
26%	9%	3%	3%	59%

Determine the probability that a randomly selected student in grades 9–12 is

   a. studying Spanish   **b.** not studying French

   c. studying French or German

8. The land area of Canada is 9,094,000 square kilometers. Of this land, 4,176,000 square kilometers is forested. A square kilometer of land in Canada is randomly selected.

    a. What is the probability that the selected area of land is forested?

    b. What is the probability that it is not forested?

9. During World War II, the British determined that the probability that a bomber was lost through enemy action on a mission over occupied Europe was 0.05.

    a. What is the probability that a bomber survived its mission?

    b. What is the probability that a bomber survived 2 missions? Assume the events are independent.

    c. What is the probability that a bomber returns safely after 10 missions?

10. Each of the six joints in the *Challenger* space shuttle's booster rocket had a 0.977 reliability. The six rocket joints worked independently.

    a. What does it mean that the six rocket joints work independently?

    b. A 0.977 reliability means that the probability of a rocket joint working successfully is 0.977. What is the probability that all six rocket booster joints work successfully?

11. The following table gives the number of degrees (in thousands) earned in the United States in a recent academic year.

	BACHELOR'S	MASTER'S	PROFESSIONAL	DOCTORATE	TOTAL
Female	784	276	39	20	
Male	559	197	44	25	
Total					

    a. Complete the table.

    b. If a student receiving a degree is randomly selected, what is the probability that the person selected is female?

c. What is the probability that the degree recipient is a male who earned a Doctorate?

d. What is the probability that the person is a female, given that the person received a Master's degree?

e. What is the probability that the person selected received a bachelor's degree, given the recipient is male?

12. In a recent year, the percentage of computer games sold is summarized in the following table:

	GAME TYPE					
	STRATEGY	FAMILY AND CHILDREN'S	SHOOTERS	ROLE PLAYING	SPORTS	OTHER
Percentage	26.9%	20.3%	16.3%	10.0%	5.4%	21.1%

a. What is the probability that a computer game sold was a strategy or shooters game?

b. What is the probability that the computer game is not a family game?

c. What is the probability that the computer game sold is a strategy game, given it is not a sports game?

13. If a license plate consists of three letters (excluding I and O) followed by three digits (excluding 0 and 1), how many different plates are possible?

14. You have ten books to arrange on your bookshelf. How many different orderings of all ten books are possible?

15. There are 96 boys in the freshman class. If five boys are selected at random to form a new rock band, how many possible bands are there?

16. If the band members in the previous problem are to be assigned specific instruments (drums, bass, keyboard, guitar, and vocals), how many bands are possible?

17. How many five-digit numbers are possible, if no digit can be repeated (and 0 as the first digit is allowed)?

18. If a coin is flipped 30 times, how many possible ways could there be exactly 10 heads?

19. If a fair coin is flipped ten times, and heads comes up every time, what is the probability that on the next flip the coin will come up tails?

20. What is the probability that in a family with 12 children, exactly half of them are boys?

21. Suppose you know 2% of all switches are defective. If two switches are used in a device:

    a. What is the probability that both switches are good?

    b. What is the probability that exactly one switch is good?

    c. What is the probability that at least one switch is good?

    d. Could you also determine your answer to part c by subtracting the probability of getting two defective switches from 1?

    e. What is the minimum number of switches that need to be selected so the probability of getting at least one good switch is greater than 0.999999?

22. In playing Monopoly, rolling doubles three times in a row sends you to jail.

    a. What is the probability of rolling three consecutive doubles?

    b. If you roll the dice 100 times over the course of a game, what is the probability that you will have rolled doubles, at any point, exactly three times?

23. A baseball player has a .300 batting average (this is the probability of getting a hit during one at bat).

    a. If the player has four at bats in a game, find the probability that the player goes hitless until his last at bat.

    b. Suppose the player has five at bats. What is the probability that he gets no hits until the third at bat?

Chapter 7   Summary

The bracketed numbers following each concept indicate the activity in which the concept is discussed.

CONCEPT/SKILL	DESCRIPTION	EXAMPLE
Relative frequency [7.1]	Frequency of a particular value expressed as a fractional part of all possible values.	If a coin is flipped 100 times, and heads comes up 43 times, its relative frequency is 43%.
Event [7.1]	A particular outcome or result for which a probability can be defined.	The event of getting an even number when a six-sided die is rolled.
Experimental probability [7.1]	Probability of event $E = P(E)$ $= \dfrac{\text{Number of observed occurrences of event}}{\text{Total number of observations}}$	Results of rolling a die ten times: $\{1, 4, 3, 6, 5, 2, 1, 4, 1, 5\}$ $P(\text{Even}) = \dfrac{4}{10} = 0.4.$
Theoretical probability [7.1]	$P(E) =$ $\dfrac{\text{Number of outcomes in event}}{\text{Total number of equally likely possible outcomes}}$	In a single toss of a fair coin, $P(\text{Heads}) = \dfrac{1}{2} = 0.5$
Sample space [7.1], [7.2]	The collection of all possible individual outcomes.	The sample space for rolling a single die is $\{1, 2, 3, 4, 5, 6\}$.
Probability properties [7.1]	• The probability of any event is always between 0 and 1.	$0 \le P(E) \le 1$
	• The probability of an impossible event is 0.	The probability of getting 7 on a single die roll is zero.
	• The probability of a certainty is 1.	The probability of getting less than 7 on a single die roll is one.
	• The sum of the probabilities of all possible outcomes in a sample space is equal to 1.	For a single die roll, $P(1) + P(2) + P(3) + P(4) + P(5) + P(6)$ $= \dfrac{1}{6} + \dfrac{1}{6} + \dfrac{1}{6} + \dfrac{1}{6} + \dfrac{1}{6} + \dfrac{1}{6} = 1.$
Law of large numbers [7.1]	The larger the number of observations, the closer the experimental probability of an event will get to the theoretical probability of that same event.	When a coin is flipped many times, the relative frequency of getting heads is most likely not exactly 0.5, but gets closer to 0.5 as the number of flips gets larger.
Simulation [7.1]	When some experiment or other real world situation is modeled using mathematics, and usually computers, to observe patterns and make predictions about the real world.	With a computer or calculator, flipping a coin can be simulated, to observe many more flips than could actually be done.
Fundamental Counting Principle [7.2]	If some choice can be made in M ways and a subsequent choice can be made in N ways, then there are M times N ways these two choices can be made in succession.	If you have five sweaters and four hats, there are $5 \cdot 4 = 20$ different ways for you to wear a sweater and hat.

CONCEPT/SKILL	DESCRIPTION	EXAMPLE
Tree diagram [7.2]	Displays all possible outcomes for a sequence of choices, one outcome for each branch of the tree.	Tree for flipping a coin twice. 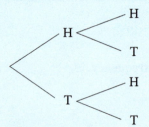
Complement of Event A [7.2]	The sum of the probability that an event will occur and the probability that an event will not occur is 1. The event "not A," denoted by $\overline{A}$, is called the complement of A. Stated symbolically, $P(A) + P(\overline{A}) = 1$ or $P(A) = 1 - P(\overline{A})$	If A is the event that you get a 3 on the roll of a die, then the complement of A is the event that you roll a 1, 2, 4, 5, or 6. $$P(3) + P(\text{not }3) = \frac{1}{6} + \frac{5}{6} = 1$$
Independent events [7.3]	Events A and B are independent events if the occurrence of either event does not affect the probability of the occurrence of the other event.	The outcomes of flipping a coin two times are independent.
Dependent events [7.3]	Events A and B are dependent events if the occurrence of either event affects the probability of the occurrence of the other event.	The outcomes of selecting two cards from a deck of cards without replacement are dependent.
Probability of two independent events both occurring [7.3]	The probability that two independent events both occur is equal to the product of the probabilities of the individual events. Stated symbolically, if A and B are independent events, then $P(A \text{ and } B) = P(A) \cdot P(B)$	Selecting two cards with replacement: $P(\text{1st ace and 2nd ace}) =$ $P(\text{1st ace}) \cdot P(\text{2nd ace})$ $$= \frac{4}{52} \cdot \frac{4}{52} = \frac{1}{169}$$
Probability of two dependent events both occurring [7.3]	If A and B are dependent events, then $P(A \text{ and } B) = P(A) \cdot P(B, \text{ given } A \text{ has occurred})$.	Selecting two cards without replacement: $P(\text{1st ace and 2nd ace})$ $= P(\text{1st ace}) \cdot P\left(\begin{array}{l}\text{2nd ace,}\\\text{given 1st}\\\text{ace selected}\end{array}\right)$ $$= \frac{4}{52} \cdot \frac{3}{51} = \frac{1}{221}$$
Mutually exclusive events [7.3]	Two events A and B are said to be mutually exclusive if it is not possible for both events to occur together.	Selecting one card from a standard deck: the outcomes card is a king and card is an ace are mutually exclusive.

CONCEPT/SKILL	DESCRIPTION	EXAMPLE
Probabilities of mutually exclusive events [7.3]	If A and B are mutually exclusive events, then **i.** $P(A \text{ and } B) = 0$ **ii.** $P(A \text{ or } B) = P(A) + P(B)$	**i.** a card is drawn: $P(\text{ace and king}) = 0$ **ii.** a card is drawn: $P(\text{club or heart}) =$ $P(\text{club}) + P(\text{heart})$ $= \dfrac{13}{52} + \dfrac{13}{52} = \dfrac{1}{2}$
Probability of events that are not mutually exclusive [7.3]	If A and B are not mutually exclusive, then $P(A \text{ or } B) =$ $P(A) + P(B) - P(A \text{ and } B)$	A card is drawn: $P(\text{club or ace})$ $= P(\text{club}) + P(\text{ace}) -$ $P(\text{club and ace}).$ $= \dfrac{13}{52} + \dfrac{4}{52} - \dfrac{1}{52} = \dfrac{16}{52} = \dfrac{4}{13}$
Conditional probability [7.4]	The **conditional probability** of an event A is when the probability of the event is affected by the knowledge that other event has already occurred. $P(A \mid B)$ denotes the probability of event A, given that event B has occurred.	Two fair dice are rolled. The probability that the sum of the dots on the dice is 5, given first die is a 2.
Calculating conditional probabilities [7.4]	The formula used to calculate conditional probabilities is $P(A \mid B) = \dfrac{P(A \text{ and } B)}{P(B)},$ where $P(B) \neq 0$. The probabilities $P(A \text{ and } B)$ and $P(B)$ are determined using the original sample space.	Two fair dice are rolled: $P(\text{total } 5 \mid \text{first is } 2)$ $= \dfrac{P(\text{total } 5 \text{ and first is } 2)}{P(\text{first is } 2)}$ $= \dfrac{\frac{1}{36}}{\frac{6}{36}} = \dfrac{1}{6}$
Permutations [7.6]	Arrangements of objects, from first to last, where the order in which the objects are selected is important.	There are 6 permutations of the letters abc: $abc, acb, bac, bca, cab, cba$
Factorial notation [7.6]	The product of any positive integer with all smaller positive integers down to one has a special name and notation.	6 factorial (denoted 6!) is $6! = 6 \cdot 5 \cdot 4 \cdot 3 \cdot 2 \cdot 1 = 720.$
Permutations of n objects taken r at a time [7.6]	The first r factors of $n!$, denoted by $_nP_r = n(n-1)(n-2) \cdots$ $(n - r + 1) = \dfrac{n!}{(n-r)!}.$	Different ways to arrange 3 books from a collection of 5 books is $_5P_3 = 5 \cdot 4 \cdot 3 = 60.$
Combinations [7.6]	Collections of objects selected from a larger collection, where the order of selection is not important.	The ways to select 6 lottery numbers are combinations, since the order of selection is not important.

CONCEPT/SKILL	DESCRIPTION	EXAMPLE
Combinations of n objects taken r at a time [7.6]	The number of ways to select r objects from a collection of n objects, denoted by $$_nC_r = \frac{_nP_r}{_rP_r} = \frac{n!}{(n-r)!\,r!}.$$	The number of ways of selecting a committee of 4 from a group of 20 people is $$_{20}C_4 = \frac{20 \cdot 19 \cdot 18 \cdot 17}{4 \cdot 3 \cdot 2 \cdot 1} = 4845.$$
Binomial coefficients [7.7]	Another name for combinations, usually denoted by $\binom{n}{r} = {_nC_r}.$	The number of ways to get 3 heads on 7 flips of a coin is $$\binom{7}{3} = {_7C_3} = 35.$$
Binomial probabilities [7.7]	When an experiment is repeated n times (called n trials), each trial having only two possible outcomes (success or failure), and all trials being independent, then the probability of having x successes in n trials is called a binomial probability, given by the formula $P(x) = \binom{n}{x} p^x q^{n-x}$, where n is the number of trials, x is the random variable, what is being counted, p is the probability of a success (the outcome x is counting), q is the probability of a failure (the other outcome).	The probability of getting 3 heads on 7 flips of a coin is an example of a binomial probability, where $n = 7, x = 3$ and $p = q = 0.5.$ $$P(3) = \binom{7}{3} 0.5^3 \cdot 0.5^4$$ So, $$= \frac{35}{128} = 0.2734375.$$
Geometric probabilities [7.7]	When an experiment has trials with only two possible outcomes (success or failure), and all trials are independent, then the probability that the first success occurs on the nth trial is called a geometric probability, given by the formula $P(x = n) = q^{n-1} \cdot p$, where p is the probability of success, and q is the probability of failure.	The probability of getting the first two on the 4th roll of a fair die. $$n = 4$$ $$p = \frac{1}{6}$$ $$q = \frac{5}{6}$$ $$P(x = 4) = \left(\frac{5}{6}\right)^{4-1} \cdot \left(\frac{1}{6}\right)$$ $$= \frac{125}{1296} \approx 0.0965$$

Chapter 7 Gateway Review

1. The following table summarizes characteristics of registered voters in a small town district:

	REPUBLICAN	DEMOCRAT	REFORM	UNAFFILIATED	TOTAL
Female	156	192	23	41	412
Male	177	160	45	28	410
Total	333	352	68	69	822

If one voter is selected at random, determine the following probabilities:

a. P(Democrat)

b. P(male)

c. P(female Republican)

d. P(male Reform)

e. P(not Republican)

f. Given the person selected is male, P(Reform)

2. A box of candies contains 25 red, 40 green, 30 yellow, and 50 brown candies. If one candy is selected at random, determine the following probabilities:

a. P(red)

b. P(yellow or brown)

c. P(not green)

d. P(blue)

3. A fair coin is flipped three times.

a. Display the sample space for this experiment.

b. Let the input variable x be the number of heads. Complete this table for the corresponding probability distribution.

NUMBER OF HEADS, x	$P(x)$
0	
1	
2	
3	

4. The six most popular colors for new cars or light trucks (including SUVs and minivans) in North America in a given year are listed in the following table.

	COLOR					
	SILVER	WHITE	GRAY	BLACK	RED	BLUE
Percent of New Cars and Light Trucks	18%	17%	15%	11%	11%	12%

A new car or light truck is randomly selected.

a. What is the probability that the selected vehicle is not red?

b. What is the probability that the selected vehicle is silver or black?

c. What is the probability that the selected vehicle has a color not listed in the table?

5. Select the statement that best completes the following: When simulating the number of heads that result from flipping a fair coin, the law of large numbers means that as the number of flips gets very large,

   a. the probability of getting a head gets closer to one-half.
   b. the relative frequency of heads must eventually equal one-half.
   c. the percentage of heads will eventually equal the percentage of tails.
   d. the relative frequency of tails should eventually get very close to one-half.

6. Suppose the probability that a resident of the United States has traveled to Canada is 0.18, to Mexico is 0.09, and to both countries is 0.04. What is the probability that a randomly selected American traveled to either Canada or Mexico?

7. What is the sample space for the number of boys in a family with five children? Are the outcomes equally likely?

8. In a recent poll, 85% of the season ticket holders for Yankee Stadium surveyed favored testing professional baseball players for steroids and other performance-enhancing drugs. If four season ticket holders are randomly selected, what is the probability that all four will be in favor of drug testing? Assume independence.

9. You live in a large city and commute to a part-time job by subway or by taxi. You take the subway 85% of the time because it costs less. You take a taxi the remaining 15% of the time. When taking the subway, you arrive at work on time 75% of the time. You make it on time 90% of the time when traveling by taxi.

   a. What is the probability that on a given day, you took the subway and arrived on time?

   b. What is the probability that you arrived on time, given you took a taxi?

10. A random sample of 200 people were asked whether they bought a new product and whether they saw an advertisement for the product before the purchase. The results are given in the table.

	SAW ADVERTISEMENT	DID NOT SEE THE ADVERTISEMENT	TOTAL
Purchased	105	20	125
Did Not Purchase	15	60	75
Total	120	80	200

   a. What is the probability that a randomly selected person from the survey purchased the product?

   b. What is the probability that a person purchased the product given the person saw the advertisement?

   c. Does the probability in part b support the claim that advertising influences buying? Explain.

   d. Determine $P(\text{purchase the product} \mid \text{did not see ad})$.

11. The quality control department of the Starr Communications Company, the manufacturer of video-game cartridges, has determined from records that 1.5% of the cartridges sold have video defects, 0.8% have audio defects, and 0.4% have both audio and video defects. What is the probability that a cartridge purchased by a customer will have neither a video nor an audio defect? *Hint:* Use a Venn diagram.

12. You have narrowed your vacation reading list to ten books. But you really only have time to read three of them. In how many ways can you choose the three books, assuming that you will also pick the order in which you will read them?

13. How many eight-digit serial numbers are possible if they cannot start with 0 or 1?

14. The portion of a city street map below shows city blocks packed with buildings. In walking from point A to point B someone needs to walk ten city blocks. How many different ten-block walks are possible? (*Hint:* Start with a shorter walk.)

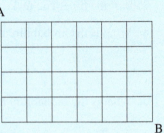

15. Determine the probability of getting all even numbers when a single die is rolled 12 times.

16. In how many ways can the letters of the word *PERMUTATION* be rearranged to form a unique word?

17. A family has five children, all girls. What is the probability that the next child born will be a boy?

18. A coin is tossed 15 times. What is the probability of exactly ten heads?

19. If you know 42% of the registered voters in your county are Democrats, what is the probability of selecting 10 voters at random and getting at least eight Democrats?

20. A vase contains 6 blue marbles, 9 white marbles, and 5 yellow marbles. One marble is taken at random from the vase. What is the probability that:

    a. the marble is not white?

    b. the marble is either blue or white?

    c. the marble is black?

21. By mistake, a manufacturer of CD mini-rack systems includes three defective systems in a shipment of 10 going out to a small retailer.

    a. If two systems are randomly selected, what is the probability that both systems are defective?

    b. The retailer has decided to accept the shipment of CD systems only if none is found to be defective. Upon receipt of the shipment, the retailer examines only four of the CD systems. What is the probability that the shipment will be accepted?

22. In a particular state, 34% of all registered voters are Republican, 39% are Democrats, and 9% are members of other parties. All other voters are considered Independent. Three registered voters are called at random. What is the probability that:

a. all three are Republicans?

b. none of the three are Independent?

c. at least one is a Democrat?

23. The following table contains a record of party affiliation of registered voters in a small town.

GENDER	REPUBLICAN	DEMOCRAT	INDEPENDENT	TOTAL
Male	190	250	40	480
Female	230	150	140	520
Total	420	400	180	1000

If one voter is selected at random, identify the possible events as follows.

R = a Republican is selected.
D = a Democrat is selected
I = an Independent is selected
M = a male is selected
F = a female is selected

Determine the following probabilities.

$P(M) =$ $P(F) =$ $P(R) =$ $P(D) =$ $P(I) =$
$P(\overline{M}) =$ $P(\overline{F}) =$ $P(\overline{R}) =$ $P(\overline{D}) =$ $P(\overline{I}) =$
$P(M\ or\ R) =$ $P(F\ or\ I) =$ $P(M\ or\ F) =$ $P(M\ and\ R) =$
$P(F\ and\ I) =$ $P(F\ and\ M) =$

24. Two standard dice are rolled.

a. What is the probability that the dots on the upper faces total nine?

b. If two standard dice are rolled repeatedly, what is the probability that the first total of nine occurs on the fourth roll?

Problem Solving with Financial Models

You are starting your first job as a salesperson at a sporting goods store. You receive a base salary of $12.50 per hour with time-and-a-half for hours in excess of 40 hours per week. You also receive a 3% commission on your total sales.

1. Determine your annual gross base salary based on working 8 hours per day, 5 days per week, and 52 weeks per year.

2. You are paid biweekly. What is your gross base salary per pay period?

You need to deduct Social Security, Medicare, and federal income tax withholdings from your gross salary to determine your net pay for each pay period. Note that your state does not have a state income tax.

3. a. Social Security tax is 6.2% of the gross pay. What is your Social Security deduction for the gross pay calculated in Problem 2?

b. The Medicare tax is 1.45% of the gross pay. What is your Medicare deduction for the gross pay calculated in Problem 2?

The federal income tax withholding is determined from tax tables found in IRS Publication 15-A, Employer's Supplemental Tax Guide. The biweekly table is printed at the top of page 954. The table is based on your taxable income. Your taxable income is your gross pay minus $150 (for biweekly) for each allowance claimed on the W-4 form you completed when you started the job.

4. a. Assume you claimed one withholding allowance on your W-4 form. What is your taxable income?

b. Look up your taxable income on the biweekly table for a single person. Your biweekly taxable income is over $428 but not over $1,479. The instructions to calculate the income tax withheld are given as: 15% of your taxable income minus $29.90. Let x represent your taxable income, and write an equation given the amount, t, of income tax withheld.

c. Use the equation from part b to determine the amount of tax withheld.

Table B(1)—BIWEEKLY PAYROLL PERIOD (Amount for each allowance claimed is $150.00)

Single Person				Married Person			
If the wage in excess of allowance amount is:		The income tax to be withheld is:		If the wage in excess of allowance amount is:		The income tax to be withheld is:	
Over—	But not over—	Of such wage—	From product	Over—	But not over—	Of such wage—	From product
$0	—$85..............	0%	$0	$0	—$319..............	0%	$0
$85	—$428..............	10% less	$8.50	$319	—$1,006............	10% less	$31.90
$428	—$1,479.............	15% less	$29.90	$1,006	—$3,108............	15% less	$82.20
$1,479	—$3,463.............	25% less	$177.80	$3,108	—$5,950............	25% less	$393.00
$3,463	—$7,133.............	28% less	$281.69	$5,950	—$8,898............	28% less	$571.50
$7,133	—$15,406...........	33% less	$638.34	$8,898	—$15,640...........	33% less	$1,016.40
$15,406	—$15,469...........	35% less	$946.46	$15,640	—$17,627...........	35% less	$1,329.20
$15,469	—	39.6% less	$1,658.03	$17,627	—	39.6% less	$2,140.04

Data Source: Internal Revenue Service

5. Calculate your net pay by deducting the Social Security tax and the Medicare deduction (see Problem 3) and the federal withholding amount (see Problem 4) from your gross pay.

6. You accrue 8 hours of overtime and have $12,000 in sales during one two-week pay period.

   a. Calculate your gross pay for this pay period.
      Gross pay = base pay + overtime pay + commission.

   b. Calculate your net pay for this pay period by completing the following table.

Social Security Tax	
Medicare Deduction	
Taxable Income	
Federal Withholding	
Net Pay	

7. You want to set up a budget to control your expenses. The first step is assessing your current monthly expenditures. You determine the following expense categories: (1) household, (2) auto, (3) entertainment, and (4) miscellaneous. The actual expenses for last month are given below. The number in the parentheses indicates the expense category.

car payment (2), $210	rent (1), $360
car insurance (2), $80	utilities (1), $160
cell phone (1), $60	movies (3), $30
groceries (1), $200	gas (2), $160
dining out/fast food (3), $90	clothing (4), $100
toiletries (4), $40	music (3), $30

a. Complete the following table to determine your expenses for last month.

					TOTAL
Household (1)					
Auto (2)					
Entertainment (3)					
Miscellaneous (4)					
Total					

b. What were your total expenses last month?

c. Assume last month is a typical month in terms of expenses. Will your base salary (without overtime or commission) cover your expenses? Explain your answer.

8. You want to purchase a new tablet that costs $549 plus 7% sales tax. You plan to save your overtime pay to purchase the tablet outright and not pay monthly installments on the bill. Estimate the number of hours you need to work to yield enough net overtime pay to purchase the tablet.

(Use $15\% = 0.15$ federal income tax, plus $6.2\% = 0.062$ Social Security and $1.45\% = 0.0145$ Medicare.)

Activity 8.2

Banking Options

Objective

Analyze data to make decisions about banking, including online banking, checking accounts, overdraft protection, processing fees, and debit/ATM fees.

Congratulations! You just received word that you have the job you interviewed for last week. It will pay $12.50 per hour, and most weeks there will be a 40-hour workweek. You decide to crunch some numbers to see how much money you will have to live on.

1. You will be subject to 15% federal withholding tax, 6.2% for Social Security, and 1.45% for Medicare. If you work 40 hours this week, what will be your net pay after all of the withholding taxes are taken out?

Your employer has offered three payment options:

    Option 1: The traditional paper paycheck

    Option 2: Pay deposited directly into a bank account

    Option 3: Pay directly transferred to a debit card

Debit Card

The debit card option is intriguing, because it means you would never have to worry about signing and depositing your paycheck, nor would you have to set up a bank account.

There are some important conditions on the debit card that you need to consider:

- A $0.50 charge every time the card is swiped for cash or for a purchase

- A $2.50 charge for each cash withdrawal from a company machine and an additional $2.75 if cash is withdrawn from a non-company machine

- A $25 charge for every transaction that is unsuccessful because of insufficient funds

2. You estimate that you will use the card on average 10 times per week and will probably make an average of 5 cash withdrawals every 4 weeks with 4 of these from non-company machines. Although you are going to try to keep track of all of your transactions, you will probably overspend your account once a month. Determine the total of these bank charges every 4 weeks.

You are now concerned that the debit card option may be a risk, so you decide to review your expenses for an average month. Among other things, you want to begin putting aside $500 each month for upcoming college expenses. Additional expenses are summarized in the following table.

ITEM	EXPENSE
Share of the Rent	$350
Food	$125
Cell Phone	$85
Share of the Cable Bill	$35
Share of the Utilities	$55
Car Insurance	$120
Gas	$50
School Expenses	$500
Miscellaneous	$100

3. a. What are your total monthly expenses?

   b. Using the results from Problem 1, how much income will you receive in an average month?

   c. Would you be able to afford the estimated debit card charges from Problem 2?

Online Banking vs. Economy Checking Account

Based on the work above, you decide to explore some other banking options including online banking and an economy checking account. You go to several banks and come home with a wealth of information. A great deal of it is rather confusing. You pick out the options that are most appealing and summarize their features.

Online Banking: Your paycheck will be deposited directly. You can pay all bills on line, but there is a $5.00 per month service charge. It rises to $15 per month if the average balance is less than $5000. There is a $25.00 charge for transactions made with insufficient funds (NSF). A $14 per month fee is charged if you want paper statements. There is an $8.95 fee if you use a teller at a bank. You also may opt for an overdraft protection of $12.50 per day in which the bank pays for the transaction but debits your account $12.50 per day for each day you have insufficient funds. The bank will pay 0.2% interest on the average balance.

Economy Checking Account: Your paycheck will be deposited directly. There is an $8 fee for each 100 checks. You must have a minimum balance of $500 to avoid fees. If the balance ever drops below $500 during the month, the fees are activated. If your account's balance is below $500, then the monthly fee is $7.00 per month. The fee for insufficient funds is $29 per transaction. This account is also eligible for the same overdraft protection as online banking.

Because there is a good possibility of your account being overdrawn, you decide to **investigate overdraft protection**.

   4. a. Suppose you write a check for $65.00, and there are insufficient funds. With overdraft protection, the bank will pay the bill. How much money would you need to deposit the following day to cover the total cost of this transaction if you include the overdraft fee?

b. What is the total cost of this transaction if you cannot make a sufficient deposit for 5 days?

c. You write the same check without the overdraft protection. For the economy checking account, the bank charges you $29 for the bad check. The company you are trying to pay charges 1.25% interest for the 30 days it takes to get them a good check. What is the cost of the transaction if you include the check charges and the interest?

d. Based on the outcome of parts a–c, do you believe that this kind of overdraft protection is a good idea? Explain.

You decide against the overdraft protection and vow to watch any potential overdrafts very carefully. Now, you want to compare the online option to the economy checking.

5. a. You estimate your average balance per month to be $250. You decide there is no need for paper statements. You expect to visit a teller once a month on average. Knowing your budget is tight, you anticipate one overdraft per month, but you are not looking forward to the interest payment. Your bank explains that the interest payment can be estimated closely using simple interest on your average balance. Determine the monthly bank charges less the interest earned.

b. Do the same for the economy checking account, assuming you also have to purchase 100 checks.

Although it may be more work, you decide to go with the checking account.

6. You are looking into two different checking accounts:

Account 1: $7.00 per month service fee and $29.00 per transaction for every check returned for insufficient funds

Account 2: $16.00 per month service fee and $20.00 per transaction for every check returned for insufficient funds

a. Write an equation for Account 1 that relates the total fees, F, to the number of returned checks, n.

b. Write an equation for Account 2 that relates the total fees, F, to the number of returned checks, n.

c. Complete the following table.

NUMBER OF RETURNED CHECKS, n	TOTAL FEES ON ACCOUNT 1	TOTAL FEES ON ACCOUNT 2
0		
1		
2		
3		

   d. Determine the minimum number of returned checks necessary for the fees on Account 2 to equal the fees of Account 1.

   e. If on average you have less than 1 returned check per month, which account would you choose? Explain.

After choosing the checking account, the following information appears on your first monthly statement:

   Opening balance: $232.56

   5 deposits: $386.75, $386.75, $386.75, $386.75, $100.00

   10 checks: $350.00, $84.36, $35.00, $53.97, $120.05, $256.33, $37.55, $63.47, $331.45, $13.12

   Service fee: $7.00

   2 NSF fees: $29.00, $29.00

In addition, you know there are three checks you have written that are not currently accounted for: $101.72, $43.33, and $27.02.

   7. a. From the information given in the statement, what is the current balance in your account?

   b. Taking into account the outstanding checks, how much money do you currently have available in your account?

SUMMARY: ACTIVITY 8.2

When making decisions about banking, there are multiple options to consider:

1. Online banking vs. teller service

2. Overdraft protection

3. Processing fees

4. ATM and debit card fees

Consider them all very carefully, and make decisions that make sense for you.

1. Your employer offers you two options to have your paycheck deposited directly to a debit card. The options are with two different companies, but both come with fees attached.

Option 1: A fixed fee of $5.00 per month plus a swipe fee (a charge every time you use your card) of $0.30 per swipe.

Option 2: A fixed fee of $10.00 per month plus a swipe fee of $0.20 per swipe

a. Write an equation for the fees attached to Option 1 that relate the total fees, F, to the number of swipes, n.

b. Write an equation for the fees attached to Option 2 that relate the total fees, F, to the number of swipes, n.

c. Complete the following table.

NUMBER OF SWIPES, n	TOTAL FEES FOR OPTION 1	TOTAL FEES FOR OPTION 2
0		
10		
20		
30		
40		
50		
60		

d. If you use your card an average of 30 times per month, which option would you choose? Explain.

e. For what number of swipes would the fees on both options be equal?

2. You recently converted to an online banking account. It also came with a debit card. Upon reading the conditions of the account, you discover that in addition to a $7 monthly fee, there is a $0.25 charge for each electronic check and $0.25 for each swipe of the debit card. Therefore, you consider acquiring a separate credit card. The credit card comes with a $108 annual fee.

a. Let x represent your number of monthly transactions, i.e., both electronic checks and credit card swipes. Write a linear function that represents the total fees assigned to your online banking account each month.

b. What is the monthly cost of the credit card you are considering?

c. How many transactions do you need to average each month to justify the additional credit card?

3. You have opened a checking account at a local bank. The fee for insufficient funds is $25.00 per transaction, but you can apply for overdraft protection at $10.00 per day. Last month, you had two unexpected bills: $650 to replace the water pump on your car and $75 to remove a virus from your laptop. You wrote checks for both bills. Your checking account balance was $432.74 at the time. You forgot to deposit money to cover the checks. The bank called a week later to let you know that you were overdrawn. You had to attend to your balance immediately or another day would have added to your charges beyond the one already in the system.

a. Because you did not have overdraft protection, how much money did you need to deposit to bring your balance back to $0.00?

b. If you had overdraft protection, how much money would you have needed to deposit to bring your balance back to $0.00?

Activity 8.3

Time Is Money

Objectives

1. Distinguish between simple and compound interest.

2. Use the compound interest formula to determine the future value of a lump-sum investment earning compound interest.

3. Determine the future value using technology.

4. Determine the effective interest rate.

5. Use the present value formula in a given situation involving compound interest.

6. Determine the present value using technology.

Simple Interest

1. Suppose $10,000 is deposited in a bank at 3.5% annual interest. What is the interest earned after one year?

If you have money invested at **simple interest**, the interest earned each year is based on the original investment, called the **principal**, and not on interest earned in previous years. This means that the interest amount earned remains the same each year.

The formula for the interest on money invested with simple interest is

$$I = Prt,$$

where

$I = $ the amount of interest

$P = $ the principal (the original investment)

$r = $ the annual rate of interest written in decimal form

$t = $ the number of years the principal earns interest.

2. a. What is the total interest on a principal investment of $10,000 at 3.5% simple interest in five years?

b. What is the balance in the account in five years?

You determined the balance by adding the interest earned to the principal. The formula for the balance, A, in an account earning simple interest is

$$A = P + I = P + Prt = P(1 + rt).$$

Compound Interest

The interest paid on savings accounts in banks generally is not simple interest. Interest on money in a bank is compounded. With **compound interest**, the interest earned in each period is added to the balance. The interest earned in the next period is based on this new balance. In other words, the interest earns interest.

If $10,000 is deposited in an account at 3.5% interest compounded annually, the balance after one year is

$$\text{Balance} = 10,000 + 10,000(0.035) = \$10,350.$$

The new balance of $10,350 is now used to compute the interest at the end of the second year. Therefore, the balance at the end of the second year is

$$\text{Balance} = 10,350 + 10,350(0.035) = \$10,712.25$$

3. How much interest was earned on the $10,000 principal in the second year?

The interest in Problem 3 was compounded annually (once a year). Interest can also be compounded more than once a year.

INTEREST IS COMPOUNDED	NUMBER OF TIMES PER YEAR, n
Annually	1
Quarterly	4
Monthly	12
Daily	365

If interest is compounded, the balance, A, in t years is given by the following **compound interest formula**.

> If interest is compounded, then the current balance is given by the formula
>
> $$A = P\left(1 + \frac{r}{n}\right)^{nt},$$
>
> where A is the current balance or compound amount in the account,
>
> P is the principal (the original amount deposited),
>
> r is the annual interest rate (in decimal form),
>
> n is the number of times per year that interest is compounded, and
>
> t is the time in years the money has been invested.

Example 1   *You invest $2000 at 3.5% compounded quarterly. How much money do you have after five years?*

SOLUTION

The principal is $2000; so $P = 2000$. The annual interest rate is 3.5%; so $r = 0.035$. Interest is compounded quarterly; that is, four times per year, so $n = 4$. The money is invested for five years; so $t = 5$. Substituting numbers for the variables in the preceding formula, you have

$$A = 2000\left(1 + \frac{0.035}{4}\right)^{4\cdot5} = \$2380.68 \text{ (nearest cent)}$$

The amount $A = \$2380.68$ in Example 1 is called the **future value** of the investment.

4. In Problem 2, you were asked about $10,000 deposited in an account with a 3.5% annual interest rate. Use the compound interest formula in parts a–d to determine the future value in five years with different compounding periods.

    a. Determine the balance in five years if the interest is compounded annually. What is the total interest earned?

a.

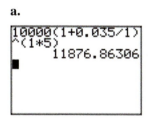

    b. Determine the balance in 5 years if the interest is compounded quarterly. What is the total interest earned?

c. Determine the balance in 5 years if the interest is compounded monthly. What is the total interest earned?

d. Compare these results for compound interest to the result from Problem 2 for simple interest. Why do the compound interest results yield a greater balance than the simple interest result?

Future Value Using Technology

Your TI-83/84 Plus graphing calculator has several built-in functions to help you solve many of the finance problems encountered in this chapter. To access the finance applications menu, press (APPS),

followed by (ENTER).

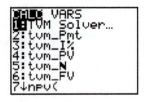

Next, press (ENTER) to select option 1: TVM Solver . . . The following screen will appear:

where the time-value-of-money (TVM) variables represent the following:

N   = total number of compounding periods ($N = nt$);

I%   = r, where $r\%$ is the annual interest rate;

PV   = present value (amount invested or principal);

PMT   = payment;

FV   = future value;

P/Y   = number of payments per year;

C/Y   = number of compounding periods per year;

PMT:END   = payment due at the end of the period; and

PMT:BEGIN = payment due at the beginning of the period.

> **Example 2** *Determine the future value of $2000 invested at 3.5% compounded quarterly for five years (see Example 1).*

SOLUTION

Use the up and down arrow keys to enter the following information into the TVM Solver:

$$N = 20, I\% = 3.5, PV = -2000, PMT = 0, P/Y = 4, C/Y = 4$$

The value of N is 20 because 5 years $\times$ 4 periods/year = 20 compounding periods. The present value is entered as a negative value since it is a cash outflow; 0 is entered for PMT to specify no payments; P/Y and C/Y are set to 4 to indicate 4 periods/year. Now, arrow up to FV to highlight FV = $\blacksquare$ and press (ALPHA) followed by (ENTER) to select [SOLVE]. The future value appears next to $\blacksquare$FV.

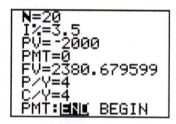

```
N=20
I%=3.5
PV=-2000
PMT=0
FV=2380.679599
P/Y=4
C/Y=4
PMT:END BEGIN
```

The future value is $2380.68, which is the same result as in Example 1.

5. Use the TVM Solver to redo Problem 4. That is, determine the future value in 5 years of $10,000 deposited in an account with 3.5% annual interest rate with different compounding periods. Round your answers to the nearest cent.

a. Determine the balance in five years if the interest is compounded annually.

b. Determine the balance in five years if the interest is compounded quarterly.

c. Determine the balance in five years if the interest is compounded monthly.

d. How do the results in parts a, b, and c compare to the results in the corresponding parts a, b, and c of Problem 4?

Effective Annual Yield

As you can see from the results of Problems 4 and 5, the balance in the account is larger with more compounding periods per year. If interest is compounded more than once per year, the balance actually grows at a higher rate than the stated annual rate. This is called the **effective annual yield** or the **annual percentage yield (APY)**. The effective interest yield can be determined using several different methods. Example 3 demonstrates the use of the graphing calculator.

Example 3   *Money is deposited in an account with an annual interest rate (APR) of 4% compounded monthly. Use the graphing calculator to determine the annual percentage yield (APY).*

SOLUTION

In the APPS menu select option 1, Finance; then option C, Eff(; enter 4 (for 4% annual interest rate) followed by a comma, followed by 12 (for monthly compounding) followed by a closing parenthesis,) ; then press enter.

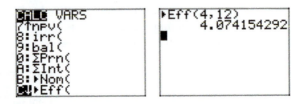

The APY is 4.074%. So a balance in an account that pays 4% interest compounded monthly yields 4.074% annually.

6. Your local bank is offering a five-year certificate of deposit (CD) at an annual interest rate of 4.8% compounded daily ($n = 365$). Determine the annual percentage yield (APY).

Present Value

You want to deposit enough money into the CD described in Problem 6 to have a balance of $6000 at the end of the five years. The amount you need to deposit now to obtain a future balance of $6000 is called the **present value** of the investment.

A formula for the present value can be determined by using the compound interest formula

$$A = P\left(1 + \frac{r}{n}\right)^{nt}$$

In the given formula, A represents the **future value** and P represents the **present value**.

7. a. Solve the compound interest formula for P by dividing both sides of the formula by $\left(1 + \frac{r}{n}\right)^{nt}$.

   b. Replace P, the present value, with PV. Replace A, the future value, with FV. The result is a formula for present value.

7. c.

c. Use the present value formula in part b to determine the amount needed to be deposited (the present value) in the five-year CD that earns 4.8% compounded daily to have $6000 at the end on the five years. Your calculator screen should appear as displayed to the left.

8.

N=1825
I%=4.8
• PV=-4719.841642
PMT=0
FV=6000
P/Y=365
C/Y=365
PMT: **END** BEGIN

The calculator TVM Solver can be used to determine the present value.

8. Redo Problem 7c using the TVM Solver. Enter the following values:
N = 365·5 = 1825, I% = 4.8, PMT = 0, FV = 6000, P/Y = 365, C/Y = 365
and solve for PV. How does your answer compare to the result in Problem 7c?

9. Beginning in 2008 the Federal Reserve, the nation's central bank, has kept its lending rate to member banks very low. This has resulted in low interest rates on both deposits and loans. In 2013, the interest rate paid on a "premium" savings account was 0.85%.

9. a.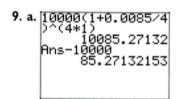

a. Determine the interest created by a $10,000 deposit after one year, if the money was deposited at 0.85% compounded quarterly. Note: 0.85% = 0.0085.

b. Determine the interest created by a $10,000 deposit after five years, if the money was deposited at 0.85% compounded monthly.

9. b.

10000(1+0.0085/1
2)^(12*5)
       10434.00358
Ans-10000
       434.0035832

c. Do you believe that keeping interest rates low encourages savings? Explain.

For the remainder of this chapter, when necessary, round all dollar amounts to the nearest cent.

SUMMARY: ACTIVITY 8.3

1. Simple interest is interest paid on the principal only. The simple interest formula for the accumulated amount in an account, A, is

$$A = P + I = P + Prt = P(1 + rt), \text{ where}$$

I is the amount of interest.

P is the principal (the original investment).

r is the annual rate of interest written in decimal form.

t is the number of years the principal earns interest.

2. Compound interest is interest paid on interest as well as the principal.

3. The **compound amount**, also called **future value**, is the total amount of compound interest and principal at the end of the investment.

4. The compound amount (future value) can be calculated using the **compound interest formula**:

$$A = P\left(1 + \frac{r}{n}\right)^{nt},$$

where A is the compound amount, or future value, in the account,

P is the principal (the original amount deposited),

r is the annual interest rate (annual percentage rate in decimal form),

n is the number of times per year that interest is compounded, and

t is the time in years the money has been invested.

5. The compound amount (future value) can be calculated using the TVM Solver feature of the TI-83/84 Plus graphing calculator.

6. **Effective interest yield** on an investment (called **annual percentage yield** or **APY**) is the true rate of return on an investment.

7. **Present value** (or **principal**) is the amount that must be invested now at compound interest to reach a given future value.

8. **Present value formula** is

$$PV = \frac{FV}{\left(1 + \frac{r}{n}\right)^{nt}},$$

where PV is present value (principal),

FV is future value (amount accumulated),

r is the annual interest rate (annual percentage rate in decimal form),

n is the number of times per year that interest is compounded, and

t is the time in years the money has been invested.

EXERCISES: ACTIVITY 8.3

1. a. You deposit $2000 earned at a summer job in an account that pays 4.2% simple interest. What is the balance in the account in three years?

   b. Determine the balance in three years if the $2000 is deposited in an account that pays 4.2% compounded quarterly.

2. Your high school freshman class had a total of $1200 from fund-raisers during the year. To help pay for a class trip at the end of your senior year, the class deposits the money into a 30-month CD paying 4.2% interest compounded monthly.

   a. Determine the amount the class will receive after 30 months.

   b. How much interest did the class earn on the investment?

3. a. You have just graduated from high school and received $1500 in gifts from family and friends. You also received scholarships in the amount of $800. If you deposit the total amount received into a 24-month CD at 5.5% compounded daily, how much will you receive at the end of 24 months?

b. How much interest did you earn on the investment?

4. You started a new job and received a $2000 signing bonus. In order to save for a new car, you decide to deposit the money into an account earning 6.75% compounded monthly. How much will you have saved after five years?

5. Determine the effective annual yield for $100 invested for one year at 6.75% compounded monthly.

6. An advertisement in the local newspaper claims that the APY on a CD that paid 4.5% interest compounded quarterly was 4.58%. Is this claim accurate?

7. a. You are the beneficiary of a trust fund established by your grandparents 21 years ago, when you were born. The original amount of the fund was $15,000. If the money earned interest at the rate of 6% compounded annually, what is the current amount in the fund?

b. How much money would be in the fund if it had been invested at 6% compounded monthly?

8. The parents of a child have just received a large inheritance and want to put a portion of the money into a college fund. They estimate that they will need $75,000 in 12 years. If they invest the money at 7% annual interest compounded quarterly, how much should they deposit into the fund?

9. You have two investment options:

Option 1: Invest $10,000 at 10% compounded semiannually.

Option 2: Invest $10,000 at 9.5% compounded daily.

Which option has the greater return after five years?

10. When you graduate from law school in four years, you estimate that you will need $20,000 to set up a law office.

a. How much money must you invest now at 9% interest compounded quarterly to achieve this goal?

b. How much interest will you have earned on your investment?

11. You are a member of the town planning board. The town intends to pay off a $4,000,000 bond issue that comes due in 6 years. How much must the town set aside now in order to accumulate the necessary amount of money? Assume the money is invested in an account that earns 8% compounded quarterly.

12. Your car is for sale. You have received two offers:

Offer 1: The first buyer offers you $6000. He wants to pay you $2000 now, $2000 in six months, and $2000 in one year.

Offer 2: The second buyer offers to pay you $5600 in cash now.

a. What is the present value of offer 1? Assume you can earn 10% interest compounded monthly on your money.

b. Which offer is the better deal? Explain.

Continuous Compounding

You could extend the compounding problems in Activity 8.3 so that the compounding periods become shorter and shorter (compounding every hour, every minute, every second), n gets larger and larger. If you consider the period to be so short that it's essentially an instant in time, you have what is called **continuous compounding**. Some banks use this method for compounding interest.

The compound interest formula $A = P\left(1 + \dfrac{r}{n}\right)^{nt}$ is no longer used when interest is compounded continuously. The following steps develop a formula for continuous compounding.

Step 1. Rewrite the given formula as indicated using properties of exponents.

$$A = P\left(1 + \frac{r}{n}\right)^{nt} = P\left[\left(1 + \frac{r}{n}\right)^{n/r}\right]^{rt} \text{ because } \frac{n}{r} \cdot rt = nt$$

Step 2. Let $\dfrac{n}{r} = x$. It follows that $\dfrac{r}{n} = \dfrac{1}{x}$. Note that as the number of compounding periods n gets very large, the value of x also gets very large.

Step 3. Substituting x for $\dfrac{n}{r}$ and $\dfrac{1}{x}$ for $\dfrac{r}{n}$ in the rewritten formula in step 1, you have

$$A = P\left[\left(1 + \frac{r}{n}\right)^{n/r}\right]^{rt} = P\left[\left(1 + \frac{1}{x}\right)^{x}\right]^{rt}.$$

1. a. Now take a closer look at the expression $\left(1 + \dfrac{1}{x}\right)^{x}$. Enter $\left(1 + \dfrac{1}{x}\right)^{x}$ into your calculator as a function of x. Display a table that starts at 0 and is incremented by 100. The results are displayed below.

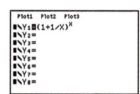 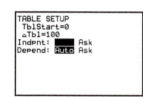 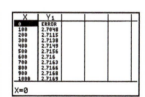

b. In the table of values, why is there an error at $x = 0$?

c. Determine the output for $x = 4000, 5000,$ and $10{,}500$. What happens to the output, $\left(1 + \dfrac{1}{x}\right)^{x}$, as the input, x, gets very large?

The result in Problem 1c may look familiar. Recall from Activity 4.8 that the number $2.7182\ldots$ is represented by the letter e.

The letter e is used to represent the number that $\left(1 + \dfrac{1}{x}\right)^{x}$ approaches as x gets very large.

This notation was devised by mathematician Leonhard Euler (pronounced oiler) (1707–1783). Euler used the letter e to denote this number.

You are now ready to complete the compound interest formula for continuous compounding.

Substituting e for $\left(1 + \dfrac{1}{x}\right)^{x}$ in $A = P\left[\left(1 + \dfrac{1}{x}\right)^{x}\right]^{rt}$, you obtain the following continuous compounding formula.

$$A = Pe^{rt},$$

where A is the current amount, or balance, in the account;

P is the principal;

r is the annual interest rate (annual percentage rate in decimal form);

t is the time in years that your money has been invested; and

e is the base of the continuously compounded exponential function.

Example 1   *You invest $100 at a rate of 4% compounded continuously. How much money will you have after five years?*

SOLUTION

The principal is $100, so $P = 100$. The annual interest rate is 4%, so $r = 0.04$. The money is invested for five years, so $t = 5$. Because interest is compounded continuously, you use the formula for continuous compounding as follows:

$$A = 100e^{0.04 \cdot 5} \approx \$122.14$$

2. a. Calculate the balance of a $10,000 investment in ten years with an annual interest rate of 3.5% compounded continuously.

3. In 2013, the interest rate paid on a "premium" savings account was 0.85%.

 a. Determine the interest created by a $10,000 deposit after 1 year if the money was deposited at 0.85% compounded quarterly.

 b. Determine the value of a $10,000 deposit after five years if the money was deposited at 0.85% compounded monthly.

 c. Determine the value of a $10,000 deposit after ten years if the money was deposited at 0.85% compounded continuously.

4. a. Historically, investments in the stock market have yielded an average rate of 11.7% per year. Suppose you invest $10,000 in an account at an 11% annual interest rate that compounds continuously. Use the formula $A = Pe^{rt}$ to determine the balance after 35 years.

 b. What is the balance after 40 years?

 c. Your grandfather claimed that $10,000 could grow to more than half a million dollars by the time you retire (40 years). Is your grandfather correct in his claim?

SUMMARY: ACTIVITY 8.4

The formula for **continuous compounding** is $A = Pe^{rt}$.

EXERCISES: ACTIVITY 8.4

1. You inherit $25,000 and deposit it in an investment account that earns 4.5% annual interest compounded quarterly.

   a. Write an equation that gives the amount of money in the account after t years.

   b. How much money will be in the account after 10 years?

   c. If the interest were to be compounded continuously at 4.5%, how much money would be in the account after 10 years?

2. Your friend deposits $1900 in an account that earns 3% compounded continuously.

   a. What will be her balance after two years?

   b. Estimate how long it will take your friend's investment to double.

Activity 8.5

Saving for Retirement

Objectives

1. Distinguish between an ordinary annuity and an annuity due.

2. Determine the future value of an ordinary annuity using a formula.

3. Determine the future value of an annuity due using technology.

4. Determine the present value of an ordinary annuity using technology.

5. Solve problems involving annuities.

You have a discussion in your business class regarding a retirement plan, including Social Security, company pension plans, 401(k) accounts, and individual accounts such as Roth IRA. According to a recent report by the Federal Reserve, an individual will need 85% of his income at retirement in order to maintain the same standard of living. Currently, the average 401K account of 60- to 62-year-olds in the United States provides only 25% of the income needed to maintain their standard of living at retirement. Social Security only covers about 45%. Even if you are fortunate to have a pension, there is still a possibility that you will face a retirement "income gap."

An **annuity** is an account into which, or out of which, a series of scheduled payments are made. Annuities are often used to save for long-term goals such as retirement. A contract is made for a series of scheduled payments to be deposited into an account. Then, upon retirement, the lump sum accumulated can be annuitized, and you will receive monthly payments from the annuity.

There are many different types of annuities. One of the most basic is an **ordinary (or fixed) annuity**.

Definition

An annuity into which equal payments are made at the end of each regular periodic interval, with interest compounded at the end of each period, and with a fixed interest rate for each compounding period, is called an ordinary or fixed annuity. The payment is called an annuity payment. The number of annuity payments made each year is the same as the number of compounding periods each year.

The following problem demonstrates how an ordinary annuity grows over time.

1. A deposit of $1000 is made at the end of every six months (semiannually) for two years at 4% annual interest compounded semiannually.

   a. Determine the amount in the account at the end of the first six-month period.

   b. Determine the amount in the account at the end of the second six-month period. Note that the period interest rate is 4% ÷ 2 = 2%.

   c. Enter your results from parts a and b into the following table. Then complete the table.

PERIOD	END-OF-PERIOD VALUE IN THE ACCOUNT
1	
2	
3	
4	

   d. What is the future value of the annuity described in part a in two years?

In an ordinary annuity, the payment is made at the end of each period. If the payment is made at the beginning of each period, the annuity is called an **annuity due**. Therefore, the first annuity payment in an annuity due will earn interest. The difference in the future value of an ordinary annuity and an annuity due is one additional period's worth of interest for the length of the loan period.

2. Suppose the $1000 annuity payment in Problem 1 is made at the beginning of every six months for two years at 4% compounded semiannually.

    a. Determine the amount in the account at the end of period 1.

    b. Determine the amount in the account at the end of period 2.

    c. Record your results from part a and b into the following table. Then complete the table by filling the information for periods 3 and 4.

PERIOD	END-OF-PERIOD VALUE IN THE ACCOUNT
1	
2	
3	
4	

    d. What is the future value of this annuity due account at the end of two years?

    e. Compare the future value of the ordinary annuity in Problem 1d to your result in Problem 2d. Is this what you expected?

Future Value of an Annuity

The future value of an annuity is the total accumulation of the payments and interest earned. The formulas to determine the future value of an annuity are:

Future Value of an Ordinary Annuity	**Future Value of an Annuity Due**
$$FV = P \cdot \frac{(1 + i)^N - 1}{i}$$	$$FV = P \cdot \frac{(1 + i)^N - 1}{i} \cdot (1 + i),$$

where FV = future value,

        P = annuity payment,

        i = interest rate per period (written as a decimal), and

        N = total number of periods.

3. Consider the ordinary annuity in Problem 1: A deposit of $1000 made at the end of every six months for two years at 4% annual interest compounded semiannually.

   a. Determine the interest rate-per-period and the number of periods.

   b. Use the appropriate formula above to determine the future value of this annuity.

You may be wondering if the TVM Solver feature of your graphing calculator can be used to determine the future value. The answer is yes.

4. Redo Problem 3 using the TI-84 Plus graphing calculator. (See screen on left.)

   a. Enter the values for the TVM Solver variables into the calculator.

4. a.

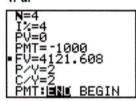

```
N=4
I%=4
PV=0
PMT=-1000
▪FV=4121.608
P/Y=2
C/Y=2
PMT:END BEGIN
```

Be careful. The value for I% is the annual interest rate; the PMT (payment) value is entered as a negative number since the payment is a cash outflow. Since the annuity is ordinary, the payment is made at the end of the period. Therefore, the END *must* be highlighted. Simply move the cursor to END and press (ENTER).

   b. Determine the future value of the annuity. How does your result compare to the answer in Problem 3b using a formula?

5. a. Suppose $1000 is deposited at the beginning of every six months for two years at 4% compounded semiannually. Determine the future value of this annuity by using the appropriate formula.

   b. Redo Problem 5a using the TVM calculator application. How does the result compare with the future value of the annuity due in Problem 2d?

6. How much money will accumulate if you deposit $500 at the beginning of each year for 40 years at 6% compounded annually?

Present Value of an Annuity

One of the requirements in your business class is to write a report describing your retirement plans. You estimate that you will need $20,000 a year at the end of each year for 20 years to supplement your pension and Social Security during retirement. How much money (lump sum) will you need at the time of retirement so that you can make the $20,000 annual payment for 20 years? You assume that the lump-sum amount needed will earn 5% annual interest compounded annually during the payout period.

Until now, you have considered annuities in which you start with a zero balance. Then, by making equal annuity payments for a given period of time at compound interest, a future value or accumulation is reached. This new situation starts with a lump-sum amount (unknown) so that equal payments ($20,000 per year) can be made over a specified time (20 years).

The starting lump-sum amount will earn interest (5% compounded annually) while annuity payments are being made. However, the starting amount will continue to decline to a zero balance at the end of 20 years. The lump sum that must be present in the beginning is called the **present value of an annuity**.

> **Definition**
>
> The **present value of an annuity** is a lump sum that is put into a fund in order for the fund to payout a specified regular payment for a specified amount of time.

7. Now, let us review the details of the problem just given.

As part of your retirement plan, you want to set up an annuity in which a regular payment of $20,000 is made at the end of each year. You need to determine how much money must be deposited (present value) in order to make the annuity payment for 20 years. Assume the money is earning 5% compounded annually.

a. What is the annuity payment? What is the number of payments?

b. What is the annual interest rate?

c. What is the future value?

d. Is the annuity ordinary or due?

7. e.

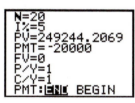

```
N=20
I%=5
PV=249244.2069
PMT=-20000
FV=0
P/Y=1
C/Y=1
PMT:END BEGIN
```

e. Use the TVM feature of your graphing calculator to determine the present value of the annuity.

Formulas to determine the present value of an annuity are given in Exercise 8 of this activity.

Applications

8. Your business teacher asks you to consider 2 different retirement plans for someone who retires after 40 years of work. Assume money in each plan earns 4.75% compounded annually and that each plan is an ordinary annuity. If payments into the plan stop before retirement, the accumulated amount will continue to accrue interest at 4.75% compounded annually until retirement.

Plan 1: Payments of $500 per year are made into the account for the first 25 years and then stop. No payments are made for the next 15 years.

Plan 2: No payments are made into the account for the first 15 years, and then payments of $500 per year are made for the next 25 years.

a. How much money will be in the account for Plan 1 at retirement?

b. How much money will be in the account for Plan 2 at retirement?

    c. Each plan has payments of $500 per year for 25 years. Why does Plan 1 have a greater balance at retirement?

9. When you were born, your parents began depositing $500 at the beginning of each year for ten years into a college fund, paying 4% per year compounded annually. They then increased the annual payments to $1000 for the next eight years, but the interest rate dropped to 2.5% compounded annually. Determine the balance in the college fund after the 18 years of payments.

SUMMARY: ACTIVITY 8.5

1. An **annuity** is the payment (or receipt) of equal cash payments per period for a given period of time.

2. Ordinary annuity is an annuity in which the payments are made at the end of the period.

3. An **annuity due** is an annuity in which the payments are made at the beginning of the period.

4. The **future value** of an annuity is the total accumulation of the payments and interest earned.

5. Future value of an ordinary annuity

$$FV = P \cdot \frac{(1 + i)^N - 1}{i}$$

Future value of an annuity due

$$FV = P \cdot \frac{(1 + i)^N - 1}{i} \cdot (1 + i),$$

where FV = future value,

      P = annuity payment,

      i = interest rate per period (written as a decimal), and

     N = total number of periods.

6. The **present value** of an annuity is the lump sum that is put into a fund in order for the fund to pay out a specified regular payment for a specified amount of time.

7. The future value (and present value) of an annuity can be determined using the TVM solver feature of the TI-83/84 Plus graphing calculator.

EXERCISES: ACTIVITY 8.5

1. A college fund is established in which $100 is deposited at the end of every month at 4% per year compounded monthly. If the account started when you were five years old, how much will be in the account when you turn 18?

2. After making a $1500 down payment on a truck, you paid $210 per month for 36 months for a loan. The interest rate for the loan was 6% compounded monthly.

 a. Use the TVM feature of your graphing calculator to determine the amount of the loan (a present value).

 b. What was the original cost of the truck?

3. As part of your retirement plan, at the beginning of each year you want to invest $2000 into an IRA. If you plan to retire in 25 years, how much will be in the IRA if it earns 6% compounded annually?

4. Suppose you have $100,000 in a retirement account at your current place of employment. Since you are beginning work at a new company, you plan to roll over the $100,000 into an IRA account earning 5.5% per year compounded quarterly. You plan to deposit an additional $2000 at the end of each quarter into a 401k account until you retire. The 401k account also pays 5.5% per year compounded quarterly.

 a. If you plan to retire in 20 years, what amount of money will be in the IRA account?

 b. How much money will be in the 401k account?

5. The winner of the $2,000,000 state lottery will receive 20 payments of $100,000 per year. The payments are to be made as follows:

 1. The first payment of $100,000 is to be made immediately.
 2. The next 19 payments of $100,000 will be made at the end of each year.

 Determine the amount of money that the state needs to have on deposit in order to make the payments. Assume that the balance on deposit will earn 5% per year compounded annually.

6. As part of your retirement plan, you have decided to deposit $2500 at the beginning of each year into an account paying 5.5% interest compounded annually.

 a. How much will be in the account after ten years?

 b. How much will be in the account after 20 years?

 c. If you retire in 30 years, what will be the total value of the account?

 d. Suppose you had invested the $2500 at the beginning of each year into an account earning 6.5% compounded annually. How much more would you have in this account after 30 years?

7. You plan to have $500,000 available when you retire. You want to liquidate that amount in retirement, so that you will receive equal amounts for 25 years. At the end of 25 years, the balance in the account will be $0.00.

If the retirement fund earns 6.25% interest compounded annually, how much will you receive each year? Assume an ordinary annuity.

8. The formulas to determine the present value of an annuity are:

Present Value of an Ordinary Annuity	Present Value of an Annuity Due
$PV = P \cdot \dfrac{1 - (1 + i)^{-N}}{i}$	$PV = P \cdot \dfrac{1 - (1 + i)^{-N}}{i} \cdot (1 + i)$

where PV = present value, P = annuity payment, i = interest rate per period, N = number of periods

   a. You want to receive $2000 at the end of each year for the next ten years. How much should be on deposit now at 6.75% compounded annually to accomplish this goal? Use the appropriate present value formula.

   b. What technological tool would you use to verify your results in part a?

   c. Use the tool selected in part b to verify your result in part a.

9. Two other types of annuities are variable annuities and immediate annuities. Using the Internet, financial magazines, books, and other resources, write a report that describes each annuity and how each is different from or similar to an ordinary annuity. Be sure to include the advantages and disadvantages of each type. Be prepared to give an oral presentation of your findings.

10. Other options for saving for retirement include IRAs (traditional and Roth), 401(k)s, and 403(b)s. Using the Internet, financial magazines, and other resources, write a report in which you explain the differences among the different savings options and describe the advantages of each option over the others.

Activity 8.6

Buy or Lease?

Objectives

1. Determine the amortization payment on a loan using a formula.

2. Determine the amortization payment on a loan using technology.

3. Solve problems involving repaying a loan or liquidating a sum of money by amortization model.

You are interested in purchasing a new car. You have decided to buy a Honda Civic EX for $21,605. The credit union requires a 10% down payment and will finance the balance with a 4.5% interest loan for 36 months. The sales tax is 8.25%, and there is a document fee of $75.

1. a. What is the total purchase price of the car including the sales tax and document fee?

b. What is the amount of the down payment?

c. What is the total amount of the loan?

Many large-sum loans, such as business loans for new equipment, home mortgages, and car loans are repaid through **amortization**. Generally, a loan is repaid by a series of equal periodic payments over a specific period of time. Each payment typically includes an amount for interest and a principal reduction (amortization).

When amortizing a loan, the original amount of the loan is known (present value). Amortization is actually an application of an annuity, in which the annuity payment is actually the amortization payment. Therefore, you can use the present value formula for annuities (see Exercise 8 in Activity 5.4) to develop a new formula for determining the amortization payment.

2. Solve the following present value formula from Exercise 8 in Activity 8.5 for P.

$$PV = P \cdot \frac{1 - (1 + i)^{-N}}{i}$$

Let *Amt* represent the amortization payment. Substituting *Amt* for P in the formula obtained in Problem 2 leads to the following.

Calculating Amortization Payments by Formula

The formula to determine the amount of the payment, *Amt*, is

$$Amt = PV \cdot \frac{i}{1 - (1 + i)^{-N}}$$

or equivalently

$$Amt = PV \cdot \frac{i}{1 - \dfrac{1}{(1 + i)^{N}}}$$

where   *Amt* = amortization payment (principal and interest payment),

   PV = amount of loan (present value),

   i = interest rate per period, and

   N = total number of payments.

Example 1   *What is the monthly amortization payment required to pay off an $8000 loan at 6% interest in 2 years?*

SOLUTION

The rate per compounding period is $i = \dfrac{r}{n} = \dfrac{6\%}{12} = 0.5\% = 0.005$.

The total number of compounding periods is

$N = 2$ years $\times$ 12 periods per year $= 24$ periods.

Use the amortization payment formula with $PV = 8000$, $N = 24$, and $i = 0.005$.

$$Amt = PV \cdot \frac{i}{1 - (1 + i)^{-N}}$$

$$= 8000 \cdot \frac{0.005}{1 - (1 + 0.005)^{-24}}$$

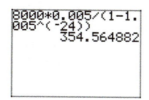

```
8000*0.005/(1-1.
005^(-24))
          354.564882
```

The monthly payment is $354.56.

3. Use the amortization payment formula to calculate the monthly payment on your car loan in Problem 1.

4. Redo Problem 3 using the TVM Solver feature of your calculator. Compare your results.

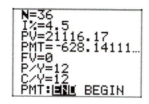

```
N=36
I%=4.5
PV=21116.17
PMT=-628.14111...
FV=0
P/Y=12
C/Y=12
PMT:END BEGIN
```

You can use this technology to create amortization models to investigate automobile financing, including amortization models to help compare buying versus leasing a vehicle.

Leasing a Car

Rather than purchasing the car, you have the option of **leasing**. The car salesperson explains to you that a lease is an agreement in which you make equal monthly payments for a specific period of time. At the end of this period, you return the car to the leasing dealer. You do not have ownership of the car and, therefore, have no equity or asset at the end of the leasing period. You have the option of purchasing the car at a predetermined price. There can be end-of-lease termination fees as well as charges for excess mileage or damage.

She gives you the following advertisement that the dealership just placed in the local newspaper:

Lease a New Honda Civic Ex

Only $299 per month for 36 months

WITH NO SECURITY DEPOSIT

$2500 at signing plus tax and document fee with approved credit

5. How does the monthly leasing payment compare to the monthly purchase payment determined in Problem 3?

Since the monthly leasing payment is much lower than the monthly amortization payment, it seems like you are getting more for your money. Rather than comparing the monthly payments, you decide to compare the total cost of purchasing (Problem 6) to the total cost of leasing the car (Problem 7).

6. a. What is the down payment if you purchase the car?

b. What is the total amount of the 36 payments if you purchase the car?

c. Your down payment could have earned interest at 3.25% compounded annually for 3 years. How much interest would have been earned?

When the loan is paid off in 3 years, the **residual value** is the amount the dealer thinks the car will be worth at the end of the lease time. It is your equity in the car. The dealer tells you the residual value is $12,625.

d. Use parts a–c and the residual value to determine the total cost of purchasing the car.

7. a. What is the total of the 36 monthly lease payments?

b. What is the down payment if you lease the car?

c. The 8.25% sales tax is based on the amount of the car you use in 3 years. This is the purchase price minus the residual value. $(21{,}605 - 12{,}625 = 8980)$. What is the sales tax amount if you lease the car?

d. What is the total amount due at signing if you lease the car?

e. The amount in part d could have earned interest at 3.25% compounded annually for 3 years. How much interest is lost?

f. Assume there are no end-of-lease termination fees for damage or excess mileage. What is the total cost of leasing the car?

8. Is it better to lease or buy your new car? Explain your answer.

SUMMARY: ACTIVITY 8.6

1. Amortization is the process of repaying a loan by a series of equal periodic payments over a specific period of time. Amortization can also be used to liquidate a fund down to a zero balance.

2. The **amortization payment** can also be determined using the following formula:

$$Amt = PV \cdot \frac{i}{1 - (1 + i)^{-N}}$$

or equivalently

$$Amt = PV \cdot \frac{i}{1 - \dfrac{1}{(1 + i)^{N}}},$$

where Amt = amortization payment,
   PV = amount of loan,
   i = interest rate per period, and
   N = number of payments.

EXERCISES: ACTIVITY 8.6

1. a. The price for a new Ford Focus is $20,595.00 including tax and fees. You can put $2,595.00 down so you need to finance $18,000.00. You find a 48-month loan at 4% interest. Use technology to create an amortization model to investigate automobile financing.

i. Use the TVM Solver to determine the amortization monthly payment.

ii. Using the amortization monthly payment, determine the total amount of the 48 loan payments?

    iii. You could have deposited the $2,595.00 down payment at 2% interest compounded monthly. Use technology to determine the amount of interest that could have been earned over the 4-year period?

    iv. At the end of the 48-month loan, the car's residual value will be $11,400. Calculate the total cost of buying the car.

b. Instead of buying the Ford Focus, you can lease it for only $300 per month with $2500.00 due at signing. The $300 payments include taxes and fees.

    i. What is the total amount of the 48 lease payments?

    ii. You could have deposited the $2,500.00 down payment at 2% interest compounded monthly. What amount of interest could have been earned?

    iii. What is the total cost of leasing the car?

c. Use the results from parts a. iv and b. iii to determine if it is better to buy or lease the car.

2. Your grandmother borrows $5000 from the credit union to purchase a car. She agrees to repay the loan in 36 equal monthly payments. The credit union charges 6% per year compounded monthly.

a. What is the monthly payment? Use the amortization formula and then verify your result using your graphing calculator.

b. How much interest was charged for the loan?

3. Your parents buy a new high-definition (HD) TV for $2000 and finance the purchase at 9.5% interest compounded monthly for 24 months.

a. What is the monthly payment?

b. What was the total finance charge?

4. A dealership offers two options to finance $22,500 for a new truck:

Option 1: 0% financing for 48 months
Option 2: A $3000 rebate

If a customer chooses the rebate, she can obtain a loan for the balance at 4.5% compounded annually for 48 months.

a. If the customer chooses option 1, what is the monthly payment?

b. If the customer chooses option 2, what is the monthly payment?

c. Which option should the buyer choose? Explain.

5. The outstanding balance on your credit card is $4000. The bank issuing the credit card is charging 18.5% per year compounded monthly. You decide you must pay off the balance before you get married in 18 months. If you make equal monthly payments for the next 18 months, determine the monthly payment. Assume no additional charges are made.

6. Your cousin makes a profit of $30,000 on the sale of her small business. She deposits the money into an account paying 4.2% interest compounded annually. She wants to make equal annual withdrawals from the account over the next five years. At the end of the five years, there will be no funds in the account.

a. Determine the amount of each withdrawal. Assume the withdrawal is made at the beginning of each year.

b. Determine the amount of each withdrawal if the money must last 10 years.

7. Your parents are starting to make a budget for their retirement. They have been contributing to an account that will provide a portion of their retirement income. Money in this account is expected to earn interest at the annual rate of 5%. They determine that they will need $30,000 per year for 20 years from this account. How much money needs to be in the account at retirement to meet this goal?

8. Find an advertisement in your local newspaper for a lease offer on a vehicle you would like to own. Obtain the information necessary to determine whether it is better to lease or purchase the vehicle. Use your local dealership, banks, and the Internet. Submit a report of your findings and conclusions.

Objectives

1. Determine the amount of the down payment and points in a mortgage.

2. Determine the monthly mortgage payment using a table.

3. Determine the total interest on a mortgage.

4. Prepare a partial amortization schedule of a mortgage.

5. Determine if borrowers qualify for a mortgage.

You are the loan officer at a bank that specializes in long-term loans, called **mortgages**, in which property is used as security for a debt. A young couple, recently married, has decided to purchase a townhouse at a negotiated price of $120,000. They have applied for a conventional mortgage in which the interest rate on the loan remains fixed for the duration of the loan. Because the townhouse is older, the bank is requiring a 20% down payment. The bank is currently offering a fixed interest rate of 4.25%.

The bank is also requiring 2 points for their loan at the time of closing (the final step in the sale process). You explain to the couple that charging points enables the bank to reduce the amount of interest on the mortgage loan and, therefore, will reduce the amount of monthly payment.

1. Determine the amount of the down payment. This amount is paid to the seller at closing.

2. a. If the couple agrees to pay the premium for private mortgage insurance (PMI), the bank can lend up to 95% of the value of the property. What would be the down payment if the bank required a 5% down payment?

b. Assuming they agree to purchasing the PMI, what is the amount to be mortgaged?

c. A point is 1% of the amount to be borrowed. Determine the amount of the points. This amount is paid to the bank at closing.

Calculating the Monthly Mortgage Payment

In Activity 8.6, you learned that amortization is a special type of annuity in which a large loan is paid off with equal periodic payments over a period of time. Mortgages are loans that are paid off using amortization. The amount of the mortgage is actually the present value of an annuity.

The amount of the monthly mortgage payments can be calculated by using a table. Since mortgages often run for a period of 5 to 30 years, there is a special present value table in which the periods are listed in years. Also, since mortgage loans involve large sums of money, the table values represent the monthly payment per $1000 of mortgage.

Example 1 illustrates the procedure to calculate the monthly mortgage payment using Table 8.1 on page 989.

Example 1 *Determine the monthly payment and total interest on a $250,000 mortgage at 4% for 20 years.*

SOLUTION

Step 1. Divide the amount to be mortgaged by 1000. This gives the number of thousands of dollars to be mortgaged.

$$\frac{250,000}{1000} = 250$$

Step 2. Using Table 8.1, locate the table value, monthly payment per $1000 financed, at the intersection of the number of years row (20) and annual interest rate column (4%).

Table value is 6.06.

Step 3. Multiply the amount to be mortgaged (step 1) by the table value (step 2) to determine the monthly mortgage payment.

$$250 \cdot 6.06 = \$1515$$

Table 8.1 Monthly Principal and Interest per $1000 of Mortgage

INTEREST RATE (%)	5 YEARS	10 YEARS	15 YEARS	20 YEARS	25 YEARS	30 YEARS
3%	17.97	9.66	6.91	5.55	4.75	4.22
3.25%	17.97	9.78	7.03	5.68	4.88	4.36
3.5%	18.2	9.89	7.15	5.8	5.01	4.5
3.75%	18.31	10.01	7.28	5.93	5.15	4.64
4%	18.42	10.13	7.4	6.06	5.28	4.78
4.25%	18.53	10.25	7.53	6.2	5.42	4.92
4.5%	18.65	10.37	7.65	6.33	5.56	5.07
4.75%	18.76	10.49	7.78	6.47	5.71	5.22
5%	18.88	10.61	7.91	6.6	5.85	5.37
5.25%	18.99	10.73	8.04	6.74	6	5.53
5.5%	19.11	10.86	8.18	6.88	6.15	5.68
5.75%	19.22	10.98	8.31	7.03	6.3	5.84
6%	19.34	11.11	8.44	7.17	6.45	6
6.25%	19.45	11.23	8.58	7.31	6.6	6.16
6.5%	19.57	11.36	8.72	7.46	6.76	6.33
6.75%	19.69	11.49	8.85	7.61	6.91	6.49
7%	19.81	11.62	8.99	7.76	7.07	6.66
7.25%	19.92	11.75	9.13	7.91	7.23	6.83
7.5%	20.04	11.88	9.28	8.06	7.39	7

Using Table 8.1, the monthly mortgage payment in Example 1 is $1515. The TVM Solver feature of your TI-83/84 Plus can be used to determine the monthly mortgage payment. Let N = 240, I% = 4.0, PV = 250,000, FV = 0, P/Y = C/Y = 12, PMT:END, you obtain PMT = −1514.95. Your screen should appear as follows:

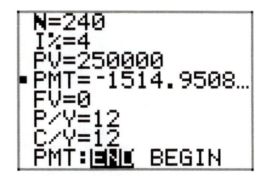

Note that there is a small difference between the results obtained from the graphing calculator and the result using the mortgage table. This is a result of rounding numbers in the calculations.

3. a. Recall from Problem 2b that the couple is applying for a mortgage amount of $114,000 at 4.25% APR. Determine the monthly payment for a 30-year mortgage.

b. What is the total interest paid? Use the monthly payment obtained by the TVM Solver.

4. a. The couple would like to compare the monthly payment and total interest paid for different lengths of time, in years. Complete the following table by using an amortization schedule calculator on the Internet (www.myamortizationchart.com):

$114,000 MORTGAGE AT 4.25% ANNUAL INTEREST		
LENGTH OF MORTGAGE	MONTHLY PAYMENT	TOTAL INTEREST
15		
20		
25		
30		

b. What is the effect on the monthly payment and total interest paid as the length of the mortgage increases?

Amortization Schedule

The couple decides on a 30-year mortgage in order to have the lowest monthly payments. You explain to them that the monthly payment is applied to pay interest on the loan and to reduce the amount owed. The couple asks you if the $560.81 monthly payment is divided equally between reducing the amount owed (principal) and paying the interest. They are surprised that this is not the case.

Example 2 *The calculated monthly payment for the couple's 30-year $114,000 mortgage at 4.25% annual interest is $560.81 (using TVM). Determine the outstanding balance at the start of the second month.*

SOLUTION

Step 1. Determine the amount of interest in the first month using $I = Prt$, where P is the outstanding balance of the loan for given month, r is the annual interest rate (fixed at 4.25%), and t is time $\left(\text{fixed at } \dfrac{1}{12} \right)$.

For the first payment, $P = 114{,}000$, $r = 0.0425$, $t = \dfrac{1}{12}$

$$I = Prt = 114{,}000(0.0425)\left(\frac{1}{12}\right) = \$403.75$$

Step 2. Determine the portion of the payment used to reduce the amount of the principal by subtracting the amount of interest owed (from step 1) from the monthly payment.

$$560.81 - 403.75 = \$157.06$$

Step 3. Calculate the outstanding balance for the next month by subtracting the amount to reduce the principal (step 2) from the outstanding balance of the current month.

$$114,000 - 157.06 = \$113,842.94$$

5. a. Repeat steps 1–3 in Example 2 to complete the following table:

PAYMENT NUMBER	MONTHLY PAYMENT	INTEREST	AMOUNT USED TO REDUCE PRINCIPAL	END-OF-MONTH BALANCE OF LOAN
1	$560.81	$403.75	$157.06	$113,842.94
2				
3				
4				

b. Using the results in part a, determine whether each of the following gradually increases, gradually decreases, or remains the same, as monthly payments are made over the years.

    i. monthly payment   _____

    ii. interest   _____

    iii. amount that reduces principal   _____

    iv. end-of-month balance   _____

The table in Problem 5a is a partial **amortization schedule** for the first four payments. An amortization schedule for the entire loan would require 360 lines. (Why?) Therefore, such schedules are usually done by spreadsheet programs and calculator Web sites. For example, see the monthly mortgage payment site at www.myamortizationchart.com.

Qualifying for a Mortgage

An important responsibility of a mortgage loan officer is to determine the borrower's ability to repay the loan. To determine if the couple is qualified for the mortgage, you obtain the following information:

Couple's combined gross annual income: $60,000

Estimated annual property taxes on house: $ 2200

Annual house insurance: $ 670

6. a. Determine the total monthly housing expenses by adding the monthly mortgage payment and the monthly expenses for property taxes and house insurance.

b. Determine the couple's combined gross monthly income.

Most lenders of conventional loans use the qualifying rule that the monthly housing expenses should not be more than 28% of the monthly gross income. The ratio

$$\frac{\text{monthly housing expense}}{\text{monthly gross income}}$$

is called the **housing expense ratio**.

7. a. Using the results in Problem 6, determine the housing expense ratio for the couple applying for the loan.

b. Do they qualify for the loan?

8. The couple has just closed on the townhouse. Determine the total cost of the house by summing the down payment, the principal, the interest, and the points.

There are usually many other expenses every home buyer needs to pay when purchasing a house. These expenses include the cost of a lawyer, loan application fees, state and local real estate taxes, and title search fees.

SUMMARY: ACTIVITY 8.7

1. A **mortgage** is a long-term loan in which the property is used as security for the debt.

2. **Amortization** is a special type of annuity in which a large loan is paid off with equal regular payments over a specified time.

3. The monthly mortgage payment can be determined by using a table (see Example 1) or the TVM Solver feature of the TI-83/84 Plus graphing calculator.

4. An **amortization schedule** is a list containing the payment number, payment on the interest, payment on the principal, and balance of the loan.

5. A borrower qualifies for a mortgage if the **monthly housing expense** is no more than 28% of the monthly gross income.

EXERCISES: ACTIVITY 8.7

1. A house is purchased at a selling price of $285,000. The bank requires a 15% down payment. The current mortgage rate is 3.5%.

    a. Determine the amount of the down payment.

    b. What is the amount to be financed?

    c. What is the monthly payment for a 20-year mortgage on the amount in part b?

    d. What is the total interest paid?

2. Your aunt is buying a new condominium for $210,000. A 20% deposit is required. She secured a mortgage at 4.25% for 30 years.

    a. What is the amount of the down payment?

    b. What is the amount to be financed?

    c. What is the monthly payment for a 30-year mortgage on the amount in part b?

    d. What is the total interest paid?

3. a. How much money in interest would your aunt save in mortgaging the condominium in Problem 2 if she obtained a 20-year mortgage?

    b. By how much would her monthly payments increase if she chose the 20-year mortgage?

4. The monthly payment for a 30-year, $185,000 conventional mortgage at 3.75% is $856.76.

    a. Determine the amount of interest in the first payment.

    b. How much of the first payment is applied to the principal?

    c. What is the outstanding balance at the start of the second month of the loan?

d. Enter the results from parts a–c into the table below for payment 1. Then, repeat the steps to complete the table for payments 2 through 4.

PAYMENT NUMBER	MONTHLY PAYMENT	INTEREST	AMOUNT USED TO REDUCE PRINCIPAL	END-OF-MONTH BALANCE OF LOAN
1				
2				
3				
4				

e. Use the Amortization Schedule Calculator at www.myamortizationchart.com to create an amortization model to investigate home financing. Use your model to verify your results for the amortization schedule in part d.

5. You are looking for a mortgage to purchase a house selling for $300,000. You have gathered the following information about mortgage offers at two local banks:

First Bank: 10% down payment, an interest rate of 3.75%, a 30-year conventional mortgage, and 2 points to be paid at the time of closing

Second Bank: 20% down payment, an interest rate of 4.5%, a 25-year conventional mortgage, and no points

a. Determine the total cost of the house if the first bank is selected.

b. What is the total cost of the house if the mortgage offer from the second bank is chosen?

c. Based on the results from parts a and b, which bank should you select? Explain.

6. For new homeowners, banks often require that $\frac{1}{12}$ of the estimated property taxes and house insurance premiums be included in the monthly payment. Each month, the taxes and insurance portions of the payment are deposited into a type of savings account called an **escrow account**. The bank pays the property taxes and insurance premiums when they are due.

A new townhouse is purchased with a mortgage of $125,000 at 4.75% for 30 years.

a. What is the monthly payment needed to pay principal and interest?

b. If the property taxes are $2100 per year and the insurance premium is $875.70, what amount must be added to the monthly payment?

c. Determine the total monthly payment for principal, interest, taxes, and insurance (called the PITI payment).

7. Your parents sold their house for a profit of $65,000 and have just received $30,000 from your grandparents' estate. They offer $310,000 on a new house, which has been accepted by the seller.

a. Your parents plan to make a large down payment, including the monies from the sale of their old house, your grandparents' estate, and $15,000 from savings. What will be the amount of their down payment?

b. What is the amount they will be financing?

c. The lender has offered a 30-year mortgage at 4.5% with no points. The estimated property tax is $4500 and hazard insurance costs $1200. Determine the PITI payment.

8. You have enough money saved for the down payment for a new home. The monthly mortgage payment will be $850.40. All that is left is whether you qualify for the mortgage. Your annual gross earnings are $45,000. The estimated annual property taxes on the home are $1500 and the annual house insurance is $750.

a. Determine your total housing monthly expenses including the monthly mortgage payment and the property taxes and the insurance.

b. Determine your combined gross monthly income.

c. Determine your housing expense ratio.

d. Do you qualify for the loan?

Renting versus Buying a House

Objective

Compare renting versus buying a house.

In Activity 8.7, Home Sweet Home, a couple purchased a townhouse for $120,000. The couple had the opportunity to rent the townhouse for $650 per month. Would it have been better for the couple to rent the townhouse than to buy it? Comparing buying to renting is actually a fairly complicated endeavor. There are several Web sites that feature a rent versus buy calculator in which you are asked to provide certain information. You can use this technology to create amortization models. The analysis in Problems 1–4 demonstrates many of the calculations and much of the information that is necessary in making a rent versus buy comparison.

1. a. Recall that the monthly payment for the couple's 30-year $114,000 mortgage at 4.25% annual interest is $560.81. Record the total mortgage payments for specified years in column 1 of the table below.

 b. Assume that the annual property tax is initially 1% of the selling price ($120,000) and increases by 3% per year. Complete column 2 of the table.

 c. The total interest portion of mortgage payments for the year is recorded in column 3. These values were obtained from the amortization schedule for a 30-year $114,000 mortgage at 4.25% at www.hsh.com.

 d. Mortgage interest and property taxes are an itemized deduction with a tax savings equal to the total amount paid multiplied by the couple's marginal tax rate. Assume the couple's marginal tax rate is 30%. Complete column 4 of the table.

 e. The end-of-the-year balance on the loan is recorded in column 5 of the table. Complete column 5 of the Table. Recall, the total of the mortgage payments for the year minus the total interest paid for the year gives the amount applied toward reducing the principal (mortgage balance). These values could also be obtained from the amortization table in part c.

END OF YEAR	(1) MORTGAGE PAYMENTS ($)	(2) PROPERTY TAXES ($)	(3) MORTGAGE INTEREST ($)	(4) TAX SAVINGS ($)	(5) MORTGAGE BALANCE ($)	(6) APPRECIATED HOUSE VALUE ($)
1			4807.85			
2			4724.56			
3			4637.66			
4			4546.99			
5			4452.40			

 f. Due to inflation, the value of the townhouse increases in value. Inflation rates are difficult to predict. Assume an inflation rate of 3% per year of the original selling price of $120,000. Complete column 6 of the table.

 g. Determine the totals in columns 1, 2, and 4.

 h. Subtract the totals of column 4 from the sum of the totals in columns 1 and 2. What does this number represent?

2. a. As homeowners make payments on a mortgage, equity is built up in the value of the house. Equity is the difference between the selling price and the balance owed on the mortgage. Assuming the townhouse was sold after five years at the appreciated value (see column 6), determine the equity in the townhouse.

   b. What is the total expense to the couple over the five-year period when buying the townhouse? Recall that a $6000 down payment on the townhouse was required.

3. a. The rent is initially $650 a month. Assuming a 3% increase per year, use technology to complete the following table.

YEAR	1	2	3	4	5
Total rent for the year ($)					

   b. What is the total amount spent on rent in the five-year period?

4. a. Using the results from Problems 2b and 3b, compare the costs in renting the townhouse for five years versus buying the townhouse.

   b. In this situation, which is better: renting or buying?

   c. Search the Internet for a web site featuring a rent versus buy calculator. Use this technology to determine if it is better to rent or buy in the townhouse situation. How does it compare to your results?

5. There were many assumptions made in the analysis of this situation in Problems 1–4. A change in one or more of these assumptions could dramatically change the results.

   a. Give some reasons why, in general, it is better to buy than to rent.

   b. Give some reasons why, in certain situations, it could possibly be better to rent than to buy.

6. Use the amortization model developed in Problems 1–4 to help you create an amortization model to compare buying versus renting a home in the following situation.

a. Your aunt and uncle are moving to the suburbs of San Diego, California. They want to compare buying versus renting a home. The price of a certain house is $400,000.

- They are qualified for a 30-year mortgage at 4.75% interest rate.
- A 20% down payment is required with no points.
- A similar home in the area is available to rent at $2000 per month.
- The marginal tax rate is 35%. The initial property tax is $4000 and is expected to grow at 2% per year.
- Assume that the rent and housing prices each grow at a rate of 4% per year.

Use the above information to complete a table similar to the table in Problem 1e. Use technology to complete the amortization portion (columns 1, 3, 5).

END OF YEAR	(1) MORTGAGE PAYMENTS ($)	(2) PROPERTY TAXES ($)	(3) MORTGAGE INTEREST ($)	(4) TAX SAVINGS ($)	(5) MORTGAGE BALANCE ($)	(6) APPRECIATED HOUSE VALUE ($)

b. Using the results in part a, determine the total expense over the five-year period when buying the house?

c. Determine the total expenses for renting the house for five years.

d. Use the results from parts b and c to determine if it is better to rent or to buy if your aunt and uncle plan to sell and move to Oregon in five years.

Buy Now, Pay Later

Objectives

1. Determine the amount financed, the installment price, and the finance charge of an installment loan.

2. Determine the installment payment.

3. Determine the annual percentage rate (APR) using the APR formula and using a table.

4. Determine the unearned interest on a loan if paid before it is due.

5. Determine the interest on a credit card account using the average daily balance method.

Significant price cuts have recently taken place in the cost of high-definition televisions. Your parents have decided that now is the time to take the plunge. After researching the features of different types of HD TVs, they have selected a 60-inch HD LED, 1080P TV at a cost of $1500.

The electronics store salesperson informs them that the store is offering a 30-month installment plan to finance the TV. They are interested and discuss the details of the plan with the salesperson.

Fixed or Closed-End Installment Loan

The installment loan that is being offered by the electronics store is an example of a **fixed** or **closed-end installment loan**. The amount borrowed is repaid plus interest in equal payments (usually monthly) over a certain period of time. Such loans are commonly used for college tuition loans and loans to purchase furniture, appliances, and computers. These loans are generally repaid in 24, 36, 48, or 60 equal monthly payments.

1. The store requires no down payment. The salesperson tells your parents that they can finance the HD LED TV with 30 monthly payments of $56.02.

    a. Determine the total amount paid. This total is called the **installment price**.

    b. Determine the finance charge (interest) of the installment loan. Remember that the monthly payment includes payment of interest as well as repayment of some of the amount borrowed (principal).

Annual Percentage Rate (APR)

The results from the following Problem 2 can be used to demonstrate a hidden cost when borrowing money on an installment plan.

2. a. Suppose you borrowed $1000 for a year and had an interest charge of $45. If you repaid the amount borrowed plus interest at the end of the year, what is the annual interest rate on the loan?

    b. Suppose you paid the amount borrowed plus interest on a 12-month installment plan, what is the monthly payment?

In Problem 2a, you would have the use of the $1000 for the entire year. In part b, you are losing the use of some of the $1000 because you are using some of it to pay $87.08 each month. Therefore, when paying a loan on an installment plan, the actual interest rate charged is more than 4.5% (the actual rate charged if you paid off the loan at the end of the year). The true annual interest rate charged is called the **annual percentage rate (APR)**.

Definition

The **annual percentage rate (APR)** is the *true rate of interest* charged for the loan.

The APR can be approximately calculated using the formula:

$$\text{APR} = \frac{72i}{3P(n+1) + i(n-1)},$$

where   APR = annual percentage rate,

      i = interest (finance) charge on the loan,

      P = principal or amount borrowed, and

      n = number of months of the loan.

3. a. In Problem 2, $1000 was borrowed at an interest rate of 4.5%. If the loan is paid off in 12 monthly payments and $45 interest was charged, determine the APR using the APR formula.

3. b.

```
N=12
I%=8.205195864
PV=1000
PMT=-87.083333...
FV=0
P/Y=12
C/Y=12
PMT:END BEGIN
```

    b. Verify your result in part a using the TVM Solver feature of the TI-83/84 Plus graphing calculator. The screen appears on the left.

4. Let us return to your parents' TV loan. They are borrowing $1500 to purchase a 60-inch HD LED TV using a 30-month installment plan. Determine the annual percentage rate (APR) using the APR formula. Round your answer to the nearest whole percent. Recall that the interest charge was $180.60 (see Problem 1b).

In 1969, Congress passed the **Truth in Lending Act**. This legislation required the lender to provide the borrower with the finance charge and APR of the loan. In order to help lending institutions and businesses to provide this information, the Federal Reserve Board has published APR tables. A portion of one of these tables appears in Table 8.2. More complete APR tables are available on the Internet or at your local bank or credit union.

Table 8.2   Annual Percentage Rate Table for Monthly Payment Plans

NUMBER OF PAYMENTS	ANNUAL PERCENTAGE RATE												
	4.0%	4.5%	5.0%	5.5%	6.0%	6.5%	7.0%	7.5%	8.0%	8.5%	9.0%	9.5%	10.0%
	(FINANCE CHARGE PER $100 OF AMOUNT FINANCED)												
6	1.17	1.32	1.46	1.61	1.76	1.90	2.05	2.20	2.35	2.49	2.64	2.79	2.93
12	2.18	2.45	2.73	3.00	3.28	3.56	3.83	4.11	4.39	4.66	4.94	5.22	5.50
18	3.20	3.60	4.00	4.41	4.82	5.22	5.63	6.04	6.45	6.86	7.28	7.69	8.10
24	4.22	4.75	5.29	5.83	6.37	6.91	7.45	8.00	8.54	9.09	9.64	10.19	10.75
30	5.25	5.92	6.59	7.26	7.94	8.61	9.30	9.98	10.66	11.35	12.04	12.74	13.43
36	6.29	7.09	7.90	8.71	9.52	10.34	11.16	11.98	12.81	13.64	14.48	15.32	16.16
48	8.38	9.46	10.54	11.63	12.73	13.83	14.94	16.06	17.18	18.31	19.45	20.59	21.74
60	10.50	11.86	13.23	14.61	16.00	17.40	18.81	20.23	21.66	23.10	24.55	26.01	27.48

The following example demonstrates the process for using APR Table 8.2 to determine the APR for a loan.

Example 1   *In Problems 2 and 3, you financed $1000 for a year and agreed to repay the loan with 12 monthly payments of $87.08. The interest charge was $45. Use the APR Table 8.2 to determine the annual percentage rate (APR).*

SOLUTION

Step 1. Determine the total finance charge (interest) on the loan. The finance charge is given as $45.

Step 2. Determine the finance charge per $100 of the amount financed.

$$\frac{\text{finance charge}}{\text{amount financed}} \cdot 100 = \frac{45}{1000} \cdot 100 = 4.5$$

Therefore, you pay $4.50 for each $100 being financed.

Step 3. Using APR Table 8.2, determine the row for the given number of payments (12) and then move across to locate the number nearest to 4.5. Next, move to the top of the column to determine the APR of the loan.

Therefore, the APR is 8%.

5. Redo Problem 4 (TV loan) using APR Table 8.2. How does your answer compare to the APR determined in Problem 4?

Calculating Monthly Installment Payment

If you know the APR and the number of monthly payments, the APR Table 8.2 can be used to determine the monthly payment.

6. Suppose the salesperson at a competing electronics store offers the same model 60-inch HD LED TV, but at a lower price of $1350. With no down payment required, your parents can borrow $1350 at 7.5% APR for 24 months.

   a. Use the APR Table 8.2 to determine the finance charge per $100 financed.

   b. Determine the total finance charge of the loan. Note that if you divide the table value of 8.00 by 100, you get the finance charge for $1 borrowed. Multiplying this number by the amount financed will give the total finance charge of the $1350 loan.

   c. What is the total of the amount financed and finance charge for the HD LED TV at this competing store?

   d. What is the monthly payment on the loan?

Using the TVM Solver feature, where $N = 24, I\% = 7.5, PV = 1350, FV = 0, P/Y = 12, C/Y = 12$, PMT: END, you obtain $60.75 for the monthly payment.

   e. Complete the following table to summarize the information you have calculated for each of the installment plans to purchase the 60-inch HD LED TV.

	INSTALLMENT PLAN 1	INSTALLMENT PLAN 2 AT COMPETING STORE
Purchase Price		
Down Payment		
Monthly Payment		
Number of Monthly Payments		
APR		
Finance Charge		
Total Cost		

   f. Which installment plan should your parents choose? Explain.

Paying an Installment Loan before It Is Due

Your parents purchase the HD LED TV at the second electronics store (installment plan 2). Recall from Problem 6b that the total finance charge for plan 2 is $108. Suppose your parents pay the first 18 payments and then decide to pay off the loan. Would you have to pay the total finance charge of $108?

The answer is no. Your parents are entitled to a finance charge rebate. At the time of the loan payoff, the lender must return any unearned interest that is saved by paying off the loan early.

The most commonly used method to determine the unearned interest is the **actuarial method**.

The actuarial method formula is

$$u = \frac{npv}{100 + v},$$

where u = unearned interest,

n = number of remaining monthly payments,

p = monthly payment, and

v = value from the APR table that corresponds to the APR for the number of remaining payments.

7. Recall that your parents' monthly payment for the 24-month installment plan 2 to purchase the HD LED TV is \$60.75.

   a. If you pay off the loan after 18 payments, how many payments remain?

   b. Determine the value v in the actuarial method formula. *Hint:* Using the APR table, find the intersection of the number of remaining payments now (6) with the APR column headed by the loan APR (7.5%).

   c. Substitute the appropriate values for n, p, and v into the actuarial formula to determine the unearned interest u.

   d. Determine the total amount due to pay off the loan after 18 payments.

Open-end Installment Loans

You are enjoying your new HD LED TV. Your parents purchase a \$450 cabinet to contain the TV equipment and store your large collection of DVDs. They charge the cabinet on their credit card.

Use of a credit card to purchase goods is a type of **open-end installment loan**. Rather than making fixed equal payments over a specific period time, you make variable monthly payments. There is generally no specific period of time to pay off the loan. As a matter of fact, you can actually borrow additional money to purchase merchandise while you still have unpaid loans in the account.

The Credit Card Accountability Responsibility and Disclosure (CARD) Act of 2009 required credit card companies to provide additional information to their customers on their monthly statements.

Interest rates for open-ended accounts are generally given as annual rates. The interest on most open-end accounts is calculated using the **average daily balance method**. In this method, a balance is determined for each day of the billing period and then the total is divided by the number of days in that billing period. This gives an average of all the daily balances.

The following example demonstrates the process:

Example 2   *The balance on your credit card on March 1, the billing date, is $242.50. The following transactions were made during the month of March.*

March 6	Payment	$50.00
March 15	Charge: clothing	37.00
March 20	Charge: bookstore	85.00
March 25	Charge: music CDs	41.50

Determine the balance due on April 1.

SOLUTION

i. Determine the average daily balance for the month of March as follows:

Step 1. Determine the balance due for each transaction date:

March 1	242.50
March 6	$242.50 - 50.00 = 192.50$
March 15	$192.50 + 37.00 = 229.50$
March 20	$229.50 + 85.00 = 314.50$
March 25	$314.50 + 41.50 = 356.00$

Step 2. Determine the number of days that the balance did not change between each transaction.

March 1–March 5	5
March 6–March 14	9
March 15–March 19	5
March 20–March 24	5
March 25–March 31	7

Step 3. Multiply the balance due times the number of days it did not change.

$$5(242.50) = 1212.50$$
$$9(192.50) = 1732.50$$
$$5(229.50) = 1147.50$$
$$5(314.50) = 1572.50$$
$$7(356.00) = 2492.00$$

Step 4. Determine the sum of the products in step 3 and divide by the number of days in the billing cycle (month).

$$\frac{1212.50 + 1732.50 + 1147.50 + 1572.50 + 2492.00}{31} = \$263.13$$

Therefore, the average daily balance is $263.13.

ii. Determine the finance charge for the month. Assume that the interest rate is 1.25% per month. The finance charge is

$$0.0125 \times 263.13 = 3.29$$

iii. Determine the balance due on April 1.

$$\text{finance charge} + \text{balance} = 3.29 + 356 = \$359.29$$

SUMMARY: ACTIVITY 8.9

1. A **closed-end installment** loan is one in which you pay a fixed amount of money for a specified number of payments.

2. The **finance charge** is the total amount of money that the borrower must pay for its use.

3. The **finance charge** is the amount borrowed subtracted from the total monthly payments.

4. The **annual percentage rate (APR)** is the true rate of interest charged for the loan.

5. The APR can be calculated using the following formula

$$\text{APR} = \frac{72i}{3P(n + 1) + i(n - 1)},$$

where   APR = annual percentage rate,

         i = interest (finance) charge on the loan,

        P = principal or amount borrowed, and

       n = number of months of the loan.

The APR can also be calculated using Table 8.2 or the TVM Solver on the TI-83/84 Plus.

6. The **installment payment** is the amount that is paid (including interest) in regular payments.

7. The total **installment price** is the down payment plus the total of monthly payments.

8. An **open-end installment loan** is one in which you make variable payments each month.

9. The **actuarial method** is used to determine the amount of the unearned interest of an installment loan if it is paid off before it is done. The actuarial method formula is

$$u = \frac{npv}{100 + v},$$

where u = unearned interest,

       n = number of remaining monthly payments,

       p = monthly payment, and

       v = value from the APR table that corresponds to the APR for the number of remaining payments.

10. The interest on most credit card accounts is calculated using the **average daily balance method** (see Example 2).

EXERCISES: ACTIVITY 8.9

1. The high school's computer lab purchases a new laptop computer for $1200. They pay a 5% down payment and $55 per month on a 24-month purchase plan.

  a. What is the amount of the down payment?

  b. Determine the amount financed.

    c. Determine the total finance charge.

    d. Use the APR formula to determine the annual percentage rate to the nearest half percent.

    e. Use the TVM Solver feature of the TI-84 Plus to verify your result in part d.

2. Ever Green Landscaping wants to finance a used dump truck for $8000. The dealership requires a $1000 down payment and offers a 36-month installment plan to finance the balance. If the monthly payment is $235, use the APR formula to determine the APR. Verify your result using the calculator.

3. The senior class is planning a fund-raiser to purchase a new refrigerator for a local daycare center. The refrigerator's total cost is $950. In order to finance the purchase, the appliance store requires a 10% down payment, with the balance being financed with a 24-month installment plan having an APR of 10%.

    a. What is the amount of the down payment?

    b. What is the amount to be financed?

    c. Determine the finance charge. Use the APR table.

    d. Determine the monthly payment. Verify using the TVM Solver.

4. Your uncle bought a new recreational vehicle for $34,000. He received $8000 for his trade-in and used that money as a down payment. He financed the vehicle at 10% APR over 48 months. He received a bonus check at work and paid off the loan after making the 30th payment.

    a. Determine the total interest that would have been paid if he had made all 48 payments.

    b. What were his monthly payments?

    c. How many payments remained when the loan was paid off?

d. Use the actuarial method formula to determine the amount of unearned interest.

e. Determine the total amount due to pay off the loan after 30 payments.

5. You want to buy a wireless music system from Home Entertainment Center so you can stream Internet radio and your music library. The total cost of the system is $800. You have an option of paying for the system with the store's credit card or taking advantage of a fixed installment loan offered by the Home Entertainment Center.

    a. The credit card charges 1.5% interest per month. You plan to pay $200 a month until the debt is paid off. You will not make any other purchases with the credit card.

        i. The first credit card bill will have an $800 balance. You will make a $200 payment. The balance on the next bill will include 1.5% interest on the unpaid balance. What will the balance on the next bill?

        ii. If you make another $200 payment, what will be the balance on the next bill?

        iii. Again, if you make another $200 payment, what will be the balance on the next bill?

        iv. If you make a final $200 payment, what will be the balance on the next bill?

        v. If you pay the total balance on the next bill, how much total interest will you have paid using the credit card?

    b. The fixed installment plan has an annual interest rate of 6% with 6 equal monthly payments. Use Table 8.2 to determine the finance charge per $100 of the amount financed. Then, determine the interest, or finance charge on the $800 loan.

    c. Which option would you choose? Explain.

Activity 8.10

Insuring the Future

Objectives

1. Distinguish between term life insurance and permanent life insurance.

2. Determine the annual life insurance premium for different types of policies using a table.

3. Calculate the value of each of the nonforfeiture options for a cancelled permanent life insurance policy.

An actuary visits your school on career day. She is a statistician who works for an insurance company. Her job is to calculate the probability of certain insurable events (such as death of an individual, a home or business fire, or automobile accident) occurring. These probabilities are then used to determine insurance rates for individuals, groups, and businesses for a variety of types of insurances, including life, homeowner's, health, and automobile.

Investigate buying life insurance by searching "life insurance" on the following Web site:

www.ambest.com

You are surprised to learn that there are several types of life insurance policies available. You note that the two most common types are term insurance and permanent life insurance (whole-life and universal).

1. Visit the given Web site and write a brief description of each of the following types of insurance.

   a. Term Insurance

   b. Permanent Life Insurance

   c. Whole-Life (Ordinary Life) Insurance

   d. Universal Life Insurance

Term Life Insurance

An **insurance premium** is the amount paid by the insured for the protection provided by the policy. Life insurance rates are generally determined by the age, gender, and health of the individual seeking the insurance. The **face value** of a life insurance policy is the dollar amount of protection provided by the policy. Many rate calculators are available on the Internet that can be used for personalized rate quotes. Table 8.3 gives typical annual life insurance premiums per $1000 of face value for ten-year and 20-year term insurance. The word **term** refers to the length of time over which premiums are paid and the insurance is payable in the event of death.

Table 8.3 Annual Life Insurance Premium Rates per $1000 of Face Value

| | 10-YEAR LEVEL TERM | | | | | | 20-YEAR LEVEL TERM | | | | | |
| | MALE | | | FEMALE | | | | MALE | | | FEMALE | | |
AGE	PREF	NT	T	PREF	NT	T	AGE	PREF	NT	T	PREF	NT	T
20	0.87	1.27	2.28	0.75	1.10	1.88	20	1.09	1.50	2.86	0.91	1.31	2.56
25	0.87	1.27	2.28	0.75	1.10	1.88	25	1.09	1.50	2.86	0.91	1.31	2.56
30	0.87	1.36	2.49	.075	1.16	2.06	30	1.12	1.61	3.24	0.96	1.42	2.67
35	0.87	1.44	2.73	0.75	1.26	2.23	35	1.17	1.73	3.62	1.02	1.53	2.78
40	1.13	1.96	3.78	1.00	1.57	2.93	40	1.49	2.36	5.38	1.26	2.00	3.77
45	1.51	2.69	5.33	1.38	2.12	4.08	45	2.23	3.73	8.42	1.75	2.92	5.57
50	2.03	3.76	8.08	1.72	2.98	5.78	50	3.45	5.99	12.90	2.59	4.40	8.02
55	2.95	5.61	12.48	2.27	4.44	8.33	55	5.38	9.52	19.15	3.96	6.66	11.45
60	4.61	9.07	20.07	3.46	6.95	12.57	60	8.46	15.15	29.14	6.17	10.36	16.74

PREF = Preferred; NT = Non-tobacco; T = Tobacco usage
Data Source: Quick Reference Tables to Accompany Business Math 10e by Cheryl Cleaves, Margie Hobbes, and Jeffrey Noble

Life insurance companies use several factors to determine your rate classification. The primary determinants are tobacco/nicotine use, your weight/height ratio, and your family health history. Only those individuals that meet an insurer's strictest standards are eligible for the very best (preferred) rates. This is typically less than 25% of all applicants.

2. a. As the age of the insured increases, do the premiums increase, decrease, or remain the same? Explain why.

b. Do males or females pay higher rates? Explain why.

c. Use Table 8.3 to determine the annual premium for a $100,000 ten-year term policy for a 25-year-old nonsmoking male.

d. If the insured passes away in a car accident during the ten-year period, what amount would the beneficiary receive?

e. At the end of the ten-year period, you can renew the term policy, assuming you are still in good health. What is the new annual premium for another $100,000 ten-year term policy?

3. What is the annual premium for a $100,000 20-year term policy for the 25-year-old nonsmoking male?

4. Renewable term insurance enables the policyholder the option of renewing the policy for a five- or ten-year period, regardless of the health of the insurer. Since the premiums on this type of policy are higher than nonrenewable policies, why would someone choose the renewable option?

Rather than paying an annual premium, you can elect to make payments semiannually, quarterly, or monthly. The following table shows how the premiums are affected if paid for periods less than one year:

PREMIUM RATES FOR PERIODS LESS THAN ONE YEAR	
PERIOD	PERCENT OF ANNUAL PREMIUM
Semiannually	51%
Quarterly	26%
Monthly	8.75%

Therefore, if you pay on a semiannual basis, the premium paid is not 50% of the annual premium, but 51% of the annual premium.

5. a. If the annual premium for the $100,000 ten-year term policy of a 25-year-old nonsmoking male is $127, what is the premium if it is paid semiannually?

b. Determine the quarterly and monthly premium for the same policy.

c. How much more will you have to pay per year if you choose the monthly payments rather than the annual payments?

d. Why do you have to pay more per year if you choose to pay the premiums for a period less than one year?

Permanent Life Insurance

Whole-life and universal life are examples of permanent life insurance. If you purchase this type of life insurance, you agree to either

i. pay premiums for your entire life (whole life),

ii. pay premiums for a certain period of time (such as 20 years) to completely pay for the policy, and then you are covered for the rest of your life (limited life), or

iii. pay premiums for a certain period of time (such as 20 years) and then it pays you the face amount if you are still alive at the end of the specified time (endowment).

Table 8.4 gives typical annual premiums per $1000 of face value for permanent life insurance.

Table 8.4   Annual Life Insurance Premiums per $1000 of Face Value

	WHOLE LIFE						UNIVERSAL LIFE					
	MALE			FEMALE				MALE			FEMALE		
AGE	PREF	NT	T	PREF	NT	T	AGE	PREF	NT	T	PREF	NT	T
20	8.39	9.02	10.55	7.55	8.12	9.95	20	5.25	5.78	7.17	4.53	4.65	5.97
25	9.51	10.22	12.70	8.65	9.30	11.59	25	6.21	6.69	8.61	5.37	5.61	7.17
30	10.86	11.68	14.77	9.82	10.56	13.48	30	7.41	8.13	10.29	6.45	6.81	8.73
35	12.37	13.30	16.59	10.97	11.80	14.85	35	9.09	9.93	12.69	7.89	8.25	10.77
40	14.42	15.50	19.31	12.54	13.48	16.69	40	11.25	12.33	15.69	9.69	10.17	13.41
45	17.65	18.98	24.03	15.14	16.28	20.43	45	14.01	15.45	19.65	12.09	12.69	16.77
50	22.45	24.14	30.65	19.09	20.53	25.39	50	17.61	19.17	24.45	15.09	16.05	21.33
55	28.67	30.82	39.00	24.11	25.92	30.97	55	22.41	24.57	31.41	18.93	20.25	26.97
60	36.06	38.77	49.39	29.75	31.99	37.25	60	28.77	34.53	43.65	23.97	28.29	37.53

PREF = Preferred; NT = Non-tobacco; T = Tobacco usage
Data Source: Quick Reference Tables to Accompany Business Math 10e by Cheryl Cleaves, Margie Hobbes, and Jeffrey Noble

2. a. What is the annual premium for a 55-year-old male who buys a ten-year term insurance policy with a face value of $175,000. Use the nontobacco rate.

b. What is the annual premium for a 55-year-old female for the same policy in part a?

3. A 35-year-old male who does not smoke is in good health. Determine the annual premium for each of the following policies having a $220,000 face value (use Table 8.4):

a. Whole Life   **b.** Universal Life

4. A 40-year-old female can afford about $500 per year for a life insurance premium. She wants protection for the next ten years for her family. An insurance agent suggests either a ten-year term policy or a whole life policy (nonsmoker).

How much insurance coverage can she purchase under each policy?

5. A 20-year-old female purchased a whole life insurance policy having a face value of $300,000. At age 50, the policy has a cash value of $19,340. She decides to convert the policy to an extended term policy having a face value of $300,000. Estimate the number of years the extended term insurance will last using the 20-year level term non-smoker rates.

The Stock Market

Objectives

1. Distinguish between a stock and a bond.

2. Read a stock listing.

3. Read a bond listing.

4. Calculate the price of bonds.

5. Solve problems involving stocks and bonds.

Stocks

A share of stock represents partial ownership in a corporation. Each share has a particular value called the *face value (par value)*. When purchased, the value of the stock can fluctuate from day to day. Since a share of stock is not a loan, interest is not paid to the shareholder. However, some corporations may give their shareholders *dividends*. Dividends are generally paid on a quarterly basis. Money from the corporation's profits is used to pay dividends.

There are two types of stocks that a corporation may offer:

1. **Common Stock:** Dividends may or may not be paid, depending on the success of the company and the discretion of senior management. Once purchased, the price (value) of the stock can rise or fall, depending on many factors. Common stockholders have voting rights in the company.

2. **Preferred Stock:** Dividends are fixed and are paid to the stockholder, whether the corporation is doing well or not. In the event that the company goes bankrupt, preferred stockholders have priority over common stockholders in having some or all of their investment returned. Preferred stockholders do not have voting rights.

Once shares of stocks have been issued, stockholders buy and sell their shares in the stock market. The price of a company's stock is affected by many factors including success of the company (profits), overall economic and political climate, interest and tax rates, and unemployment. As a result, investing in stocks is more risky than other types of investment. There is the potential to make a great deal of money as well as to lose a great deal of money.

The daily prices of stocks are available in local newspapers and on the Internet.

The following is a listing of a few stocks from *The New York Times*.

Example 1   *New York Stock Exchange stock listing.*

52-WEEK		STOCK (TICKER)	DIV	YLD%	P/E	VOL 100s	CLOSE	NET CHG
HIGH	LOW							
15.98	10.98	Bank of Am (BAC)	0.04	0.30	23.78	759968	15.60	−0.15
99.95	84.70	Exxon Mobi (XOM)	2.52	2.60	12.89	233338	98.68	−0.75
55.89	22.67	Facebook I (FB)	—	—	140.97	925385	55.05	0.07
27.50	20.26	General El (GE)	0.88	3.20	20.57	309524	27.36	0.04

Data Source: The New York Times

- Columns 1 and 2 give the high and low price of the stock in the last 52 weeks. Stock prices are listed in dollars and cents.

- In the Stock (Ticker) column, Bank of Am is the abbreviated name of the company (Bank of America®) issuing the stock; BAC is the ticker symbol.

- The Div column shows the annual dividend paid per share. The BAC stock paid $0.04 per share for the year.

- The Yld % column gives the previous year's dividends as a percent of the current price per share (Close). For BAC,

$$\text{Yld\%} = \frac{\text{Div}}{\text{Close}} = \frac{0.04}{15.60} = 0.0025 \approx 0.3\%$$

- The column headed P/E gives the price-earnings ratio.

$$P/E = \cfrac{\text{current price per share}}{\cfrac{\text{company's total earnings (past 12 months)}}{\text{number of stockholder shares}}}$$

  The P/E ratio generally ranges between 3 and 50. In general, the P/E ratio is comparable to the company's growth rate. Companies that are expected to grow and have higher future earnings should have a higher P/E.

- The VOL 100s column gives the volume (number) of hundreds of shares traded on the given day. For Bank of America®, the volume was $759968(100) = 75{,}996{,}800$ shares.

- The CLOSE column shows the price of the stock (in dollars and cents) at the end of the day's closing of trading.

- The NET CHG represents the increase (or decrease) of the closing price from the previous day's closing price. Therefore, the price of a share of stock of Bank of America® was 15 cents lower than the closing price the day before.

1. Using the stock listing in Example 1, answer the following questions for the stock General El (General Electric).

  a. How many shares of General Electric were traded on this day?

  b. What was the closing price of the stock?

  c. Is the closing price larger or smaller than the closing price the previous day?

  d. During the previous 52 weeks, what was the high price of the stock?

  e. Did General Electric give a dividend to its shareholders? If so, what is the dividend?

  f. If you own 45 shares of General Electric stock, how much will you receive in dividends for the year?

2. Suppose you purchased 40 shares of Bank of America® stocks at $11.25 per share. You paid $5 discounted fee for the transaction using an online discount broker service.

  a. What is the total amount of money spent on the stock transaction?

  b. If you hold onto the stock for one year, what is the total amount of dividends you receive? Use the stock listing dividend in Example 1.

  c. If you sell the stock at the current price, what are your total earnings (or losses)? You paid an additional $5 commission to the brokerage service to sell the stocks.

As an investor, you purchase stocks in order to increase the value of the investment. You want to buy at a low price and sell at a higher price.

The amount of money that you receive after selling the stock is called the **proceeds**. The gain (or loss) from the stock investment is the difference between the cost of buying the stock and the proceeds.

$$\text{Gain (or loss) on stock} = \text{Proceeds} - \text{total cost of buying}$$

Stocks are generally bought and sold using a stockbroker (full service) or an online discount brokerage service. A broker charges a commission or transaction fee for this service, usually a percent (1% to 3%) of the total cost of buying or selling the shares.

3. You purchase 50 shares of common stock at \$35.50 per share. Several months later, you sell the shares for \$43.10. No dividends were paid. Your stockbroker charges 3% commission on purchases and sales of less than 100 shares (odd lot).

a. Determine the total cost of buying the shares.

b. Determine the proceeds. Remember, you pay a commission on selling shares.

c. Determine the gain (or loss) of the stock transaction.

The return on investment, ROI, is a measure of the performance of the investment. The ROI is the quotient of the total gain (or loss) divided by the total cost of the investment. It is expressed as a percentage or ratio.

$$\text{ROI} = \frac{\textit{total gain (or loss)}}{\textit{total cost of investment}}$$

d. Determine the ROI of the investment in this problem.

Bonds

Basically, a **bond** is a type of loan. When cities or corporations need money, they (the issuer) can borrow money from investors by issuing (selling) bonds. The city or corporation agrees to pay the money back at a certain rate of interest in a stated period of time, generally from 5 to 20 years. The maturity date is the date the issuer of the bonds agrees to repay the loan. Interest is paid to the bondholder on a regular basis during the loan period. The interest paid by the issuer to the bondholder is referred to as the **coupon**.

There are three types of bonds that are available to the public:

i. Treasury Bonds: Issued by the U.S. government and considered to be an extremely safe investment because treasury bonds are backed by the government. The drawback is that the interest rate is often low. Treasury bond interest is exempt from state and local taxes, but generally not from federal income taxes.

ii. Municipal Bonds: Issued by state and local governments. The amount of risk can vary a great deal. Municipal bond interest is generally exempt from federal income taxes and sometimes from state and local taxes.

iii. Corporate Bonds: Issued by corporations. The amount of risk can vary greatly, depending on the stability of the corporation issuing the bonds.

Bonds are generally purchased by investors at a **face value (par value)** of \$1000.

4. Suppose a corporation issues a $1000 par value bond at 7% simple interest (called **coupon rate**) that has a maturity date in ten years. You buy one of these bonds through a broker. Determine the interest that is paid each year for the next ten years. What amount is due to the bond holder when the bond matures in ten years?

There are millions of individual bonds in the municipal bond market alone. Listing all of them would take over 100 pages. Since most bondholders buy the bond with the intent of keeping the bond until it matures, bonds are not traded with the same frequency as stocks. Most newspapers or financial media (such as *The Wall Street Journal* or *Barron's*) will give some bond listings to show current trading prices. Bond listings are readily available on the Internet.

Bond price tables are somewhat different than stock listings. Although the bond listing table format may vary from one source to another, the same basic information is given. Corporate bond listings look like the following:

Example 2   *A typical corporate bond listing.*

ISSUER NAME (SYMBOL)	COUPON	MATURITY	RATING MOODY'S/ FITCH	CLOSE	NET CHANGE	YIELD %
Bank of America® (BAC.IOP)	5.750%	Aug 2016	A2/A$^+$	110.87	0.521	1.503

- The first column gives the name of the issuing company and its corresponding symbol.

- The COUPON column gives the interest, or coupon, rate (5.750%) expressed as a percent of the face value.

- The MATURITY column gives the month and year when the bond will mature.

- The RATING column provides the bond rating from two primary rating services.

- The CLOSE column gives the closing price of the bond. Note that bond prices are listed as a percent of the face value of $1000. Therefore, the price 110.87 means that the bond is selling for 110.87% of $1000.

$$110.87\% \text{ of } \$1000 = 1.1087 \times 1000 = \$1108.70$$

- The NET CHANGE column gives the difference in price from the previous close. Therefore, the closing price of 110.87 is 0.521% higher than the previous closing price. In terms of dollars, the closing price was up

$$\$1000 \times 0.521\% = 1000 \times 0.00521 = \$5.21$$

Therefore, the closing price at the previous closing was $5.21 less than today's closing price.

$$1108.70 - 5.21 = \$1103.49$$

- The YIELD % column gives the yield to maturity, a measure of how profitable the bond issue is for the life of the bond.

5. Use the following bond listing for General Electric Capital to answer the following:

ISSUER NAME (SYMBOL)	COUPON	MATURITY	RATING MOODY'S/ FITCH	CLOSE	NET CHANGE	YIELD%
General Electric Capital (GE.HMX)	3.150%	March 2024	AA/A	97.88	0.028	3.398

a. What is the coupon rate and maturity date for this bond?

b. What is the amount of interest paid each year?

c. What was the closing price per bond? What was the dollar amount?

d. The current yield of a bond is

$$\frac{(Coupon\ Rate\ as\ a\ decimal)(\$1000)}{Closing\ Price\ of\ Bond}$$

Determine the current yield of the General Electric Capital bond.

e. Compare the current yield with the yield to maturity from the table.

6. You purchase six municipal bonds on the first of the year. The current market price is 103. The commission charge is $5 per bond. What is the cost of purchasing these bonds?

Accrued Interest

An additional charge due to the seller is **accrued interest**. The interest earned on a bond is generally paid semiannually on January 1 and July 1. If the bond is traded between interest payments, the interest earned from the last payment date must be paid to the seller. The accrued interest is calculated using the simple interest formula

$$I = prt,$$

where I = accrued interest,

p = face value of the bond,

r = coupon rate, and

t = number of days since the last payment date divided by 360 (or number of months since the last payment date divided by 12).

7. Suppose you buy eight corporate bonds having a coupon rate of 7.500 and a current market price of $1,000. You purchase the bonds on May 1. The commission charge is $5.00 per bond.

a. If the bond pays interest semiannually on January 1 and July 1, how much interest was paid on January 1?

b. How many months since the last payment was made?

c. Determine the amount of accrued interest that is due the seller.

d. Determine the price per bond.

e. What is the total amount of money paid for the eight corporate bonds?

SUMMARY: ACTIVITY 8.11

1. A **stock** is an ownership share of a corporation.

2. A **share** is one unit of stock.

3. A **dividend** is an amount of money paid to the owners of the stock (shareholders) from the corporation's profits.

4. **Face value (par value)** is the value of one share of stock.

5. **Common stock** is a type of stock that gives the stockholder voting rights. The stockholders may or may not be paid dividends.

6. **Preferred stock** is a type of stock that guarantees a specific dividend to the stockholders. The preferred stockholder does not have voting rights.

7. Stock listings contain information about the price of a share of stock and some historical information that is published in newspapers and the Internet.

8. **Proceeds** refers to the amount of money you receive after selling the stock.

9. A **bond** is a type of loan to a company or municipality to raise money.

10. The **face value (par value)** is the original value of the bond, usually $1000.

11. There are three types of bonds: 1) **treasury bonds** issued by the federal government; 2) **municipal bonds** issued by local and state governments; and 3) **corporate bonds** issued by corporations.

12. In a bond listing, the price of the bond is listed as a percent of the face value of $1000.

13. When a bond is traded on the market, the interest earned from the last payment, called **accrued interest**, is paid to the seller.

EXERCISES: ACTIVITY 8.11

1. Using the stock listings in Example 1, answer the following questions for ExxonMobi (ExxonMobil):

   a. How many shares of ExxonMobil were traded on this day?

   b. What was the closing price of the stock?

   c. Is the closing price larger or smaller than the closing price the previous day?

   d. During the previous year, what was the high price of the stock?

   e. Did ExxonMobil give a dividend to its shareholders? If so, what is the dividend?

   f. If you own 35 shares of ExxonMobil stock, how much will you receive in dividends for the year?

2. Using the stock listings in Example 1, answer the following questions for Facebook:

   a. How many shares of Facebook were traded on this day?

   b. What was the closing price of the stock?

   c. Is the closing price larger or smaller than the closing price the previous day?

   d. During the previous year, what was the high price of the stock?

   e. Did Facebook give a dividend to its shareholders? If so, what is the dividend?

   f. If you own 50 shares of Facebook stock, how much will you receive in dividends for the year?

3. The following stock listing appeared recently in *The New York Times:*

52-WEEK		STOCK	DIV	YLD %	P/E	SALES 100s	LAST	CHG
HIGH	LOW							
81.37	67.37	Wal-Mart® (WMT)	1.88	2.40	14.88	250,558	77.43	0.19

a. How many shares of stock of Wal-Mart® were traded on this day?

b. What was the closing price of the stock?

c. Is the closing price larger or smaller than the closing price the previous day?

d. During the previous year, what was the high price of the stock?

e. If you own 25 shares of Wal-Mart®, how much will you receive in dividends for the year?

4. You purchase 20 shares of common stock at 19.86 per share. Several months later, you sell the shares for 31.50. Your stockbroker charges a flat fee of $25 commission on transactions of less than 100 shares (odd lot).

a. Determine the total cost of buying the shares.

b. Determine the proceeds.

c. Determine the gain (or loss) of the investment.

d. Determine the ROI.

5. Your current CD matures in a few days. You would like to find an investment with a higher rate of return than the CD. Stocks historically have a rate of return between 10% and 12%, but you do not like the risk involved. You have been looking at bond listings online. A friend wants you to look at the following corporate bonds as a possible investment.

BOND	COUPON	MATURITY	RATING MOODY'S/FITCH	CLOSE	NET CHANGE	YIELD %
ABC	7.5%	May 2018	A2/A	104.75	—	4.784
XYZ	7.75%	Sept 2020	A1/A⁺	100.5	0.25	5.836

a. What is the annual interest you would earn on each bond?

b. What price would you pay for each bond if you purchased one of them today? *Note:* Face value is $1000.

c. If you buy three of the ABC bonds with $10 commission for each, how much will it cost? Assume you do not owe any accrued interest.

d. Determine the current yield of the XYZ bond.

6. Determine the current bond yield for a bond having a current bond price of 98.431% and a coupon rate of 6.375%.

7. Determine the ROI for 2000 shares of a stock having a purchase price of $15.83 per share and selling at $18.72 per share. The A dividend of $0.87 was paid per share during the period of ownership.

Project 8.12

Which Is the Best Option?

Objective

Use the financial models developed in this chapter to solve problems.

Each of the following problems presents a financial situation in which you must select the best option. Work with a group and be prepared to give an oral presentation of your findings to the class. The presentation should include visual displays showing any tables, equations, or graphs used in the problem solving process.

1. You are 25 years old and begin to work for a large company that offers you two different retirement options.

   Option 1:   You will be paid a lump sum of $20,000 for each year you work for the company.

   Option 2:   The company will deposit $10,000 annually into an account that will pay you 12% compounded monthly. When you retire, the money will be given to you.

   Which option is better? Explain.

2. Congratulations! You have just won $100,000 in the state lottery. You decide to invest half of the money into a savings account, in order to start your own business when you graduate from college. You have three investment options:

   Option 1:   Open a savings account that pays 6.5% simple interest.

   Option 2:   Invest in a long term certificate of deposit (CD) that pays 3% compounded annually.

   Option 3:   Open a savings account that pays 2.7% compounded daily.

   Which option will result in having the most money in six years? Explain.

3. You have just purchased a new desktop computer for $2400. You have two options to pay off the purchase:

   Option 1:   Apply for a store credit card and make payments of $150 per month until paid off. Assume you will make no additional purchases using the card and that there is no annual fee. The interest on the card is 1.5% per month. Payment is due at the beginning of each month.

   Option 2:   Apply for a fixed installment loan of $2400 at 6.55% APR for two years.

   Which option is better? Explain.

Activities 8.1–8.12   **What Have I Learned?**

I. You invest $1000 at 4% for three years.

   a. Use the simple interest formula to determine the amount accumulated.

   b. If the interest is compounded annually, use the compound interest formula to determine the amount accumulated.

2. Which option would result in a greater return on your investment?

   Option 1:   Invest at 5% interest compounded quarterly.
   Option 2:   Invest at 5% interest compounded monthly.

   Explain.

3. If you invest $100 every month for five years, will you earn more interest if you made the payments at the beginning of each month or at the end of each month? Explain.

4. Banking regulations require that the effective interest rate (APR for loans, APY for deposits) be stated on all loan or investment contracts. Explain why.

5. A recent promotion for a credit card made the following introductory offer:

   • 0% fixed APR on all purchases and balance transfers for the first 12-month period
   • a low 8.5% APR thereafter

   However, if the minimum payment is not received by the due date on the billing statement or if the outstanding balance exceeds your credit limit, then your interest rate becomes a so-called default rate of 24.99%.

   a. What is the interest for the month on an average daily balance of $8000 at 8.5% APR?

   b. What is the interest for the month on an average daily balance of $8000 at 24.99% APR?

   c. How much more would you pay for the month at the default rate of 24.99% APR?

6. For a mortgage having a given amount and interest rate,

   a. What happens to the total amount of interest paid if the number of years of the mortgage increases?

   b. What happens to the monthly payments as the number of years increase?

7. How are bonds different from certificates of deposit or savings accounts?

8. How are bonds different from stocks?

9. Why is whole life insurance more expensive than term insurance?

Activities 8.1–8.12   How Can I Practice?

1. You just started a new job and received a $5000 bonus. You decide to invest this money so that you can purchase a new car in four years. Your local credit union offers a CD paying 4% annual interest compounded semiannually. How much money will you have at the end of four years?

2. You are a freshman in high school and plan to attend the local community college in four years. Your father said he would pay for your college tuition, fees, and books for the first year. You estimate that $4500 will be needed to cover these expenses. Your father has done some research and has decided on a four-year CD with an interest rate of 3.75% compounded monthly. How much will he have to invest now to have $4500 in 48 months?

3. You plan to start a consulting business for computer networking. You deposit $200 at the beginning of each month into an account paying 4.75% compounded monthly.

   a. How much will be in the account after five years?

   b. How much interest was earned?

4. You notice the following advertisement in the local newspaper:

> A 24-foot motorboat for sale
> price of $30,000; $6000 down
> and $555 per month for 60 months

   a. Determine the total finance charge.

   b. What is the APR of the loan?

5. You are ready to purchase your first home. This home sells for $160,000. You must put a 10% down payment and you are required to pay 3 points at the time of closing. The conventional mortgage will be for 30 years at an interest rate of 4.5%. You wonder if you can afford this home.

   a. Determine the amount of the down payment.

   b. What is the amount of the mortgage?

   c. Determine the cost of the three points.

   d. Determine the monthly mortgage payment.

e. What is the total cost of the house including the down payment and points?

f. What is the total interest that you will pay over the life of the loan?

g. Determine how much of the first payment on the loan is applied to the principal.

6. You deposited $10,000 in an account that pays 3% annual interest compounded monthly.

a. Write an equation to determine the amount, A, you will have in t years.

b. How much will you have in five years?

c. Write an equation to determine the amount, A, you will have in t years if the interest is compounded continuously.

d. Use the equation in part c to determine how much you will have in five years. Compare your answer with your answer in part b.

The bracketed numbers following each concept indicate the activity in which the concept is discussed.

CONCEPT/SKILL	DESCRIPTION	EXAMPLE
Banking Options [8.2]	When making decisions about banking, there are multiple options to consider including online banking vs. teller service, overdraft protection, processing fees, and ATM and debit card fees.	Consider all options very carefully. See Problems 2–6 in Activity 8.2.
Compound amount [8.3]	The compound amount, also called future value, is the total amount of principal and compound interest at the end of an investment.	The compound amount of a $1000 investment at 6% interest compounded annually for two years:

Original principal	$1000.00
Interest first year	+60.00
Balance after first year	1060.00
Interest (0.06×1060)	+63.60
Balance after 2 years	$1123.60

CONCEPT/SKILL	DESCRIPTION	EXAMPLE
Calculating compound amount (future value) using the compound interest formula [8.3]	The compound interest formula is $A = P\left(1 + \dfrac{r}{n}\right)^{nt}$, where A is the compound amount, P is the principal, r is the annual interest rate (decimal form), n is the number of compounding periods per year, and t the number of years money invested.	See Example 1 in Activity 8.3.
Calculating compound amount (future value) using technology [8.3]	The TI-83/84 Plus graphing calculator has a finance applications feature (TVM Solver).	See Example 2 in Activity 8.3.
The annual percentage yield (APY) or effective interest rate [8.3]	The APY is the true rate of return on an investment. One method to calculate the APY is the Eff(feature of the TI-84 Plus.	See Example 3 in Activity 8.3.
Calculating the present value of a compound amount (future value) using a formula [8.3]	Present value is the amount that must be invested now at a compound interest to reach a given future value. The present value formula is $$PV = \frac{FV}{\left(1 + \dfrac{r}{n}\right)^{nt}}.$$	The amount that must be invested now in order to have $6000 in five years, at 5.7% compounded monthly is $$P = \frac{6000}{\left(1 + \dfrac{0.057}{12}\right)^{5 \cdot 12}}$$ $$= \frac{6000}{(1 + 0.00475)^{60}}$$ $$= \$4515.13.$$

CONCEPT/SKILL	DESCRIPTION	EXAMPLE
Calculating the present value using technology [8.3]	The present value for a given future value of an investment can be calculated using the TVM Solver feature of the TI-83/84 Plus.	See Problem 8 in Activity 8.3.
Continuous compounding [8.4]	The formula for continuous compounding is $A = Pe^{rt}$.	Example 1, Activity 8.4.
Calculating the future value of an ordinary annuity using a formula [8.5]	An annuity is the payment (or receipt) of equal cash payments per period for a given period of time. For an ordinary annuity, the payments are made at the end of each period. The future value of ordinary annuity formula is $$FV = P \cdot \frac{(1 + i)^N - 1}{i},$$ where FV is future value, P is annuity payment, i is interest rate per period (written as a decimal), and N is total number of periods.	The future value of a deposit of $1000 made at the end of every six months for two years at 8% compounded semiannually is $$FV = 1000 \cdot \frac{(1 + 0.04)^4 - 1}{0.04}$$ $$= \$4246.46.$$
Calculating the future value of an annuity due using a formula [8.5]	For an annuity due, the payments are made at the beginning of each period. The future value of an annuity due formula is $$FV = P \cdot \frac{(1 + i)^N - 1}{i} \cdot (1 + i).$$	The future value of a deposit of $1000 made at the beginning of every six months for two years at 8% compounded semi-annually is $$FV = 1000 \cdot \frac{(1 + 0.04)^4 - 1}{0.04}(1 + 0.04)$$ $$= \$4416.32.$$
Calculating the future value of an annuity using technology [8.5]	The future value of an ordinary annuity or annuity due can be calculated using the TVM Solver feature of the TI-83/84 Plus graphing calculator.	What is the future value of an ordinary annuity of $500 per month, for three years, at 10% compounded monthly? Using $N = 36, I\% = 10,$ $PMT = -500, PV = 0,$ $P/Y = C/Y = 12,$ $PMT{:}END,$ you have $FV = 20{,}890.91$
Calculating the present value of an annuity using technology [8.5]	The present value of annuity is a lump sum that is deposited now to yield a pay-out of equal periodic payments for a given time. Present value can be calculated using the TVM Solver feature of TI-83/84 Plus.	How much must be deposited now, at 6% compounded annually, to yield an annuity payment of $5000 at the end of each year, for five years? Using $N = 5, I\% = 6,$ $PMT = -5000, FV = 0,$ $P/Y = C/Y = 1,$ $PMT{:}END,$ $PV = 21{,}061.82.$

CONCEPT/SKILL	DESCRIPTION	EXAMPLE
Calculating the amortization payment by formula [8.6]	Amortization is repaying a loan by a series of equal periodic payments over a specified period of time. The amount *Amt* of the amortization payment can be calculated using the formula $$Amt = PV \cdot \frac{i}{1 - (1 + i)^{-N}}$$ or, equivalently $$Amt = PV \cdot \frac{i}{1 - \dfrac{1}{(1 + i)^{N}}}$$ where *Amt* is amortization payment, *PV* is amount of loan, i is interest rate per period, and *N* is number of periods.	See Example 1 in Activity 8.6.
Calculating the amount of the equal period payments to amortize (liquidate) a lump sum with equal periodic payments over specified period of time [8.6]	The TVM Solver feature of the TI-83/84 Plus graphing calculator can be used to determine the amount of the payments to liquidate a lump sum.	What payment is required at the end of each month, at 8% interest compounded monthly, to liquidate $2000 in two years? Using N = 24, I% = 8 PV = 2000, FV = 0, P/Y = C/Y = 12, PMT:END, you obtain PMT = −90.45.
Calculating the monthly mortgage payment using a table [8.7]	The monthly mortgage payment can be determined using Table 8.1.	See Example 1 in Activity 8.7.
Calculating the monthly mortgage payment using technology [8.7]	The TVM Solver feature of the TI-83/84 Plus can be used to determine the monthly mortgage payment.	What is the monthly payment on a $200,000 mortgage at 5.5% APR for 25 years? Let N = 300, I% = 5.5, PV = 200,000, FV = 0, P/Y = C/Y = 12, PMT:END, you obtain PMT = −1228.17
Amortization schedule [8.7]	An amortization schedule displays, for given monthly payment, the amount of interest paid and the amount used to reduce principal.	See Example 2 in Activity 8.7 for a partial amortization schedule.
Housing expense ratio [8.7]	A borrower qualifies for a mortgage if the monthly housing expenses do not exceed 28% of the borrower's monthly gross income. The housing expense ratio is $$\frac{\text{monthly housing expense}}{\text{monthly gross income}}.$$	The housing expense ratio for borrowers having $910 monthly housing expenses and $4800 monthly gross income is $$\frac{910}{4800} = 0.190 = 19\% < 28\%.$$ The borrower qualifies.

CONCEPT/SKILL	DESCRIPTION	EXAMPLE
Finance charge [8.9]	The finance charge on an installment loan is the total amount that the borrower must pay for its use.	The finance charge on an installment loan of $2000 with 24 monthly payments of $95 is the total amount of payments minus the amount financed: $$24(95) - 2000 = \$280$$
Annual percentage rate (APR) [8.9]	The APR is the true rate of interest charged for the loan. The APR formula is $$APR = \frac{72i}{3P(n + 1) + i(n - 1)},$$ where APR is annual percentage rate, i is interest (finance) charge on the loan, P is principal or amount borrowed, and n is number of months of the loan.	The APR on a 12-month installment loan of $1000 having finance charge of $100 is $$APR = \frac{72(100)}{3(1000)(12 + 1) + 100(12 - 1)}$$ $$= \frac{7200}{39000 + 1100}$$ $$\approx 0.18 \text{ or } 18\%.$$
Calculating the APR using tables [8.9]	The APR of a loan can be determined using APR Table 8.2 in Activity 8.9.	See Example 1 in Activity 8.9.
Calculating the monthly installment payment using tables [8.9]	If you know the APR and the number of monthly payments, the APR Table 5.2 can be used to determine the monthly payment.	See Problem 6, parts a–d, in Activity 8.9.
The amount of unearned interest when an installment loan is paid off before it is due [8.9]	The unearned interest can be determined by the actuarial method formula $$u = \frac{npv}{100 + v},$$ where u is unearned interest, n is number of remaining monthly payments, p is monthly payment, and v is value from the APR table that corresponds to the APR for the number of remaining payments.	See Problem 7, parts a–c, in Activity 8.9. The answer in part c is: Let $n = 12, p = 142.50$. $v = 7.74$, $$u = \frac{npv}{100 + v}$$ $$= \frac{12(142.50)(7.74)}{100 + 7.74}$$ $$= \$122.85$$
Interest on an open-ended installment loan, such as a credit card [8.9]	The interest on most credit card accounts is calculated using the average daily balance method.	See Example 2 in Activity 8.9.
Life insurance [8.10]	Life insurance is a policy (contract) that pays a specified amount of money (face value) to the beneficiary upon the death of the person who is insured.	Types of life insurance include: **a.** Term Insurance **b.** Permanent Life Insurance **c.** Whole-Life Insurance **d.** Universal Life Insurance

CONCEPT/SKILL	DESCRIPTION	EXAMPLE
Life insurance premium [8.10]	An insurance premium is the amount paid by the insured, in regular intervals, for a life insurance policy. Life insurance rates are generally determined by age, gender, and health.	A typical rate for a five-year term policy for a healthy 24-year-old male is $2.69 for each $1000 of insurance. The annual premium for a $50,000 policy is $$2.69(50) = \$134.50.$$
Stocks [8.11]	A share of stock represents partial ownership in a corporation. Each share has a particular value called the face value (par value). Types of stocks include common and preferred.	On a given day, the value of a share of General Electric stock was $14.25.
Stock listings [8.11]	Stock listings give information about the price of a share of stock and some historical information that is published in newspapers and the Internet.	See Example 1 in Activity 8.11.
Calculating the gain (or loss) on the sale of shares of stock [8.11]	The gain (or loss) on the sale of shares of stock is the proceeds (amount of money you receive after selling the shares) minus the total cost of buying the shares.	If the proceeds for selling 30 shares of a particular stock is $2200, and the cost of buying the shares is $1750, then the gain is $$2200 - 1750 = \$450.$$
Bonds [8.11]	A bond is a type of loan to a company (corporate bonds), a city (municipal bonds), or federal government (treasury bonds) to raise money.	GE Capital 3.15% coupon bond maturing in March 2024.
Bond price [8.11]	Bond prices are reported as a percent of the face value of $1000.	A corporate bond having price 103.25 is selling for 103.25% of its face value of $1000. $$103.25\% \times 1000$$ $$= 1.0325 \times 1000$$ $$= \$1032.50 \text{ per bond}$$
Bond listings [8.11]	Bond listings give the price of bonds and historical information regarding the performance of the bond.	See Example 2 in Activity 8.11.
Accrued interest [8.11]	When a bond is traded on the market, the interest earned from the last interest payment, called accrued interest, is paid to the seller.	A bond with a coupon rate (interest rate) of 6.375% is sold 50 days since the last interest payment. The accrued interest is $$1000 \times .06375 \times \frac{50}{360} = \$8.85.$$

1. You have a part-time job in your junior year of high school. You need to borrow $2,000 to purchase a used car to drive to work. The local bank will lend you the money for two years at a rate of 9% compounded quarterly. Your uncle offers to lend you the money in the form of a promissory note for two years with an annual simple interest rate of 6%. How much money will you save by borrowing the money from your uncle?

2. If $100 is invested in a bank in the year the *Declaration of Independence* was signed, how much would it be worth in the year 2010? Assume the investment earned 3.5% compounded monthly.

3. You deposit $2000 at the end of each year in a Roth IRA that earns 4.85% compounded annually. After ten years, you stop making payments, but the money in the account continues to earn interest.

   a. Determine the amount in the Roth IRA after ten years.

   b. If you withdraw the money ten years after you stopped making payments, how much money did you receive?

4. Your older brother and his wife just had a baby boy. They want to establish a college fund for their son. If they decide to deposit $2000 at the beginning of each year at 4.5% compounded annually, how much money will be in the fund after 18 years?

5. Your parents are planning to retire and want to set aside a lump sum, earning 4.25% compounded quarterly, in order to have a pay out of $5000 per quarter for 20 years. What lump sum should your parents put aside when they retire? Assume payments are made at the beginning of each quarter.

6. a. What lump sum of money must a 20-year-old invest today at 3.75% compounded monthly in order for the investment to have a value of $1,000,000 at age 55?

   b. For this same 20-year-old, how much would their monthly payments be in an ordinary annuity at 3.75% compounded monthly in order to accumulate $1,000,000 at age 55?

7. A landscaper needs to finance $27,000 in order to purchase a 4-by-4 truck. The dealer offers him the option of 0% financing for 60 months or a $4500 rebate.

   a. What is the monthly payment if he chooses the 0% financing?

b. If the rebate is chosen, the buyer can finance the balance for 60 months at 6% compounded monthly. What is the monthly payment? Assume payments are made at the end of each month.

c. Which option is the best? Explain.

8. Your passion is flying remote control model airplanes. You find the ultimate plane in a hobby shop for $675. The terms of the sale include a down payment of $175 and 12 monthly payments of $44.83.

   a. Determine the total amount paid or the installment price.

   b. Determine the finance charge of the installment loan.

   c. Use the APR formula and determine the APR.

9. You are going to college in the fall. Your roommate asks you to bring the stereo system for the dorm room. You find the perfect system for $1750. You have to put 25% down and you can pay the balance in 18 equal installments. The clerk tells you that you will pay a total finance charge of $112. You wonder if that is a good rate and if you can afford the monthly payments.

   a. How much is the down payment?

   b. What is the total amount financed including the finance charge?

   c. What is the monthly payment on the loan?

   d. Use the formula and determine the APR.

10. A house is selling for $250,000. The bank requires a 15% down payment, but no points. The current fixed mortgage rate is 3.75%.

   a. Determine the amount of the down payment.

   b. What is the amount of the mortgage?

c. If a 25-year mortgage is obtained, what is the monthly mortgage payment?

d. What is the total interest that will be paid on the loan?

e. How much of the first payment on the loan is applied to principal?

11. You just inherited $5000. You can invest the money at a rate of 4% compounded continuously. In eight years, how much will be in the bank. Use the equation $A = A_0 e^{rt}$.

Appendix A

The TI-83/84 Plus Graphing Calculator

A Basic Primer for Necessary Features and Routines

Contents

Getting Started with the TI-83/84 Plus Family of Calculators

ON-OFF

To turn on the TI-83/84 Plus, press the ⟨ON⟩ key. To turn off the calculator, press ⟨2nd⟩ and then ⟨ON⟩.

Most keys on the calculator have multiple purposes. The number or symbolic function/command written directly on the key is accessed by simply pressing the key. The symbolic function/commands written above each key are accessed with the aid of the ⟨2nd⟩ and ⟨ALPHA⟩ keys. The command above and to the left is color coded to match the ⟨2nd⟩ key. That command is accessed by first pressing the ⟨2nd⟩ key and then pressing the key itself. Similarly, the command above and to the right is color coded to match the ⟨ALPHA⟩ key and is accessed by first pressing the ⟨ALPHA⟩ key and then pressing the key itself.

Contrast

To adjust the contrast on your screen, press and release the ⟨2nd⟩ key and hold ⟨▲⟩ to darken and ⟨▼⟩ to lighten.

Mode

The ⟨MODE⟩ key controls many calculator settings. The activated settings are highlighted. For most of your work in this course, the settings in the left-hand column should be highlighted.

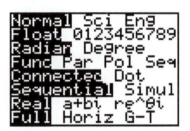

To change a setting, move the cursor to the desired setting and press ⟨ENTER⟩.

The Home Screen

The home screen is used for calculations.

You may return to the home screen at any time by using the QUIT command. This command is accessed by pressing ⟨2nd⟩ ⟨MODE⟩. All calculations in the home screen are subject to the order of operations.

Enter all expressions as you would write them. Always observe the order of operations. Once you have typed the expression, press ⟨ENTER⟩ to obtain the simplified result. Before you press ⟨ENTER⟩, you may edit your expression by using the arrow keys, the delete command ⟨DEL⟩, and the insert command ⟨2nd⟩ ⟨DEL⟩.

Three keys of special note are the reciprocal key ⟨X^{-1}⟩, the caret key ⟨^⟩, and the negative key ⟨(−)⟩.

Typing a number and then pressing the reciprocal command key $\boxed{X^{-1}}$ will give the reciprocal of the number. The reciprocal of a nonzero number, n, is $\frac{1}{n}$. As noted in the screen below, when performing an operation on a fraction, the fraction MUST be enclosed in parentheses before accessing this command.

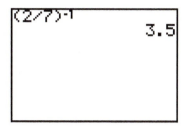

The caret key $\boxed{\wedge}$ is used to raise numbers to powers

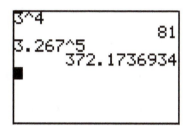

The negative key $\boxed{(-)}$ on the bottom of the keyboard is different from the subtraction key $\boxed{-}$. They cannot be used interchangeably. The negative key is used to change the sign of a single number or symbol; it will not perform a subtraction operation. If you mistakenly use the negative key in attempting to subtract, you will likely obtain an ERROR message.

```
15-6
              9
15- -6
             21
```

A table of keys and their functions follows.

KEY	FUNCTION DESCRIPTION
ON	Turns calculator on or off.
CLEAR	Clears the current line on the text screen.
ENTER	Executes a command.
(−)	Calculates the additive inverse.
MODE	Displays current operating settings.
DEL	Deletes the character at the cursor.
∧	Symbol used for exponentiation.
ANS	Storage location of the last answer, or calculation.
ENTRY	Retrieves the previously executed expression.

ANS and ENTRY

The last two commands in the table can be real time savers. The result of your last calculation is always stored in a memory location known as ANS. It is accessed by pressing [2nd] [(−)] or it can be automatically accessed by pressing any operation button.

Suppose you want to evaluate $12.5\sqrt{1 + 0.5 \cdot (0.55)^2}$. It could be evaluated in one expression and checked with a series of calculations using ANS.

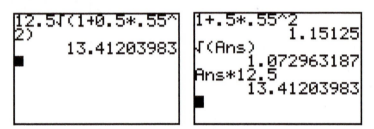

After you have keyed in an expression and pressed [ENTER], you cannot move the cursor back up to edit or recalculate this expression. This is where the ENTRY ([2nd] [ENTER]) command is used. The ENTRY command retrieves the previous expression and places the cursor at the end of the expression. You can use the left and right arrow keys to move the cursor to any location in the expression that you wish to modify.

Suppose you want to evaluate the compound interest expression $P\left(1 + \dfrac{r}{n}\right)^{nt}$, where P is the principal, r is the interest rate, n is the number of compounding periods annually, and t is the number of years, when $P = \$1000$, $r = 6.5\%$, $n = 1$, and $t = 2, 5$, and 15 years.

Using the ENTRY command, this expression would be entered once and edited twice.

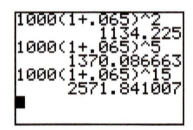

Note that there are many last expressions stored in the ENTRY memory location. You can repeat the ENTRY command as many times as you want to retrieve a previously entered expression.

Functions and Graphing

"Y =" Menu

Functions of the form $y = f(x)$ can be entered into the TI-83/84 Plus using the "Y =" menu. To access the "Y =" menu press the [Y=] key. Type the expression $f(x)$ after Y_1 using the [X,T,θ,n] key for the variable x and press [ENTER].

For example, enter the function $f(x) = 3x^5 - 4x + 1$.

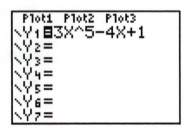

Note the $=$ sign after Y_1 is highlighted. This indicates that the function Y_1 is active and will be graphed when the graphing command is executed and will be included in your table when the table command is executed. The highlighting may be turned on or off by using the arrow keys to move the cursor to the $=$ symbol and then pressing (ENTER). Notice in the screen below that Y_1 has been deactivated and will not be graphed nor appear in a table.

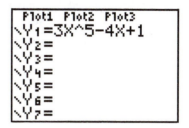

Once the function is entered in the $Y =$ menu, function values may be evaluated in the home screen.

For example, given $f(x) = 3x^5 - 4x + 1$, evaluate $f(4)$. In the home screen, press (VARS).

Move the cursor to Y-VARS and press (ENTER).

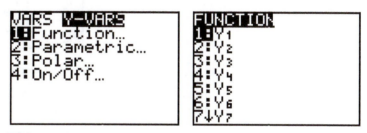

Press (ENTER) again to select Y_1. Y_1 now appears in the home screen.

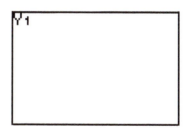

To evaluate $f(4)$, press (I) (4) (I) after Y_1 and press (ENTER).

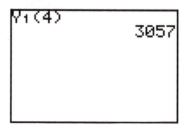

Tables of Values

If you are interested in viewing several *y*-values for the same function, you may want to construct a table.

Before constructing the table, make sure the function appears in the "Y =" menu with its "=" highlighted. You may also want to deactivate or clear any functions that you do not need to see in your table. Next, you will need to check the settings in the Table Setup menu. To do this, use the TBLSET command (2nd WINDOW).

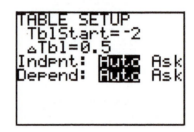

As shown in the screen above, the default setting for the table highlights the Auto options for both the independent (*x*) and dependent (*y*) variables. Choosing this option will display ordered pairs of the function with equally spaced *x*-values. TblStart is the first *x*-value to be displayed, and here is assigned the value -2. ΔTbl represents the equal spacing between consecutive *x*-values, and here is assigned the value 0.5. The TABLE command (2nd GRAPH) brings up the table displayed in the screen below.

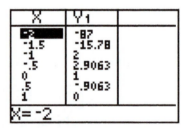

Use the (▲) and (▼) keys to view other values in the table.

If the *x*-values of interest are not evenly spaced, you may want to choose the ask mode for the independent variable from the Table Setup menu.

The resulting table is blank, but you can fill it by choosing any values you like for *x* and pressing (ENTER) after each.

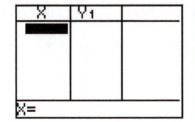 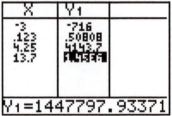

Note that the number of digits shown in the *y*-column is limited by the table width, but if you want more digits, move the cursor to the desired output and more digits appear at the bottom of the screen.

Graphing a Function

Once a function is entered in the "Y =" menu and activated it can be displayed and analyzed. For this discussion, we will use the function $f(x) = -x^2 + 10x + 12$. Enter this as Y_1, making sure to use the negation key ((-)) and not the subtraction key $(-)$.

The Viewing Window

The viewing window is the portion of the rectangular coordinate system that is displayed when you graph a function.

Xmin defines the left edge of the window.

Xmax defines the right edge of the window.

Xscl defines the distance between horizontal tick marks.

Ymin defines the bottom edge of the window.

Ymax defines the top edge of the window.

Yscl defines the distance between vertical tick marks.

In the standard viewing window, Xmin $= -10$, Xmax $= 10$, Xscl $= 1$, Ymin $= -10$, Ymax $= 10$, and Yscl $= 1$.

To select the standard viewing window, press (ZOOM) and (6).

```
ZOOM MEMORY
1:ZBox
2:Zoom In
3:Zoom Out
4:ZDecimal
5:ZSquare
6:ZStandard
7↓ZTrig
```

If you press the (GRAPH) key now, you will view the following:

Is this an accurate and or complete picture of your function, or is the window giving you a misleading impression? You may want to use your table function to view the *y*-values for $x = -10$ to 10.

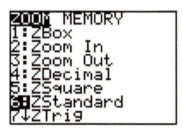

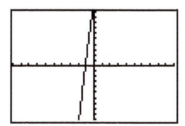

The table indicates that the minimum y-value on the interval from $x = -10$ to $x = 10$ is -188 and the maximum y-value is 37. Press [WINDOW] and reset the settings to approximately the following;

$$Xmin = -10, Xmax = 10, Xscl = 1, Ymin = -190, Ymax = 40, Yscl = 10$$

 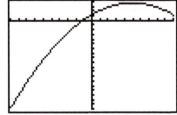

The new graph gives us a much more complete picture of the behavior of the function for $-10 \leq \times \leq 10$.

The coordinates of specific points on the curve can be viewed by activating the trace feature. While in the graph window, press [TRACE]. The function equation will be displayed at the top of the screen, a flashing cursor will appear on the curve at the middle of the screen, and the coordinates of the cursor location will be displayed at the bottom of the screen.

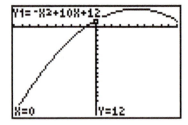

The left arrow key, [◀], will move the cursor toward smaller x-values. The right arrow key, [▶], will move the cursor toward larger x-values. If the cursor reaches the edge of the window and you continue to move the cursor, the window will adjust automatically.

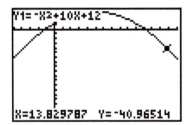

Zoom Menu

The Zoom menu offers several options for changing the window very quickly.

The features of each of the commands are summarized in the following table:

ZOOM COMMAND	DESCRIPTION
1: ZBox	Draws a box to define the viewing window.
2: Zoom In	Magnifies the graph near the cursor.
3: Zoom Out	Increases the viewing window around the cursor.
4: ZDecimal	Sets a window so that Xscl and Yscl are 0.1.
5: ZSquare	Sets equal size pixels on the x and y axes.
6: ZStandard	Sets the window to standard settings.
7: ZTrig	Sets built-in trig window variables.
8: ZInteger	Sets integer values on the x and y axes.
9: ZoomStat	Sets window based on the current values in the stat lists.
0: ZoomFit	Replots graph to include the max and min output values for the current Xmin and Xmax.

Solving Equations Graphically

The Intersection Method

This method is based on the fact that solutions to the equation $f(x) = g(x)$ are values of x that produce the same y-values for the functions f and g. Graphically, these are the x-coordinates of the intersection points of $y = f(x)$ and $y = g(x)$.

The following procedure illustrates how to use the intersection method to solve $x^3 + 3 = 3x$ graphically:

Step 1. Enter the left-hand side of the equation as Y_1 in the "Y =" editor and the right-hand side as Y_2.

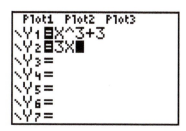

Step 2. Examine the graphs to determine the number of intersection points.

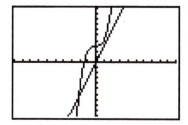

You may need a couple of windows to be certain of the number of intersection points.

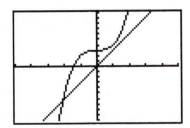

Step 3. Access the Calculate menu by pushing 2nd TRACE, then choose option 5: intersect.

 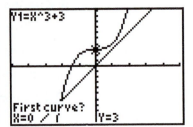

The cursor will appear on the first curve in the center of the window.

Step 4. Move the cursor close to the desired intersection point and press ENTER.

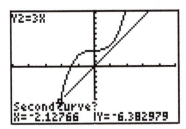

The cursor will now jump vertically to the other curve.

Step 5. Repeat step 4 for the second curve.

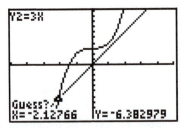

Step 6. To use the cursor's current location as your guess, press ENTER in response to the question on the screen that asks Guess? If you want to move to a better guess value, do so before you press ENTER.

The coordinates of the intersection point appear below the word *Intersection*.

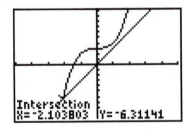

The *x*-coordinate is a solution to the equation.

If there are other intersection points, repeat the process as necessary.

Determining the Zeros of a Function

Graph the function and size the window so the x-intercept is visible.

For example: The zero of $y = x^3 + 2x^2 - x - 5$ is visible in the standard window, between $x = 1$ and $x = 2$. To determine the zero approximately, follow these steps.

1. Press 2nd CALC to go to the CALCULATE menu, then select 2:zero.

2. Move the cursor on the graph until it is clearly to the left of the x-intercept, then press enter.

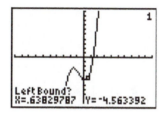

3. Move the cursor on the graph until it is clearly to the right of the x-intercept, then press enter.

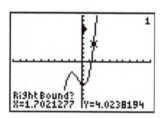

4. For the guess you can simply press ENTER. (If you move the cursor outside of the [left, right] interval you will get an error message).

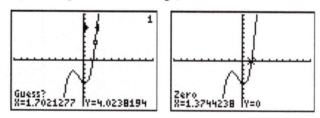

The zero is $x \approx 1.3744$.

If there are more zeros, each must be determined separately by following the same procedure.

Alternative Method to Determine a Zero

If the constant function $y = 0$ is entered as a second function, the zeros can be found by using the intersection method discussed previously.

Determining the Linear Regression Equation for a Set of Data

Example:

x	y
2	2
3	5
4	3
5	7
6	9

Enter the data into the calculator as follows:

I. Press (STAT) and choose EDIT.

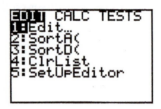

2. The TI-83/84 Plus has six built-in lists, L1, L2, . . . , L6. If there is data in L1, clear the list as follows:

a. Use the arrows to place the cursor on L1 at the top of the list. Press (CLEAR) followed by (ENTER), followed by the down arrow.

b. Follow the same procedure to clear L2 if necessary.

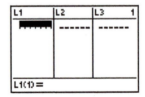

c. Enter the x-values into L1 and the y-values into L2.

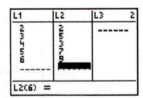

To see a scatterplot of the data proceed as follows.

I. STAT PLOT is the 2nd function of the (Y=) key. You must press (2nd) before pressing (Y=) to access the STAT PLOT menu.

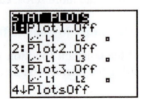

2. Select Plot 1 and make sure that Plots 2 and 3 are Off. The screen shown below will appear. Select On and then choose the scatterplot option (first icon) on the Type line. Confirm that your x and y values are stored, respectively, in L_1 and L_2. The symbols L_1 and L_2 are 2nd functions of the ⬚1⬚ and ⬚2⬚ keys, respectively. Finally, select the small square as the mark that will be used to plot each point.

3. Press ⬚Y=⬚ and clear or deselect any functions currently stored.

4. To display the scatterplot, have the calculator determine an appropriate window by pressing ⬚ZOOM⬚ and then ⬚9⬚ (ZoomStat).

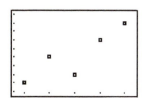

Calculate the linear regression equation as follows.

1. Press ⬚STAT⬚ and right arrow to highlight CALC.

2. Choose 4: LinReg($ax + b$). LinReg($ax + b$) will be pasted to the home screen. To tell the calculator where the data is, press ⬚2nd⬚ and ⬚1⬚ (for L1), then ⬚,⬚, then ⬚2nd⬚ and ⬚2⬚ (for L2) because the x list and y list are stored in L_1 and L_2 respectively. The display looks like this:

3. Press ⟨,⟩ and then press ⟨VARS⟩.

4. Right arrow to highlight Y-VARS.

5. Choose 1, FUNCTION.

6. Choose 1 for Y_1 (or 2 for Y_2, etc.).

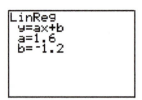

7. Press ⟨ENTER⟩.

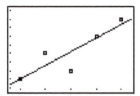

The linear regression equation for this data is $y = 1.6x - 1.2$.

8. To display the regression line, press ⟨GRAPH⟩.

9. Press the $\boxed{Y=}$ key to see the equation.

```
Plot1 Plot2 Plot3
\Y1■1.6X+-1.2
\Y2=
\Y3=■
\Y4=
\Y5=
\Y6=
\Y7=
```

Determining the Correlation Coefficient and Residuals

10. If the correlation coefficient, r, does not appear with the linear regression equation, as in step 7 on page A-14, press $\boxed{2nd}$ and $\boxed{0}$ (CATALOG), scroll down to select DiagnosticOn, press $\boxed{ENTER}$ twice. This will allow you to see r whenever you find a linear regression equation.

```
LinReg
 y=ax+b
 a=1.6
 b=-1.2
 r²=.7804878049
 r=.8834522086

■
```

11. Whenever a regression equation is calculated, the residuals are automatically stored in a list. To see the residuals, go to $\boxed{STAT}$ $\boxed{EDIT}$ and highlight L3 at the top of the list. Press $\boxed{2nd}$ $\boxed{DEL}$ (for INSERT), which will open a new list. At the flashing cursor, press $\boxed{2nd}$ $\boxed{STAT}$ (for LIST) and scroll down to select RESID. Press $\boxed{ENTER}$ twice and the list of residuals will appear.

```
L1      L2      RESID  4
 2       2       0
 3       5       1.4
 4       3       -2.2
 5       7       .2
 6       9       .6
------  ------  ------
RESID =(0,1.4,-2,...
```

Histograms, Boxplots, and Statistics

Entering a Collection of Data

Press $\boxed{STAT}$, then choose $\boxed{EDIT}$. You will see a table with heading L1, L2, L3, (L is for List). If there is a list of numbers already stored under L1, move the cursor to the L1 heading, then press $\boxed{CLEAR}$ and $\boxed{ENTER}$. When the list is empty start entering the data one number at a time. If you wish to store more than one set of data values, simply repeat the process for any of list L2 through L6.

Example:

COLLECTION OF DATA				
18	20	19	23	20
19	25	21	18	27
20	35	23	19	21
28	18	21	28	33

1. Press ⟨STAT⟩ and choose ⟨EDIT⟩.

2. To enter one variable data, you will need only one list.

   a. If there is data in L1, clear the list (move the cursor to the top, highlighting L1, press ⟨CLEAR⟩ and then ⟨ENTER⟩).

   b. Enter the data into L1.

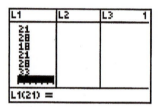

Calculating One-Variable Statistics

Press ⟨STAT⟩, then choose ⟨CALC⟩. From the calculate menu, press ⟨ENTER⟩ to choose the one-variable statistics. When you see 1-Var Stats on your home screen, enter the name of the list where your data is stored (the second function of 1 through 6).

1. To calculate the one variable statistics using the example data in L1, press ⟨STAT⟩, highlight CALC, then choose option 1: 1-Var Stats.

2. When 1-Var Stats appears on the home screen, enter ⟨2nd⟩ and ⟨1⟩ (for L1). The following basic statistics appear. Scroll down to see the second screen.

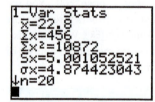 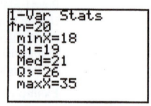

The statistics you should note are:

   mean, $\bar{x} = 22.8$

   standard deviation, $\sigma_x = 4.874423043$

   number of data values, $n = 20$

   median, med $= 21$

Creating a Histogram for One-Variable Data

1. Using the example data in L1, press ⌨2nd⌨ and ⌨Y=⌨ to get the STAT PLOT menu. Choose any of Plot1, Plot2, or Plot3, making sure the two you *don't* choose are Off.

2. In the plot menu, highlight and select On, histogram type of plot, L1 for the Xlist, and 1 for the Freq, as shown:

3. To see a histogram, press ⌨ZOOM⌨ and choose option 9: ZoomStat. The following histogram shows the domain of the distribution, grouping data values into classes.

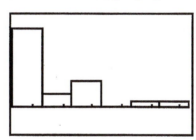

4. To show the histogram where each rectangle represents one data value, press ⌨WINDOW⌨ and set Xscl to 1, then press ⌨GRAPH⌨.

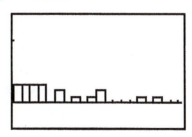

5. Pressing ⌨TRACE⌨ allows you to see the frequency for each data value.

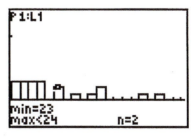

Creating a Boxplot for One-Variable Data

1. In the plot menu, highlight and select On, boxplot type of plot, L1 for the Xlist, and 1 for the Freq, as shown below:

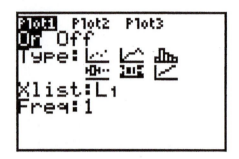

2. To see a boxplot, press ⌈ZOOM⌉ and choose option 9: ZoomStat.

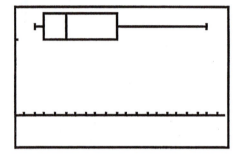

3. Pressing ⌈TRACE⌉ allows you to see the five-number summary for this distribution.

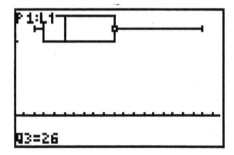

Factorials, Permutations, and Combinations

All are found by selecting the probability menu.

Press ⌈MATH⌉ and choose the PRB menu.

1. To calculate ten factorial, enter the integer followed by ⌈MATH⌉, choose the PRB menu, and select option 4: !, then ⌈ENTER⌉.

2. To calculate the number of permutations of ten things taken four at a time, enter 10 followed by (MATH), choose the PRB menu, and select option 2: nPr, followed by 4, then (ENTER).

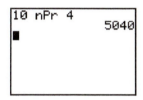

3. To calculate the number of combinations of ten things taken four at a time, or the binomial coefficient $\binom{10}{4}$, enter 10 followed by (MATH), choose the PRB menu, and select option 3: nCr, followed by 4, then (ENTER).

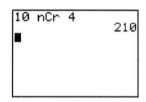

Generating a List of Random Numbers

1. Press (STAT) and choose (EDIT).

2. Clear list L1 and highlight the L1 heading. Press (MATH), choose the PRB menu, and select option 5: randInt(.

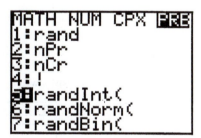

3. To randomly generate integers between 0 and 1, enter this domain, followed by the quantity of numbers desired, separated by commas. The following command generates a list of 50 zeros and ones. (Note that the command is too long to fit on the screen at once.)

4. To quickly add the numbers in a list, from the home screen, press (2nd) and (STAT) (for LIST), choose the MATH menu, and select option 5: sum(. Enter the name of the list, L1, and the sum will be displayed, as shown.

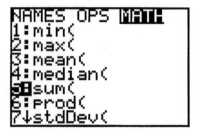 

In this example, 22 of the 50 integers were ones.

Binomial Probabilities

Simulation of a Binomial Experiment

Example: Tossing a fair coin five times and recording the number of heads, 0 through 5.

To generate a random integer from a binomial distribution, with five trials and probability of success 0.5, press (MATH) and choose the PRB menu, and select option 7: randBin(, (ENTER) followed by 5, 0.5) (ENTER).

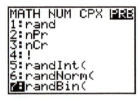 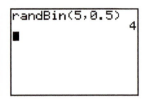

In this case a number between 0 and 5 (the number of heads) will be generated.

To generate a list of such random numbers, to create a simulation, a third number is included after the probability of success, the number of simulations. If not included, as in the previous example, the default is one simulation.

To simulate this experiment 500 times, storing the results in a list, start by pressing (STAT), select EDIT, and highlight the name of a list (L1 in this example) and press (CLEAR) and (ENTER) to empty L1. Highlight L1 again and press (MATH), choose the PRB menu, select option 7: randBin(, (ENTER) followed by 5, 0.5, 500) (ENTER). After a short while, the 500 randomly generated results will appear in L1.

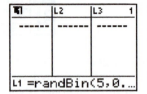

The results can best be seen by generating a histogram for the list of data in L1. (Refer to Creating a Histogram for One-Variable Data on page A-17). The following window settings will allow you to trace the histogram and record the frequency for each possible number of heads (successes).

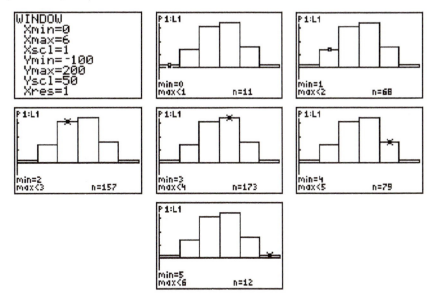

Calculating Binomial Probabilities

A single binomial probability can be quickly determined by using the probability distribution feature on the TI-83/84 Plus. To find the probability of getting three heads when a fair coin is tossed five times ($n = 5, p = 0.5, x = 3$), in the home screen, press ⌈2nd⌉ and ⌈VARS⌉ and select option 0:binompdf(, followed by 5, 0.5, 3) ⌈ENTER⌉, to get the exact theoretical probability.

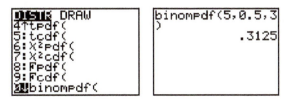

Geometric Probabilities

A geometric probability can be quickly determined using the probability distribution feature on the TI-83/84 Plus. To find the probability that you get your first head on the fourth toss of a coin ($p = 0.5, n = 4$), in the home screen, press ⌈2nd⌉ and ⌈VARS⌉ and select option E:geometpdf(, then enter 0.5, 4). Hit ⌈ENTER⌉ to get the exact theoretical probability.

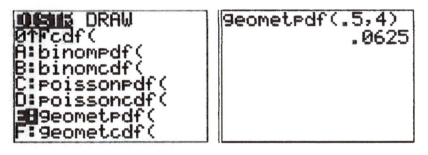

Metric–U.S. Conversions

Metric–U.S. Conversions

Length

1 meter $=$ 3.28 feet

1 meter $=$ 1.094 yards

1 centimeter $=$ 0.394 inches

1 kilometer $=$ 0.6214 miles

1 foot $=$ 0.305 meters

1 yard $=$ 0.914 meters

1 inch $=$ 2.54 centimeters

1 mile $=$ 1.6093 kilometers

Temperature

Celsius (C) to Fahrenheit (F)

$$F = \frac{9}{5}C + 32$$

Fahrenheit (F) to Celsius (C)

$$C = \frac{5}{9}(F - 32)$$

Weight

1 gram $=$ 0.03527 ounces
1 kilogram $=$ 2.205 pounds
1 gram $=$ 0.002205 pounds

1 ounce $=$ 28.35 grams
1 pound $=$ 0.454 kilograms
1 pound $=$ 454 grams

Volume

1 liter $=$ 1.057 quarts
1 liter $=$ 0.2642 gallons

1 quart $=$ 0.946 liters
1 gallon $=$ 3.785 liters

U.S. System of Measurement

Length

1 foot $=$ 12 inches
1 yard $=$ 3 feet
1 mile $=$ 5280 feet

Weight

1 pound $=$ 16 ounces
1 ton $=$ 2000 pounds

Volume

1 cup $=$ 8 fluid ounces
1 pint $=$ 2 cups
1 quart $=$ 2 pints
1 gallon $=$ 4 quarts

Metric System of Measurement

Length

1 kilometer (km) $=$ 1000 meters (m)

1 hectometer (hm) $=$ 100 m

1 dekameter (dam) $=$ 10 m

1 decimeter (dm) $= \frac{1}{10}$ m $=$ 0.1 m

1 centimeter (cm) $= \frac{1}{100}$ m $=$ 0.01 m

1 millimeter (mm) $= \frac{1}{1000}$ m $=$ 0.001 m

Mass

1 kilogram (kg) = 1000 grams (g)

1 hectogram (hg) = 100 g

1 dekagram (dag) = 10 g

1 decigram (dg) = $\dfrac{1}{10}$ g = 0.1 g

1 centigram (cg) = $\dfrac{1}{100}$ g = 0.01 g

1 milligram (mg) = $\dfrac{1}{1000}$ g = 0.001 g

Volume

1 kiloliter (kl) = 1000 liters (L)

1 hectoliter (hl) = 100 L

1 dekaliter (dal) = 10 L

1 deciliter (dl) = $\dfrac{1}{10}$ L = 0.1 L

1 centiliter (cl) = $\dfrac{1}{100}$ L = 0.01 L

1 milliliter (ml) = $\dfrac{1}{1000}$ L = 0.001 L

Glossary of Geometric Formulas, Interval Notation, Glossary of Functions

Glossary of Geometric Formulas

Perimeter and Area of a Triangle and Sum of Measures of the Angles

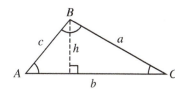

$$P = a + b + c$$
$$A = \frac{1}{2}bh$$
$$A + B + C = 180°$$

Pythagorean Theorem

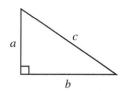

$$a^2 + b^2 = c^2$$

Perimeter and Area of a Rectangle

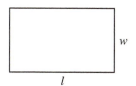

$$P = 2l + 2w$$
$$A = lw$$

Perimeter and Area of a Square

$$P = 4s$$
$$A = s^2$$

Area of a Trapezoid

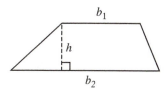

$$A = \frac{1}{2}h(b_1 + b_2)$$

Circumference and Area of a Circle

$$C = 2\pi r$$
$$A = \pi r^2$$

Volume and Surface Area of a Rectangular Solid

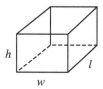

$$V = lwh$$
$$SA = 2lw + 2lh + 2wh$$

Volume and Surface Area of a Sphere

$$V = \frac{4}{3}\pi r^3$$
$$SA = 4\pi r^2$$

Volume and Surface Area of a Right Circular Cylinder

$$V = \pi r^2 h$$
$$SA = 2\pi r^2 + 2\pi rh$$

Volume and Surface Area of a Right Circular Cone

$$V = \frac{1}{3}\pi r^2 h$$
$$SA = \pi r^2 + \pi rl$$

Interval Notation

$(a, b) = a < x < b$ $(-\infty, a) = x < a$

$(a, b] = a < x \leq b$ $(-\infty, a] = x \leq a$

$[a, b) = a \leq x < b$ $(a, \infty) = x > a$

$[a, b] = a \leq x \leq b$ $[a, \infty) = x \geq a$

Glossary of Functions

Constant Function	Linear Function	Quadratic Function	Cubic Function
$f(x) = b$	$f(x) = mx + b$	$f(x) = ax^2 + bx + c,$ $a \neq 0$	$f(x) = x^3$

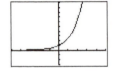 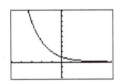 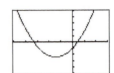 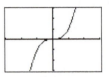

Exponential Function	Exponential Function	Logarithmic Function	Logarithmic Function
$f(x) = ab^x$ $a > 0, b > 1$	$f(x) = ab^x$ $a > 0, 0 < b < 1$	$f(x) = a(\log_b x)$ $a > 0, b > 1$	$f(x) = a(\log_b x)$ $0 < b < 1, a > 0$

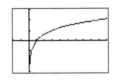 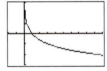

Index

Credits

Cover: Hareluya/Shutterstock **page 676:** MedioImages/Alamy **page 684:** Photoneye/Shutterstock **page 726:** Sorincolac/Fotolia **page 727T:** Heritage Image Partnership Ltd/Alamy **page 727C:** Christian Delbert/Shutterstock **page 727B:** Martin Pugh/Hubble Legacy Archive, ESA/NASA **page 728:** Ian 2010/Fotolia **page 729:** Steven Wynn/iStock/Getty Images **page 860:** Courtesy of Statistics Canada, 2001